## Metal Forming Analysis

The introduction of numerical methods, particularly finite-element (FE) analysis, represents a significant advance in metal forming operations. Numerical methods are used increasingly to optimize product design and deal with problems in metal forging, rolling, and extrusion processes. *Metal Forming Analysis* describes the latest and most important numerical techniques for simulating metal forming operations. The first part of the book describes principles and procedures and includes numerous examples and worked problems. The remaining chapters focus on applications of numerical analysis to specific forming operations. Most of these results are drawn from the authors' research in the areas of metal testing, sheet-metal forming, forging, extrusion, and similar operations. Sufficient information is presented so that readers can understand the nonlinear finite-element method as applied to forming problems without a prior background in structural finite-element analysis. Graduate students, researchers, and practicing engineers will welcome this thorough reference to state-of-the-art numerical methods used in metal forming analyses.

R. H. Wagoner is Distinguished Professor of Engineering in the Department of Materials Science and Engineering, The Ohio State University. He is a Life Member of the National Academy of Engineering, a Fellow of ASM International, and a Member and Past President of The Minerals, Metals, and Materials Society.

J.-L. Chenot is Professor and Head of the Material Forming Center, Ecole des Mines de Paris. He is a Founding Member and Chairman of ESAFORM (European Scientific Association for Material Forming) and a Member of SFM (Société Française de Mécanique) and SF2M (Société Française de Mitallengri et ds Matenary).

# METAL FORMING ANALYSIS

**R. H. WAGONER**
The Ohio State University

**J.-L. CHENOT**
Ecole des Mines de Paris

CAMBRIDGE UNIVERSITY PRESS
Cambridge, New York, Melbourne, Madrid, Cape Town, Singapore, São Paulo

Cambridge University Press
The Edinburgh Building, Cambridge CB2 2RU, UK

Published in the United States of America by Cambridge University Press, New York

www.cambridge.org
Information on this title: www.cambridge.org/9780521642675



First published 2001
This digitally printed first paperback version 2005

*A catalogue record for this publication is available from the British Library*

*Library of Congress Cataloguing in Publication data*

Wagoner, R. H. (Robert H.)
Metal forming analysis / R.H. Wagoner, J.-L. Chenot.
p. cm.
ISBN 0-521-64267-1
1. Metal-work – Mathematical models. 2. Finite element method. 3. Numerical analysis.
I. Chenot, J. L. II. Title.
TS213.W32 2001
671 – dc21

00-031249

ISBN-13 978-0-521-64267-5 hardback
ISBN-10 0-521-64267-1 hardback

ISBN-13 978-0-521-01772-5 paperback
ISBN-10 0-521-01772-6 paperback

To our wives

# Contents

# Preface

*Metal Forming Analysis* has two purposes: (a) to acquaint the advanced graduate student with numerical principles and procedures used in the modern analysis of industrial forming operations, and (b) to provide reference material for those performing such an analysis in industrial settings, government laboratories, and academia. In both cases, an understanding of the most important methods and their respective characteristics is the goal.

The first seven chapters focus on principles and procedures, which are derived and presented in an intuitive, informal manner. Exercises appear throughout these chapters, proposing and then solving illuminating problems related to the subject. Extensive problems are provided in three categories at the end of each chapter: proficiency, depth, and numerical, to solidify the information presented.

The last five chapters focus on applications of the numerical analysis to specific forming operations in order to illustrate the lessons learned from these simulations. Most of these results are drawn from the authors' research in this area, using programs developed over many years at their laboratories. Exercises are presented where appropriate and practical, and a limited number of problems are provided at the end of some chapters.

It should be noted that this advanced text and reference volume does not provide a detailed treatment of the underlying physical equations or principles necessary to understand metal deformation itself. This material is limited to Chapter 1, which is a very brief review of the physical descriptions and equations. For a thorough treatment of the physical fundamentals leading to the numerical treatment, we recommend that the interested reader refer to *Fundamentals of Metal Forming* (Wiley, 1997, ISBN 0-471-57004-4), which we wrote for this purpose.

Because of the nature of metal forming, the challenges to a numerical analysis lie predominantly in the large deformation experienced by materials and the nonlinear aspects of the finite-element method. With this required focus, it is impossible to introduce in any systematic way the broad field of linear finite-element modeling for structural applications. There are many excellent and exhaustive texts on these subjects that may be consulted. However, sufficient information is provided to understand the nonlinear finite-element method as applied to forming problems without this breadth of background.

# Acknowledgments

This book is made possible by the selfless contributions of close friends, whom we imposed upon to read, write, edit, criticize, and give up other activities so that we could do these same things. Robyn K. Wagoner, without an iota of technical training, typed much of this manuscript and made nearly all of the many changes required in figures, equations, and text. Robert H. Wagoner (senior), also without technical training, read and edited every chapter several times, finding all of the inevitable misreferenced equations and figures, the improper English, and the many subtle errors that occur when technical people from two countries try to write. Beatrice Chenot and Robyn Wagoner gave up weekends with their husbands over a period of years in order to make this book possible.

The first author would like to acknowledge the support of The Ohio State University and the Ecole des Mines de Paris, both of which provided support for a sabbatical leave during which this book began to take shape. Some of the travel facilitated the collaboration of the authors and was provided by the National Science Foundation (DMR-8814926).

The second author would like to acknowledge the Ecole des Mines de Paris and his Director, J. Lévy, for creating the favorable conditions to develop research and engineering applications with industrial companies, upon which this work is based. Most of the computational examples in Chapters 9, 10, and 12 were kindly provided by colleagues and students of the CEMEF laboratory: they will find here my deep gratitude.

Chapters 8 and 11 were drafted by Dr. Dajun Zhou and the equations were proofed by Dr. Kaiping Li. Many of the original figures were drafted by Weili Wang, wife of Dajun Zhou, who provided criticism for the technical content, especially in the area of numerical methods.

Without these many services, provided without complaint or compensation, this book would have not existed, or would have been even later, by years, than it is now.

Robert H. Wagoner, Columbus, Ohio
Jean-Loup Chenot, Sophia-Antipolis, France

CHAPTER ONE

# Mathematical Background

This book assumes a background in the fundamentals of solid mechanics and the mechanical behavior of materials, including elasticity, plasticity, and friction. A previous book by the same authors[1] covers these topics in detail, including derivation or explanation of the most important concepts. It is beyond the scope of the current book to reproduce all of this important information.

In this chapter, the essential equations from this background are reproduced. This serves two purposes: to introduce the notation that will be used throughout the remaining chapters, and to list the principal background equations in one place. Frequent reference to the equations presented in this chapter will be made. However, it should be kept in mind that the full context for these equations is found in *Fundamentals of Metal Forming*.[1]

## 1.1 Notation

There are many alternate forms of notation used in solid mechanics and finite-element modeling. In some cases, it is clearer to use a form that has become a de facto standard in the area, even though such usage might not be rigorous. In other cases, there is no consensus on notation, so it is less confusing to be consistent with other equations.

In general, scalars are denoted by plain Roman or Greek letters, with or without subscripts or superscripts: a, A, $\alpha$, t, T, $a_1$, $a_{12}$, ....

Vectors (whether physical or numerical ones, which are generalized one-dimensional arrays of numbers) are typically represented by lower-case or upper-case bold letters to emphasize the vector nature of the variable, with alternate notations used to refer to the components of the vector:

$$\mathbf{a} = a_1\hat{e}_1 + a_2\hat{e}_2 + a_3\hat{e}_3 = |\mathbf{a}|\hat{\mathbf{g}} \leftrightarrow a_1, a_2, a_3 \leftrightarrow \begin{bmatrix} a_1 \\ a_2 \\ a_3 \end{bmatrix} = [a_i] = [a] \leftrightarrow a_i, \qquad (1.1)$$

where $\hat{e}_1$, $\hat{e}_2$, $\hat{e}_3$, are the Cartesian orthogonal unit vectors, $|\mathbf{a}|$ is the norm of vector $\mathbf{a}$, and $\hat{\mathbf{g}}$ is the unit vector with direction $\mathbf{a}$. The symbol $\leftrightarrow$ is used here in order to treat the differences in the forms rigorously. However, this convention will often be dropped and the various forms of such a quantity will be used interchangeably, depending on convenience and clarity.

[1] R. H. Wagoner and J.-L. Chenot, *Fundamentals of Metal Forming* (Wiley, New York, 1997).

Notation for tensors of rank higher than 1 ("vector") follows vector usage, although an attempt will be made to use bold upper-case letters when there is not a conventional usage of another symbol. Tensors are sometimes expressed in matrix form to illustrate the required manipulation:

$$\mathbf{A} \leftrightarrow [A] = [A_{ij}] = \begin{bmatrix} A_{11} & A_{12} & A_{13} \\ A_{21} & A_{22} & A_{23} \\ A_{31} & A_{32} & A_{33} \end{bmatrix} \leftrightarrow A_{ij}. \tag{1.2}$$

Of course, the matrix shown above need not correspond to any tensor (such as $\mathbf{A}$ shown at the left side of the chain). Several other common notations for matrices are as follows:

$$\begin{aligned} &\mathbf{A}^T \leftrightarrow [A]^T\text{: transpose of } \mathbf{A}, \\ &\mathbf{A}^{-1} \leftrightarrow [A]^{-1}\text{: inverse of } \mathbf{A}, \\ &\det(\mathbf{A}) = \det[A]\text{: determinant of } \mathbf{A}, \\ &\text{trace}(\mathbf{A}) = \text{trace}[A]\text{: trace of } \mathbf{A}, \\ &\mathbf{C} = \mathbf{AB} \leftrightarrow [C] = [A][B]\text{: matrix multiplication}, \\ &\mathbf{I} \leftrightarrow [I]\text{: identity matrix.} \end{aligned} \tag{1.3}$$

Each form can also be modified to show the subscript indices to emphasize the components.

The indicial form of matrix equations makes use of standard rules. Any repeated index within a term is a "dummy index," following Einstein's summation convention in which any repeated index is summed. Other indices are "free indices," which may independently adopt certain values. Two standard operators are used to complete indicial equations.

One is the Kronecker delta, with the property that

$$\begin{aligned} &\delta_{ij} = 0 \qquad \text{if } i \neq j, \\ &\delta_{ij} = 1 \qquad \text{if } i = j. \end{aligned} \tag{1.4}$$

The other is the permutation operator epsilon, with the property that

$$\begin{aligned} &\varepsilon_{ijk} = 0 \qquad \text{if } i = j, \text{ or } j = k, \text{ or } k = i, \\ &\varepsilon_{ijk} = 1 \qquad \text{if } ijk = 1,2,3, \text{ or } 2,3,1 \text{ or } 3,1,2, \\ &\varepsilon_{ijk} = -1 \qquad \text{if } ijk = 3,2,1, \text{ or } 1,3,2, \text{ or } 2,1,3. \end{aligned} \tag{1.5}$$

## 1.2 Stress

Throughout this book, reference to stress will always mean Cauchy stress, which relates real force intensities on planes and areas defined in a current deformation state. This standard stress measure is always symmetric by equilibrium considerations applied to a continuum and thus may be written as follows:

$$\boldsymbol{\sigma} \leftrightarrow [\sigma] = \begin{bmatrix} \sigma_{11} & \sigma_{12} & \sigma_{13} \\ \sigma_{21} & \sigma_{22} & \sigma_{23} \\ \sigma_{31} & \sigma_{32} & \sigma_{33} \end{bmatrix}. \tag{1.6}$$

The actual tractions, or stress vector, may be obtained from the stress tensor as follows:

$$\mathbf{T} = \boldsymbol{\sigma}\mathbf{n} \leftrightarrow [T_i] = \sum_j \sigma_{ij} n_j \leftrightarrow \begin{bmatrix} T_1 \\ T_2 \\ T_3 \end{bmatrix} = \begin{bmatrix} \sigma_{11} & \sigma_{12} & \sigma_{13} \\ \sigma_{21} & \sigma_{22} & \sigma_{23} \\ \sigma_{31} & \sigma_{32} & \sigma_{33} \end{bmatrix} \begin{bmatrix} n_1 \\ n_2 \\ n_3 \end{bmatrix}. \tag{1.7}$$

The force acting on an elementary surface can be calculated similarly by

$$\mathbf{df} = \boldsymbol{\sigma}\mathbf{da} \leftrightarrow \mathrm{d}f_i = \sum_j \sigma_{ij} \mathrm{d}a_j \leftrightarrow \begin{bmatrix} \mathrm{d}f_1 \\ \mathrm{d}f_2 \\ \mathrm{d}f_3 \end{bmatrix} = \begin{bmatrix} \sigma_{11} & \sigma_{12} & \sigma_{13} \\ \sigma_{21} & \sigma_{22} & \sigma_{23} \\ \sigma_{31} & \sigma_{32} & \sigma_{33} \end{bmatrix} \begin{bmatrix} \mathrm{d}a_1 \\ \mathrm{d}a_2 \\ \mathrm{d}a_3 \end{bmatrix}, \tag{1.8}$$

where $\mathbf{da} = |\mathbf{da}|\mathbf{n}$ is the elementary surface vector and $\mathbf{n}$ is the normal vector.

The principal stresses are obtained as the eigenvalues $\lambda$ of $\boldsymbol{\sigma}$ in the usual way:

$$\boldsymbol{\sigma}\mathbf{n} = \lambda\mathbf{n} \leftrightarrow \begin{bmatrix} \sigma_{11} & \sigma_{12} & \sigma_{13} \\ \sigma_{21} & \sigma_{22} & \sigma_{23} \\ \sigma_{31} & \sigma_{32} & \sigma_{33} \end{bmatrix} \begin{bmatrix} n_1 \\ n_2 \\ n_3 \end{bmatrix} = \begin{bmatrix} \lambda n_1 \\ \lambda n_2 \\ \lambda n_3 \end{bmatrix}, \tag{1.9}$$

or with indicial form:

$$\begin{bmatrix} \sigma_{11} - \lambda & \sigma_{12} & \sigma_{13} \\ \sigma_{21} & \sigma_{22} - \lambda & \sigma_{23} \\ \sigma_{31} & \sigma_{32} & \sigma_{33} - \lambda \end{bmatrix} \begin{bmatrix} n_1 \\ n_2 \\ n_3 \end{bmatrix} = 0. \tag{1.10}$$

This is solved by noting that the determinant of the tensor on the left-hand side must be identically equal to zero (if $\mathbf{n}$ is not to be a null vector):

$$\det(\boldsymbol{\sigma}) = 0 \leftrightarrow \det \begin{bmatrix} \sigma_{11} - \lambda & \sigma_{12} & \sigma_{13} \\ \sigma_{21} & \sigma_{22} - \lambda & \sigma_{23} \\ \sigma_{31} & \sigma_{32} & \sigma_{33} - \lambda \end{bmatrix} = 0, \tag{1.11}$$

where the result of this calculation gives a cubic equation, with real roots $\lambda_i$ corresponding to the principal stresses (eigenvalues):

$$\lambda^3 - J_1\lambda^2 - J_2\lambda - J_3 = 0, \tag{1.12}$$

where $J_1$ is the first stress invariant,

$$J_1 = \mathrm{trace}(\boldsymbol{\sigma}) = \sigma_{11} + \sigma_{22} + \sigma_{33}, \tag{1.13}$$

$J_2$ is the second stress invariant (quadratic invariant),

$$J_2 = -(\sigma_{11}\sigma_{22} + \sigma_{22}\sigma_{33} + \sigma_{33}\sigma_{11}) + \sigma_{23}^2 + \sigma_{31}^2 + \sigma_{12}^2, \tag{1.14}$$

and $J_3$ is the third stress invariant,

$$J_3 = \det(\boldsymbol{\sigma}) = \begin{vmatrix} \sigma_{11} & \sigma_{12} & \sigma_{13} \\ \sigma_{21} & \sigma_{22} & \sigma_{23} \\ \sigma_{31} & \sigma_{32} & \sigma_{33} \end{vmatrix}. \tag{1.15}$$

For the numerical formulation of elastoplastic constitutive equations, it is frequently convenient to perform an additive decomposition of the stress tensor to obtain the spherical or pressure part (which is insensitive to plastic deformation) and the deviatoric part. The hydrostatic pressure is taken to be:

$$p = -\sigma_p = -\frac{\sigma_{11} + \sigma_{22} + \sigma_{33}}{3} = -\frac{J_1}{3}, \tag{1.16}$$

which allows further decomposition to obtain the deviatoric part:

$$\sigma = \sigma_p \mathbf{I} + \mathbf{s} = (-p)\mathbf{I} + \mathbf{s}, \tag{1.17}$$

where s is the deviatoric stress tensor.

## 1.3 Strain

The work-conjugate (actually power-conjugate) quantity to the Cauchy stress is the rate of deformation (**D**), which may be referred to without ambiguity as the strain rate $\dot{\varepsilon}$ (epsilon with overdot). In nearly every instance throughout this book, an updated Lagrangian formulation will be used, in which the preceding and current steps are considered sufficiently close that infinitesimal deformation theory may be used, such that, for example, $\Delta\varepsilon = \dot{\varepsilon}\Delta t$. Thus, there will be no distinction between power and work formulations. With this in mind, a few important equations may be presented, starting with the Lagrangian view of continuum deformation.

The coordinate vector is

$$\mathbf{x} = \chi(\mathbf{X}, t), \tag{1.18}$$

where each material point is labeled by its position **X** at some time $t_0$, and its current position at time $t$ is given by **x**.

The material velocity is

$$\mathbf{v} = \left(\frac{\partial \mathbf{x}}{\partial t}\right)_{\mathbf{X}}. \tag{1.19}$$

The material acceleration is

$$\gamma = \left(\frac{\partial \mathbf{v}}{\partial t}\right)_{\mathbf{X}}. \tag{1.20}$$

The displacement vector is

$$\mathbf{U}(\mathbf{X}, t) = \mathbf{x}(\mathbf{X}, t) - \mathbf{X}. \tag{1.21}$$

In the updated Lagrangian sense, **X** represents the position of a material point at the previous time step, which is considered only infinitesimally removed from the current position of the same material element, **x**.

The deformation gradient defines the transformation between corresponding material vectors at the two instants,

$$\mathrm{d}\mathbf{x} = \frac{\partial \mathbf{x}}{\partial \mathbf{X}}\mathrm{d}\mathbf{X} = \mathbf{F}\,\mathrm{d}\mathbf{X}, \tag{1.22}$$

with the component form

$$\begin{bmatrix} \mathrm{d}x_1 \\ \mathrm{d}x_2 \\ \mathrm{d}x_3 \end{bmatrix} = \begin{bmatrix} F_{11} & F_{12} & F_{13} \\ F_{21} & F_{22} & F_{23} \\ F_{31} & F_{32} & F_{33} \end{bmatrix} \begin{bmatrix} \mathrm{d}X_1 \\ \mathrm{d}X_2 \\ \mathrm{d}X_3 \end{bmatrix} \leftrightarrow \mathrm{d}x_i = \sum_j \frac{\partial x_i}{\partial X_j}\mathrm{d}X_j = \sum_j F_{ij}\mathrm{d}X_j. \tag{1.23}$$

It may be shown that the determinant of **F**, also called the Jacobian of the transformation, relates the initial differential volume $\mathrm{dvol}_0$ (or density $\rho_0$) at a point with the corresponding one dvol (or $\rho$) after deformation:

$$\frac{\mathrm{dvol}}{\mathrm{dvol}_0} = \frac{\rho}{\rho_0} = J = \det(\mathbf{F}). \tag{1.24}$$

A similar transformation is written for the relative displacements of the head and tail of such a vector:

$$\mathbf{du} = \frac{\partial \mathbf{u}}{\partial \mathbf{X}} = \mathbf{J}\,\mathrm{d}\mathbf{X}. \tag{1.25}$$

With the indicial forms we have

$$\begin{bmatrix} du_1 \\ du_2 \\ du_3 \end{bmatrix} = \begin{bmatrix} \frac{\partial u_1}{\partial X_1} & \frac{\partial u_1}{\partial X_2} & \frac{\partial u_1}{\partial X_3} \\ \frac{\partial u_2}{\partial X_1} & \frac{\partial u_2}{\partial X_2} & \frac{\partial u_2}{\partial X_3} \\ \frac{\partial u_3}{\partial X_1} & \frac{\partial u_3}{\partial X_2} & \frac{\partial u_3}{\partial X_3} \end{bmatrix} \begin{bmatrix} \mathrm{d}X_1 \\ \mathrm{d}X_2 \\ \mathrm{d}X_3 \end{bmatrix} = \begin{bmatrix} J_{11} & J_{12} & J_{13} \\ J_{21} & J_{22} & J_{23} \\ J_{31} & J_{32} & J_{33} \end{bmatrix} \begin{bmatrix} \mathrm{d}X_1 \\ \mathrm{d}X_2 \\ \mathrm{d}X_3 \end{bmatrix}. \tag{1.26}$$

$$\leftrightarrow \mathrm{d}u_i = \sum_j \frac{\partial u_i}{\partial X_j}\mathrm{d}X_j = \sum_j J_{ij}\mathrm{d}X_j$$

**J** is called the displacement gradient matrix, or Jacobian matrix, which can be written in terms of **F** with the help of Eq. (1.21):

$$\mathbf{J} = \mathbf{F} - \mathbf{I}. \tag{1.27}$$

In order to ignore pure rotations of a vector, the stretch of a material element is considered, starting from **F**:

$$\mathrm{d}s^2 = \mathrm{d}\mathbf{X}^T\mathbf{C}\,\mathrm{d}\mathbf{X}, \tag{1.28}$$

with

$$\mathbf{C} = \mathbf{F}^T\mathbf{F}, \quad C_{ij} = \sum_k \frac{\partial x_k}{\partial X_i}\frac{\partial x_k}{\partial X_j}, \tag{1.29}$$

where d$s$ is the final length of such an element or vector, and **C** is known as the deformation tensor, or the Cauchy deformation tensor.

Whereas **C** transforms the length of a vector from one state to another, it is most frequently of interest to focus on the change of length, in which case the strain tensor **E** is used:

$$\mathrm{d}s^2 - \mathrm{d}S^2 = \mathrm{d}\mathbf{X}^T(2\mathbf{E})\,\mathrm{d}\mathbf{X}. \tag{1.30}$$

Taking into account Eq. (1.28), we get

$$\mathbf{E} = \frac{1}{2}(\mathbf{C} - \mathbf{I}) = \frac{1}{2}(\mathbf{F}^T\mathbf{F} - \mathbf{I}), \tag{1.31}$$

the components of which are

$$E_{ij} = \frac{1}{2}\left(\sum_k \frac{\partial x_k}{\partial X_i}\frac{\partial x_k}{\partial X_j} - \delta_{ij}\right), \tag{1.32}$$

and with the help of Eq. (1.27) we can also write

$$E = \frac{1}{2}(\mathrm{J} + \mathrm{J}^T + \mathrm{J}^T\mathrm{J}). \tag{1.33}$$

The factor of 1/2 is used conventionally such that the infinitesimal strain components (which are historically older) become the small limit of **E**.

For small deformation and rotation (i.e., where the components of J are much less than one, and can be considered in the limit of approaching zero), it may be seen that the last term of Eq. (1.33) is vanishingly small relative to the first term. Elimination of the second-order term produces the definition of the small-strain tensor:

$$\varepsilon = \frac{1}{2}(\mathrm{J} + \mathrm{J}^T) \quad \text{or } \varepsilon_{ij} = \frac{1}{2}\left(\frac{\partial u_i}{\partial x_j} + \frac{\partial u_j}{\partial x_i}\right). \tag{1.34}$$

The small strain is clearly the symmetric part of **J**, whereas the antisymmetric part corresponds to the rigid-body rotation:

$$\omega = \frac{1}{2}(\mathrm{J} - \mathrm{J}^T) \quad \text{or } \omega_{ij} = \frac{1}{2}\left(\frac{\partial u_i}{\partial x_j}\frac{\partial u_j}{\partial x_i}\right). \tag{1.35}$$

Although there is no formal distinction between infinitesimal displacements and velocities (aside from a homogeneous factor of $dt$), conventional notation is often based on rate or velocity forms. In this case, the relative velocity **dv** of the head to tail of a vector **dx** is related to **dx** as follows[2]:

$$\mathbf{dv} = \mathbf{L}\,\mathbf{dx}, \tag{1.36}$$

where **L** is defined by

$$\mathbf{L} = \frac{\partial \mathbf{v}}{\partial \mathbf{x}}, \quad L_{ij} = \frac{\partial v_i}{\partial x_j}, \tag{1.37}$$

and **L** is the velocity gradient or the spatial gradient of velocity. **L** may be decomposed additively, analogous to the decomposition of **J** (because **L** is simply $\dot{\mathbf{J}}$) for infinitesimal differences between the start and end of deformation:

$$\mathbf{L} = \dot{\varepsilon} + \dot{\omega}. \tag{1.38}$$

Here $\dot{\varepsilon}$, often denoted by **D**, is the strain rate, or rate of deformation tensor:

$$2\dot{\varepsilon} = \mathbf{L} + \mathbf{L}^T, \quad \dot{\varepsilon}_{ij} = \frac{1}{2}\left(\frac{\partial v_i}{\partial x_j} + \frac{\partial v_j}{\partial x_i}\right), \tag{1.39}$$

and $\dot{\omega}$ is the spin tensor:

$$2\dot{\omega} = \mathbf{L} - \mathbf{L}^T, \quad \dot{\omega}_{ij} = \frac{1}{2}\left(\frac{\partial v_i}{\partial x_j} - \frac{\partial v_j}{\partial x_i}\right). \tag{1.40}$$

[2] Note that the lower case has been used for **dx**, even though this is a material vector at the beginning of the deformation step. Because the focus is on infinitesimal steps, the distinction between **dx** and **dX** is lost in this context.

For small strain, denoted here by $\Delta$, the relationship between $\Delta\varepsilon$ and $\dot{\varepsilon}$, or $\Delta\omega$ and $\dot{\omega}$, is thus:

$$\Delta\varepsilon = \dot{\varepsilon}\Delta t, \tag{1.41}$$

$$\Delta\omega = \dot{\omega}\Delta t. \tag{1.42}$$

It is sometimes useful to use relationships among the various deformation measures:

$$\mathbf{L} = \dot{\mathbf{F}}\mathbf{F}^{-1} \quad \text{or } \dot{\mathbf{F}} = \mathbf{L}\mathbf{F}, \tag{1.43}$$

$$\mathbf{E} = \frac{1}{2}(\mathbf{F}^T\mathbf{F} - \mathbf{I}), \quad E_{ij} = \frac{1}{2}\left(\frac{\partial x_k}{\partial X_i}\frac{\partial x_k}{\partial X_J} - \delta_{ij}\right). \tag{1.44}$$

It is also useful to perform a polar decomposition of the deformation gradient, as follows:

$$\mathbf{F} = \mathbf{R}\mathbf{U}, \tag{1.45}$$

where **R** is orthogonal and called the rotational operator and where **U** is symmetric positive definite and is called the right stretch tensor. **R** is often used to estimate the rigid-body rotation of a large deformation, whereas **U** is used to find the stretch ratios for a large deformation:

$$\mathbf{U}^2 = \mathbf{C} = (2\mathbf{E} + \mathbf{I}), \quad \lambda_i = U_i = \sqrt{C_i} = \sqrt{2E_i + 1}. \tag{1.46}$$

Here $\lambda_i$ is the $i$th stretch ratio corresponding to the $i$th principal value (eigenvalue) of **U**. Note that raising a tensor to a power signifies a tensor with the same principal directions but with principal values raised to that power.

## 1.4 Mechanical Principles

There are many alternate, but equivalent, ways to formulate the mechanical equations governing continuum motion. The continuity equation states that mass cannot be lost or gained, and it implies that velocity fields must be well behaved:

$$\frac{d\rho}{dt} + \rho\,\mathrm{div}(\mathbf{v}) = 0, \quad \frac{\partial\rho}{\partial t} + \mathrm{div}(\rho\mathbf{v}) = 0. \tag{1.47}$$

Similarly, Newton's laws must be obeyed for each material element. Including dynamic (inertial) and static effects internally, and gravity as an external body force, we can write the equation of motion as

$$\rho\gamma = \mathrm{div}(\boldsymbol{\sigma}) + \rho\mathbf{g}. \tag{1.48}$$

For the cases that dominate the examples in this book, only static equilibrium need be considered, and gravity forces may be neglected as much smaller than other forces:

$$\mathrm{div}(\boldsymbol{\sigma}) = 0. \tag{1.49}$$

By considering a virtual displacement field $\delta\mathbf{u}$ (which is infinitesimal and has the property that $\delta\mathbf{u}$ is zero wherever the displacement is specified), we can equate the internal work absorbed by the deformation of the material to the external work done by outside forces acting on the body:

$$\int_{\partial\Omega} \mathbf{T}^d \delta\mathbf{u}\, \mathrm{d}S + \int_{\Omega} \rho \mathbf{g} \delta\mathbf{u}\, \mathrm{d}V = \int_{\Omega} \rho \gamma \delta\mathbf{u}\, \mathrm{d}V + \int_{\Omega} \sigma \delta\varepsilon\, \mathrm{d}V,$$

$$\text{external work increment} = \text{internal work increment} \tag{1.50}$$

where $\mathbf{T}^d$ is defined on the whole boundary $\partial\Omega$ of $\Omega$. It is equal to the external stress vector where it is prescribed, and it is equal to zero elsewhere; $\delta\mathbf{u}$ is any virtual displacement equal to zero on the part of $\partial\Omega$ where the displacement is imposed.

As before, if static equilibrium is sought and gravitational forces may be neglected as second order, this reduces to

$$\int_{\partial\Omega} \mathbf{T}^d \delta\mathbf{u}\, \mathrm{d}S = \int_{\Omega} \sigma \delta\varepsilon\, \mathrm{d}V.$$

$$\text{external work increment} = \text{internal work increment} \tag{1.51}$$

Variational principles may be used to establish functional formulations whenever the virtual work principle can be integrated exactly. Then, instead of solving for a root of a function where the net force equals zero [e.g., the function can be the left-hand side of Eq. (1.50) less the right-hand side of Eq. (1.51)], we seek the minimum of a functional, generally homogeneous to the net work, or to the rate of work.

Although it is always possible to derive the virtual work statement from the functional (by differentiation), the converse is not always possible. Where the variational principle or functional exists, it ensures that the stiffness matrix is symmetric, which has advantages for numerical solution. Thus, when such a principle can be written, it is often useful to do so. Although there is no convenient elastoplastic functional, both elastic and purely plastic or viscoplastic functionals can be derived. Examples of such functionals are shown as follows, under the appropriate constitutive equations.

## 1.5 Elasticity

Hooke's law may be written simply as follows:

$$\sigma = \mathbf{c} : \varepsilon, \qquad \text{or } \varepsilon = \mathbf{S} : \sigma. \tag{1.52}$$

With the components notation it is also written as

$$\sigma_{ij} = \sum_{k,l} c_{ijkl} \varepsilon_{kl}, \qquad \text{or } \varepsilon_{ij} = \sum_{k,l} S_{ijkl} \sigma_{kl}, \tag{1.53}$$

where $\mathbf{c}$ is known as the elastic constant tensor (often denoted by $\mathbf{D}$) and $\mathbf{S}$ represents the compliance tensor.

For an isotropic material, $\mathbf{c}$ and $\mathbf{S}$ take special forms ensuring that the material has the same properties in every direction. The terms in $\mathbf{c}$ and $\mathbf{S}$ are often written in terms of conventional elastic constants, as follows:

$$c_{ijkl} = \lambda \delta_{ij} \delta_{kl} + \mu (\delta_{ik} \delta_{jl} + \delta_{il} \delta_{jk}), \tag{1.54}$$

$$S_{ijkl} = \frac{1+\nu}{E} \delta_{ij} \delta_{kl} - \frac{\nu}{E} (\delta_{ik} \delta_{jl} + \delta_{il} \delta_{jk}), \tag{1.55}$$

where the conventional elastic constants $\lambda$, $\mu$ are the Lamé coefficients, $E$ is the Young modulus, and $\nu$ is the Poisson coefficient.

Using Eq. (1.54), we write Hooke's equation in the usual form:

$$\sigma = \lambda\theta \mathrm{I} + 2\mu\varepsilon, \tag{1.56}$$

where $\theta = \sum_i \varepsilon_{ii}$ is the dilatation. The component form is obviously

$$\sigma_{ij} = \lambda\theta\delta_{ij} + 2\mu\varepsilon_{ij}. \tag{1.57}$$

If Eq. (1.56) is inverted, we get

$$\varepsilon = \frac{3\nu}{E} p\mathrm{I} + \frac{1+\nu}{E}\sigma \tag{1.58}$$

or

$$\varepsilon_{ij} = \frac{3\nu}{E} p\delta_{ij} + \frac{1+\nu}{E}\sigma_{ij}. \tag{1.59}$$

Other elastic constants are also used commonly and are related as follows:

$$\mu = G = \frac{E}{2(1+\nu)}, \tag{1.60}$$

$$\lambda = \frac{\nu E}{(1+\nu)(1+2\nu)} = \frac{2\mu\nu}{1-2\nu}. \tag{1.61}$$

For the bulk modulus,

$$B = \lambda + \frac{2}{3}\mu = \frac{E}{3(1-2\nu)} = \kappa, \tag{1.62}$$

$$\nu = \frac{\lambda}{2(\lambda+\mu)} = \frac{3B-2\mu}{2(3B+\mu)}, \tag{1.63}$$

$$E = \frac{\mu(3\lambda+2\mu)}{\lambda+\mu} = \frac{9\mu B}{3B+\mu}. \tag{1.64}$$

For this material constitutive equation (i.e., an isotropic, linearly elastic material), a functional exists and may be written as follows:

$$\Pi(\varepsilon) = \int_\Omega \left(\frac{1}{2}\lambda\theta^2 + \mu\varepsilon : \varepsilon\right) dV - \int_{\partial\Omega} \mathbf{T}^d \mathbf{u}\, dS. \tag{1.65}$$

Here $\mathbf{T}^d$ is defined on the whole boundary $\partial\Omega$ of $\Omega$. It is equal to the external stress vector where it is prescribed, and it is equal to zero elsewhere; $\delta\mathbf{u}$ is any virtual displacement equal to zero on the part of $\partial\Omega$ where the displacement is imposed.

## 1.6 Plasticity

Unless otherwise stated, in this book we consider only flow theory plasticity laws, which obey the following principles.

$f(\sigma) < 0$: no plastic deformation,

$f(\sigma) = 0$: plastic deformation is possible,

$f(\sigma) > 0$: is forbidden.

- $f(\boldsymbol{\sigma})$ is a yield function with the following properties:
  - There is no plastic deformation in the elastic (or rigid) region enclosing $\boldsymbol{\sigma} = 0$, which is defined by

$$f(\boldsymbol{\sigma}) < 0, \quad \text{or} \left( f(\boldsymbol{\sigma}) = 0 \quad \text{and} \ \frac{\partial f}{\partial \boldsymbol{\sigma}} : \dot{\boldsymbol{\sigma}} < 0 \right). \tag{1.66}$$

  - The plastic region corresponds to

$$f(\boldsymbol{\sigma}) = 0 \quad \text{and} \ \frac{\partial f}{\partial \boldsymbol{\sigma}} : \dot{\boldsymbol{\sigma}} \geq 0. \tag{1.67}$$

  - The surface defined in the stress space by $f(\boldsymbol{\sigma}) = 0$ is convex.
- The yield surface changes size but not shape (isotropic hardening).
- The normality condition (associated flow) holds for plastic loading:

$$\dot{\varepsilon} = \dot{\lambda}^p \frac{\partial f}{\partial \boldsymbol{\sigma}}, \quad \dot{\varepsilon}_{ij} = \dot{\lambda}^p \frac{\partial f}{\partial \sigma_{ij}}, \qquad \dot{\lambda}^p > 0, \tag{1.68}$$

or

$$\mathrm{d}\varepsilon = \mathrm{d}\lambda^p \frac{\partial f}{\partial \boldsymbol{\sigma}}, \quad \mathrm{d}\varepsilon_{ij} = \mathrm{d}\lambda^p \frac{\partial f}{\partial \sigma_{ij}}, \qquad \mathrm{d}\lambda^p > 0. \tag{1.69}$$

- Plastic flow is time independent (unless viscoplasticity is noted).

The von Mises yield function may be expressed in stress components as follows:

$$\begin{aligned} f(\boldsymbol{\sigma}) &= \frac{1}{2}[(\sigma_1 - \sigma_2)^2 + (\sigma_2 - \sigma_3)^2 + (\sigma_3 - \sigma_1)^2] - \bar{\sigma}_0^2 \\ &= \frac{1}{2}\left[(\sigma_{11} - \sigma_{22})^2 + (\sigma_{22} - \sigma_{33})^2 + (\sigma_{33} - \sigma_{11})^2 + 6\sigma_{12}^2 + 6\sigma_{23}^2 + 6\sigma_{31}^2\right] - \bar{\sigma}_0^2. \end{aligned} \tag{1.70}$$

If the deviatoric stress tensor s is used, we obtain the equivalent expression,

$$f(\mathrm{s}) = \frac{3}{2} \sum_{i,j} s_{ij}^2 - \bar{\sigma}_0^2, \tag{1.71}$$

which is also

$$f(\mathrm{s}) = J_2' - \bar{\sigma}_0^2, \tag{1.72}$$

where $J_2'$ is the second invariant of the deviatoric stress tensor [see Eq. (1.14)] and $\bar{\sigma}_0$ is the yield stress in tension. We also define $\sigma_i$ and $s_i$ as the principal values of the stress tensor and deviatoric stress tensor, respectively.

It is convenient to define the effective stress (or equivalent stress) as the tensile stress corresponding to any state of stress by means of a yield surface passing through the state of stress:

$$\bar{\sigma} = f(\boldsymbol{\sigma}). \tag{1.73}$$

For the von Mises flow, the effective stress takes the usual form:

$$\begin{aligned} \bar{\sigma} &= \left\{ \frac{1}{2}\left[(\sigma_{11} - \sigma_{22})^2 + (\sigma_{22} - \sigma_{33})^2 + (\sigma_{33} - \sigma_{11})^2 + 6\sigma_{12}^2 + 6\sigma_{23}^2 + 6\sigma_{31}^2\right] \right\}^{1/2} \\ &= \left\{ \frac{1}{2}[(\sigma_1 - \sigma_2)^2 + (\sigma_2 - \sigma_3)^2 + (\sigma_3 - \sigma_1)^2] \right\}^{1/2} = \left( \frac{3}{2} \sum_{i,j} s_{ij}^2 \right)^{1/2}. \end{aligned} \tag{1.74}$$

Thus, the yield surface separates states of stress based on the relationship of two scalars, the sigma-bar and the sigma-zero-bar scalars, where

$\bar{\sigma} < \bar{\sigma}_0$ is the elastic region enclosing $\boldsymbol{\sigma} = 0$, and
$\bar{\sigma} = \bar{\sigma}_0$ is elastic or plastic depending on the orientation of $\mathrm{d}\sigma$.

The remaining equations pertinent to the von Mises flow may be derived from these principles, as follows:

$$\begin{aligned}
\mathrm{d}\varepsilon_{11} &= \mathrm{d}\lambda^p(2\sigma_{11} - \sigma_{22} - \sigma_{33}) = \dot{\varepsilon}_{11}\mathrm{d}t,\\
\mathrm{d}\varepsilon_{22} &= \mathrm{d}\lambda^p(2\sigma_{22} - \sigma_{33} - \sigma_{11}) = \dot{\varepsilon}_{22}\mathrm{d}t,\\
\mathrm{d}\varepsilon_{33} &= \mathrm{d}\lambda^p(2\sigma_{33} - \sigma_{11} - \sigma_{22}) = \dot{\varepsilon}_{33}\mathrm{d}t,\\
\mathrm{d}\varepsilon_{12} &= 3\mathrm{d}\lambda^p\sigma_{12} = \dot{\varepsilon}_{12}\mathrm{d}t,\\
\mathrm{d}\varepsilon_{23} &= 3\mathrm{d}\lambda^p\sigma_{23} = \dot{\varepsilon}_{23}\mathrm{d}t,\\
\mathrm{d}\varepsilon_{31} &= 3\mathrm{d}\lambda^p\sigma_{31} = \dot{\varepsilon}_{31}\mathrm{d}t,
\end{aligned} \tag{1.75}$$

or

$$\mathrm{d}\boldsymbol{\varepsilon} = \frac{3}{2\bar{\sigma}_0}\mathbf{s}\,\mathrm{d}\bar{\varepsilon}, \qquad \dot{\boldsymbol{\varepsilon}} = \frac{3}{2}\frac{\dot{\bar{\varepsilon}}}{\bar{\sigma}_0}\mathbf{s}, \tag{1.76}$$

with the effective strain increment defined by

$$\mathrm{d}\bar{\varepsilon} = \left(\frac{2}{3}\sum_{i,j}\mathrm{d}\varepsilon_{ij}^2\right)^{1/2} = \left(\frac{2}{3}\sum_{i,j}\dot{\varepsilon}_{ij}^2\right)^{1/2}\mathrm{d}t = \dot{\bar{\varepsilon}}\,\mathrm{d}t. \tag{1.77}$$

Note that the notations $\mathrm{d}\varepsilon$ and $\dot{\varepsilon}$ are considered equivalent because the plasticity equations are homogeneous in $\mathrm{d}t$. That is, the equations are not changed by making the substitution $\mathrm{d}\varepsilon = \dot{\varepsilon}\mathrm{d}t$. The true time rate of change of $\varepsilon$ is not important except in the case of viscoplasticity.

Hill[3] introduced a quadratic yield function that provides a simple form of anisotropy (orthotropy) suitable for rolled sheet metal. The anisotropy constants are usually expressed in terms of the so-called plastic anisotropy parameter, $r$, defined as follows in a sheet tensile test:

$$r = \varepsilon_w/\varepsilon_t. \tag{1.78}$$

Here $\varepsilon_w$ is the plastic strain in the width direction and $\varepsilon_t$ is the plastic strain in the thickness direction of the sheet. The conventional axes in sheet forming are taken to be the rolling direction (RD, or 0°), the long-transverse direction (TD or 90°), and the thickness direction. For the full orthotropic symmetry parameters, $r_{90}$ and $r_0$, to be obtained, sheet tensile tests are performed in the 0° and 90° directions, respectively. Hill's quadratic yield function and auxiliary equations may be expressed as follows:

$$f(\sigma) = \frac{1}{r_{\mathrm{TD}}(1 + r_{\mathrm{RD}})}[r_{\mathrm{RD}}(\sigma_2 - \sigma_3)^2 + r_{\mathrm{TD}}(\sigma_3 - \sigma_1)^2 + r_{\mathrm{RD}}r_{\mathrm{TD}}(\sigma_1 - \sigma_2)^2] - \sigma_0^2, \tag{1.79}$$

[3] R. Hill, *Mathematical Theory of Plasticity* (Oxford University, London, 1950), Chap. 12.

so that the normality condition is

$$d\varepsilon_1 = \frac{d\bar{\varepsilon}}{\sigma_0} \frac{1}{r_{TD}(1 + r_{RD})} [r_{TD}(\sigma_1 - \sigma_3) + r_{RD} r_{TD}(\sigma_1 - \sigma_2)],$$
$$d\varepsilon_2 = \frac{d\bar{\varepsilon}}{\sigma_0} \frac{1}{r_{TD}(1 + r_{RD})} [r_{RD}(\sigma_2 - \sigma_3) - r_{RD} r_{TD}(\sigma_1 - \sigma_2)],$$
$$d\varepsilon_3 = \frac{d\bar{\varepsilon}}{\sigma_0} \frac{1}{r_{TD}(1 + r_{RD})} [-r_{RD}(\sigma_2 - \sigma_3) - r_{TD}(\sigma_1 - \sigma_3)]. \tag{1.80}$$

Then the effective strain increment is expressed as

$$d\bar{\varepsilon}^2 = \frac{r_{TD}(1 + r_{RD})}{(r_{RD} r_{TD} + r_{RD}^2 r_{TD} + r_{RD} r_{TD}^2)^2} [r_{RD}(r_{TD} d\varepsilon_2 - r_{RD} r_{TD} d\varepsilon_3)^2 + r_{TD}(r_{RD} d\varepsilon_1 - r_{RD} r_{TD} d\varepsilon_3)^2 + r_{RD} r_{TD}(r_{RD} d\varepsilon_1 - r_{TD} d\varepsilon_3)^2]. \tag{1.81}$$

Note that Hill's quadratic yield function is identical to that of von Mises for all $r$ values equal to unity.

It is often convenient to consider all directions lying in the sheet as equivalent, with the through-thickness direction different. This condition is called normal anisotropy (with planar isotropy only), in which case all of the $r$ values determined by sheet tensile tests will be equal. In fact, because this is seldom found precisely, an average $r$ value ($\bar{r}$, or sometimes just $r$) is determined from the individual measurements as follows:

$$\bar{r} = \frac{r_0 + 2r_{45} + r_{90}}{4}. \tag{1.82}$$

For normal anisotropy, Hill's quadratic yield function becomes

$$f(\sigma) = \frac{1}{(1 + r)} [(\sigma_2 - \sigma_3)^2 + (\sigma_3 - \sigma_1)^2 + r(\sigma_1 - \sigma_2)^2] - \bar{\sigma}_0^2 = \bar{\sigma}^2 - \bar{\sigma}_0^2, \tag{1.83}$$

and the auxiliary equations are

$$d\varepsilon_1 = \frac{d\bar{\varepsilon}}{\sigma_0} \frac{1}{(1 + r)} [(\sigma_1 - \sigma_3) + r(\sigma_1 - \sigma_2)],$$
$$d\varepsilon_2 = \frac{d\bar{\varepsilon}}{\sigma_0} \frac{1}{(1 + r)} [(\sigma_2 - \sigma_3) + r(\sigma_1 - \sigma_2)], \tag{1.84}$$
$$d\varepsilon_3 = \frac{d\bar{\varepsilon}}{\sigma_0} \frac{1}{(1 + r)} [(\sigma_3 - \sigma_2) + (\sigma_3 - \sigma_1)],$$

and

$$d\bar{\varepsilon}^2 = \frac{1 + r}{(1 + 2r)^2} [(d\varepsilon_2 - r\, d\varepsilon_3)^2 + (r\, d\varepsilon_3 - d\varepsilon_1)^2 + r(d\varepsilon_1 - d\varepsilon_2)^2]. \tag{1.85}$$

In later work,[4] Hill introduced a nonquadratic form of yield function that allows more adjustment of the shape while maintaining the measured $r$ values in tensile tests.

[4] R. Hill, Math. Proc. Camb. Phil. Soc. 85, 179–191 (1979); also see A. Parmer and P. B. Mellor, Int. J. Mech. Sci, 1978, vol. 20, p. 385 and p. 707, and A. Parmer and P. B. Mellor, *Mechanics of Sheet Metal Forming* (Plenum, New York, 1978), p. 53.

The quadratic yield function is recovered by setting the new anisotropy parameter $M$ equal to 2, and the von Mises function is recovered by setting $M = 2$ and $r = 1$. The nonquadratic yield function, which is usually used in plane stress cases (where $\sigma_3 = 0$), and auxiliary equations may be expressed as follows:

$$f(\boldsymbol{\sigma}) = \frac{2r+1}{2(r+1)}(\sigma_1 - \sigma_2)^M + \frac{1}{2(r+1)}(\sigma_1 + \sigma_2)^M - \sigma_0^M, \tag{1.86}$$

$$d\varepsilon_1 = \frac{d\bar{\varepsilon}}{\sigma_0^{M-1}}\left[\frac{2r+1}{2(1+r)}(\sigma_1 - \sigma_2)^{M-1} + \frac{1}{2(1+r)}(\sigma_1 + \sigma_2)^{M-1}\right],$$

$$d\varepsilon_2 = \frac{d\bar{\varepsilon}}{\sigma_0^{M-1}}\left[-\frac{2r+1}{2(1+r)}(\sigma_1 - \sigma_2)^{M-1} + \frac{1}{2(1+r)}(\sigma_1 + \sigma_2)^{M-1}\right],$$

$$d\varepsilon_3 = \frac{d\bar{\varepsilon}}{\sigma_0^{M-1}}\left[-\frac{1}{1+r}(\sigma_1 + \sigma_2)^{M-1}\right], \tag{1.87}$$

$$d\bar{\varepsilon}^2 = \frac{[2(1+r)]^{1/M}}{2}\left[\frac{1}{(1+r)^{1/(M-1)}}(\varepsilon_1 - \varepsilon_2)^{M/(M-1)} + (\varepsilon_1 + \varepsilon_2)^{M/(M-1)}\right]^{(M-1)/M}. \tag{1.88}$$

## 1.7 Viscoplasticity

For the case of a general plastic or viscoplastic potential $\varphi$, where $\varphi$ is convex and positive, and $\varphi(0) = 0$, the stresses are obtained as follows:

$$\sigma = \frac{\partial\varphi(\dot{\varepsilon})}{\partial\dot{\varepsilon}}, \quad \sigma_{ij} = \frac{\partial\varphi(\dot{\varepsilon})}{\partial\dot{\varepsilon}_{ij}}, \tag{1.89}$$

and the viscoplastic functional may be written as

$$\Phi(\mathbf{v}) = \int_\Omega \varphi(\dot{\varepsilon})dV - \int_{\partial\Omega} \mathbf{T}^d\mathbf{v}\,dS. \tag{1.90}$$

A useful form of the viscoplastic potential is known as the Norton–Hoff law:

$$\varphi(\dot{\varepsilon}) = \frac{K}{m+1}(\sqrt{3}\dot{\bar{\varepsilon}})^{m+1}, \tag{1.91}$$

which, according to Eq. (1.87), corresponds to the following constitutive equation:

$$\mathbf{s} = 2K(\sqrt{3}\dot{\bar{\varepsilon}})^{m-1}\dot{\varepsilon}. \tag{1.92}$$

The Norton–Hoff functional is a special case of viscoplastic functionals:

$$\Phi(\mathbf{v}) = \int_\Omega \frac{K}{m+1}(\sqrt{3}\dot{\bar{\varepsilon}})^{m+1}dV - \int_{\partial\Omega} \mathbf{T}^d\mathbf{v}\,dS. \tag{1.93}$$

A functional for a rigid-plastic material (i.e., one without an elastic component of the material response) can be constructed by utilizing Eq. (1.90). The plastic potential $\varphi$ differs from the yield function by only a constant. The stresses may be obtained from the plastic potential according to Eq. (1.89). For example, the von Mises law corresponds to Eqs. (1.91)–(1.93) with $m = 1$. It should be noted that rigid plasticity has limited usefulness without modification because the stresses inside the yield function are indeterminate with respect to strain increments. Thus, for small strain increments, a unique stress cannot be obtained. In order to circumvent this numerical problem,[5] the stresses are usually "regularized," such that very small strains produce unique stresses inside the yield function.

A viscoplastic potential may be used for the regularization of rigid plasticity by allowing small strain rates (or strain increments) to be assigned unique stress values. An example of such a viscoplastic potential is

$$\mathrm{s} = 2K\left[\sqrt{3}\left(\dot{\varepsilon}_0^2 + \dot{\bar{\varepsilon}}^2\right)\right]^{(m-1)/2} \dot{\varepsilon}, \tag{1.94}$$

where $\dot{\varepsilon}_0$ is a small positive constant. For $m = 1$ and $\dot{\bar{\varepsilon}} \gg \dot{\varepsilon}_0^2$, Eq. (1.94) tends to the von Mises flow rule, Eq. (1.76), with $K = \bar{\sigma}_0/\sqrt{3}$.

## 1.8 Elastoplasticity

Elastoplastic behavior is considered by decomposing the strain increment (or strain rate) into a part corresponding to elastic behavior (which is thus recoverable) and a plastic part (which is permanent):

$$\dot{\varepsilon} = \dot{\varepsilon}^e + \dot{\varepsilon}^p. \tag{1.95}$$

Then the isotropic elastic constitutive equation may first be written as

$$\dot{\sigma} = \lambda \ \mathrm{trace}(\dot{\varepsilon}^e)\mathrm{I} + 2\mu\dot{\varepsilon}^e, \tag{1.96}$$

or, with the Jauman stress rate to ensure objectivity, as

$$\mathrm{d}_J\sigma/\mathrm{d}t = \dot{\sigma} - \dot{\omega}\sigma + \sigma\dot{\omega} = \lambda \ \mathrm{trace}(\dot{\varepsilon}^e)\mathrm{I} + 2\mu\dot{\varepsilon}^e, \tag{1.97}$$

where $\dot{\omega}$ is the spin tensor.

Furthermore, it is convenient to decompose the elastic strain rate or increment into a dilatational part $\dot{\theta}$ (which has no plastic counterpart),

$$\dot{\theta} = \mathrm{trace}(\dot{\varepsilon}^e), \tag{1.98}$$

and a deviatoric part $\dot{\mathrm{e}}^e$,

$$\dot{\mathrm{e}}^e = \dot{\varepsilon}^e - \frac{1}{3}\dot{\theta}\mathrm{I}. \tag{1.99}$$

Then the elastic law becomes

$$\dot{\mathrm{s}} = 2\mu\dot{\mathrm{e}}^e - \dot{\omega}\sigma + \sigma\dot{\omega} \tag{1.100}$$

and

$$\frac{1}{3}\mathrm{trace}(\dot{\sigma}) = -\dot{p} = \kappa \ \mathrm{trace}(\dot{\varepsilon}), \tag{1.101}$$

with the bulk modulus, or compressibility coefficient, $\kappa = \lambda + \frac{2}{3}\mu$.

The plastic part of the strain rate obeys the following equations in term of the plastic yield criterion $f$.

[5] The problem is purely a numerical one because of the rigid-plasticity idealization. Nature solves the problem by introducing either elastoplasticity or viscoplasticity for real materials.

- Deformation is only elastic when the yield function is negative:

$$\dot{\varepsilon}^p = 0 \qquad \text{if } f(\sigma) < 0, \quad \text{or } f(\sigma) = 0 \ \text{ and } \frac{\partial f}{\partial \sigma} : \dot{\sigma} < 0. \tag{1.102}$$

- It is elastoplastic if the yield function is nil; more precisely,

$$\dot{\varepsilon}^p = \dot{\lambda}^p \frac{\partial f}{\partial \sigma} \qquad \text{if } f(\sigma) = 0 \ \text{ and } \frac{\partial f}{\partial \sigma} : \dot{\sigma} \geq 0. \tag{1.103}$$

- The plastic multiplier must be positive $\dot{\lambda}^p \geq 0$, and the additional scalar equation holds:

$$\frac{\partial f}{\partial \sigma} : \dot{\sigma} + \frac{\partial f}{\partial \bar{\varepsilon}} \dot{\bar{\varepsilon}} = 0. \tag{1.104}$$

There is no convenient functional form of elastoplastic mechanical formulation, so the virtual work principal is used and discretized directly for finite-element applications.

## 1.9 Elastoviscoplasticity

The drawback to the Norton–Hoff viscoplastic law is that nonreversible straining can occur at very small stresses. That is, there is no true yield function, as usually observed in room-temperature metal deformation. An alternate form, introduced by Perzyna, combines the idea of a plastic yield function with a strain-rate-dependent flow stress:

$$\dot{\varepsilon}^p = \frac{\partial \varphi'}{\partial \sigma}, \tag{1.105}$$

where $\varphi'$ is the complementary viscoplastic potential, which is often defined by

$$\varphi'(\sigma) = \frac{m}{m+1} K \left\langle \frac{\sigma_{\text{eq}} - R}{K} \right\rangle^{(m+1)/m}. \tag{1.106}$$

Here $R$ represents the zero-strain-rate yield function and we use the notation

$$\langle x \rangle = 0 \qquad \text{if } x < 0, \quad \langle x \rangle = x \qquad \text{if } x \geq 0. \tag{1.107}$$

## 1.10 Friction

Friction is one of the largest sources of error and uncertainty in the modeling of metal-forming operations. The problem usually reduces to knowing a friction coefficient or friction factor under true forming conditions (including strain rates, displacement rates, pressure, surface roughness, lubricant quantities and conditions, local temperature, and so on). Thus, typically only two friction formulations are used.

Coulomb's friction law (also known as Amonton's law) states that the friction force (or friction stress) resisting relative motion is proportional to the normal force (or contact pressure):

$$\tau_F = \mu |\sigma_N|, \tag{1.108}$$

where $\mu$ is known as the friction coefficient. Whenever the unbalanced force at a point is less than the friction force, the point does not move, or is considered "sticking." Whenever the internal force equals or exceeds the friction force, the friction force impedes relative motion. In the latter case, when the sliding velocity between the part and the tool is denoted $\Delta \mathbf{v}$, Eq. (1.108) becomes

$$\tau_F = -\mu |\sigma_N| \frac{\Delta \mathbf{v}}{\|\Delta \mathbf{v}\|}. \tag{1.109}$$

The second common friction formulation is variously known as sticking friction or Tresca friction, which may be written as follows:

$$\tau_F = -m\tau_{\max} \frac{\Delta \mathbf{v}}{\|\Delta \mathbf{v}\|}, \tag{1.110}$$

where $m$ is known as the friction factor, and $\tau_{\max}$ is the shear strength of the material. For a Tresca yield function we have

$$\tau_{\max} = \frac{\bar{\sigma}_0}{2}, \tag{1.111}$$

and for a von Mises yield criterion it is

$$\tau_{\max} = \frac{\bar{\sigma}_0}{\sqrt{3}}. \tag{1.112}$$

Here $m$ can be interpreted simply as an undetermined constant to be measured (as in the Coulomb case), or can be taken to be unity in the case in which no sliding takes place, except by shearing the base material. (Such conditions are seldom encountered in important metal-forming operations.)

CHAPTER TWO

# Introduction to the Finite-Element Method

There are many books devoted to a general introduction to the finite-element method. It is impossible to list all of them here, but those referenced in the footnotes[1,2,3,4] appear in increasing order of difficulty. Interested readers could select one or two for study, depending on their backgrounds. In this chapter a very simple introduction is given, mainly in view of applications to the numerical modeling of metal-forming processes.

A first introduction is given in Section 2.1, with a comparison with another popular numerical method: the finite-difference method. The selected thermal problem allows a complete "hand calculation" by the two methods, and a comparison of the results with an analytical solution. The aim of this first overview is to summarize very briefly the main aspects (and vocabulary) of the method. It shows that a quasi-intuitive approach is conceivable and could help to demythologize the finite-element method.

A more systematic approach follows in Sections 2.2 to 2.5, in order to separate as clearly as possible the steps that lead to the rational development of the method. In most occasions the basic concepts are first explained in detail by using one-dimensional examples, and the generalization is only mentioned, as more detail will be given in Chapters 3 and 4 for two- and three-dimensional approaches.

The basic idea is to follow four steps:

1. Establish the physical equations of the model (see Chapter 1, or an earlier work[5] on this subject).
2. Find an appropriate mathematical formulation for a finite-element treatment.
3a. Discretize the mathematical problem spatially by choosing the mesh and the shape functions (Sections 2.2 and 2.3).
3b. That is, convert the partial differential equations into a set of ordinary equations (Sections 2.4 and 2.5).
4. Solve the resulting set of ordinary equations.

[1] R. D. Cook, *Concepts and Applications of Finite Element Analysis* (Wiley, New York, 1974).
[2] O. C. Zienkiewicz and R. L. Taylor, *The Finite Element Method* (McGraw-Hill, New York, 1989).
[3] T. J. R. Hughes, *The Finite Element Method. Linear Static and Dynamic Finite Element Analysis* (Prentice-Hall, Englewood Cliffs, N.J., 1987).
[4] R. E. White, *An Introduction to the Finite Element Method with Application to Nonlinear Problems* (Wiley, New York, 1985).
[5] R. H. Wagoner and J.-L. Chenot, *Fundamentals of Metal Forming* (Wiley, New York, 1997).

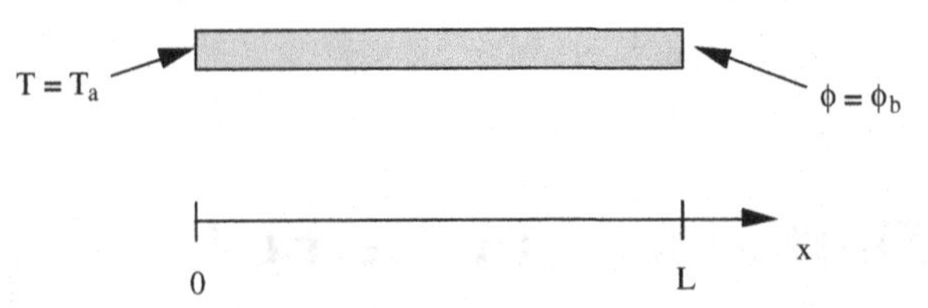


Figure 2.1. The one-dimensional thermal problem for a beam.

## 2.1 Comparison of Finite-Difference and Finite-Element Methods with Analytical Solutions

A simple example is introduced in order to distinguish the finite-difference method (FDM) from the finite-element method (FEM); the two appear very similar at first sight. Consider a one-dimensional thermal problem defined on a beam of section $S$, length $L$, and thermal conductivity $k$. At the ends of the beam a constant temperature $T_a$ is imposed at the left, and a constant heat flux $\phi_b$ on the right. We consider only the steady-state case, without internal heat sources or sinks.

As indicated in Fig. 2.1, the O$\hat{\mathbf{x}}$ axis is chosen parallel to the beam, with an origin marked "O" corresponding to the left side of the beam. With these conventions, and if the heat conduction perpendicular to O$\hat{\mathbf{x}}$ is assumed negligible, the heat equation can be written as

$$\frac{\mathrm{d}}{\mathrm{d}x}\left(kS\frac{\mathrm{d}T}{\mathrm{d}x}\right) = 0. \tag{2.1a}$$

This equation means that the heat flux through each cross section $S$ is constant, corresponding to the steady-state case. The boundary conditions correspond to a prescribed temperature on the left,

$$T(0) = T_a, \tag{2.1b}$$

and to a heat flux on the right:

$$-k\frac{\mathrm{d}T}{\mathrm{d}x}(L) = \phi_b. \tag{2.1c}$$

If $k$ and $S$ are constants, the solution is very simple. Equation (2.1a) reduces to

$$\frac{\mathrm{d}^2T}{\mathrm{d}x^2} = 0, \tag{2.2}$$

which shows that $T$ is simply linear on the interval. With the end conditions given by Eqs. (2.1a) and (2.1b), we obtain

$$T = T_a - \frac{\phi_b}{k}x. \tag{2.3}$$

---

**Exercise 2.1: Calculate the analytical solution of the problem corresponding to Eqs. (2.1a)–(2.1c), when the cross-sectional area is not constant and is instead given by**

$$\mathbf{S} = \mathbf{S}_0\left(1 + \alpha\frac{\mathbf{x}}{\mathbf{L}}\right), \tag{2.1–1}$$

**where $\alpha$ is sufficiently small with respect to unity so that the approximation of unidirectional heat conduction is valid.**

Substituting Eq. (2.1–1) into Eq. (2.1a) and integrating with respect to $x$, we obtain

$$kS_0\left(1+\frac{\alpha x}{L}\right)\frac{dT}{dx}=c, \tag{2.1–2}$$

where $c$ is a constant that can be evaluated by using Eq. (2.1c):

$$k\frac{dT}{dx}(L)=\frac{c}{S_0(1+\alpha)}=-\phi_b, \tag{2.1–3}$$

so that

$$\frac{dT}{dx}=-\frac{\phi_b(1+\alpha)}{k}\frac{1}{1+(\alpha x/L)}, \tag{2.1–4}$$

which is again integrated with the help of Eq. (2.1b):

$$T=T_a-\frac{L}{\alpha}\frac{\phi_b(1+\alpha)}{k}\ln\left(1+\frac{\alpha x}{L}\right). \tag{2.1–5}$$

---

For a simple illustration of a numerical treatment of the problem, we will take a length $L=2$, and divide it into two segments of unit length, as indicated in Fig. 2.2. The three points of subdivision will be called *nodes*, and the temperatures $T_1$, $T_2$, and $T_3$ are introduced as the unknowns. They correspond to the nodes numbered 1, 2, and 3, with coordinates $x_1=0$, $x_2=1$, and $x_3=2$.

### A Finite-Difference Example

In the finite-difference method, Eqs. (2.1a)–(2.1c) will be approximated in terms of unknown nodal values. At the left end of the beam, the boundary condition can be written as

$$T_1=T_a \tag{2.4}$$

and, for the right end, the derivative of $T$ is approximated by the simplest first-order Taylor expansion[6]:

$$-k\frac{T_3-T_2}{L/2}=\phi_b. \tag{2.5}$$

After application of the boundary conditions, the only inner (nonboundary) node is node 2, for which the heat equation, Eq. (2.1a), reduces to Eq. (2.3). As is usual in the FDM, a second-order Taylor expansion is used to approximate the second

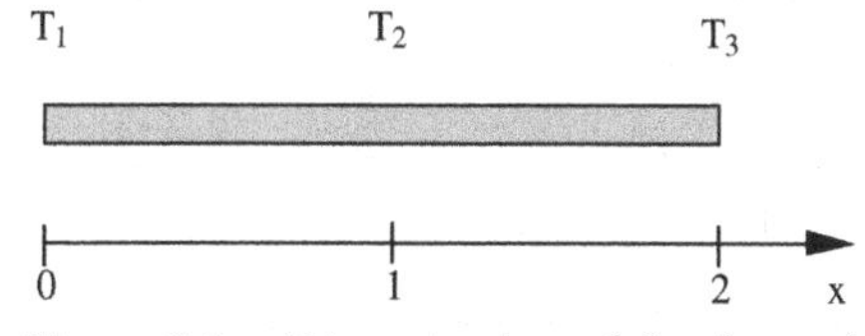


Figure 2.2. Discretization of the thermal beam problem.

[6] That is, the derivatives are interpreted as finite slopes based on the finite differences in $x$ and $T$; hence the name for the method. Note that as $\Delta x$ tends to zero, the limiting form of the true derivative is automatically recovered. This property is fundamental to all properly formulated discrete numerical methods.

derivative, the general form of which is written as

$$T(x+h) \cong T(x) + h\frac{\mathrm{d}T}{\mathrm{d}x}(x) + \frac{h^2}{2}\frac{\mathrm{d}^2T}{\mathrm{d}x^2}(x). \tag{2.6}$$

We put $x = x_2$, and $h = L/2 = 1$ into Eq. (2.6) to obtain the approximated temperature at $x_3$:

$$T_3 \cong T_2 + \frac{\mathrm{d}T}{\mathrm{d}x}(x_2) + \frac{1}{2}\frac{\mathrm{d}^2T}{\mathrm{d}x^2}(x_2). \tag{2.6a}$$

Similarly, the approximated temperature at $x_1$ is written as

$$T_1 \cong T_2 - \frac{\mathrm{d}T}{\mathrm{d}x}(x_2) + \frac{1}{2}\frac{\mathrm{d}^2T}{\mathrm{d}x^2}(x_2). \tag{2.6b}$$

If Eq. (2.6a) is added to Eq. (2.6b), the classical approximation of Eq. (2.2) is obtained[7]:

$$\frac{\mathrm{d}^2T}{\mathrm{d}x^2}(x_2) = T_3 + T_1 - 2T_2 = 0. \tag{2.7}$$

Equations (2.4), (2.5), and (2.7) can be put into matrix form:

$$\begin{bmatrix} 1 & 0 & 0 \\ 1 & -2 & 1 \\ 0 & -1 & 1 \end{bmatrix} \begin{bmatrix} T_1 \\ T_2 \\ T_3 \end{bmatrix} = \begin{bmatrix} T_a \\ 0 \\ -\phi_b/k \end{bmatrix}. \tag{2.8}$$

The solution of Eq. (2.8) can be found by simple substitution:

$$T_1 = T_a, \quad T_2 = T_a - \frac{\phi_b}{k}, \quad T_3 = T_a - 2\frac{\phi_b}{k}.$$

A comparison with the analytical result [Eq. (2.3)] shows that the FDM obtains the exact solution in this simple case, because the linearity of the solution matches the assumed form used to relate nodal values and spatial derivatives. In all other cases, the exact agreement holds only for infinitesimal distances between nodes. A more general problem is illustrated in the following exercise.

---

**Exercise 2.2: Calculate the FDM solution of the thermal problem in a beam, for a nonconstant section S, given by the linear expression**

$$\mathbf{S = S_0(1 + 0.1x),} \tag{2.2–1}$$

**and compare it with the exact solution determined in Exercise 2.1 when $\alpha = 0.2$ and $L = 2$.**

Equation (2.1a) gives

$$kS_0(1+0.1x)\frac{\mathrm{d}^2T}{\mathrm{d}x^2} + 0.1kS_0\frac{\mathrm{d}T}{\mathrm{d}x} = 0, \tag{2.2–2}$$

which (after simplification by $kS_0$) is discretized at node $x = 1$ into

$$1.1(T_3 + T_1 - 2T_2) + 0.1\left(\frac{T_3 - T_2}{2}\right) = 0. \tag{2.2–3}$$

[7] The same result can also be obtained more simply from the definition of the second derivative as the "derivative of the derivative," which involves the difference of expressions similar to Eq. (2.5).

The other conditions, Eqs. (2.1b) and (2.1c), are the same as the constant section case. The matrix form of the resulting equations are:

$$\begin{bmatrix} 1 & 0 & 0 \\ 1.05 & -2.2 & 1.15 \\ 0 & -1 & 1 \end{bmatrix} \begin{bmatrix} T_1 \\ T_2 \\ T_3 \end{bmatrix} = \begin{bmatrix} T_a \\ 0 \\ -\phi_b/k \end{bmatrix}. \tag{2.2–4}$$

The linear system can be solved by successive substitution, leading to:

$$T_1 = T_a, \quad T_2 = T_a - (1.15/1.05)\phi_b/k, \quad T_3 = T_a - (2.2/1.05)\phi_b/k. \tag{2.2–5}$$

With the use of Exercise 2.1, the analytical solution is

$$T = T_a - 12\ln(1 + 0.1x)\phi_b/k, \tag{2.2–6}$$

and a comparison of the two solutions gives

|  | *analytical solution* | *FDM solution* |
|---|---|---|
| $T_1$ | $T_a$ | $T_a$ |
| $T_2$ | $T_a - 1.144\phi_b/k$ | $T_a - 1.095\phi_b/k$ |
| $T_3$ | $T_a - 2.188\phi_b/k$ | $T_a - 2.095\phi_b/k$ |

(2.2–7)

The results are no longer exact because the solution is no longer linear. However, the result is a good approximation in spite of the small number of nodes, as the variation from linearity is modest.

### A Finite-Element Example

In the finite-element method, an approximate solution is built from the values of the temperature at selected nodes by introducing a piecewise linear interpolation $\bar{T}$, as pictured in Fig. 2.3.[8] The linear interpolation was chosen here for simplicity; other examples are given in Section 2.3.

Assuming linearity in each element, the analytical expression of $\bar{T}$ over the two intervals[9] can be easily expressed in term of the nodal values $T_1$, $T_2$, and $T_3$:

$$\text{interval } [0, 1]: \quad \bar{T} = T_1 + (T_2 - T_1)x, \quad \frac{\mathrm{d}T}{\mathrm{d}x} = T_2 - T_1, \tag{2.9a}$$

$$\text{interval } [1, 2]: \quad \bar{T} = T_2 + (T_3 - T_2)(x - 1), \quad \frac{\mathrm{d}T}{\mathrm{d}x} = T_3 - T_2. \tag{2.9b}$$

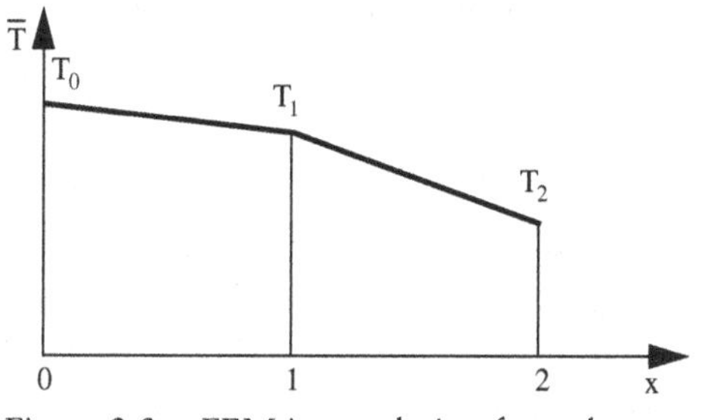


Figure 2.3. FEM interpolation for a three-node discretization.

[8] Note the basic difference from the FDM, in which there is no assumed approximation to the solution *between* nodes. In the FEM, all values are expressed between nodes according to assumed functions that may not always exit at nodes.

[9] These intervals will be called elements in Section 2.2. In this example, for the sake of brevity, only the most important ideas are outlined.

Now a *variational principle*[10] is introduced, which can be shown to be equivalent to the governing differential equations, by the method presented in Chapter 5. The variational principle states that the exact solution of the problem defined by Eqs. (2.1a)–(2.1c) is the function $T$, defined on the segment [0, 2], which minimizes the functional

$$I(T) = \frac{1}{2}\int_0^2 kS\left(\frac{\mathrm{d}T}{\mathrm{d}x}\right)^2 \mathrm{d}x + ST(2)\phi_b, \tag{2.10}$$

subject to the boundary condition $T(0) = T_1 = T_a$. Now if the approximate solution $\bar{T}$ is introduced in Eq. (2.10), we obtain the corresponding discrete functional:

$$I(\bar{T}) = \frac{1}{2}kS\{(T_2 - T_1)^2 + (T_3 - T_2)^2\} + ST_3\phi_b. \tag{2.11}$$

We obtain the numerical solution by minimizing $I$ with respect to $T_2$ and $T_3$ only (as the left-hand side condition is $T_1 = T_a$):

$$\frac{\partial I(\bar{T})}{\partial T_2} = kS\{T_2 - T_1 + T_2 - T_3\} = 0,$$
$$\frac{\partial I(\bar{T})}{\partial T_3} = kS\{T_3 - T_2\} + S\phi_b = 0. \tag{2.12}$$

We obtain a linear system, where the coefficient matrix is called the *stiffness matrix*:

$$\begin{bmatrix} 1 & 0 & 0 \\ -kS & 2kS & -kS \\ 0 & kS & -kS \end{bmatrix} \begin{bmatrix} T_1 \\ T_2 \\ T_3 \end{bmatrix} = \begin{bmatrix} T_a \\ 0 \\ S\phi_b \end{bmatrix}. \tag{2.13}$$

After dividing the second and third row by $-kS$ in Eq. (2.7), we find that the new form is identical to Eq. (2.8), so that the numerical solution is identical to the FDM solution and to the analytical solution[11] given by Eq. (2.3).

---

**Exercise 2.3: Repeat Exercise 2.2, using the FEM, with a piecewise linear interpolation for a nonconstant cross section (as in Exercise 2.1).**

Using the expression of $\bar{T}$, given by Eqs. (2.9a) and (2.9b), and the general expression of the functional defined by Eq. (2.10) (with a nonconstant section), we can write the approximate form of the variational principle as

$$I(\bar{T}) = \frac{1}{2}kS_0\int_0^1 (1 + 0.1x)(T_2 - T_1)^2\,\mathrm{d}x,$$
$$+\frac{1}{2}kS_0\int_1^2 (1 + 0.1x)(T_3 - T_2)^2\,\mathrm{d}x + 1.2S_0T_3\phi_b. \tag{2.3–1}$$

After integration with respect to $x$, we obtain the explicit form:

$$I(\bar{T}) = kS_0\left[\frac{1.05}{2}(T_2 - T_1)^2 + \frac{1.15}{2}(T_3 - T_2)^2\right] + 1.2S_0T_3\phi_b. \tag{2.3–2}$$

[10] The concepts and usage of variational principles, weak forms, and functionals are presented briefly in Chap. 1 and in more detail in Chap. 5 of R. H. Wagoner and J.-L. Chenot, *Fundamentals of Metal Forming* (Wiley, New York, 1997). Note that there are several ways to formulate the governing equations, which give equivalent results when they can be used.

[11] As we found with the FDM, the FEM solution is exact when the solution is linear because the assumed form of $T$ between nodes happened to be precisely correct. This agreement is fortuitous.

The solution is obtained by the minimization of $I$ (subject to the boundary condition):

$$T_1 = T_a$$

$$\frac{dI}{dT_2} = kS_0[1.05(T_2 - T_1) + 1.15(T_3 - T_2)] = 0$$

$$\frac{dI}{dT_3} = kS_0 1.15(T_3 - T_2) + 1.2S_0\phi_b = 0. \tag{2.3–3}$$

The second and third equations are divided by $kS_0$, and we get successively:

$$1.05T_1 - 2.2T_2 + 1.15T_3 = 0,$$

$$1.15(-T_2 + T_3) = 1.2\phi_b/k. \tag{2.3–4}$$

The matrix form of the three equations is obviously

$$\begin{bmatrix} 1 & 0 & 0 \\ 1.05 & -2.2 & 1.15 \\ 0 & -1.15 & 1.15 \end{bmatrix} \begin{bmatrix} T_1 \\ T_2 \\ T_3 \end{bmatrix} = \begin{bmatrix} T_a \\ 0 \\ -1.2\phi_b/k \end{bmatrix}, \tag{2.3–5}$$

which is not exactly the same as the matrix equation originating from the FDM. The numerical result is easily derived:

$$T_1 = T_a,$$

$$T_2 = T_a - (1.2/1.05)\phi_b/k = T_a - 1.143\phi_b/k,$$

$$T_3 = T_a - 2.2{}^*1.2/(1.05{}^*1.15)\phi_b/k = T_a - 2.186\phi_b/k. \tag{2.3–6}$$

The numerical solution is very close to the analytical solution and provides here a better accuracy than the FDM solution of Exercise 2.2.[12] The comparison of the difference $\Delta T_D$ between the analytical solution and the FDM one, and of the difference $\Delta T_E$ between the analytical solution and the FEM one, is made in the figure below, in which we have assumed $\phi_b/k = 1$.

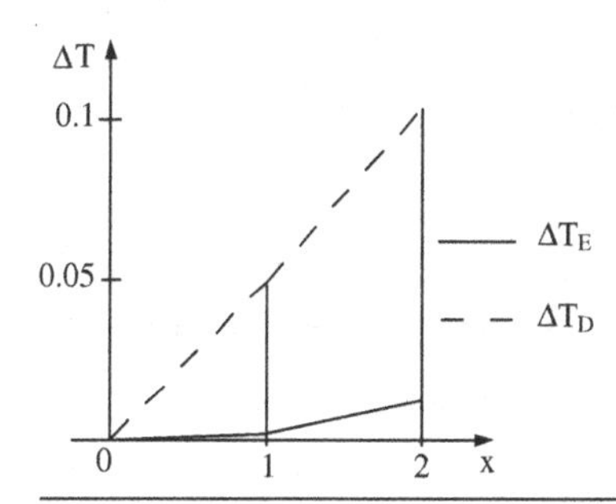


## 2.2 Spatial Discretization

In the previous section we treated a very simple case in one dimension, with two elements. Thus we did not face any serious problem of node or element numbering. For more complex problems, when a realistic solution must be computed, we generally need a much greater number of nodes and elements, and the precise correspondence between nodes and elements requires a systematic approach.

In this chapter we will restrict ourselves to static or stationary problems, for which time does not play an explicit role. We suppose that the physical problem to be solved is defined in a geometric domain $\Omega$. The basic idea in the finite-element method is

[12] The apparently poor accuracy of the FDM solution is caused by the right-hand boundary condition, which is imposed with a poor approximation. A better approximation of this boundary condition results in an improvement of the solution.

to *discretize* the domain $\Omega$, by cutting it into a finite number of nonoverlapping subdomains. These are called *element domains* or, more simply, *elements*. If in this process of discretization, $n_e$ elements are introduced, we label them as

$$\Omega_e \qquad e = 1, 2, \ldots, n_e.$$

In doing so, we must at the same time create $n_n$ geometrical points in the domain $\Omega$, called *nodes*, which are numbered in a given order; this is *global numbering*. Within an element there is a smaller number of nodes, which are again numbered, with a *local numbering*.

### Examples of One-Dimensional Elements

For a one-dimensional problem, the domain $\Omega$ will be a segment $[a, b]$, with ends corresponding to coordinates $x = a$ and $x = b\,(a < b)$. To define the elements, we introduce $n_n$ geometrical points or nodes in the segment, so that their coordinates are

$$x_1(= a), x_2, \ldots, x_{n_n-1}, x_{n_n}(= b).$$

The global numbering is clearly indicated: it often corresponds to the order of increasing $x$ coordinates (for a one-dimensional problem). The nodes can be unequally spaced.

The simplest elements that can be considered are the segments

$$\Omega_e = [x_e, x_{e+1}] \qquad e = 1, 2, \ldots, n_e. \tag{2.14}$$

In order to recognize the elements to which a node belongs, the *element node array* lg is introduced, which is also called the *connection table*. lg allows the transfer of local numbering, in an element, to global node numbers. The definition of lg by

$$\lg(m, e) = n$$

means that the $m$th local node, in element number $e$, is the node numbered $n$ in the meshing of the global domain. More specifically, in the one-dimensional example with the previous elements it is easy to verify that

| | | |
|---|---|---|
| for element number 1 | $\lg(1, 1) = 1$ | $\lg(2, 1) = 2$ |
| for element number 2 | $\lg(1, 2) = 2$ | $\lg(2, 2) = 3$ |
| for element number $n_e$ | $\lg(1, n_e) = n_e$ | $\lg(2, n_e) = n_e + 1 = n_n$ |

In fact, even in this very simple example, we see most clearly the two kinds of node numbering that are used. The first is the local numbering within each element. Here the index $m$ can take the values 1 or 2. The second is the global numbering within the entire domain. Here the $n$ index runs from 1 to $n_n$. We note also in this example that for each *internal node*,[13] there are two *connected elements* specified, which can be determined by using the connection table in a reverse way.

The general process of domain decomposition into a set of elements is schematically represented in Fig. 2.4, for the two-node line element. Of course, the elements can be defined differently, even for a given distribution of nodes. For example,

[13] Here the internal node is one that is not a boundary node of the global domain. The term "internal node" is also used to designate a node within a single element.

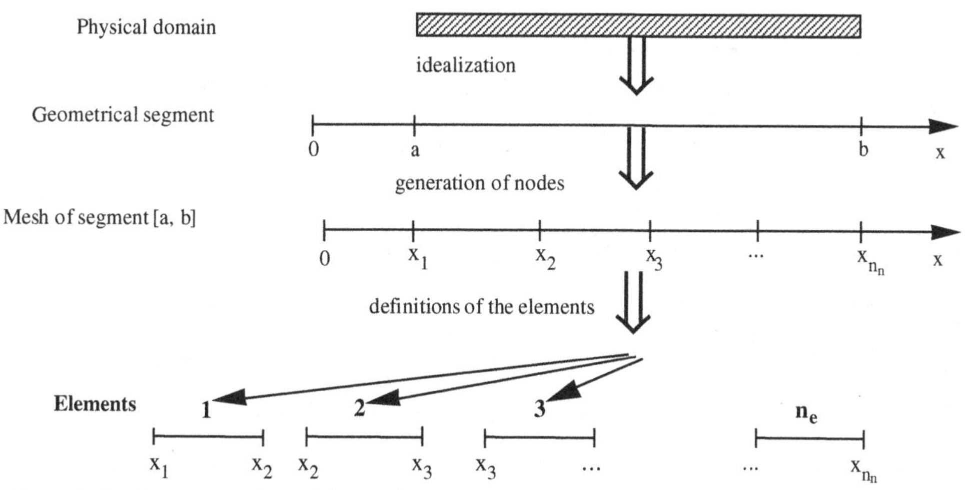


Figure 2.4. Representation of the meshing process.

each element can include three nodes, and provided that $n_n$ is odd, the elements would be

$$\Omega_e = [x_{2e-1}, x_{2e+1}] \qquad \text{for } e = 1, 2, \ldots, n_e = 1/2(n_n - 1). \tag{2.15}$$

For this element the local nodes (or internal nodes) will be $x_{2e-1}, x_{2e}, x_{2e+1}$.

---

**Exercise 2.4: Write the element node array for the three-node elements defined by Eq. (2.15).**

| *element* | | *element node array* | |
|---|---|---|---|
| 1 | $lg(1, 1) = 1$ | $lg(2, 1) = 2$ | $lg(3, 1) = 3$ |
| 2 | $lg(2, 1) = 3$ | $lg(2, 2) = 4$ | $lg(3, 2) = 5$ |
| $e$ | $lg(1, e) = 2e - 1$ | $lg(2, e) = 2e$ | $lg(3, e) = 2e + 1$ |
| $n_e$ | $lg(1, n_e) = n_n - 2$ | $lg(2, n_e) = n_n - 1$<br>with $n_e = 1/2(n_n - 1)$ | $lg(3, n_e) = n_n$ |

---

### A Two-Dimensional Element

The simplest generalization of the concept of meshing in two dimensions involves *triangulation*. We suppose first that our problem is defined within a two-dimensional domain $\Omega$, which has been approximated by a polygon $\bar{\Omega}$. The approximate domain is then divided into triangular elements $\Omega_e$, with the index $e$ ranging from 1 to $n_e$, where $n_e$ is the total number of elements. The triangulation is *consistent* if no triangles are allowed to overlap and if no "artificial hole" is present in the meshing. This is illustrated in Figs. 2.5(a) and 2.5(b).

The procedure can be performed "by hand," but this becomes tedious when a large number of nodes is involved in two-dimensional problems, and it is nearly impossible for complicated three-dimensional cases. A more rational method will be discussed in Chapter 12; it is based on Delauney triangulation, which allows fully automatic meshing.

The triangles are defined from a set of $n_n$ nodes, distributed in the discretized domain. All the nodes are numbered globally, and the vector coordinate of node

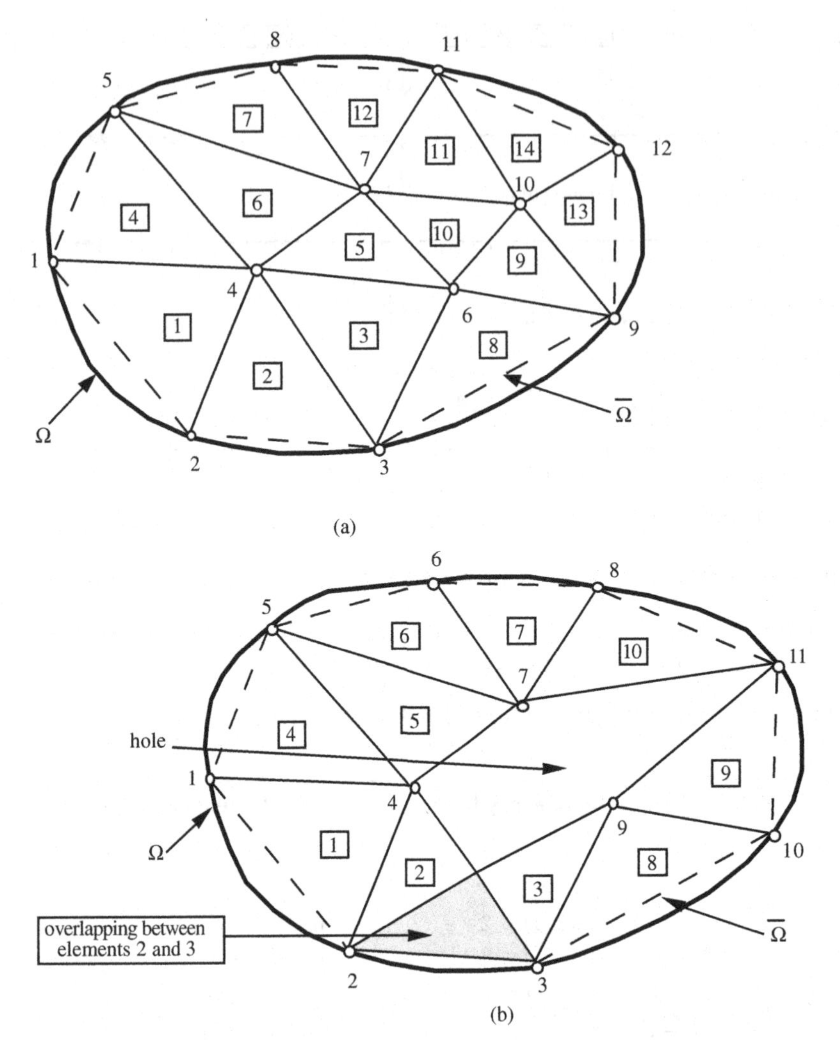


Figure 2.5. Examples of meshes: (a) consistent and (b) inconsistent.

$n$ is denoted by $\mathbf{X}_n$ ($n$ ranging from 1 to $n_n$), with components $X_{1n}$ and $X_{2n}$. The global node vector is a mathematical vector defined in a $2n_n$ abstract space, whose components are the nodal coordinates, listed in a coherent order:

$$\mathbf{X} \rightarrow [X] = \begin{bmatrix} X_{11} \\ X_{21} \\ X_{12} \\ X_{22} \\ \cdots \\ X_{1n_n} \\ X_{2n_n} \end{bmatrix} \begin{matrix} \left.\begin{matrix} \\ \\ \end{matrix}\right\} \text{node } 1 \\ \left.\begin{matrix} \\ \\ \end{matrix}\right\} \text{node } 2 \\ \\ \left.\begin{matrix} \\ \\ \end{matrix}\right\} \text{node } n_n \end{matrix}$$

More complex two- or three-dimensional element shapes are sometimes preferred for the practical computation of metal deformation. They will be discussed later in Chapter 4.

**Exercise 2.5: Put values to the element node array for the first three elements in Fig. 2.5(a), in such a way that the nodes for a triangle are numbered in a counterclockwise direction.**

The solution is simply

$$\begin{aligned}
&\lg(1,1) = 1, \quad \lg(2,1) = 2, \quad \lg(3,1) = 4,\\
&\lg(1,2) = 2, \quad \lg(2,2) = 3, \quad \lg(3,2) = 4,\\
&\lg(1,3) = 3, \quad \lg(2,3) = 6, \quad \lg(3,3) = 4.
\end{aligned}$$

## 2.3 Shape Functions

In the example of Section 2.1, we built an approximate solution which was piecewise linear, that is, that could be expressed only element by element with an analytic expression (i.e., continuously differentiable up to any order within the element). The objective of this section is to provide a basis for generalizing this approach to a more complex interpolation within elements. In doing so, we will observe that the concept of shape functions can be introduced naturally. We proceed in several steps. First we present a more detailed analysis of the one-dimensional linear element, and then we present two other one-dimensional elements in order to illustrate the possibility of increasing the order of the polynomials with $C^0$ or $C^1$ continuity. Finally we give an introductory example for a multidimensional analysis.

The *shape functions* can be viewed as *basis functions*, a linear combination of which is used to build the approximate solution of our physical problem. The shape functions relate nodal values (the primary, discrete unknowns) to the function values (which approximate the exact solution) inside an element. In these functions lies the core of the difference between the FEM and the FDM. The FEM shape functions express the partial differential equations in a form that can be viewed as an "average" in some sense, and therefore an integral form. With the FDM, the partial differential equation is expressed only at given (nodal) points.

The shape functions can be defined *locally*, that is, at the element level, or globally on the whole discretized domain. In this presentation, the local definition will be used at the beginning, and the global one will be introduced later.

A *finite element* is the association of the element subdomain with the local shape functions defined in this subdomain. That is, a finite-element model has two essential characteristics: it fills space corresponding to a physical part of the body being modeled, and it has an associated functional form (or an integral form) in its interior that approximates the variation of a quantity sought in the solution.

### Linear One-Dimensional Shape Functions

Consider the two-noded one-dimensional elements expressed by Eq. (2.14) in Section 2.2, and let us suppose that a linear interpolation of a given unknown $T$ is assumed within each element $\Omega_e$, with the end values $T_e$ and $T_{e+1}$. It is easy to see that the only linear interpolation in element $e$ having the appropriate values at the extremities of the element is expressed as

$$\bar{T}(x) = T_e + (T_{e+1} - T_e)\frac{x - x_e}{x_{e+1} - x_e}, \tag{2.16}$$

which can be rewritten as

$$\bar{T}(x) = T_e \frac{x - x_{e+1}}{x_e - x_{e+1}} + T_{e+1} \frac{x - x_e}{x_{e+1} - x_e}. \tag{2.17}$$

This is equivalent to the general expression

$$\bar{T}(x) = T_e N_e(x) + T_{e+1} N_{e+1}(x), \tag{2.18}$$

where the shape functions $N_e$ and $N_{e+1}$ are obtained by identification of Eq. (2.18) with Eq. (2.17):

$$N_e(x) = \frac{x - x_{e+1}}{x_e - x_{e+1}}, \quad N_{e+1} = \frac{x - x_e}{x_{e+1} - x_e}. \tag{2.19}$$

It is important to remark here that a shape function takes the value 1 at one node of an element, and is 0 at all other nodes. In element number $e$ we have

$$N_e(x_f) = \delta_{ef} \qquad \text{with } f = e \text{ or } e + 1. \tag{2.20}$$

This property ensures that summing nodal values multiplied by the corresponding shape functions always returns the specified nodal value of $\bar{T}$ at each nodal position. The functions $\bar{T}$, $N_e$, and $N_{e+1}$ are represented in Fig. 2.6.

It is important to note that the mathematical expression of Eq. (2.17) is only valid for element number $e$. In this element a *local* numbering of the nodal coordinates, nodal unknowns and shape functions is often used. We use the superscript $e$ to distinguish local numbering, and the $e$ may be replaced by a particular element number to distinguish all of the local numbers. For the $e$th element we have

$$\begin{aligned} x_e &= x_1^e, & x_{e+1} &= x_2^e; \\ T_e &= T_1^e, & T_{e+1} &= T_2^e; \\ N_e &= N_1^e, & N_{e+1} &= N_2^e; \end{aligned} \tag{2.21}$$

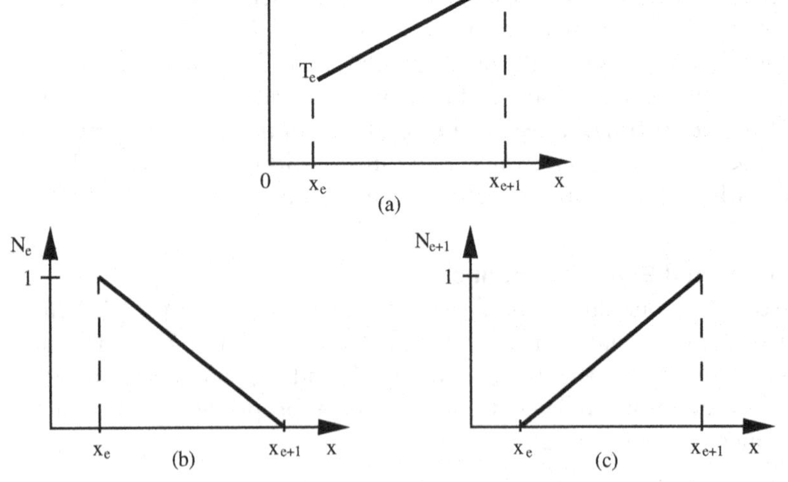


Figure 2.6. FE linear interpolation inside the $e$th element: (a) the unknown function; (b) and (c) shape functions.

and with these new notations, Eq. (2.18), corresponding to element number $e$, becomes

$$\bar{T}(x) = T_1^e N_1^e(x) + T_2^e N_2^e(x). \tag{2.22}$$

A formula analogous to Eq. (2.20) still holds for local shape functions:

$$N_i^e(x_j^e) = \delta_{ij}. \tag{2.23}$$

Thus the procedure can be summarized: a given functional *shape* is assumed within each element. This functional shape is then deconvoluted into a linear combination of nodal values times shapes functions, defined on the element domain.

### Quadratic One-Dimensional Shape Functions

More sophisticated shape functions can be built: the choice depends on various mathematical criteria. The first criterion imposes the use of shape functions that guarantee a continuity of the approximate function, and of its partial derivatives, up to at least one order less than the order of the partial derivatives present in the *integral form*. For example, the heat equation is a *second-order partial differential equation*, and the functional we use involves only *first-order derivatives*, so the shape functions need only be *continuous*. Another important requirement is the precision of the finite-element calculation; it can be improved by refinement of the mesh, or by use of more sophisticated shape functions, or both. Finally, we shall see that certain kinds of material behavior could lead to the elimination of some kinds of shape functions; this is true, for example, for the incompressible viscoplastic laws. In the following paragraphs we briefly examine three elementary cases, and in Chapters 3 and 4 we will present a more detailed analysis of two- and three-dimensional shape functions.

The first case corresponds to a piecewise quadratic interpolation. Here an element is defined by three nodes; that is, the $e$th element corresponds to the nodes $x_{2e-1}$, $x_{2e}$, and $x_{2e+1}$. The shape functions are quadratic within each element, and the same condition [Eq. (2.20)] applies. The result is graphically illustrated in Fig. 2.7.

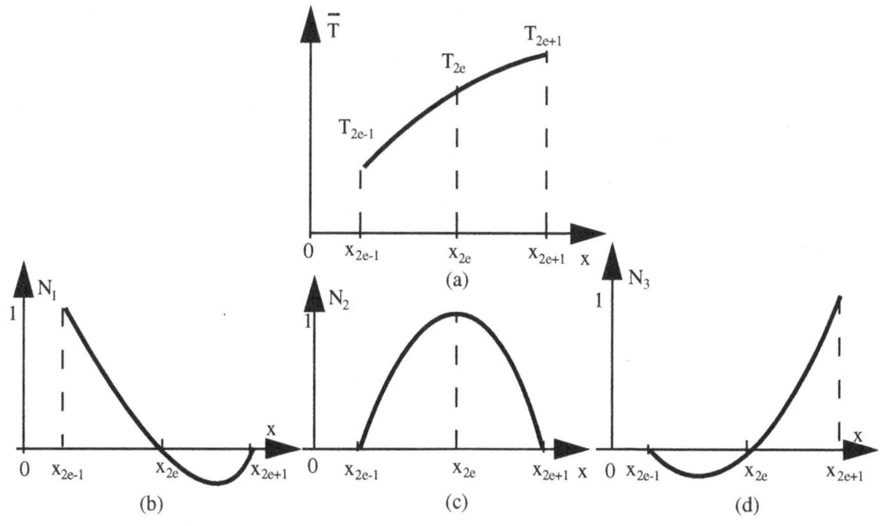


Figure 2.7. FE quadratic approximation: (a) interpolation of the unknown function and (b)–(d) shape functions.

**Exercise 2.6: Give the analytical expression of the piecewise quadratic shape functions in each three-node element defined by Eq. (2.9) in Section 2.1. Examine the special case in which $x_1 = -1$, $x_2 = 0$, and $x_3 = 1$, and give the expressions for $N_1$, $N_2$, and $N_3$.**

Given the element number $e$ with nodes at positions $x_{2e-1}$, $x_{2e}$, and $x_{2e+1}$, we need to express the shape function $N_i$ in the interval $[x_{2e-1}, x_{2e+1}]$, for $i = 2e-1, 2e$, and $2e+1$.

In the element number $e$, we seek $N_i$ as a polynomial of the general form

$$N_i(x) = a + bx + cx^2. \tag{2.6–1}$$

The first shape function corresponds to $i = 2e-1$; using Eq. (2.23), we must write

$$\begin{aligned} Ni(x_{2e-1}) &= 1 = a + bx_{2e-1} + cx_{2e-1}^2, \\ Ni(x_{2e}) &= 0 = a + bx_{2e} + cx_{2e}^2, \\ Ni(x_{2e+1}) &= 0 = a + bx_{2e+1} + cx_{2e+1}^2, \end{aligned} \tag{2.6–2}$$

which can be written in matrix form as

$$\begin{bmatrix} 1 & x_{2e-1} & x_{2e-1}^2 \\ 1 & x_{2e} & x_{2e}^2 \\ 1 & x_{2e+1} & x_{2e+1}^2 \end{bmatrix} \begin{bmatrix} a \\ b \\ c \end{bmatrix} = \begin{bmatrix} 1 \\ 0 \\ 0 \end{bmatrix}. \tag{2.6–3}$$

The required solution of the linear system of Eq. (2.6–3) is given by

$$\begin{aligned} a &= x_{2e}x_{2e+1}/D, \\ b &= -(x_{2e} + x_{2e+1})/D, \\ c &= 1/D, \\ D &= (x_{2e} - x_{2e-1})(x_{2e+1} - x_{2e-1}). \end{aligned} \tag{2.6–4}$$

The two remaining cases, when $i = 2e$ and when $i = 2e+1$, are treated with the same procedure. The linear system is quite similar; the only difference is in the right-hand side vector of Eq. (2.6–3), for which the position of 1 becomes number 2 for $i = 2e$, and number 3 for $i = 2e+1$.

For the special case in which the coordinates are $x_1 = -1$, $x_2 = 0$, and $x_3 = 1$, the left-hand side matrix is

$$[A] = \begin{bmatrix} 1 & -1 & 1 \\ 1 & 0 & 0 \\ 1 & 1 & 1 \end{bmatrix}. \tag{2.6–5}$$

We can verify that the inverse of [A] is

$$[B] = \begin{bmatrix} 0 & 1 & 0 \\ -1/2 & 0 & 1/2 \\ 1/2 & -1 & 1/2 \end{bmatrix}. \tag{2.6–6}$$

The coefficients of $N_1$ are obtained by multiplying matrix [B] by the column vector with a 1 on the first row and 0 elsewhere; we get

$$a = 0, \quad b = -1/2, \quad c = 1/2, \tag{2.6–7}$$

and the first shape function is

$$N_1(x) = 1/2x(x-1). \tag{2.6–8}$$

Similarly, the others are

$$\begin{aligned} N_2(x) &= 1 - x^2, \\ N_3(x) &= 1/2x(x+1). \end{aligned} \tag{2.6–9}$$

The shape functions $N_1$, $N_2$, and $N_3$ are respectively analogous to the functions represented in Figs. 2.7(b)–2.7(d) for the special case in which $x_1 = -1$, $x_2 = 0$, and $x_3 = 1$.

The difference between the linear and the quadratic shape functions lies in the capability of reproducing with accuracy a general function: it is clear that the quadratic approximation is potentially more precise than the linear one (for a fixed number of elements). However, it is important to see that none of these functions allows us to build a solution with a continuous derivative; only the function itself is continuous. This condition is generally denoted by the term "$C^0$ continuity," in which the superscript zero refers to the order of derivative that is required to be continuous.

### Example of One-Dimensional Shape Functions with Continuous Derivatives

To define one-dimensional shape functions with $C^1$ continuity, that is, with a continuous first-order derivative, we come back to the two-node elements of Eq. (2.8). In any element $e$, the unknown function will now be prescribed by two end values $T_e$ and $T_{e+1}$, as for the linear element, *and* two end values of the first derivative $T_e'$ and $T_{e+1}'$. For these four conditions to be satisfied, obviously at least a cubic interpolation in the element is necessary.[14]

---

**Exercise 2.7: Give the analytical expression for the cubic interpolation of the unknown function in element $e$, with given values of the function and of its first derivative at the ends of the interval. From this calculation show that, in the special case in which $x_{e+1} = x_e + 1$, it is possible to deduce explicitly the analytical expressions for the shape functions and represent graphically the variation of these functions in the element.**

The procedure can be quite similar to that which was used for the linear case, and it should give a fourth-order matrix equation. Here a simplification will be illustrated for the interval $[x_e, x_{e+1}]$. $\tilde{T}$ is a general cubic polynomial, which can be written as

$$\tilde{T}(x) = a + b(x - x_e) + c(x - x_e)^2 + d(x - x_e)^3, \tag{2.7–1}$$

without loss of generality, so that

$$\frac{\mathrm{d}\tilde{T}(x)}{\mathrm{d}x} = b + 2c(x - x_e) + 3d(x - x_e)^2. \tag{2.7–2}$$

The first two conditions on $\tilde{T}$ can be expressed very easily:

$$\tilde{T}(x_e) = T_e = a, \quad \frac{\mathrm{d}\tilde{T}}{\mathrm{d}x}(x_e) = T_e' = b. \tag{2.7–3}$$

The two other conditions are then

$$\tilde{T}(x_{e+1}) = T_{e+1} = T_e + T_e'(x_{e+1} - x_e) + c(x_{e+1} - x_e)^2 + d(x_{e+1} - x_e)^3, \tag{2.7–4}$$

$$\frac{\mathrm{d}\tilde{T}}{\mathrm{d}x}(x_{e+1}) = T_{e+1}' = T_e' + 2c(x_{e+1} - x_e) + 3d(x_{e+1} - x_e)^2. \tag{2.7–5}$$

We immediately obtain

$$c = \frac{1}{(x_{e+1} - x_e)}\left(3\frac{T_{e+1} - T_e}{x_{e+1} - x_e} - 2T_e' - T_{e+1}'\right),$$

$$d = \frac{1}{(x_{e+1} - x_e)^2}\left(-2\frac{T_{e+1} - T_e}{x_{e+1} - x_e} + T_e' + T_{e+1}'\right). \tag{2.7–6}$$

[14] Because there are four independent parameters ($T_1$, $T_2$, $T_1'$, and $T_2'$ in the local numbering system), a shape function having at least four parameters is required.

Now the explicit forms of the shape functions are detailed for the special case in which $x_{e+1} - x_e = 1$. If the values of $a$, $b$, $c$, and $d$ are put in the general cubic polynomial, we get

$$\bar{T}(x) = T_e + (x - x_e)T'_e + (x - x_e)^2(3T_{e+1} - 3T_e - T'_{e+1} - 2T'_e) + (x - x_e)^3(2T_e - 2T_{e+1} + T'_e + T'_{e+1}), \tag{2.7–7}$$

which may also be written as

$$\bar{T}(x) = T_e[1 - 3(x - x_e)^2 + 2(x - x_e)^3] + T'_e[x - x_e - 2(x - x_e)^2 + (x - x_e)^3] + T_{e+1}[3(x - x_e)^2 - 2(x - x_e)^3] + T'_{e+1}[-(x - x_e)^2 + (x - x_e)^3]. \tag{2.7–8}$$

This equation now appears in the desired form, where the approximating function is a linear combination of nodal values times shape functions with the general form

$$\bar{T}(x) = T_e N_e(x) + T'_e N'_e(x) + T_{e+1} N_{e+1}(x) + T'_{e+1} N'_{e+1}(x). \tag{2.7–9}$$

A comparison of Eq. (2.7–9) with Eq. (2.7–8) immediately gives

$$\begin{aligned} N_e(x) &= 1 - 3(x - x_e)^2 + 2(x - x_e)^3, \\ N'_e(x) &= x - x_e - 2(x - x_e)^2 + (x - x_e)^3, \\ N_{e+1}(x) &= 3(x - x_e)^2 - 2(x - x_e)^3, \\ N'_{e+1}(x) &= -(x - x_e)^2 + (x - x_e)^3. \end{aligned} \tag{2.7–10}$$

With these four formulas, it is easy to show by direct substitution (keeping in mind that $x_{e+1} = x_e + 1$) that

$$N_{e+i}(x_{e+j}) = \delta_{ij}, \quad \frac{dN_{e+i}}{dx}(x_{e+j}) = 0, \quad N'_{e+i}(x_{e+j}) = 0, \quad \frac{dN'_{e+i}}{dx}(x_{e+j}) = \delta_{ij} \quad \text{for } i, j = 0, 1. \tag{2.7–11}$$

Finally, the graphical representation of the four shape functions is given in Fig. 2.8.

**Global Shape Functions**

The definition of shape functions will now be extended to the whole discretized domain $\Omega$. Only linear shape functions will be discussed here. Equation (2.19) shows that, if $n$ is not equal to 1 or to $n_e$, any function $N_n$ is defined in both elements adjacent to node $n$, $\Omega_{n-1} = [x_{n-1}, x_n]$ and $\Omega_n = [x_n, x_{n+1}]$. Then the definition of $N_n$ is extended to all other elements by simply assuming that it is equal to zero outside $\Omega_{n-1}$ and $\Omega_n$. By doing that, we have built shape functions that possess the following two properties:

1. $N_n$ is piecewise linear in $\Omega$.
2. $N_n$ takes the value zero for any node, except for node number $n$:

$$N_n(x_m) = \delta_{nm}. \tag{2.24}$$

It is easy to verify that the set of the $n_n$ shape functions obtained by this method, with properties 1 and 2, is unique for a given mesh. Finally, if we seek an approximate function $\bar{U}(x)$, which is piecewise linear and has prescribed values $U_n$ at $x_n$ for $n = 1, 2, \ldots, n_n$, then it can be expressed as a linear combination of the global shape functions:

$$\bar{T}(x) = \sum_{n=1}^{n_e} T_n N_n(x). \tag{2.25}$$

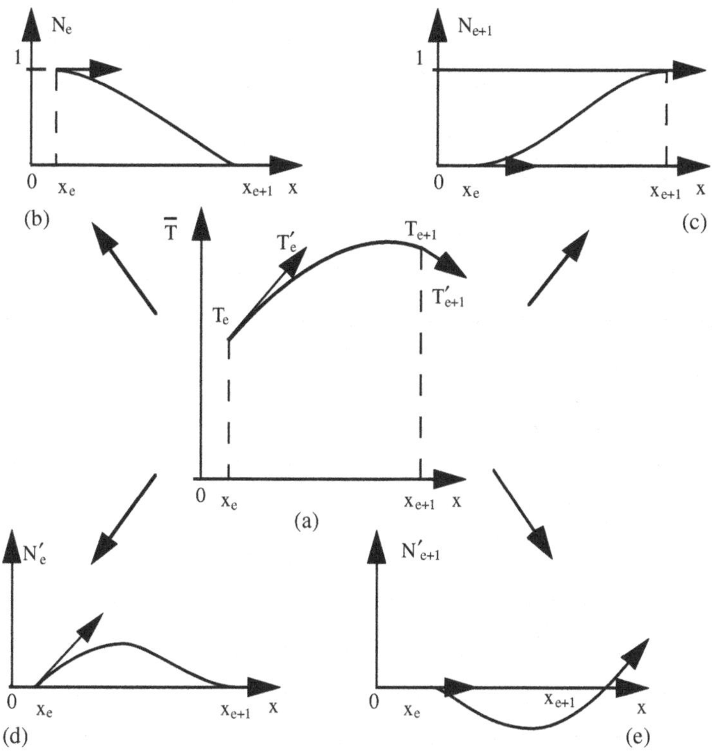


Figure 2.8. FE cubic approximation: (a) interpolation of the unknown function and (b)–(e) shape functions.

The proof of Eq. (2.25) is straightforward: if we consider any element $\Omega_e = [x_e, x_{e+1}]$, most of the shape functions are nil within this interval. According to our definition, only $N_e$ and $N_{e+1}$ have nonzero values, so that Eq. (2.25) reduces to

$$\bar{T}(x) = T_e N_e(x) + T_{e+1} N_{e+1}(x),$$

which is identical to Eq. (2.18) in element number $e$. The representation of the global shape functions on the whole domain is given in Fig. 2.9.

### Linear Two-Dimensional Shape Functions

We shall examine only briefly the generalization of one-dimensional shape functions to the two-dimensional element. The three-node triangular element shape functions are such that

- they take a value of 1 at a given node, and 0 at the two other nodes,
- they are linear functions of both coordinates.

Linear shape functions are introduced with a value of 1 on a given node, and 0 on the others. Figure 2.10 shows the three linear shape functions that can be built on a triangle.

More precisely, in element number $e$ the nodes are defined by the three local coordinates

$$\mathbf{X}_m^e = \begin{bmatrix} X_{1m}^e \\ X_{2m}^e \end{bmatrix}, \tag{2.26}$$

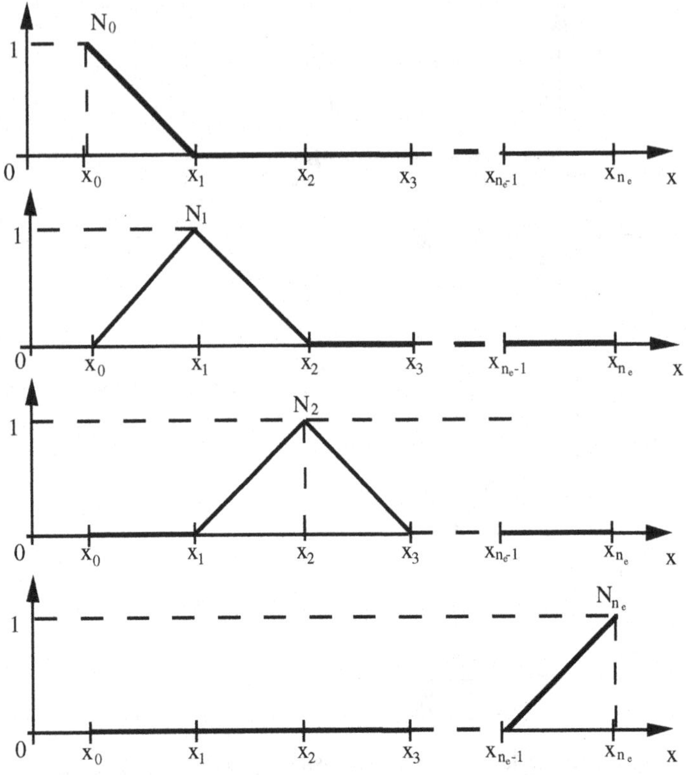

Figure 2.9. Linear global shape functions.

where $m$ ranges from 1 to 3, over the local numbering of the nodes. Every shape function must have the linear expression

$$N_m^e(\mathbf{x}) = a_m^e x_1 + b_m^e x_2^e + c_m^e, \tag{2.26a}$$

which takes values of 1 at the $m$th node and 0 at the other nodal positions:

$$N_m^e(\mathbf{X}_n^e) = \delta_{mn},$$

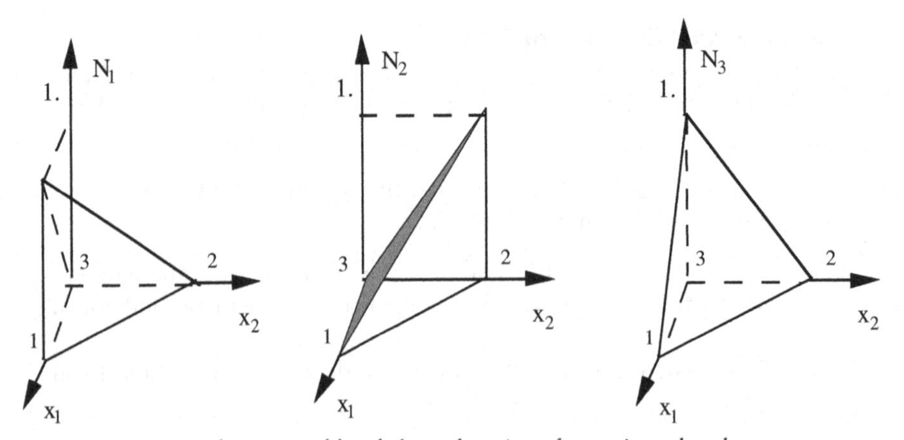

Figure 2.10. Two-dimensional local shape functions for a triangular element.

or, in a matrix form,

$$\begin{bmatrix} X^e_{11} & X^e_{21} & 1 \\ X^e_{12} & X^e_{22} & 1 \\ X^e_{13} & X^e_{23} & 1 \end{bmatrix} \begin{bmatrix} a^e_1 & a^e_2 & a^e_3 \\ b^e_1 & b^e_2 & b^e_3 \\ c^e_1 & c^e_2 & c^e_3 \end{bmatrix} = \begin{bmatrix} 1 & 0 & 0 \\ 0 & 1 & 0 \\ 0 & 0 & 1 \end{bmatrix}. \tag{2.27}$$

---

**Exercise 2.8: Calculate explicitly the coefficients $a_1, b_1, c_1, a_2, \ldots, c_3$ in terms of the nodal coordinates, and write the shape functions for a triangle (the e superscript is omitted here for simplicity).**

The coefficients of the shape functions are given by Eq. (2.27), which can be transformed into

$$\begin{bmatrix} X_{11}-X_{13} & X_{21}-X_{23} & 0 \\ X_{12}-X_{13} & X_{22}-X_{23} & 0 \\ X_{13} & X_{23} & 1 \end{bmatrix} \begin{bmatrix} a_1 & a_2 & a_3 \\ b_1 & b_2 & b_3 \\ c_1 & c_2 & c_3 \end{bmatrix} = \begin{bmatrix} 1 & 0 & -1 \\ 0 & 1 & -1 \\ 0 & 0 & 1 \end{bmatrix}, \tag{2.8–1}$$

where the third line was subtracted from the first and the second one, and the superscript $e$ was omitted.

We must calculate the determinant of the first matrix in Eq. (2.8–1), denoted by $\mathbf{A}$, appearing on the left-hand side of the above equation. The result is

$$\Delta = (X_{11}-X_{13})(X_{22}-X_{23}) - (X_{12}-X_{13})(X_{21}-X_{23}), \tag{2.8–2}$$

which is exactly two times the area of the triangle (with a positive sign if the nodes are numbered according to a right-hand rule, i.e., counterclockwise when seen along an $x_3$ direction).

The inverse of matrix $\mathbf{A}$ is

$$\mathbf{A}^{-1} = \frac{1}{\Delta}\begin{bmatrix} X_{22}-X_{23} & -(X_{21}-X_{23}) & 0 \\ -(X_{12}-X_{13}) & X_{11}-X_{13} & 0 \\ X_{23}X_{12}-X_{13}X_{22} & X_{13}X_{21}-X_{23}X_{11} & \Delta \end{bmatrix}. \tag{2.8–3}$$

By multiplying the inverse of $\mathbf{A}$ by the matrix on the right side of the above matrix equation, we immediately obtain

$$\begin{aligned} a_1 &= (X_{22}-X_{23})/\Delta, & b_1 &= (X_{13}-X_{12})/\Delta, & c_1 &= (X_{23}X_{12}-X_{13}X_{22})/\Delta, \\ a_2 &= (X_{23}-X_{21})/\Delta, & b_2 &= (X_{11}-X_{13})/\Delta, & c_2 &= (X_{13}X_{21}-X_{23}X_{11})/\Delta, \\ a_3 &= (X_{21}-X_{22})/\Delta, & b_3 &= (X_{12}-X_{11})/\Delta, & c_3 &= (X_{11}X_{22}-X_{12}X_{21})/\Delta. \end{aligned} \tag{2.8–4}$$

---

If the local nodal values of the unknown physical function, the temperature, for example, are given as $T^e_1$, $T^e_2$, and $T^e_3$ a linear interpolation can be calculated in the element by

$$\bar{T}(\mathbf{x}) = \sum_m T^e_m N^e_m(\mathbf{x}) \tag{2.28}$$

and, from the definition of the shape functions, the value of $\bar{T}$ can be evaluated at the nodal positions:

$$\bar{T}(\mathbf{X}^e_m) = T^e_m. \tag{2.29}$$

The definition of the two-dimensional shape functions can also be extended to the whole discretized domain. The numbering of the global shape functions $N_n$ will of course correspond to the global numbering of the nodal coordinate vectors $\mathbf{X}_n$

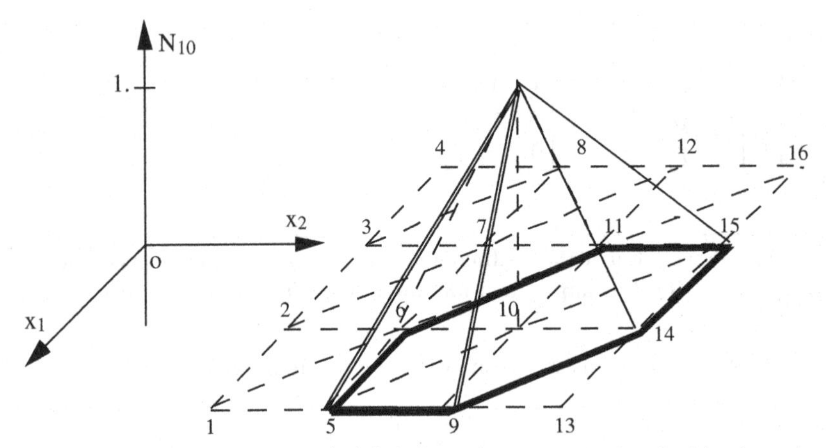


Figure 2.11. Two-dimensional global shape function associated with triangular elements.

(which was introduced in Section 2.2). The method is similar to that used in the one-dimensional case: for a given node $n$, the shape function $N_n$ is defined within each element. Two situations must be examined for element number $e$:

- the node number $n$ does not belong to the domain $\Omega_e$ of element number $e$; then for any point with vector coordinates $\mathbf{x}$ in $\Omega_e$,

  $N_n(x) = 0;$

- the node number $n$ is an apex of the domain $\Omega_e$; then the local shape function definition is used in $\Omega_e$. More precisely, we suppose that in element $e$, one local number $m$ corresponds to the global number $n$:

  $\lg(m, e) = n;$

  then for any point in $\Omega_e$ with vector coordinates $\mathbf{x}$,

  $N_n(\mathbf{x}) = N_m^e(\mathbf{x}).$

The representation of the global shape function $N_{10}$ is given in Fig. 2.11, where one can see that the shape function has nonzero values only on a subdomain. This subdomain is defined by the elements for which one node is $n = 10$: it is the polygon with nodes 5, 9, 14, 15, 11, and 6.

---

Exercise 2.9: **For a consistent triangulation, which was defined previously as one with no holes and no overlapping triangles, show that the sum function $S$ of the shape functions satisfies**

$$S(\mathbf{x}) = \sum_{n=1}^{n_n} N_n(\mathbf{x}) = 1 \tag{2.9–1}$$

**for any vector coordinate x belonging to the discretized domain.**

Here again each element is considered separately. With the use of the above definition for element $\Omega_e$ with node numbers

$$p = \lg(1, e), \qquad q = \lg(2, e), \qquad r = \lg(3, e), \tag{2.9–2}$$

most of the shape functions are seen to be identically null, except $N_p$, $N_q$, and $N_r$ and the sum reduces to

$$S(\mathbf{x}) = N_p(\mathbf{x}) + N_q(\mathbf{x}) + N_r(\mathbf{x}). \tag{2.9–3}$$

The next step makes use of the local shape functions $N^e_m$ and the sum is rewritten as

$$S(\mathbf{x}) = N^e_1(\mathbf{x}) + N^e_2(\mathbf{x}) + N^e_3(\mathbf{x}). \tag{2.9–4}$$

Now one can utilize the results of Exercise 2.7, with the explicit form of the local shape functions. It is much easier to note that, in the element number $e$, $S(\mathbf{x})$ is a linear interpolation in a triangle, which takes a value of 1 at each node and is therefore constant and equal to 1 in the whole element.

Finally, we have shown that $S(\mathbf{x})$ is equal to 1 in every element, so it is equal to 1 in the whole discretized domain.

---

With these definitions, the local basic expression of Eq. (2.28) can be extended to the whole domain. It is a generalization of the one-dimensional expression of Eq. (2.25) to two-dimensional linear triangle elements, and it is written in the same way:

$$\bar{T}(\mathbf{x}) = \sum_{n=1}^{n_n} T_n N_n(\mathbf{x}) \qquad \text{or } T_n N_n(\mathbf{x}), \tag{2.30}$$

where the summation is usually assumed over a range of $n_n$, from the first node to the $n_n$th one, and the Einstein notation is used in the second form of Eq. (2.30). It is important to note that an equation similar to Eq. (2.24) still holds, so that the only nonzero terms in the series correspond to node numbers of the elements that contain $\mathbf{x}$. Because the $N_n$ functions are continuous, the $\bar{T}$ function is also continuous and the points on the boundaries of the elements cause no problem.

A similar approach can be applied to most of the elements, for example to the one-dimensional, three-node element, or to the one-dimensional cubic $C^1$ two-node element (presented in Exercise 2.6), and to the two- and three-dimensional elements that are described in Chapters 3 and 4.

## 2.4 Stiffness Matrix

In this section we reexamine the same mathematical problem as in Section 2.1, with the use of a more formal and systematic approach that is closer to a practical use of the finite-element method. For the sake of illustration of the stiffness matrix concept, a one-dimensional example will be presented at first. This time a mechanical case is selected: the elastic deformation of a rod in tension, which is governed by the equations formally equivalent to those of the thermal problem. In Fig. 2.12 the conditions of deformation are indicated; the left-hand side is clamped, and the right-hand side is subjected to a force $F$, which is oriented along the axis of the rod.

If gravity and other body forces are neglected, it is possible to consider only the components of the displacement, the internal force, the strain tensor, and the stress

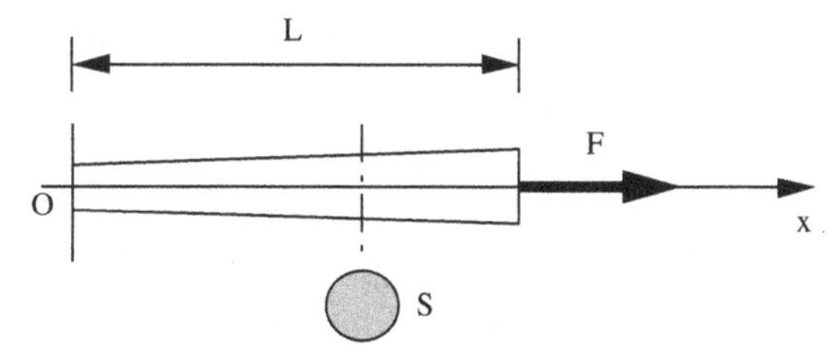


Figure 2.12. The one-dimensional elastic rod problem.

tensor, which correspond to the $Ox$ axis: they are written $u$, $f$, $\varepsilon$ (for $\varepsilon_{xx}$), and $\sigma$ (for $\sigma_{xx}$), respectively. The kinematic mechanical equations are then:

- derivation of the strain from the displacement,
$$\varepsilon = \frac{\mathrm{d}u}{\mathrm{d}x}, \tag{2.31}$$
where $u$ is the differential displacement of a material point, and the hypothesis of small elastic strain is assumed;
- internal force at the $x$ coordinate, where the cross-sectional area is $S(x)$,
$$f = S(x)\sigma; \tag{2.32}$$
- linear elastic behavior,
$$\sigma = E\varepsilon \tag{2.33}$$
(note: elasticity theory allows the calculation of the other components of the strain tensor, with the assumption of zero stress on the lateral surface of the rod; however, these strains do not affect the problem solution);
- equilibrium of any piece of the rod, which can be determined with the help of a fictitious cut perpendicular to the $Ox$ axis,
$$f \text{ uniform or } \frac{\mathrm{d}f}{\mathrm{d}x} = 0; \tag{2.34}$$
- boundary conditions,
$$u(0) = 0, \qquad f(L) = f_b. \tag{2.35}$$

If Eqs. (2.31)–(2.33) are combined appropriately, we obtain

$$\frac{\mathrm{d}}{\mathrm{d}x}\left(ES\frac{\mathrm{d}u}{\mathrm{d}x}\right) = 0, \tag{2.36}$$

which can be compared with Eq. (2.1a) to see the exact mathematical correspondence of the two quite different physical problems.

---

**Exercise 2.10: Compute the analytical solution of the elastic rod with a section given by $S = S_0(1+\alpha x)$. Formally compare the result with that obtained in Exercise 2.1.**

Starting from Eq. (2.28), we obtain by direct integration

$$ES\frac{\mathrm{d}u}{\mathrm{d}x} = c, \tag{2.10–1}$$

where $c$ is an integration constant determined by imposing the right-end boundary condition for $x = L$, which gives $c = f_b$. The next step is now the integration of the previous equation, which is transformed according to

$$\frac{\mathrm{d}u}{\mathrm{d}x} = \frac{f_b}{ES_0}\frac{1}{1+\alpha x}, \tag{2.10–2}$$

for which the solution is

$$u(x) = \frac{f_b}{\alpha ES_0}\ln(1+\alpha x). \tag{2.10–3}$$

In the last formula we have used the fact that the integration constant, involved in the second integration, is seen to be zero when the left-hand-side boundary condition is taken into account.

The present one-dimensional elastic rod can be viewed as formally equivalent to the thermal problem of Exercise 2.1, if we put

$$S(L)\phi_b = f_b, \quad k = E, \quad T = u, \quad T_a = 0. \tag{2.10–4}$$

---

Because the problem is fully one-dimensional, the stored mechanical energy can be calculated with a one-dimensional integral. It is obtained by observing first that the only nonzero component of the stress tensor is $\sigma_{xx} = \sigma$. The components of the strain tensor are computed with the help of Eq. (1.58): $\varepsilon_{xx} = \varepsilon = (\sigma/E)$, $\varepsilon_{yy} = \varepsilon_{zz} = -(\nu/E)\sigma$. Then, with Eq. (1.65), we obtain the expression

$$\Pi(\varepsilon) = \frac{1}{2}\int_0^L ES\varepsilon^2\,\mathrm{d}x - f_b u(L) = \frac{1}{2}\int_0^L ES\left(\frac{\mathrm{d}u}{\mathrm{d}x}\right)^2 \mathrm{d}x - f_b u(L). \tag{2.37}$$

According to Section 1.5, the solution of the problem can be obtained by minimizing the energy $\Pi$ with respect to the displacement functions $u$ consistent with any boundary condition on $u$ [the left-hand-side condition $u(0) = 0$]. This is the basis of the *energy method*, which is often used in conjunction with the finite-element discretization for mechanical problems, when it is possible. In fact, the energy method is a special case of the variational principle approach, for which an example in thermal analysis was presented in Section 2.1. The weak form of the energy, or the virtual work principle, can also be used. This topic is summarized in Section 3.5.

For the problem of the elastic deformation of a rod, the energy method is demonstrated with a simple mesh composed of four equally spaced nodes and three two-node linear elements. The mesh and the general appearance of the displacement, approximated with a piecewise linear interpolation, are shown in Fig. 2.13.

The nodal displacements $U_1$, $U_2$, $U_3$, and $U_4$ and the *global* piecewise linear shape functions $N_1$, $N_2$, $N_3$, and $N_4$ are introduced so that an approximation of the displacement field will be sought, in the usual way, as follows:

$$\bar{u}(x) = \sum_{n=1}^{4} U_n N_n(x). \tag{2.38}$$

If Eq. (2.38) is inserted into Eq. (2.37), the approximate elastic energy is obtained:

$$\bar{\Pi} = \frac{1}{2}\int_0^L ES\left[\sum_{n=1}^{4} U_n \frac{\mathrm{d}N_n(x)}{\mathrm{d}x}\right]^2 \mathrm{d}x - f_b U_4, \tag{2.39a}$$

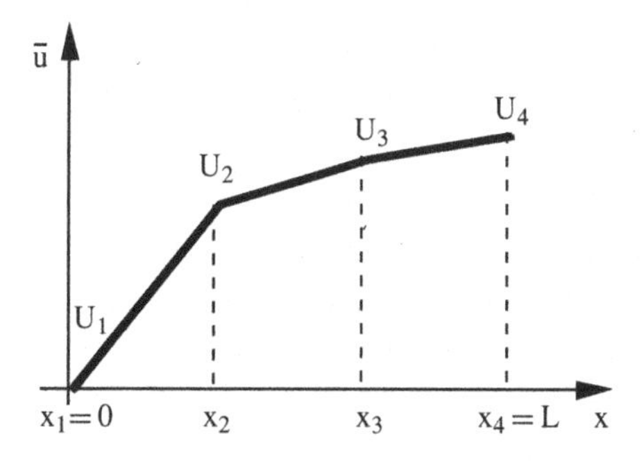


Figure 2.13. A piecewise linear discretization of the displacement function.

which can also be written as the equivalent expression:

$$\bar{\Pi} = \frac{1}{2}\int_0^L ES\left[\sum_{n=1}^{4} U_n \frac{dN_n(x)}{dx}\right]^2 \left[\sum_{m=1}^{4} U_m \frac{dN_m(x)}{dx}\right]^2 dx - f_b U_4. \tag{2.39b}$$

The resulting expression, Eq. (2.39a) [or Eq. (2.39b)], is a quadratic form of the unknowns $U_n$ ($n$ = 1–4), which must be minimized. This is done by setting the derivatives equal to zero with respect to each $U_m$:

$$\frac{d\bar{\Pi}}{dU_m} = \sum_{n=1}^{4}\left[\int_0^L ES \frac{dN_m(x)}{dx}\frac{dN_n(x)}{dx} dx\right] U_n - f_b \delta_{4m}, \tag{2.40a}$$

for $m$ = 1–4. At this stage, it is important to remark that the linear system of Eq. (2.40) must be modified to introduce the boundary condition

$$U_1 = 0. \tag{2.40b}$$

In fact, Eq. (2.39) is minimized with respect to $U_2$, $U_3$, and $U_4$ only, with the consequence that Eq. (2.40a) is written for $m = 2, 3, 4$ only. The matrix of the linear system corresponding to Eqs. (2.40b) and (2.40a) is called the *stiffness matrix*, or the *global stiffness matrix*, and it is often denoted by **K**, with components

$$K_{1n} = \delta_{1n}, \tag{2.41a}$$

$$K_{mn} = \int_0^L ES \frac{dN_m}{dx}\frac{dN_n}{dx} dx \qquad \text{for } m > 1. \tag{2.41b}$$

The array of unknowns **U** and the array of external forces $\mathbf{F}^E$ are put in vector form:

$$\mathbf{U} \to [U] = \begin{bmatrix} U_1 \\ U_2 \\ U_3 \\ U_4 \end{bmatrix}, \quad \mathbf{F}^E \to [F^E] = \begin{bmatrix} 0 \\ 0 \\ 0 \\ f_b \end{bmatrix}. \tag{2.42}$$

Thus Eq. (2.40) can be rewritten with the more compact form

$$\mathbf{KU} = \mathbf{F}^E, \tag{2.43a}$$

representing the set of equations

$$\begin{bmatrix} K_{11} & K_{12} & K_{13} & K_{14} \\ K_{21} & K_{22} & K_{23} & K_{24} \\ K_{31} & K_{32} & K_{33} & K_{34} \\ K_{41} & K_{42} & K_{43} & K_{44} \end{bmatrix} \begin{bmatrix} U_1 \\ U_2 \\ U_3 \\ U_4 \end{bmatrix} = \begin{bmatrix} F_1^E \\ F_2^E \\ F_3^E \\ F_4^E \end{bmatrix}. \tag{2.43b}$$

As we shall see in Chapter 3, from the definition of the terms of $K$, Eqs. (2.41), it is easy to see that many of them are identically zero. This is a consequence of the shape functions, which are identically zero everywhere outside of the elements connected to the node of interest. The derivatives of Eqs. (2.39)–(2.41) are always interpreted in the sense from inside the element of interest, so that the nonzero terms $K_{mn}$ correspond to an element that has nodes numbered $m$ and $n$ at positions $\mathbf{X}_m$ and $\mathbf{X}_n$. Thus, for node 2, for example, the only nonzero stiffness terms correspond to $U_{21}$, $U_{22}$, and $U_{23}$

in line 2 of K, and $U_{12}$, $U_{22}$, and $U_{32}$ in column 2. Here the symmetry of K is evident from the identical roles played by $m$ and $n$ in Eq. (2.41b).[15]

## 2.5 Assembly of the Stiffness Matrix

From a computational point of view, Eqs. (2.41) are not the most convenient form for the calculation of the stiffness matrix components, because the shape functions have an analytical form only within each element separately. The additive property of integration allows an equivalent form to Eqs. (2.40) to be written as[16]

$$\sum_{n=1}^{4}\left(\sum_{e=1}^{3}\int_{x_e}^{x_e+1} ES\frac{\mathrm{d}N_m}{\mathrm{d}x}\frac{\mathrm{d}N_n}{\mathrm{d}x}\,\mathrm{d}x\right)U_n - F_m^E = 0. \tag{2.44}$$

Now the *element stiffness matrix* $\mathbf{K}^e$ is introduced, which is defined by the compnents[17]

$$K_{mn}^e = \int_{x_e}^{x_{e+1}} ES\frac{\mathrm{d}N_m}{\mathrm{d}x}\frac{\mathrm{d}N_n}{\mathrm{d}x}\,\mathrm{d}x, \tag{2.45}$$

which allows the computation of the global stiffness matrix as the sum

$$\mathbf{K} = \sum_e \mathbf{K}^e, \tag{2.46}$$

where the first line is again modified according to Eq. (2.41a). We can see that most of the components of the $K^e$ matrices are null. This is clear when we remark that, in the domain $\Omega_e = [x_e, x_{e+1}]$ of element $e$, only $N_e$ and $N_{e+1}$ have nonzero values. The $\mathbf{K}^e$ matrix is more sparse as the number of elements increases. It is then more convenient to think of the matrix as $2 \times 2$, corresponding to only the nonzero components, which will be called the *local stiffness matrix* $\mathbf{k}^e$, for element number $e$. The local stiffness matrix is computed with the equivalent local definition of the shape functions:

$$k_{ij}^e = \int_{x_e}^{x_{e+1}} ES\frac{\mathrm{d}N_i^e}{\mathrm{d}x}\frac{\mathrm{d}N_j^e}{\mathrm{d}x}\,\mathrm{d}x \qquad \text{for} \quad i, j = 1, 2. \tag{2.47}$$

It is easy to convince oneself that Eq. (2.47) is equivalent to the gathering of the nonzero components of Eq. (2.45) with the help of Eq. (2.22).

The general procedure for the systematic derivation of the global stiffness matrix follows immediately. After initialization to zero of the K matrix, the following steps are to be performed for each element.

1. Compute the local stiffness matrix $\mathbf{k}^e$.
2. Build the global element stiffness matrix $\mathbf{K}^e$ by splitting matrix $\mathbf{k}^e$, that is, by putting the components of matrix $\mathbf{k}^e$ in the appropriate entries of matrix $\mathbf{K}^e$ with the help of the element node array lg.
3. Add matrix $\mathbf{K}^e$ to the global stiffness matrix K.

[15] In fact, K is not symmetric because of the boundary condition imposed by Eq. (2.41b); it is symmetric if the prescribed displacements are eliminated.

[16] Here we replaced the Kronecker notation of Eq. (2.40a) by its equivalent form with the help of the vector of external forces $\mathbf{F}^E$ defined by Eq. (2.42).

[17] As will be clear in the following, the element matrix was introduced for a more progressive approach. One must note that *the order of the element matrix is the same as the order of the global stiffness matrix.* In our example the order of K and of $\mathbf{K}^e$ is 4.

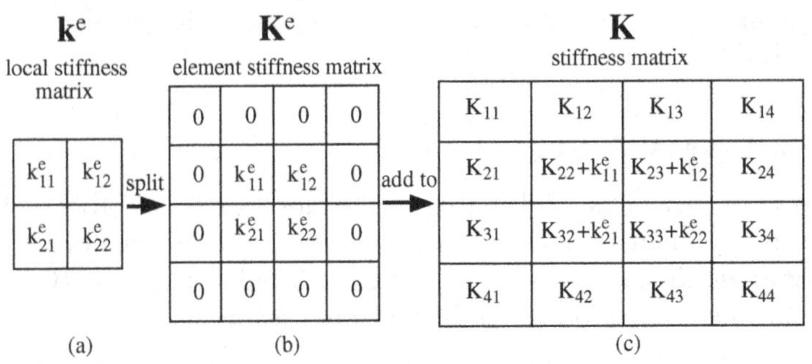


Figure 2.14. One step of the assembly of the stiffness matrix.

The whole process is called the stiffness matrix assembly, and it is illustrated in Fig. 2.14 for element number $e = 2$. At the end of the assembly, the resulting **K** matrix must be modified to take into account the boundary condition with prescribed displacement.

In fact, the explicit building of the intermediate matrix $\mathbf{K}^e$ is not necessary: this step is never performed in actual computer codes, as it would consume unnecessary time and storage. This stage was introduced here in order to separate more clearly steps 1, 2, and 3. Instead, the lg array is used to add each elemental stiffness term to the appropriate global stiffness term as assembly proceeds.

Because we chose a one-dimensional mesh and numbered the nodes sequentially, the element stiffness matrices are assembled into the global matrix without the necessity of splitting the rows and columns. This then allows the "banded" appearance of the resulting stiffness matrix. This is desirable because we can use the nature of the arrangement of the zero terms to advantage when we solve the system (see Chapter 3). However, in any problem except the simplest one-dimensional one chosen, it is impossible to have only nodes connected (by means of elements) to only nodes sequentially numbered on either side. Thus, the bandwidth (for a precise definition, see Chapter 3) increases, zero stiffness components appear inside the band, and local stiffness matrices are split in order to take their proper place in the global matrix. This situation is illustrated in the following example, with triangles as shown in Fig. 2.15. Node numbering is optimized for a minimum solution time by a pattern that minimizes the bandwidth of the global stiffness matrix.

---

**Exercise 2.11: For the problem of the elastic rod of Exercise 2.10, explicitly calculate each local stiffness matrix, if the interval $[0, L]$ is divided into $n_e = n$ equal elements. When $n = 4$, write the linear system in matrix form. The cross section of the beam is assumed to be constant here.**

The domain of element number $e$ is the segment $[x_e, x_{e+1}] = [(e-1)(L/n), e(L/n)]$, and the corresponding global shape functions are

$$N_e(x) = \frac{e(L/n) - x}{L/n}, \qquad \frac{\mathrm{d}N_e}{\mathrm{d}x} = -\frac{n}{L},$$
$$N_{e+1}(x) = \frac{x - (e-1)(L/n)}{L/n}, \qquad \frac{\mathrm{d}N_{e+1}}{\mathrm{d}x} = \frac{n}{L}. \tag{2.11–1}$$

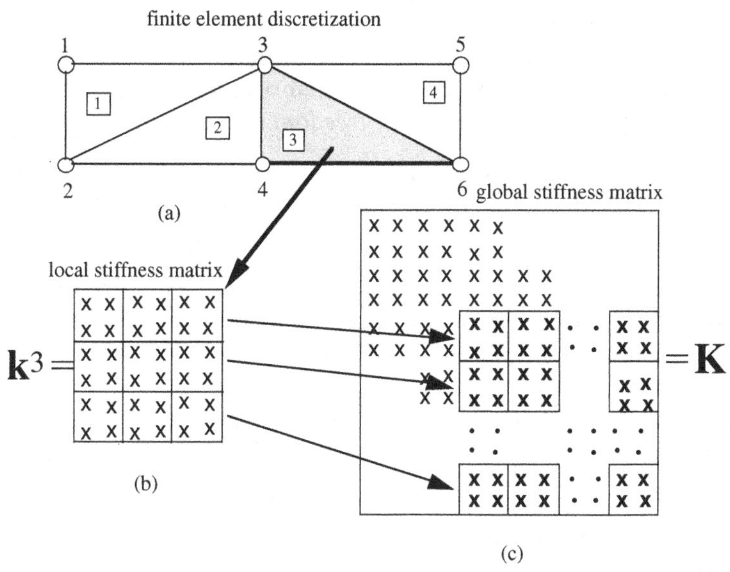


Figure 2.15. The stiffness matrix assembly for a two-dimensional problem: (a) mesh of the structure, (b) local stiffness matrix for element number 3, and (c) addition of the local stiffness matrix to the global one.

The local stiffness matrix is given by

$$k_{11}^e = \int_{x_e}^{x_{e+1}} ES\left(\frac{dN_e}{dx}\right)^2 dx = ES\frac{n}{L} = k_{22}^e,$$
$$k_{12}^e = \int_{x_e}^{x_{e+1}} ES\frac{dN_e}{dx}\frac{dN_{e+1}}{dx} dx = -ES\frac{n}{L} = k_{21}^e. \tag{2.11–2}$$

In the matrix form, we get

$$\mathbf{k}^e \to [\mathrm{k}^e] = \begin{bmatrix} ES\frac{n}{L} & -ES\frac{n}{L} \\ -ES\frac{n}{L} & ES\frac{n}{L} \end{bmatrix} \tag{2.11–3}$$

If $n = 4$ the global linear system has an order of 5 and is obtained by the same procedure as that represented schematically in Fig. 2.15. Each local stiffness matrix is identical, and we observe that after assembly the diagonal terms are doubled from row 2 to $n-1 = 4$ in the global stiffness matrix (the first row was modified according to the left-hand-side boundary condition):

$$\begin{bmatrix} 1 & 0 & 0 & 0 & 0 \\ -ES\frac{n}{L} & 2ES\frac{n}{L} & -ES\frac{n}{L} & 0 & 0 \\ 0 & -ES\frac{n}{L} & 2ES\frac{n}{L} & -ES\frac{n}{L} & 0 \\ 0 & 0 & -ES\frac{n}{L} & 2ES\frac{n}{L} & -ES\frac{n}{L} \\ 0 & 0 & 0 & -ES\frac{n}{L} & ES\frac{n}{L} \end{bmatrix} \begin{bmatrix} U_0 \\ U_1 \\ U_2 \\ U_3 \\ U_5 \end{bmatrix} = \begin{bmatrix} U_L \\ 0 \\ 0 \\ 0 \\ F \end{bmatrix}. \tag{2.11–4}$$

---

Generalization of the stiffness matrix definition and assembly to more complex physical problems are relatively easy to deduce from the previous simple case. For the finite elements we are interested in, we shall define as many local shape functions as there are nodes in the element. As a result of this assumption, the number of local unknowns can be easily calculated for the simple case of a scalar or a vector unknown field. This is often called the *local number of degrees of freedom* and is equal to[18]

- the number of nodes for a problem involving a scalar unknown function,
- two times the number of nodes for a two-dimensional unknown vector field,
- three times for a three-dimensional one.

The order of the local stiffness matrix is obviously equal to the local number of degrees of freedom. For a three-node triangle used for the discretization of a two-dimensional elastic problem in terms of an unknown displacement field with two components, the order of the local stiffness matrix is $3 \times 2 = 6$.

With regard to the total number of unknowns, called *number of degrees of freedom* (DOF) and denoted by $n_D$, the same relation holds with the number of nodes $n_n$, where

$n_D = n_n$ for a scalar function,
$n_D = 2n_n$ for a two-component unknown vector, and
$n_D = 3n_n$ for a three-component unknown vector.

Clearly the order of the global stiffness matrix must be equal to $n_D$.[19] For a graphic illustration of the principle of stiffness matrix assembly, a simple example is presented in Fig. 2.15. This example corresponds to a two-dimensional problem, where the unknown is a two-dimensional vector, and the domain is meshed with four three-node triangles. In the local and global stiffness matrices, only the positions of the nonzero entries are given by crosses; the others are blank. In the illustration, the structure of the local stiffness matrix associated with element number 3 is shown, and the positions where the entries affect the global stiffness matrix are shown with bold crosses. The dots represent entries that will be filled when the last local stiffness matrix is added.

In the example in Fig. 2.15, the element stiffness matrix is not used, and the splitting is made apparent as the node numbering is no longer consecutive.

In Chapter 4, different elements will be discussed in some detail, for their practical use in metal-forming process simulation, in two or three dimensions. The procedure for matrix assembly is essentially the same.

## PROBLEMS

### A. Proficiency Problems

1. Consider the thermal problem defined by Eqs. (2.1a)–(2.1c), where

   $$S = S_0 \exp(-\alpha x/L),$$

   with $\alpha = 0.4$ and $L = 2$. Calculate the exact solution and the FDM solution, for a discretization identical to that pictured in Fig. 2.2.

[18] Of course, more complicated cases often appear, e.g., when several fields are approximated with different shape functions within the same element subdomain.

[19] It is possible to remove the degrees of freedom specified by boundary conditions, which can reduce the order of the stiffness matrix accordingly. For purposes of our discussion, we have both specified and unspecified unknowns in $n_D$.

2. For the same case and the same discretization as in Problem 2.1, calculate the FEM solution for a piecewise linear approximation, and compare it with the exact and with the FDM solution.
3. From the consistent mesh in Fig. 2.5(a), build a new consistent mesh of four-node quadrilaterals, with each one being made of the association of two triangles. Is the solution unique? Explicitly write the element node array for the seven quadrilaterals.
4. Show that the quadratic one-dimensional shape functions, defined in Exercise 2.6, can reproduce exactly any linear function on one element.
5. Draw the seven *global* shape functions in the interval [0, 6], divided into three three-node elements, with quadratic *local* shape functions.
6. Explicitly calculate the two-dimensional shape functions defined in the three-node element in the figure below.

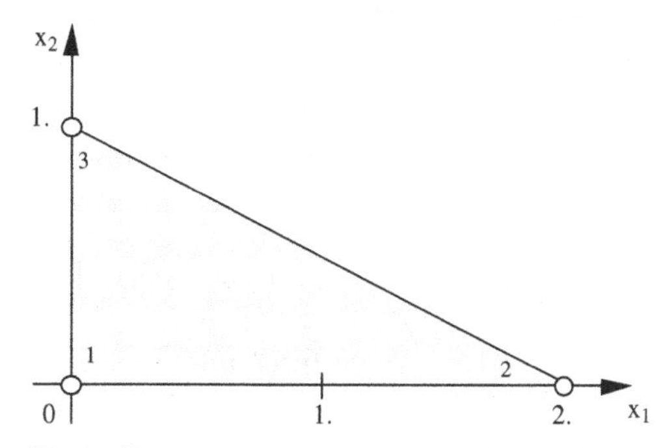

Figure P2.6.

7. A one-dimensional, three-node element is given by the node coordinates

$$x_1 = 0, \qquad x_2 = 1, \qquad x_3 = 2,$$

and the shape functions are quadratic. Give the expression of the local stiffness matrix for a linear elastic problem with constant section $S$ and constant Young modulus $E$.

8. The one-dimensional, linear elastic, three-rod problem, as presented in the figure below, is considered. The rods have the same length $L = 1$ and the same elastic modulus $E$, but their cross sections are different and equal to $S_1$, $S_2$, and $S_3$.

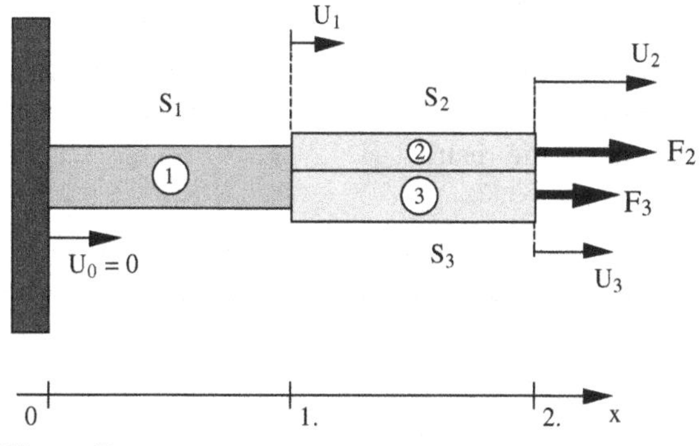

Figure P2.8.

The left-hand side of rod 1 is clamped, and the displacement of the right-hand side of rod 1, $U_1$, is supposed to be equal to the displacement of both left-hand sides of rods 2 and 3. The right-hand sides of rods 2 and 3 are submitted to axial tensions $F_1$ and $F_2$ and are allowed to have independent displacements $U_2$ and $U_3$. Each rod is discretized with one two-node linear element. Calculate the global stiffness matrix and solve the problem numerically.

## B. Depth Problems

9. A one-dimensional thermal problem is considered, for a constant section rod made of two equal parts in different materials, with conductivities $k_1$ and $k_2$, as shown in figure below.

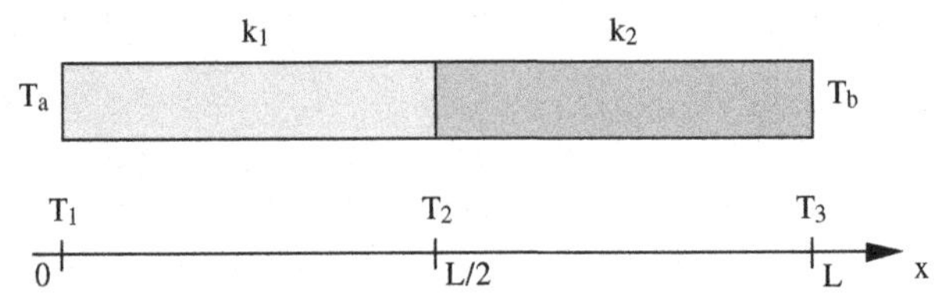


Figure P2.9.

The boundary conditions are prescribed temperatures: $T_a$ at the left-hand side and $T_b$ at the right-hand side. With only one two-node element for each part of the rod, calculate the intermediate temperature $T_2$ by the FEM. Verify that the solution is exact. The same calculation can be performed with one three-node quadratic element for the entire rod. Express $T_2$ again with this new discretization, and verify that the solution is not exact unless $k_1 = k_2$ or $T_a = T_b$. Show that the latter result was predictable.

## C. Numerical and Computational Problems

10. Write a FORTRAN computer code to solve the problem of Exercise 2.11 numerically with a variable number of nodes: $n = n_e$. Generalize the code to the non-constant section case, and show that for a linear evolution of the cross-sectional area, the numerical solution tends to the analytical one for large $n$.
11. For a two-dimensional thermal problem, with the associated functional:

$$I(T) = \frac{1}{2}\int_{\Omega}\left[\left(\frac{\partial T}{\partial x_1}\right)^2 + \left(\frac{\partial T}{\partial x_2}\right)^2\right]\mathrm{d}x_1\mathrm{d}x_2 + \int_{\partial\Omega^f} T\phi^f \mathrm{d}l$$

(where $\phi_f$ is the prescribed flux on the boundary). Explicitly calculate the local stiffness matrix $\mathbf{k}^e$, for the triangular element for the figure in Problem 6, and the corresponding eigenvalues. Comment on the results.

CHAPTER THREE

# Finite Elements for Large Deformation

The aim of this chapter is to provide the necessary numerical concepts for the practical implementation of the finite-element method. First, we review the notion of isoparametric elements; such elements are important when the elements are distorted and the shape functions cannot be easily expressed in term of the physical coordinates. This is always the case in the numerical simulation of metal-forming processes, where significant deformation of the initial mesh cannot be avoided. Second, we consider in some detail the procedure for numerical integration, which is essential for many applications, particularly when the constitutive equation is nonlinear. We resolve the resulting finite-element equation into two types: linear and nonlinear. Finally, we introduce a simple, one-dimensional mechanical example with linear material behavior. This example is treated completely and a nonlinear material is also used to illustrate the differences in the methods.

## 3.1 Isoparametric Elements

The physical problem under consideration is posed in a domain and is discretized into small subdomains, called elements, as introduced in Chapter 2. Elements are physical in the sense that they are defined themselves in real space. So far we have used the physical domains of those elements directly, by defining shape functions inside an element in terms of nodal values and positions. It is often more convenient to map the physical element domains to even simpler shapes and thus to formulate the discretized problem on these shapes. The expression of the shape functions is much simpler (for nonlinear ones). The numerical integration formulas are also easier to implement in this way.

When *isoparametric elements* are used, a mathematical space is introduced; this is called the *reference space* or *local space*, with *local coordinates* $\xi$ in one dimension, $\boldsymbol{\xi} = (\xi_1, \xi_2)$ in two dimensions, or $\boldsymbol{\xi} = (\xi_1, \xi_2, \xi_3)$ in three dimensions. In this abstract space, a reference element domain $\Omega_0$ is defined, and a mapping between each real element domain $\Omega_e$ and $\Omega_0$ is established by defining a mapping function between local and physical coordinates:

$$\mathbf{x} = \boldsymbol{\tau}^e(\boldsymbol{\xi}), \tag{3.1}$$

where $\mathbf{x}$ is the vector coordinate of a material point inside element domain $\Omega_e$ in the physical space, $\boldsymbol{\xi}$ is the vector coordinate for the corresponding point in reference space, and $\boldsymbol{\tau}^e$ is a vector function associated with element number $e$. The construction of this vector function is made, by definition of isoparametric elements, by utilizing the same shape functions[1] for interpolation of the unknown, and of the physical coordinates according to Eq. (3.1). The procedure will be clarified by the presentation of a few examples.

### One-Dimensional Isoparametric Elements

The one-dimensional linear two-node element, however trivial, will be analyzed first. The reference element will be the segment $\Omega_0 = [-1, 1]$, and the two *local shape functions*, defined on the reference space, are

$$\left.\begin{aligned} N_1^0(\xi) &= \frac{1}{2}(1-\xi) \\ N_2^0(\xi) &= \frac{1}{2}(1+\xi) \end{aligned}\right\} \quad \text{for } -1 \leq \xi \leq 1, \tag{3.2}$$

which is a special case of Eq. (2.19). That is, the reference shape function is simply a normal shape function defined over a (convenient) specified fixed interval. For a one-dimensional mesh with two-node elements, we have already seen that the domain of element number $e$ is the segment $\Omega_e = [x_e, x_{e+1}]$. The mapping between $\Omega_0$ and $\Omega_e$ must associate the ends of the segments, $-1$ to $x_e$, that is, $x_e = \tau^e(-1)$, and 1 to $x_{e+1}$, that is, $x_{e+1} = \tau^e(1)$. With these conditions, the function $\tau^e$ (which is not a vector function here, as the problem is one dimensional), must be such that

$$x = \tau^e(\xi) = x_e N_1^0(\xi) + x_{e+1}\, N_2^0(\xi) \qquad \text{for } -1 \leq \xi \leq 1. \tag{3.3a}$$

If the local coordinates $x_1^e = x_e$ and $x_2^e = x_{e+1}$ are used for element $e$, we obtain the following form, equivalent to Eq. (3.3a):

$$x = \tau^e(\xi) = x_1^e N_1^0(\xi) + x_2^e\, N_2^0(\xi) \qquad \text{for } -1 \leq \xi \leq 1. \tag{3.3b}$$

Thus, the shape functions relate the physical coordinates $x$ to the reference coordinate $\xi$. We suppose here that the unknown function is the displacement $u(x)$ and that it is approximated by the function $\bar{u}(x)$ within each element $e$, in terms of the nodal values $U_n$ and of the local shape functions. We are then led to write equations similar to Eq. (2.18) or Eq. (2.22) for the displacement function:

$$\bar{u} = U_e\, N_1^0(\xi) + U_{e+1}\, N_2^0(\xi) = U_1^e\, N_1^0(\xi) + U_2^e\, N_2^0(\xi), \tag{3.4}$$

where $U_1^e(= U_e)$ and $U_2^e(= U_{e+1})$ are the local degrees of freedom for element number $e$, as in Eq. (2.21) (for $T$).

Now the definition can be made more clear: the association of Eqs. (3.3a) or (3.3b) with Eq. (3.4) allows us to build *isoparametric finite elements*. The relation between the physical coordinate and the local coordinate on one hand, and the relation between the physical unknown and the same local coordinate on the other hand, are analogous in the sense that they are expressed with similar linear combinations of the same local shape functions.

In order to generalize Eq. (2.25) to isoparametric shape functions, the next step is to deduce a global definition of the shape functions from the previous expressions of

[1] Other parametric elements can be used, but they will not be considered here.

the local shape functions. This is always possible, at least theoretically, if we suppose that the mapping $\tau^e$ is one to one, so that it possesses an inverse $\tau^{e-1}$, which will allow consideration of the local shape functions as functions of $x$ by eliminating the local coordinate $\xi$, with the help of the equality

$$\xi = \tau^{e-1}(x). \tag{3.5}$$

With the linear element in one dimension, the calculation is rather straightforward:

$$x = x_1^e \frac{1}{2}(1-\xi) + x_2^e \frac{1}{2}(1+\xi) = \frac{x_1^e + x_2^e}{2} + \frac{x_2^e - x_1^e}{2}\xi, \tag{3.6}$$

which immediately gives

$$\xi = \frac{x - \left[(x_1^e + x_2^e)/2\right]}{(x_2^e - x_1^e)/2}. \tag{3.6a}$$

The local shape functions are transformed into

$$N_1^0[\xi(x)] = N_1^e(x) = \frac{x_2^e - x}{x_2^e - x_1^e},$$
$$N_2^0[\xi(x)] = N_2^e(x) = \frac{x - x_1^e}{x_2^e - x_1^e}. \tag{3.6b}$$

In this simple example, we observe that the isoparametric local shape functions, obtained after elimination of the local coordinate $\xi$, are identical to those defined directly with the physical (global) coordinate. This is not true for most of the elements we will describe. This property is verified only for elements with linear shape functions: a two-node segment, a three-node triangle, and four-node tetrahedrons.

The procedure is identical to that outlined in Section 2.3; that is, the global shape functions are defined element by element from the local shape functions. In general, the explicit expression of the shape functions in terms of the global coordinates is not necessary. Only the derivatives of the shape functions with respect to the physical coordinate are useful, with the first derivative being obtained in one dimension by

$$\frac{\mathrm{d}N_i^0(\xi)}{\mathrm{d}x} = \frac{\mathrm{d}N_i^0(\xi)}{\mathrm{d}\xi}\frac{\mathrm{d}\xi}{\mathrm{d}x} = \frac{\mathrm{d}N_i^0(\xi)}{\mathrm{d}\xi}\frac{1}{\mathrm{d}x/\mathrm{d}\xi}, \qquad i = 1, 2. \tag{3.7}$$

Using the local shape function definition in Eq. (3.1) and the expression of the mapping given by Eq. (3.5), we get

$$\frac{\mathrm{d}x}{\mathrm{d}\xi} = \frac{x_2^e - x_1^e}{2}. \tag{3.8}$$

Equation (3.7) immediately becomes

$$\frac{\mathrm{d}N_1^0(\xi)}{\mathrm{d}x} = -\frac{1}{x_2^e - x_1^e}, \quad \frac{\mathrm{d}N_2^0(\xi)}{\mathrm{d}x} = \frac{1}{x_2^e - x_1^e}. \tag{3.9}$$

The same result could have been obtained directly from Eq. (3.6) in this simple case, but the explicit inversion of the mapping function is not always possible, or economical from the computational point of view. Thus, by introducing the coordinate and unknown variable mapping function, we make it more convenient to find the proper derivative needed.

---

**Exercise 3.1: With the help of Exercise 2.6, give the expression of the mapping, in element number e, between the global coordinate and the local coordinate, for a three-node one-dimensional isoparametric element. The element number e has three nodes with coordinates $x_{2e-1}$, $x_{2e}$, and $x_{2e+1}$, and the local coordinates of the nodes of the reference element are $-1, 0, 1$. Compute the derivative of the local shape functions, with respect to the physical coordinate, for the intermediate node.**

In Exercise 2.6, it was proved that the three shape functions corresponding to coordinates $-1$, 0, and 1 respectively are

$$N_1^0(\xi) = \frac{1}{2}\xi(\xi - 1), \quad N_2^0(\xi) = 1 - \xi^2, \quad N_3^0(\xi) = \frac{1}{2}\xi(\xi + 1), \tag{3.1–1}$$

and that the mapping with the general expression

$$x = x_1^e N_1^0(\xi) + x_2^e N_2^0(\xi) + x_3^e N_3^0(\xi) \tag{3.1–2}$$

takes the special form

$$x = x_2^e + \frac{1}{2}\left(x_3^e - x_1^e\right)\xi + \frac{1}{2}\left(x_3^e + x_1^e - 2x_2^e\right)\xi^2. \tag{3.1–3}$$

From this last equation we notice that it is still possible to invert the mapping function, but the explicit calculation of $\xi$ in terms of $x$ is more complicated, and it involves the computation of a square root. Regarding the derivatives we have, for any $x$,

$$\frac{dx}{d\xi} = \frac{1}{2}\left(x_3^e - x_1^e\right)\xi + \left(x_3^e + x_1^e - 2x_2^e\right)\xi, \tag{3.1–4}$$

while the derivatives of the shape functions are

$$\frac{dN_1^0}{d\xi} = \xi - \frac{1}{2}, \quad \frac{dN_2^0}{d\xi} = -2\xi, \quad \frac{dN_3^0}{d\xi} = \xi + \frac{1}{2}. \tag{3.1–5}$$

As the intermediate node $x_e$ corresponds to $x = 0$, the derivatives become

$$\frac{dx}{d\xi}(0) = \frac{1}{2}\left(x_3^e - x_1^e\right), \tag{3.1–6}$$

$$\frac{dN_1^0}{d\xi}(0) = -\frac{1}{2}, \quad \frac{dN_2^0}{d\xi}(0) = 0, \quad \frac{dN_3^0}{d\xi}(0) = \frac{1}{2}. \tag{3.1–7}$$

Finally, according to Eq. (2.46) with index $i$ ranging from 1 to 3 here, the derivatives of the local shape functions with respect to the physical coordinate are, for the same node $x_e$,

$$\frac{dN_1^0}{dx}(0) = -\frac{1}{\left(x_3^e - x_1^e\right)}, \quad \frac{dN_2^0}{dx}(0) = 0, \quad \frac{dN_3^0}{dx}(0) = \frac{1}{\left(x_3^e - x_1^e\right)}. \tag{3.1–8}$$

---

**Generalization of Isoparametric Elements**

The generalization of the main equations defining isoparametric elements to two- or three-dimensional problems can be stated as follows. If

$\mathbf{X}_m^e$ is the local position vector of the node numbered $m$, in element number $e$,
$\mathbf{U}_m^e$ is the local unknown vector corresponding to node $m$ in the same element, and
$\mathbf{N}_m^0$ is the local shape function that takes a value of 1 at node $m$ (and 0 for the other nodes),

then the general form, given by Eq. (3.1), of the mapping between the physical space and the local space will be written as

$$\mathbf{x} = \sum_m \mathbf{X}^e_m N^0_m(\boldsymbol{\xi}), \tag{3.10a}$$

$$\bar{\mathbf{u}} = \sum_m \mathbf{U}^e_m N^0_m(\boldsymbol{\xi}), \tag{3.10b}$$

where the sum over $m$ extends the number of local shape functions for the element. If we introduce again the node element array lg, Eqs. (3.10a) and (3.10b) can also be written as

$$\mathbf{x} = \sum_m \mathbf{X}_{\text{lg}(m,e)} N^0_m(\boldsymbol{\xi}), \tag{3.11a}$$

$$\bar{\mathbf{u}} = \sum_m \mathbf{U}_{\text{lg}(m,e)} N^0_m(\boldsymbol{\xi}). \tag{3.11b}$$

With Eqs. (3.1), (3.10a), and (3.11a) we observe that the mapping $\tau^e$ between the local coordinate space and the global (physical) space varies from element to element, and that no unique analytical expression can be given.

It is possible to define the global shape functions $N_n$ by the values within each element domain $\Omega_e$:

- If no node of $\Omega_e$ corresponds with number $n$, then

  $N_n(\mathbf{x}) = 0.$

- If the node $n$ is mapped to the local node $m$ of the reference element domain $\Omega_o$, that is, if $n = \text{lg}(m, e)$, then

$$N_n(\mathbf{x}) = N^0_m(\boldsymbol{\xi}) \qquad \text{for } \mathbf{x} = \tau^e(\boldsymbol{\xi}). \tag{3.12}$$

This allows again the expression of the approximate finite-element solution as the fundamental linear combination:

$$\bar{\mathbf{u}}(\mathbf{x}) = \sum_{n=1}^{n_e} \mathbf{U}_n N_n(\mathbf{x}). \tag{3.13}$$

The general method for calculating the partial derivatives of the shape functions with respect to the physical coordinates can be outlined in three points.

a. Use the chain rule for the derivative of a compound function[2,3]:

$$\frac{\partial N_n}{\partial \mathbf{x}} = \frac{\partial N_n}{\partial \boldsymbol{\xi}} \cdot \frac{\partial \boldsymbol{\xi}}{\partial \mathbf{x}}, \tag{3.14}$$

which can also be written with indices:

$$\frac{\partial N_n}{\partial x_i} = \sum_k \frac{\partial N_n}{\partial \xi_k} \frac{\partial \xi_k}{\partial x_i}. \tag{3.15}$$

[2] The notation $\partial\boldsymbol{\xi}/\partial\mathbf{x}$ is used for the matrix, the components of which are $\begin{bmatrix} \partial\xi_1/\partial x_1 & \partial\xi_1/\partial x_2 \\ \partial\xi_2/\partial x_1 & \partial\xi_2/\partial x_2 \end{bmatrix}$.

[3] Note that the notation $\partial N/\partial\mathbf{x}$, where $\mathbf{x}$ is a vector and $N$ is a scalar, is a vector with components $\partial N/\partial x_1, \partial N/\partial x_2$ in two dimensions (and one more component in three dimensions).

b. Calculate the Jacobian matrix:

$$\mathbf{A} = \frac{\partial \mathbf{x}}{\partial \boldsymbol{\xi}} = \frac{\partial N_m^0(\boldsymbol{\xi})}{\partial \boldsymbol{\xi}} \mathbf{X}_m^e, \tag{3.16}$$

which is the matrix of components

$$A_{ij} = \frac{\partial x_i}{\partial \xi_j} = \sum_m \frac{\partial N_m^0(\xi)}{\partial \xi_j} X_{im}^e. \tag{3.17}$$

c. Compute $\mathbf{B}$, the inverse of $\mathbf{A}$, so that

$$\mathbf{B} = \mathbf{A}^{-1} = \frac{\partial \boldsymbol{\xi}}{\partial \mathbf{x}}, \tag{3.18}$$

and put it into Eq. (3.14).

---

**Exercise 3.2: Show that matrix $\partial \boldsymbol{\xi}/\partial \mathbf{x}$ is the inverse of matrix $\partial \mathbf{x}/\partial \boldsymbol{\xi}$.**

We start from Eq. (3.1):

$$\mathbf{x} = \tau^e(\boldsymbol{\xi}), \tag{3.2–1}$$

which is equivalent to

$$\boldsymbol{\xi} = \tau^{e-1}(\mathbf{x}). \tag{3.2–2}$$

Equation (3.1) can be rewritten as follows:

$$\mathbf{x} = \tau^e[\tau^{e-1}(\mathbf{x})] = \mathbf{x}[\boldsymbol{\xi}(\mathbf{x})]. \tag{3.2–3}$$

By differentiating the last equation with respect to $\mathbf{x}$, we get

$$\mathbf{I} = \frac{\partial \mathbf{x}}{\partial \boldsymbol{\xi}} \frac{\partial \boldsymbol{\xi}}{\partial \mathbf{x}}, \tag{3.2–4}$$

where the chain rule for differentiating compound functions was used and $\mathbf{I}$ is the unit tensor. Analogous calculations with

$$\boldsymbol{\xi} = \boldsymbol{\xi}[\mathbf{x}(\boldsymbol{\xi})] \tag{3.2–5}$$

give the result

$$\mathbf{I} = \frac{\partial \boldsymbol{\xi}}{\partial \mathbf{x}} \frac{\partial \mathbf{x}}{\partial \boldsymbol{\xi}}, \tag{3.2–6}$$

and, with the definition of the inverse, we can conclude that

$$\frac{\partial \boldsymbol{\xi}}{\partial \mathbf{x}} = \left(\frac{\partial \mathbf{x}}{\partial \boldsymbol{\xi}}\right)^{-1}. \tag{3.2–7}$$

---

## 3.2 Numerical Integration

Depending on the physical problem to solve, the finite-element method involves the evaluation of integrals in one-, two-, or three-dimensional domains for the stiffness matrix as shown in Eq. (2.45) (which represents a one-dimensional problem). As we shall see in Section 3.4, the computation of line or surface integrals is required to take into account some of the boundary conditions. Because the number of these integrals can be very high, it is desirable to find the most economical method of computation for a given accuracy. Moreover, for complex isoparametric elements, exact integration is not possible. This is also often true when the partial differential equations are nonlinear, and the functionals are thus nonquadratic.

Starting with the one-dimensional case, we have seen that the assembly of the global stiffness matrix is reduced to the successive calculation of an integral of the form

$$I_e = \underset{\text{(general)}}{\int_{\Omega_e} f(x)\mathrm{d}x} = \underset{\text{(1-D)}}{\int_{x_e}^{x_{e+1}} f(x)\,\mathrm{d}x}, \tag{3.19}$$

where $\Omega_e$ is the segment $[x_e, x_{e+1}]$ for two-node elements. For the case in which $x_e = -1$ and $x_{e+1} = 1$ we write Eq. (3.19) as

$$I_e = \int_{-1}^{1} f(x)\,\mathrm{d}x. \tag{3.20}$$

The Gaussian method of numerical integration consists of finding $p$ coordinates $x^k$, $k = 1, 2, \ldots, p$ with

$$-1 < x^1 < x^2 < \cdots < x^p < 1,$$

and the $p$ corresponding weighting factors $h^1, \ldots, h^p$ such that

$$I_e \cong \sum_{k=1}^{p} h^k f(x^k). \tag{3.21}$$

More precisely, the coordinates $x^k$, and the weighting factors $h^k$, are chosen in order to integrate *exactly* any polynomial of degree $2p - 1$. For all other functions, the integration is thus approximate.

The values of $x^k$ and $h^k$ are given in most finite-element books, but the most useful ones corresponding to $p = 1$ and $p = 2$ are treated in Exercise 3.3.

---

**Exercise 3.3: Calculate the exact values of $x^k$ and $h^k$ for $p = 1$ and for $p = 2$.**

The fundamental property is used to compute these values.

Case $p = 1$: a polynomial of degree $2p - 1 = 1$ must be integrated exactly. The general form of such a polynomial is

$$P_1(x) = a + bx. \tag{3.3–1}$$

We find immediately, by direct analytical computation,

$$\int_{-1}^{1} P_1(x)\mathrm{d}x = \left(ax + \frac{1}{2}bx^2\right)_{x=1} - \left(ax + \frac{1}{2}bx^2\right)_{x=-1} = 2a. \tag{3.3–2}$$

This must be equal to the Gaussian expression of Eq. (3.21),

$$h^1 P_1(x^1) = h^1(a + bx^1), \tag{3.3–3}$$

so that we have the following relation, which must hold for any $a$ and $b$:

$$2a = h^1 a + h^1 bx^1. \tag{3.3–4}$$

The solution is obvious:

$$h^1 = 2, \quad x^1 = 0. \tag{3.3–5}$$

Case $p = 2$: the polynomial is of degree $2p - 1 = 3$ and is written as

$$P_3(x) = a + bx + cx^2 + \mathrm{d}x^3. \tag{3.3–6}$$

The integral condition is

$$\int_{-1}^{1} P_3(x)\mathrm{d}x = h^1 P_3(x^1) + h^2 P_3(x^2), \tag{3.3–7}$$

which gives for any $a, b, c, d$:

$$2a + \frac{2}{3}c = h^1[a + bx^1 + c(x^1)^2 + \mathrm{d}(x^1)^3] + h^2[a + bx^2 + c(x^2)^2 + \mathrm{d}(x^2)^3]. \tag{3.3–8}$$

Thus, we must have

$$2 = h^1 + h^2, \tag{3.3–9}$$

$$0 = h^1 x^1 + h^2 x^2, \tag{3.3–10}$$

$$\frac{2}{3} = h^1(x^1)^2 + h^2(x^2)^2, \tag{3.3–11}$$

$$0 = h^1(x^1)^3 + h^2(x^2)^3, \tag{3.3–12}$$

for which it is easy to see that the only solution is

$$h^1 = h^2 = 1, \qquad -x^1 = x^2 = 1/\sqrt{3}. \tag{3.3–13}$$

---

When the segment $[x_e, x_{e+1}]$ is different from the reference segment $[-1, 1]$, a mapping function allows the transformation of the initial integral of Eq. (3.19) according to the usual change-of-variable rule:

$$I_e = \int_{-1}^{1} f[x(\xi)]\frac{\mathrm{d}x}{\mathrm{d}\xi}\mathrm{d}\xi. \tag{3.22}$$

This integral is transformed, using the same mapping as for the isoparametric elements [see Eq. (3.3a)], by putting

$$x(\xi) = x_e N_1^0(\xi) + x_{e+1} N_2^0(\xi) \qquad \text{for } -1 \le \xi \le 1. \tag{3.23}$$

Using Eq. (3.8) to express the derivative of the $x$ with respect to $\xi$ and putting

$$g(\xi) = f[x(\xi)]\frac{x_{e+1} - x_e}{2}, \tag{3.23a}$$

we find the initial integral of Eq. (3.22) takes the final form

$$I_e = \int_{-1}^{1} g(\xi)\,\mathrm{d}\xi, \tag{3.24}$$

which is analogous to Eq. (3.20) and can be approximated as in Eq. (3.21).

For a one-dimensional isoparametric finite element $\Omega_e$, the mapping with the reference element defined in the segment $[-1, 1]$ can be more complicated than the above linear transformation. For example, we can have a quadratic correspondence with the three-node element defined in Sections 2.4 and 3.2. However more complicated, this obviously leads to an integral that can take the same form as Eq. (3.24), by an appropriate choice of the $g$ function.

The generalization to two- or three-dimensional integrals depends on the type of element. Two cases will be discussed.

**Reference Element: Square $[-1, 1]^2$ or Cube $[-1, 1]^3$**

We suppose that the reference element is the square (with sides of length 2) defined by

$$\mathbf{x} \in \Omega_0 = [-1, 1]^2 \qquad \text{if } -1 \leq x_1 \leq 1, \quad -1 \leq x_2 \leq 1, \tag{3.25}$$

and that a mapping function is defined between the (physical) element $\Omega_e$ and the reference element $\Omega_0$. With these hypotheses the integral can be put into the following form:

$$I_e = \int_{\Omega} f(x)\mathrm{d}S = \int_{\Omega_0} f[x(\xi)]\det\left(\frac{\partial x}{\partial \xi}\right)\mathrm{d}S_0, \tag{3.26}$$

where

$\mathrm{d}S = \mathrm{d}x_1\mathrm{d}x_2$ stands for the surface element in the physical space,
$\mathrm{d}S_0 = \mathrm{d}\xi_1\mathrm{d}\xi_2$ for the surface element in the reference space, and
$\det(\frac{\partial \mathbf{x}}{\partial \boldsymbol{\xi}})$ is the Jacobian determinant of the application $\boldsymbol{\xi} \to \mathbf{x}$.

With the definition of two-dimensional integrals, we can convince ourselves that Eq. (3.26) is easily reduced to the following form:

$$I_e = \int_{-1}^{1}\left[\int_{-1}^{1} g(\xi_1, \xi_2)\mathrm{d}\xi_1\right]\mathrm{d}\xi_2, \tag{3.27}$$

provided that we put

$$g(\xi_1, \xi_2) = f\left[\mathbf{x}(\boldsymbol{\xi})\right]\det\left(\frac{\partial \mathbf{x}}{\partial \boldsymbol{\xi}}\right). \tag{3.27a}$$

With the Gaussian integration formula of Eq. (3.21) for one-dimensional integrals, we first approximate the inner integral of Eq. (3.27), with $\xi_1$ as the integration variable, by

$$I_e \approx \int_{-1}^{1}\left[\sum_{i=1}^{p} h^i g\left(\xi_1^i, \xi_2\right)\right]\mathrm{d}\xi_2, \tag{3.28}$$

where we introduced $p$ Gaussian integration points denoted by $\xi_1^i (i = 1 \text{ to } p)$. Equation (3.28) is then a one-dimensional integral, with the integration variable being $\xi_2$; using again the Gaussian approximation with the $\xi_2^j (j = 1 \text{ to } p)$ integration points, we obtain the following result:

$$I_e \cong \sum_{i,j=1}^{p} h^i h^j g\left(\xi_1^i, \xi_2^j\right). \tag{3.29}$$

A similar procedure can be used for the three-dimensional case and leads to an expression with a summation on three indices corresponding to the integration points with the three local coordinates: $(\xi_1^i, \xi_2^j, \xi_3^k)$, $(i, j, k = 1 \text{ to } p)$.

**Reference Element: Triangle or Tetrahedron**

In this case Eq. (3.29), or the three-dimensional analog, cannot be used with optimal accuracy for a given number of integration points. Special formulas were found, which are designed to respect the symmetries of the triangle, or those of the tetrahedron, and integrate exactly any complete polynomial of a given order. They can be found in finite-element books; see, for example, footnotes 2 and 3 in Chapter 2. Some examples of approximate integration formulas for triangles are given in Chapter 4.

## 3.3 Solution of Linear Finite-Element Systems

In this section we shall outline the main features of the simpler numerical methods for solving a system of linear equations, particularly those obtained after the finite-element (FE) discretization. We need first to examine the structure of the matrix, in order to take full advantage of any improvements in efficiency that are available[4]: sparsity and symmetry, if it exists. This is equally useful for nonlinear problems, as we shall see in the next section: the basic method consists of an iterative procedure, in which linear equations are solved several times at each step.

### Structure of the Finite-Element Stiffness Matrix for Linear Problems

The structure of the matrix originating from finite-element discretization is rather special. In many instances the matrix is symmetric. This property can easily be verified when the formulation is based on a variational principle, including material behavior and friction. This is true for the two problems studied so far: the linear heat equation with constant conductivity and linear elasticity.

Another interesting property is the sparsity of the matrix: for a given row corresponding to a node $n$, the only nonzero components are those with columns corresponding to nodes $n'$ for which nodes number $n$ and $n'$ belong to the same element. This property is better explained by simple examples. It is already clear from Fig. 2.16 that many components of the global stiffness matrix are equal to zero, but the sparsity increases when the number of nodes increases. This is demonstrated in Fig. 3.1: a two-dimensional linear thermal problem is discretized into three-node elements.

The stiffness matrix is $12 \times 12$, and it is built from the assembly of 10 triangular elements. In Fig. 3.2 the stiffness matrix is schematically represented with crosses for nonzero entries, and blanks elsewhere. (Note that this structure is easily determined from the connectivity, by inspection of Fig. 3.1 or by reference to the node element array lg.) One can observe that with the node numbering that was chosen in Fig. 3.2, each line has a maximum of three positions between the diagonal element, represented

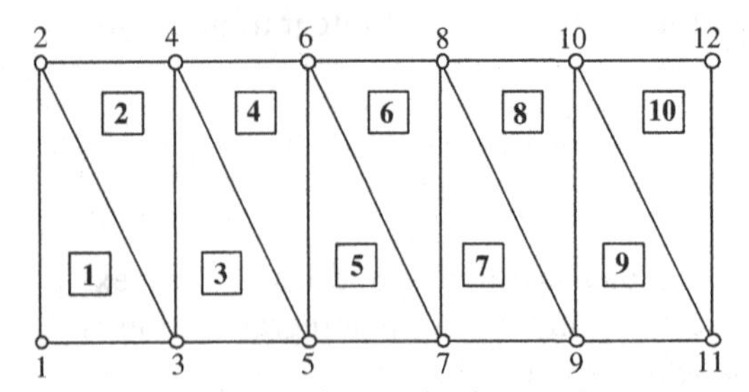

Figure 3.1. Triangular mesh of a simple structure.

[4] For any sufficiently large FE problem, that is, one with many degrees of freedom, the computation time will be dominated by the solution of the linear system. Thus, any improvement that can be made in this process is very valuable.

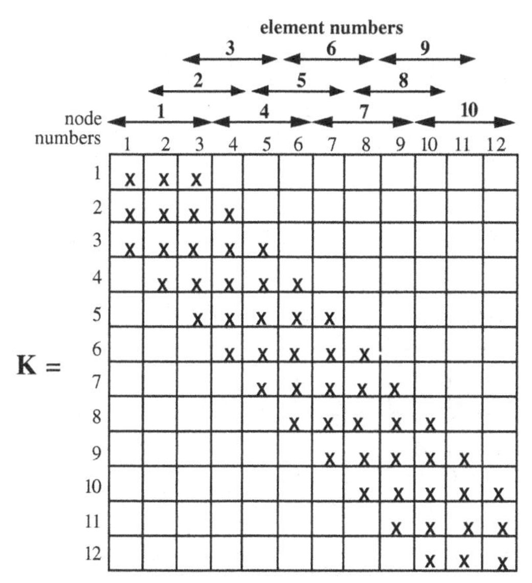


Figure 3.2. Structure of the stiffness matrix corresponding to the FE discretization of a thermal problem (defined on the mesh of Fig. 3.1, before boundary conditions are imposed).

by a cross, and a nonzero entry (including the ends). This maximum size is called the nodal half-bandwidth, denoted by $l_n$. Formally we have, for the general case,

$$l_n = \max_{e,m,m'} |\lg(m, e) - \lg(m', e)| + 1. \tag{3.30}$$

The half-bandwidth is

$$l_b = l_n \dim_u, \tag{3.31}$$

where $\dim_u$ is the dimension of the local unknown: one for the thermal problem; one, two, or three for the elasticity problem; and so on. The total bandwidth is then

$$l_w = (2l_n - 1)\dim_u. \tag{3.32}$$

In the example of Fig. 3.2 we have $l_w = 5$. (Note: the bandwidth or half-bandwidth can also be defined for each line of the stiffness matrix, and resolution algorithms can take advantage of a variable bandwidth. Furthermore, inside the bandwidth some components of the stiffness matrix are equal to zero, but in the methods we shall describe, it is difficult to exploit this additional "sparsity" effectively.)

Small bandwidths are desirable for two reasons:

- only $l_w$ elements per line must be stored for nonsymmetric systems, and $l_b$ for symmetric systems;
- the computational time for solving the set of equations by direct methods is approximately proportional to $l_{w^2}$ or $l_{b^2}$.

---

**Exercise 3.4: For a two-dimensional rectangular domain meshed into three-node triangles, as represented in the following figures, calculate the half-bandwidth for a thermal problem, and that for an elasticity problem.**

We label each node with $i$ and $j$ indices and calculate the node number according to

$$k = i + (j-1)q, \qquad i = 1 \text{ to } q, \quad j = 1 \text{ to } p.$$

We suppose the mesh is built such that any node is connected, by a maximum of six elements, to six neighbors as shown in the following figure.

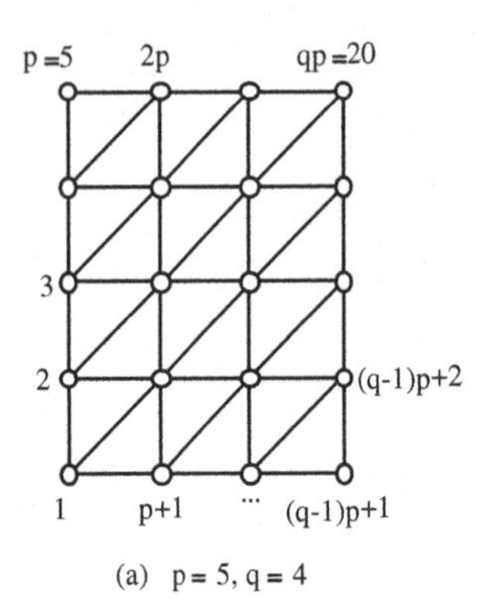


(a) p = 5, q = 4

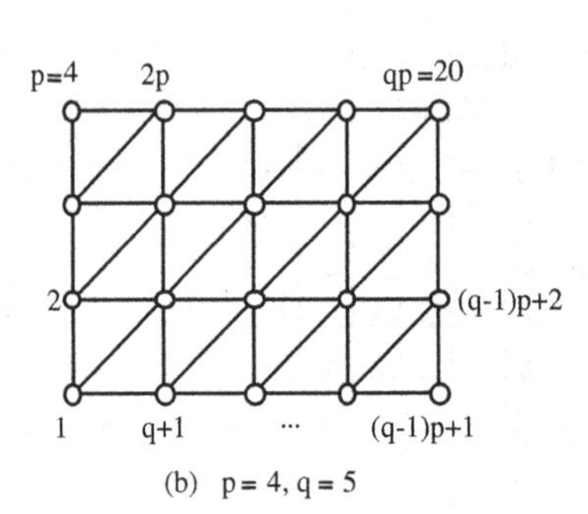


(b) p = 4, q = 5

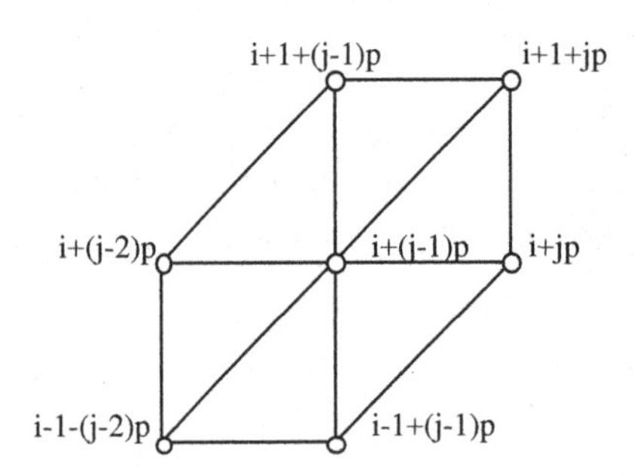


We see immediately that the maximum node number gap is

$$l_n = p + 2,$$

which gives for the half-bandwidth for the thermal problem

$$l_b = p + 2,$$

and for the elasticity problem

$$l_b = 2p + 4,$$

which is 14 for case (a) and 12 for case (b) in the figure.

---

### Gaussian Method for Solving Banded Linear Systems

The general form of the linear system is written as

$$\mathbf{KV} = \mathbf{F}, \tag{3.33}$$

where

- **V** is the unknown vector, and
- **K** is the stiffness matrix with half-bandwidth $l_b$ and dimension $n$ (note that here $n$ is used instead of $n_D$, for simplicity).

The classical Gaussian method includes $n - 1$ major steps, with the purpose of transforming the initial linear system into an equivalent one, for which all the elements

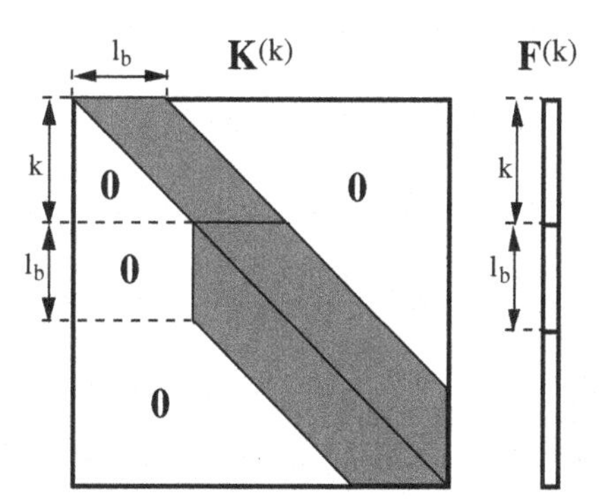


Figure 3.3. Gaussian triangulation intermediate structure of the banded stiffness matrix after the $k$th major step.

beneath the diagonal will be null. The new matrix obtained at the end of the Gaussian transformation is called *triangular*, and the resulting linear system is much simpler to solve (see Exercise 3.6).

We suppose that the $k$th major step has been already performed (with $k \geq 1$), leading to a transformed matrix $\mathbf{K}^{(k)}$ and a vector $\mathbf{F}^{(k)}$, and that the structure of the matrix is shown graphically in Fig. 3.3.

The next major step must produce zeros beneath the diagonal of the $(k+1)$th column of matrix $\mathbf{K}^{(k)}$. This is done by adding to rows numbers $(k+1)$ to $(k+l_b)$ appropriate combinations of line $k$, provided that the same operation is performed on $\mathbf{F}^{(k)}$. More precisely, we put

$$L_{ik} = \frac{K_{ki}^{(k)}}{K_{kk}^{(k)}}, \qquad i = k+1 \text{ to} \min(k+l_b, n). \tag{3.34}$$

We have supposed that $K_{kk}^{(k)} \neq 0$, and we thus obtain the new value of the matrix,

$$K_{ij}^{(k+1)} = K_{ij}^{(k)} - L_{ik}K_{kj}^{(k)}, \qquad i, j = k+1, k+2, \ldots, \min(k+l_b, n), \tag{3.35}$$

and of the right-side vector,

$$F_i^{(k+1)} = F_i^{(k)} - L_{ik}F_k^{(k)}, \qquad i = k+1, k+2, \ldots, \min(k+l_b, n). \tag{3.36}$$

---

**Exercise 3.5: Compute the number of additions and multiplications for the complete triangulation of matrix K by the Gaussian method (give only an asymptotic expression when $n \gg l_b$).**

From Eq. (3.35) we see that the number of additions and multiplications is $l_b \times l_b$ (except at the beginning when $k < l_b$ and at the end when $k > n - l_b$). In this calculation we have neglected $l_b$ divisions coming from Eq. (3.34) and $l_b$ additions and multiplications and additions from Eq. (3.36).

Finally, as we have $(n-1)$ major steps, the total number is approximately

$$n l_b^2.$$

---

This method works also for symmetric matrices, but the Cholesky or the Crout methods (described later) are generally preferred, as they are more time efficient.

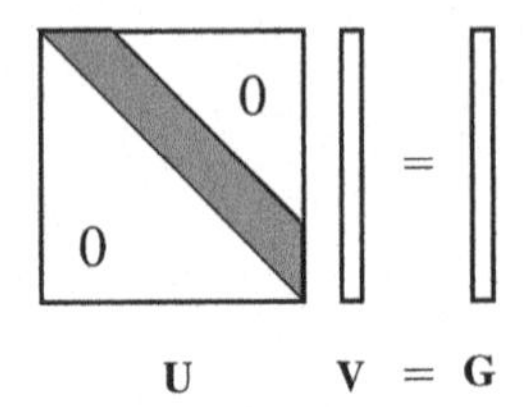


Figure 3.4. A banded upper triangular system.

The last step is to solve the linear system where **K** has been transformed into an upper triangular matrix $\mathbf{U} = \mathbf{K}^{(n-1)}$, and **F** to $\mathbf{G} = \mathbf{F}^{(n-1)}$. The linear system to solve is graphically represented in Fig. 3.4.

The algorithm is a very straightforward backsubstitution. We have first for the last row:

$$V_n = G_n / U_{nn}.$$

Then, working backward from row $n-1$ to row 1, we have

$$V_{n-i} = \frac{1}{U_{(n-i)(n-i)}} \left[ G_{n-i} - \sum_{j=n-i+1}^{n_i} U_{(n-i)j} V_j \right], \tag{3.1}$$

where $n_i = \inf(n - i + \mathrm{l}_b, n)$.

---

**Exercise 3.6: Compute the number of additions and multiplications for solving the triangular system.**

Neglecting the division, Eq. (3.37) shows that we have essentially $\mathrm{l}_b$ additions and multiplications for each component of **V**. As **V** possesses $n$ components to be computed, the total number of both arithmetic operations is

$$n\mathrm{l}_b.$$

Thus this part of the solution is second order compared with the triangulation for large bandwidth matrices. Other computations in the FE procedure involve only calculations at the element level and are therefore proportional to the number of elements.

---

### The Cholesky Method for Banded Systems

The matrix **K** is supposed to be symmetric and positive definite.[5] It is first assumed that the **K** matrix can be decomposed into the product of a lower triangular matrix **L**, and its transpose $\mathbf{L}^T$, an upper triangular matrix:

$$\mathbf{K} = \mathbf{L}\mathbf{L}^T, \tag{3.38}$$

where **L** has the same half-bandwidth as **K**. The structure of the Cholesky decomposition is illustrated in Fig. 3.5.

Then the initial problem can be transformed into the successive resolution of two triangular band systems. If we put

$$\mathbf{L}^T\mathbf{V} = \mathbf{W}, \tag{3.39}$$

[5] This is often the case when **K** derives from an energy functional, provided that there are no zero-energy modes (i.e., deformation processes that do not affect the value of the energy of the system).

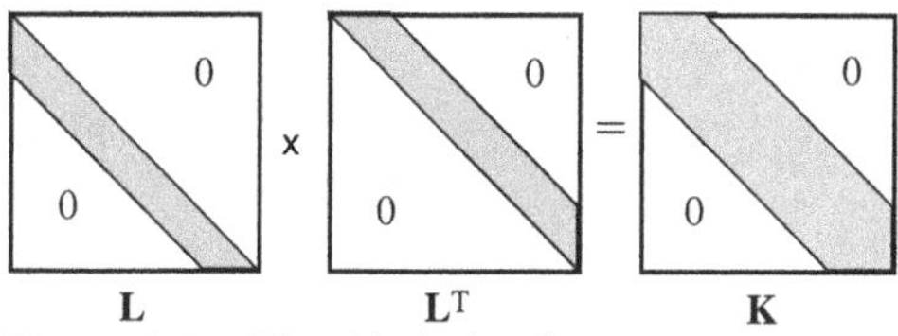

Figure 3.5. The Cholesky decomposition.

then, with the Cholesky decomposition given by Eq. (3.38), the original linear system can be written as

$$\mathbf{LW} = \mathbf{F}, \tag{3.40}$$

which is a lower triangular system and which is easily solved by using an algorithm very similar to that used for the upper triangular system. Once **W** is computed, according to Eq. (3.40), the triangular system given by Eq. (3.39) is solved.

---

**Exercise 3.7: Using Eq. (3.38), propose an algorithm for the computation of matrix L of the Cholesky decomposition.**

There are $n$ major steps, corresponding to the calculation of lines 1 to $n$ of the **L** matrix.

a. For $k = 1$ we have obviously

$$L_{11}^2 = K_{11}, \tag{3.7–1}$$

which gives

$$L_{11} = \sqrt{K_{11}}, \tag{3.7–2}$$

provided that $K_{11} > 0$.

b. Now we suppose that the $(k-1)$ first lines of matrix **L** are already computed (for $k \geq 2$). The computation of the $k$th line proceeds as follows.

- The first nonzero element of the $k$th line corresponds to the column number

$$m = \max(1, k - l_b), \tag{3.7–3}$$

and we can thus write

$$L_{km}L_{mm} = K_{km}. \tag{3.7–4}$$

If $L_{mm} \geq 0$, we get:

$$L_{km} = \frac{K_{km}}{L_{mm}} \tag{3.7–5}$$

- For any $i$ such that $m < i < k$, we suppose that the line $k$ of **L** is calculated up to element $i - 1$, and we write

$$\sum_{j=m}^{i} L_{kj}L_{ij} = K_{ki}, \tag{3.7–6}$$

which corresponds to the product of line $k$ of **L** by the column $i$ of $\mathbf{L}^T$, the latter being equal to line $i$ of **L**. Thus we obtain

$$L_{ki} = \frac{1}{L_{ii}}\left(K_{ki} - \sum_{j=m}^{i-1} L_{kj}L_{ij}\right). \tag{3.7–7}$$

- The last element of line $k$ comes from

$$\sum_{j=m}^{k} L_{kj}^2 = K_{kk}, \tag{3.7–8}$$

which is equivalent to

$$L_{kk} = \left( K_{kk} - \sum_{j=m}^{k-1} L_{kj}^2 \right)^{1/2}. \tag{3.7–9}$$

---

After the building of an algorithm for the Cholesky decomposition, the following exercise allows us to verify that this procedure is more economical than the classical Gaussian elimination.[6]

---

**Exercise 3.8: Compute the number of additions and multiplications for the complete triangulation of matrix K by the Cholesky method (give only an asymptotic expression when $n \gg l_b$).**

From Exercise 3.7 we see that, during one major step, the number of additions and multiplications is

$$1 + 2 + \cdots + (l_b - 2) + (l_b - 1) + l_b = l_b(l_b + 1)/2 \cong \frac{1}{2} l_b^2 \tag{3.8–1}$$

(except at the beginning when $k < l_b$ and at the end when $k > n - l_b$). As there are $n$ major steps, the total number of additions, or multiplications, is asymptotically equal to

$$\frac{1}{2} n l_b^2, \tag{3.8–2}$$

which corresponds to half of the corresponding number for the classical Gaussian method.

For a computer with a 50-Mflops CPU, we obtain the following figures for the Cholesky method:

| $N$ | $l_b$ | *CPU time* |
|---|---|---|
| 100 | 10 | 200 $\mu$s |
| 10,000 | 100 | 2 s |
| 1,000,000 | 1000 | 5 h, 33 min |

---

## 3.4 Numerical Solution of Nonlinear Finite-Element Systems

The nonlinear system of the unknown vector **V** may be written as

$$\mathbf{R}(\mathbf{V}) = 0, \tag{3.41}$$

where **R** is a nonlinear vector function of unknown vector **V**. The most powerful and most widely used method for solving this system is the Newton–Raphson[7] method. At iteration number 0, a first guess $\mathbf{V}^{(0)}$ of the unknown is necessary; then at an arbitrary iteration $r$, with $r \geq 1$, $\mathbf{V}^{(r-1)}$ is known and we denote $\mathbf{R}[\mathbf{V}^{(r-1)}]$ by $\mathbf{R}^{(r-1)}$. The next trial solution, at iteration $r$, may be considered as the solution at iteration $r - 1$ plus

[6] However, the Gaussian elimination can be adapted to take advantage of the symmetric matrices.

[7] The iterative method is named "Newton" for a scalar function, and "Newton–Raphson" for a system of several equations.

an increment or update:

$$\mathbf{V}^{(r)} = \mathbf{V}^{(r-1)} + \Delta\mathbf{V}^{(r)}. \tag{3.42}$$

Equation (3.42) is put into Eq. (3.41), and a first-order Taylor expansion allows the linearization of the equation about the last trial solution $\mathbf{V}^{(r-1)}$:

$$\begin{aligned}\mathbf{R}^{(r)} &= \mathbf{R}\left[\mathbf{V}^{(r-1)} + \Delta\mathbf{V}^{(r)}\right] = 0\\ &= \mathbf{R}^{(r-1)} + \frac{\partial\mathbf{R}}{\partial\mathbf{V}}\left[\mathbf{V}^{(r-1)}\right]\cdot\Delta\mathbf{V}^{(r)} + (\text{higher-order terms}).\end{aligned} \tag{3.43}$$

The first-order derivative is the stiffness matrix

$$\mathbf{K}^{(r-1)} = \frac{\partial\mathbf{R}}{\partial\mathbf{V}}\left[\mathbf{V}^{(r-1)}\right], \tag{3.43a}$$

with components

$$K_{nn'}^{(r-1)} = \frac{\partial R_n}{\partial V_{n'}}\left[\mathbf{V}^{(r-1)}\right]. \tag{3.43b}$$

$\mathbf{K}$ is symmetric if the initial Eq. (3.41) can be derived from a variational formulation. Solving Eq. (3.43) with the substitute notation of Eq. (3.43b) gives

$$\Delta\mathbf{V}^{(r)} = -\left[\mathbf{K}^{(r-1)}\right]^{-1}\mathbf{R}^{(r-1)}, \tag{3.44}$$

and the iterations can proceed until convergence. The process can be visualized for the solution of a scalar equation of one unknown, as is pictured in Fig. 3.6.

However, in some occasions, as in Fig. 3.6(b), when the initial guess is too far from the solution, this method can fail to converge and then a *subincrementation* procedure must be used. Instead of Eq. (3.42), the new solution is calculated by

$$\mathbf{V}^{(r)} = \mathbf{V}^{(r-1)} + \gamma\Delta\mathbf{V}^{(r)}, \tag{3.42a}$$

with $\Delta\mathbf{V}^{(r)}$ again expressed by Eq. (3.44). The positive scalar $\gamma$ is chosen so that the condition on the norms

$$\left|\mathbf{R}\left[\mathbf{V}^{(r-1)} + \gamma\Delta\mathbf{V}^{(r)}\right]\right| < \left|\mathbf{R}^{(r-1)}\right| \tag{3.45}$$

is fulfilled. That can be achieved in general by starting with $\gamma = 1$ and dividing it by 2 (or any number greater than 1) at each trial, until Eq. (3.45) is verified. In one variation, the *line search* method, various values are thus tried until $\mathbf{R}[\mathbf{V}^{(r-1)} + \gamma\Delta\mathbf{V}^{(r)}]$ is minimum.

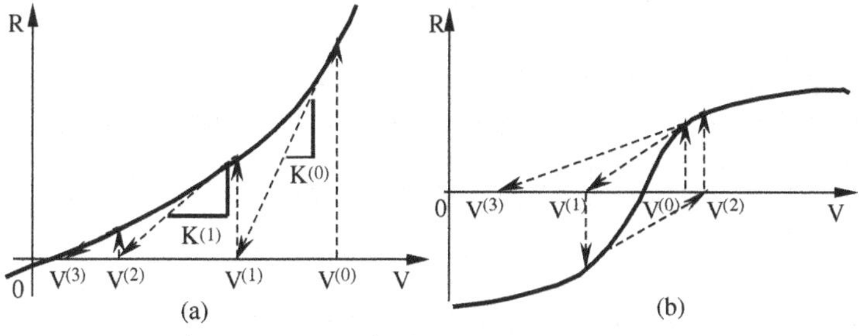


Figure 3.6. The Newton method: (a) convergence, (b) divergence.

## 3.5 Finite-Element Formulation of a Boundary-Value Problem

The stationary heat conduction problem is convenient as an illustration of the main ideas in finite-element formulation. It possesses the advantage of involving only a scalar unknown function $T$. In this section two important methods will be discussed: the variational formulation, which is attractive when it leads to a minimization problem, although it cannot be used for every system of partial differential equations; and the Galerkin method, which is based on a weak form of the equations and can be utilized virtually for any problem. For a mechanical problem, it is shown that the Galerkin method can also be viewed as a "natural" discretization of the virtual work principle.

### The Variational Formulation

We suppose that the problem is defined in a domain $\Omega$, where the stationary heat equation can be written as

$$\mathrm{div}[k\,\mathbf{grad}\,(T)] = \nabla \cdot (k \nabla T) = 0, \tag{3.46}$$

with prescribed boundary conditions on temperature,

$$T = T^d \qquad \text{on } \partial\Omega^T, \tag{3.47}$$

or on the heat flux,

$$-k\frac{\partial T}{\partial \mathbf{x}} \cdot \mathbf{n} = \phi \qquad \text{on } \partial\Omega^\phi, \tag{3.48}$$

where $\mathbf{n}$ is the normal vector to the surface. The convection form can also be introduced:

$$-k\frac{\partial T}{\partial \mathbf{x}} \cdot \mathbf{n} = \alpha(T - T^c) \qquad \text{on } \partial\Omega^c. \tag{3.49}$$

Here the radiation term has not been considered, and no internal source is taken into consideration; moreover, the scalar conduction coefficient $k$ may be space dependent, but we suppose, in this part of the section, that it does not depend on temperature itself or on the heat flux direction.

The problem being assumed, if the appropriate mathematical hypotheses are made, is equivalent to the following problem.[8]

Find the function $T$ defined in $\Omega$, so that $T = T^d$ on $\partial\Omega^T$, and which minimizes the functional:

$$I(T) = \frac{1}{2}\int_\Omega k[\mathrm{grad}(T)]^2 \mathrm{d}V + \int_{\partial\Omega^\phi} \phi T \mathrm{d}S + \frac{1}{2}\int_{\partial\Omega^c} \alpha(T - T^c)^2\, \mathrm{d}S. \tag{3.50}$$

In this form the continuous problem can be transformed into a discrete problem by restricting the space of functions $T$ to the finite-element discretized form given by Eq. (2.25). A mesh of the domain is assumed, and the choice of elements and shape functions made, such that an approximate solution will be sought in terms of the

[8] A formal justification can be obtained for a scalar function, using the same method as that developed in Section 5.5 [R. H. Wagoner and J.-L. Chenot, *Fundamentals of Metal Forming* (Wiley, New York, 1997)] for vector fields.

$n_D(= n_n)$ unknown global nodal values $T_n$. The approximate problem is obtained by putting Eq. (2.25) into Eq. (3.50). We obtain a quadratic function of the unknowns (instead of a functional depending on an infinite number of unknowns):

$$I(\bar{T}) = \int_\Omega \frac{k}{2}\left[\sum_n T_n \,\mathbf{grad}\,(N_n)\right]^2 dV + \int_{\partial\Omega^\phi} \phi \sum_n T_n N_n \,dS$$
$$+ \int_{\partial\Omega^c} \frac{\alpha}{2}\left(\sum_n T_n N_n - T^c\right)^2 dS. \qquad (3.51)$$

However, some of the nodal values are not free in general, when, on a subset of the boundary, temperatures are prescribed as in Eq. (3.47). Several sets of indices must be introduced, where

$N = \{1, 2, \ldots, n_n\}$ is the set of all node numbers of the finite element mesh,
$N^T$ is the set of the numbers of nodes lying on $\partial\Omega^T$, where temperatures are prescribed, and
$N^f$ is the set of the node numbers that correspond to nonprescribed values of the temperature.

With the classical mathematical notations for the sets, we have of course

$$N = N^T \cup N^f, \quad N^T \cap N^f = \emptyset(\text{empty set}).$$

With this additional notation introduced, the problem to solve involves $n_D = n_n$ unknowns,[9] or degrees of freedom (DOF). The approximate functional of Eq. (3.51) is minimized with respect to the unknowns, taking into account the prescribed values of the temperature:

$$\frac{\partial I}{\partial T_{n'}}(\bar{T}) = 0 \qquad \text{for } n' \in \mathcal{N}^f, \qquad (3.52)$$

$$T_{n'} = T^d(\mathbf{X}_{n'}) \qquad \text{for } n' \in \mathcal{N}^T. \qquad (3.53)$$

Equation (3.52) can be written in more detail from Eq. (3.51) as follows:

$$\sum_n \left[\int_\Omega k\,\mathbf{grad}(N_n)\cdot \mathbf{grad}(N_{n'})dV\right] T_n + \sum_n \left(\int_{\partial\Omega^c} \alpha N_n N_{n'} dS\right) T_n$$
$$- \int_{\partial\Omega^c} \alpha T^c N_{n'} dS + \int_{\partial\Omega^\phi} \phi N_{n'} dS = 0 \qquad \text{for } n' \in N^f, \qquad (3.54)$$

together with Eq. (3.53), which must be retained.

These equations can be put in matrix form:

$$\mathbf{KT} = \Phi, \qquad (3.55)$$

with the following expression for the components of the **K** matrix:

$$K_{nn'} = \int_\Omega k\,\text{grad}\,(N_n)\cdot \text{grad}(N_{n'})d\mathcal{V} + \int_{\partial\Omega^c} \alpha N_n N_{n'} d\mathcal{S} \qquad n' \in \mathcal{N}^f. \qquad (3.56)$$
$$K_{nn'} = \delta_{nn'} \qquad n' \in \mathcal{N}^p$$

[9] Of which we already know those corresponding to prescribed nodal temperatures [Eq. (3.47)].

The components of the right-hand side of Eq. (3.54) are

$$\Phi_{n'} = \int_{\partial\Omega^c} \alpha T_c N_{n'} \mathrm{d}\mathcal{S} - \int_{\partial\Omega^\phi} \phi N_{n'} \mathrm{d}\mathcal{S} = 0 \qquad n' \in \mathcal{N}^f. \tag{3.57}$$

$$\Phi_{n'} = T^d(\mathbf{X}_{n'}) \qquad n' \in \mathcal{N}^p$$

The right-hand-side vector of Eq. (3.57) is calculated by the assembly of each local vector corresponding to an element with at least one side or face on the approximate boundary, with a flux condition [given by Eq. (3.48) or (3.49)].

### The Galerkin Method for the Heat Equation

If the conductivity $k$ is a function of the temperature $T$, we cannot easily build a variational formulation analogous to Eq. (3.50). The partial differential equation in $\Omega$ can be imposed in a weak sense; that is, for any function $w$, defined in $\Omega$, it is evident from Eq. (3.46) that we must have

$$\int_\Omega \operatorname{div}[k(T)\mathbf{grad}(T)]w\,\mathrm{d}\mathcal{V} = 0. \tag{3.58}$$

If $w$ is continuously differentiable, then, by using integration by parts, we can transform Eq. (3.58) into

$$\int_{\partial\Omega} k(T)w\,\mathbf{grad}(T)\cdot\mathbf{n}\mathrm{d}\mathcal{S} - \int_\Omega k(T)\mathbf{grad}(T)\cdot\mathbf{grad}(w)\,\mathrm{d}\mathcal{V} = 0, \tag{3.59}$$

where $\mathbf{n}$ is the normal to the surface $\partial\Omega$. Now we suppose $w = 0$ on $\partial\Omega^T$, and by using the boundary conditions of Eqs. (3.48) and (3.49), we get

$$\int_\Omega k(T)\,\mathbf{grad}(T)\cdot\mathbf{grad}(w)\mathrm{d}\mathcal{V} + \int_{\partial\Omega^\phi} \phi w\,\mathrm{d}\mathcal{S} + \int_{\partial\Omega^c} \alpha(T - T^c)w\,\mathrm{d}\mathcal{S} = 0. \tag{3.60}$$

In the Galerkin method we write successively as many equations as there are non-prescribed nodal temperatures by replacing $w$ by any shape function[10] $N'_n$ for $n' \in \mathcal{N}^f$, and $T$ by the finite-element approximation to it. Finally, we obtain

$$\int_\Omega k\left(\sum_i T_i N_i\right)\mathbf{grad}\left(\sum_n T_n N_n\right)\cdot\mathbf{grad}(N_{n'})\mathrm{d}V + \int_{\partial\Omega\phi} \phi N_{n'}\,\mathrm{d}\mathcal{S}$$

$$+ \int_{\partial\Omega^c} \alpha(T - T^c)N_{n'}\,\mathrm{d}\mathcal{S} = 0. \tag{3.61}$$

Remark 1: This set of equations is nonlinear and we cannot solve it directly, because $k$ itself depends on the unknown nodal values.

Remark 2: If we refer back to the case in which $k$ is independent of temperature, we see after some calculation that Eq. (3.61) is identical to that obtained by minimization of the discretized functional [Eq. (3.54)].

[10] Generally, the shape functions are not continuously differentiable. They are continuous and differentiable by element only. The Green's theorem can be still applied for usual elements.

### The Virtual Work Principle and the Galerkin Method for the Mechanical Problem

For a general mechanical problem, defined in a domain $\Omega$, the dynamic equation is written[11] as

$$\operatorname{div}(\sigma) + \rho\gamma = 0 \text{ in } \Omega, \tag{3.62}$$

where $\mathbf{g}$ is the density of mass force (most frequently the gravity forces), and $\gamma$ is the acceleration. As usual, the boundary conditions will be expressed as

$$\mathbf{T} = \sigma \cdot \mathbf{n} = \mathbf{T}^d \tag{3.63}$$

if $\mathbf{T}^d$ is the prescribed stress vector on the part $\partial\Omega^s$ of the boundary of the domain $\Omega$. On the other part $\partial\Omega^u$ of the boundary, the displacement $\mathbf{u} = \mathbf{u}^d$ is imposed. With these hypotheses, the virtual work principle states that for any virtual velocity $\mathbf{v}^*$, with the associated virtual strain rate tensor $\dot{\varepsilon}^*$, we have the equality

$$\int_\Omega \sigma : \dot{\varepsilon}^* \, \mathrm{d}\mathcal{V} - \int_\Omega \rho(\mathbf{g} - \gamma) \cdot \mathbf{v}^* \, \mathrm{d}\mathcal{V} - \int_{\partial\Omega^s} \mathbf{T}^d \cdot \mathbf{v}^* \, \mathrm{d}\mathcal{S} = 0. \tag{3.64}$$

This equation can be obtained by multiplying Eq. (3.62) by a vector function, integrating over the domain $\Omega$, and then integrating by parts and making use of the boundary condition of Eq. (3.63).[12] Now we suppose the stress tensor depends on the displacement field, which is chosen as the unknown and is therefore discretized into finite elements. In the Galerkin method, the virtual velocity field $\mathbf{v}^*$ is also discretized with the same shape functions as the displacement $\mathbf{u}$.

## 3.6 A Simple Example of Finite-Element Calculation

In this section, we shall summarize and illustrate, on a tractable problem, the complete procedure of finite-element discretization, building the stiffness matrix and the load vector, and finally solving the resulting equations. The mechanical problem of an elastic rod presented in Section 2.4 is considered again, but with more complicated physical conditions. We suppose that the rod is rotated around a vertical axis $O_z$, with an angular velocity $\omega$, so that in addition to the external force $F$, exerted on the right end, the acceleration produces a body force per unit volume equal to $\omega^2 x$ (see Fig. 3.7).

### The Linear Elastic Rod

A one-dimensional approximation is sought in terms of the displacement function $u$, along the $ox$ axis. If the cross-sectional area $S$ is constant, Eqs. (2.31), (2.32), and (2.33) in Section 2.4 are still valid, but Eq. (2.36) contains an additional term, because of the inertia forces:

$$\frac{\mathrm{d}}{\mathrm{d}x}\left(E\frac{\mathrm{d}u}{\mathrm{d}x}\right) + \rho\omega^2 x = 0, \tag{3.65}$$

[11] The general form of the dynamic equation including the gravity force g is written as $\operatorname{div}(\sigma) + \rho(\mathbf{g} - \gamma) = 0$. In most instances the gravity forces can be neglected in a metal-forming simulation.

[12] For more details see Section 5.4 [R. H. Wagoner and J.-L. Chenot, *Fundamentals of Metal Forming* (Wiley, New York, 1997)].

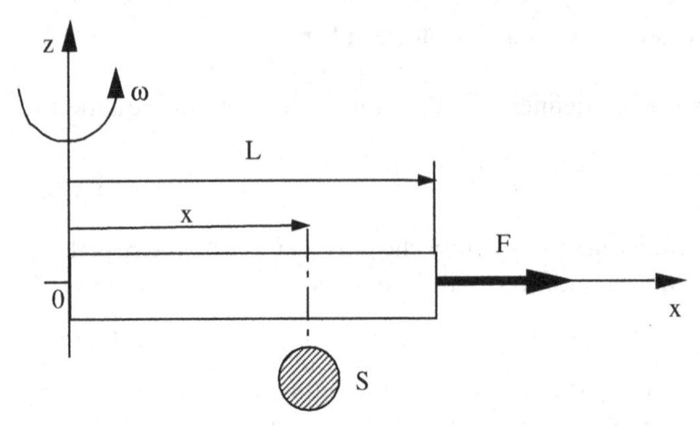


Figure 3.7. A rod submitted to a fast rotation and to an axial load.

with the same boundary conditions:

$$u(0) = 0, \tag{3.66}$$

$$\sigma(L) = F/S. \tag{3.67}$$

An analytical integration of Eq. (3.65) is possible; first it gives

$$E\frac{\mathrm{d}u}{\mathrm{d}x} = -\frac{1}{2}\rho\omega^2 x^2 + c,$$

where the integration constant $c$ is calculated from the boundary condition of Eq. (3.67), and then it gives

$$\sigma = E\frac{\mathrm{d}u}{\mathrm{d}x} = \frac{F}{S} + \frac{1}{2}\rho\omega^2(L^2 - x^2).$$

The displacement function is obtained by a new integration and, taking into account the boundary condition, Eq. (3.66), it is

$$u(x) = \frac{F}{ES}x + \frac{1}{2E}\rho\omega^2\left(L^2 x - \frac{1}{3}x^3\right). \tag{3.68}$$

Now the approximate finite-element solution will be built step by step.

**Mesh of the Rod**

Here $n_n$ equally spaced nodes are introduced, with coordinates

$$x_1 = 0, \quad x_2 = h, \quad x_3 = 2h, \ldots, \quad x_{n_n} = (n_n - 1)h,$$

where $h = L/(n_n - 1)$; the elements are the $n_e = n_n - 1$ segments:

$$\Omega_e = [(e-1)h, eh] \qquad \text{for } e = 1, 2, \ldots, n_e.$$

**Shape Functions**

Two-noded isoparametric elements are chosen, and the mapping between the reference element is done as in Section 3.2. In element number $e$ we have

$$x_1^e = (e-1)h, \qquad x_2^e = eh,$$

such that Eq. (3.6) becomes

$$x = \left(e - \frac{1}{2}\right)h + \frac{h}{2}\xi, \qquad \frac{\mathrm{d}x}{\mathrm{d}\xi} = \frac{h}{2}. \tag{3.69}$$

As explained in Section 3.2, $n_e$ global shape functions $N_n$ can be defined element by element from the local shape functions, and the displacement is sought as the linear combination

$$u(x) = \sum_{n=1}^{n_e} U_n N_n. \tag{3.70}$$

### The Galerkin Formulation

After multiplying Eq. (3.65) by any global shape function $N_m$, and integrating from $x = 0$ to $L$, we obtain:

$$U_1 = 0,$$
$$\int_0^L \left[\frac{\mathrm{d}}{\mathrm{d}x}\left(E\frac{\mathrm{d}u}{\mathrm{d}x}\right) + \rho\omega^2 x\right] N_m \,\mathrm{d}x = 0 \qquad m > 1. \tag{3.71}$$

The first term in Eq. (3.71) is integrated by parts:

$$\frac{F}{S} N_m(L) - \int_0^L \left(E\frac{\mathrm{d}u}{\mathrm{d}x}\right)\frac{\mathrm{d}N_m}{\mathrm{d}x}\,\mathrm{d}x + \int_0^L \rho\omega^2 x N_m \,\mathrm{d}x = 0, \tag{3.72}$$

in which we suppose $1 < n \cdot n_n$, so that $N_n(0) = 0$; we have made use of the right end boundary condition, with the prescribed force $F$. After putting Eq. (3.70) into Eq. (3.72), we find the linear system to solve is

$$\sum_n \left(\int_0^L E\frac{\mathrm{d}N_m}{\mathrm{d}x}\frac{\mathrm{d}N_n}{\mathrm{d}x}\right) U_n = \int_0^L \rho\omega^2 x N_m \,\mathrm{d}x + \frac{F}{S} N_m(L). \tag{3.73}$$

### Numerical Integration of the Local Stiffness Matrix and of the Load Vector

The local stiffness matrix components are given by

$$k_{ij}^e = \int_{(e-1)h}^{eh} E\frac{\mathrm{d}N_i^e}{\mathrm{d}x}\frac{\mathrm{d}N_j^e}{\mathrm{d}x}\mathrm{d}x = \int_{-1}^{1} E\frac{\mathrm{d}N_m^0}{\mathrm{d}\xi}\frac{\mathrm{d}N_n^0}{\mathrm{d}\xi}\frac{2}{h}\,\mathrm{d}\xi, \tag{3.74a}$$

where we have made use of Eq. (3.69). With the help of Eq. (3.70), this gives

$$[k^e] = \frac{E}{h}\begin{bmatrix} 1 & -1 \\ -1 & 1 \end{bmatrix}. \tag{3.74b}$$

Now the local load vector will be denoted by $\mathbf{f}^e$, with components

$$f_i^e = \int_{(e-1)h}^{eh} \rho\omega^2 x N_i^e \,\mathrm{d}x = \rho\omega^2 \int_{-1}^{1} \left[\left(e - \frac{1}{2}\right)h + \frac{h}{2}\xi\right] N_i^0 \frac{h}{2}\,\mathrm{d}\xi. \tag{3.74c}$$

For simplicity, we use only the one-point Gaussian integration formula, thus obtaining the approximate value

$$f_i^e = \frac{1}{2}\rho\omega^2\left(e - \frac{1}{2}\right)h^2 \qquad \text{if } e \neq n_e \text{ or } i \neq 2,$$
$$f_2^{n_e} = \frac{1}{2}\rho\omega^2\left(n_e - \frac{1}{2}\right)h^2 + \frac{F}{S}. \tag{3.74d}$$

**Assembly of the Stiffness Matrix and of the Load Vector**

For a more visual and tractable presentation of the numerical treatment, the following parameter values are selected: $F = 0, h = 1$, and $n_e = 4$. In the following we represent the evolution of the global stiffness matrix, after each of the four steps corresponding to the assembly of the successive local stiffness matrices given by Eq. (3.74b), for elements 1–4, and finally after the introduction of the boundary condition on the left-hand side.

- Element 1:

$$E\begin{bmatrix} 1 & -1 & 0 & 0 & 0 \\ -1 & 1 & 0 & 0 & 0 \\ 0 & 0 & 0 & 0 & 0 \\ 0 & 0 & 0 & 0 & 0 \\ 0 & 0 & 0 & 0 & 0 \end{bmatrix}, \quad \frac{1}{2}\rho\omega^2 \begin{bmatrix} 1/2 \\ 1/2 \\ 0 \\ 0 \\ 0 \end{bmatrix}.$$

- Element 2:

$$E\begin{bmatrix} 1 & -1 & 0 & 0 & 0 \\ -1 & 2 & -1 & 0 & 0 \\ 0 & -1 & 1 & 0 & 0 \\ 0 & 0 & 0 & 0 & 0 \\ 0 & 0 & 0 & 0 & 0 \end{bmatrix}, \quad \frac{1}{2}\rho\omega^2 \begin{bmatrix} 1/2 \\ 2 \\ 3/2 \\ 0 \\ 0 \end{bmatrix}.$$

- Element 3:

$$E\begin{bmatrix} 1 & -1 & 0 & 0 & 0 \\ -1 & 2 & -1 & 0 & 0 \\ 0 & -1 & 2 & -1 & 0 \\ 0 & 0 & -1 & 1 & 0 \\ 0 & 0 & 0 & 0 & 0 \end{bmatrix}, \quad \frac{1}{2}\rho\omega^2 \begin{bmatrix} 1/2 \\ 2 \\ 4 \\ 5/2 \\ 0 \end{bmatrix}.$$

- Element 4:

$$E\begin{bmatrix} 1 & -1 & 0 & 0 & 0 \\ -1 & 2 & -1 & 0 & 0 \\ 0 & -1 & 2 & -1 & 0 \\ 0 & 0 & -1 & 2 & -1 \\ 0 & 0 & 0 & -1 & 1 \end{bmatrix}, \quad \frac{1}{2}\rho\omega^2 \begin{bmatrix} 1/2 \\ 2 \\ 4 \\ 6 \\ 7/2 \end{bmatrix}.$$

An introduction of the prescribed displacement is

$$E\begin{bmatrix} 1 & 0 & 0 & 0 & 0 \\ -1 & 2 & -1 & 0 & 0 \\ 0 & -1 & 2 & -1 & 0 \\ 0 & 0 & -1 & 2 & -1 \\ 0 & 0 & 0 & -1 & 1 \end{bmatrix}, \quad \frac{1}{2}\rho\omega^2 \begin{bmatrix} 0 \\ 2 \\ 4 \\ 6 \\ 7/2 \end{bmatrix}.$$

**Resolution of the Linear System**

If we put $1/2r\omega^2/E = p$, the linear system to be solved can be written successively, by Gaussian elimination:

- Treatment of the second row:

$$\begin{bmatrix} 1 & 0 & 0 & 0 & 0 \\ 0 & 2 & -1 & 0 & 0 \\ 0 & -1 & 2 & -1 & 0 \\ 0 & 0 & -1 & 2 & -1 \\ 0 & 0 & 0 & -1 & 1 \end{bmatrix} \begin{bmatrix} U_1 \\ U_2 \\ U_3 \\ U_4 \\ U_5 \end{bmatrix} = p \begin{bmatrix} 0 \\ 2 \\ 4 \\ 6 \\ 7/2 \end{bmatrix}.$$

- Treatment of the third row:

$$\begin{bmatrix} 1 & 0 & 0 & 0 & 0 \\ 0 & 2 & -1 & 0 & 0 \\ 0 & 0 & 3/2 & -1 & 0 \\ 0 & 0 & -1 & 2 & -1 \\ 0 & 0 & 0 & -1 & 1 \end{bmatrix} \begin{bmatrix} U_1 \\ U_2 \\ U_3 \\ U_4 \\ U_5 \end{bmatrix} = p \begin{bmatrix} 0 \\ 2 \\ 5 \\ 6 \\ 7/2 \end{bmatrix}.$$

- Treatment of the fourth row:

$$\begin{bmatrix} 1 & 0 & 0 & 0 & 0 \\ 0 & 2 & -1 & 0 & 0 \\ 0 & 0 & 3/2 & -1 & 0 \\ 0 & 0 & 0 & 4/3 & -1 \\ 0 & 0 & 0 & -1 & 1 \end{bmatrix} \begin{bmatrix} U_1 \\ U_2 \\ U_3 \\ U_4 \\ U_5 \end{bmatrix} = p \begin{bmatrix} 0 \\ 2 \\ 5 \\ 28/3 \\ 7/2 \end{bmatrix}.$$

- Treatment of the last row:

$$\begin{bmatrix} 1 & 0 & 0 & 0 & 0 \\ 0 & 2 & -1 & 0 & 0 \\ 0 & 0 & 3/2 & -1 & 0 \\ 0 & 0 & 0 & 4/3 & -1 \\ 0 & 0 & 0 & 0 & 1/4 \end{bmatrix} \begin{bmatrix} U_1 \\ U_2 \\ U_3 \\ U_4 \\ U_5 \end{bmatrix} = p \begin{bmatrix} 0 \\ 2 \\ 5 \\ 28/3 \\ 21/2 \end{bmatrix}.$$

The nodal displacements are obtained by solving the upper triangular system, starting from $U_5$:

| | *FEM* | *Analytical* |
|---|---|---|
| $U_5 =$ | $42p = 42.0p$ | $42.7p$ |
| $U_4 = 3/4(28/3p + U_5)$ | $= 77/2p = 38.5p$ | $39.0p$ |
| $U_3 = 2/3(5p + U_4)$ | $= 87/3p = 29.0p$ | $29.7p$ |
| $U_2 = 1/2(2p + U_3)$ | $= 93/6p = 15.5p$ | $15.7p$ |
| $U_1 =$ | $0.0$ | $0.0$ |

It is now possible to obtain an approximation of the stresses in the rod by using the discretized form of Eqs. (2.33) and (2.31), that is, in element number $e$:

$$\sigma = E\frac{du}{dx} = E\frac{U^{e+1} - U^e}{h}.$$

Note that the stress is constant inside an element, and therefore discontinuous at the nodes: a graphical comparison of the FEM stresses and the analytical ones is given in Fig. 3.8 (where $1/2r\omega^2$ has been set equal to unity for simplicity).

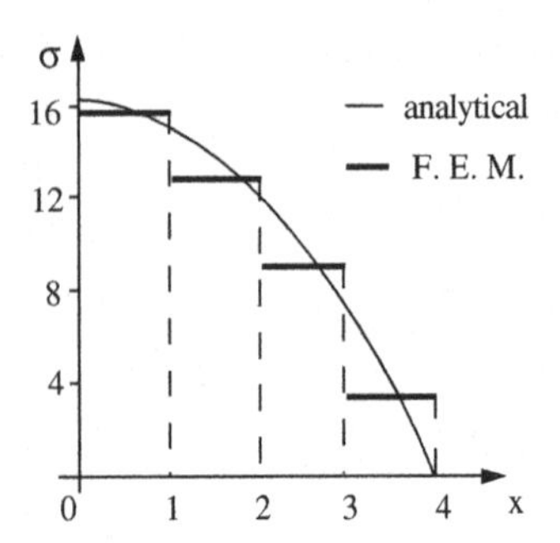


Figure 3.8. FE approximation of the stress distribution.

### A Nonlinear Elastic Rod

A quadratic elastic equation for the material behavior of the rod is defined by

$$\sigma = E\left(\varepsilon + \frac{1}{2}a\varepsilon^2\right).$$

The stress–strain relationship is pictured in Fig. 3.9, for $a = 100$.

### The Nonlinear Finite-Element Equations

With the help of the virtual work principle, we can write the continuous equation

$$\int_0^L E\left(\varepsilon + \frac{1}{2}a\varepsilon^2\right)\varepsilon^* \,\mathrm{d}x - \frac{1}{2}\int_0^L \rho\omega^2 x u^* \,\mathrm{d}x = 0, \tag{3.75}$$

which must hold for any virtual displacement field $u^*$ (and the associated virtual strain field $\varepsilon^*$). Equation (3.75) is discretized with the same elements as for the linear case and the following nonlinear equations are obtained:

$$R_m(\mathrm{U}) = \int_0^L E\left(\varepsilon + \frac{1}{2}a\varepsilon^2\right)\frac{\mathrm{d}N_m}{\mathrm{d}x}\mathrm{d}x - \frac{1}{2}\int_0^L \rho\omega^2 x N_m \mathrm{d}x = 0. \tag{3.76}$$

The solution of Eq. (3.76) requires the use of the Newton–Raphson method. For the sake of illustration one iteration is presented, starting from the linear solution with

$$p = 1/2r\omega^2/E = 10^{-4}.$$

### Computation of the Residual Vector and of the Stiffness Matrix

The residual vector components are the $R_m$ functions, and it is easy to convince ourselves that it comes only from the nonlinear term in Eq. (3.76), because we start

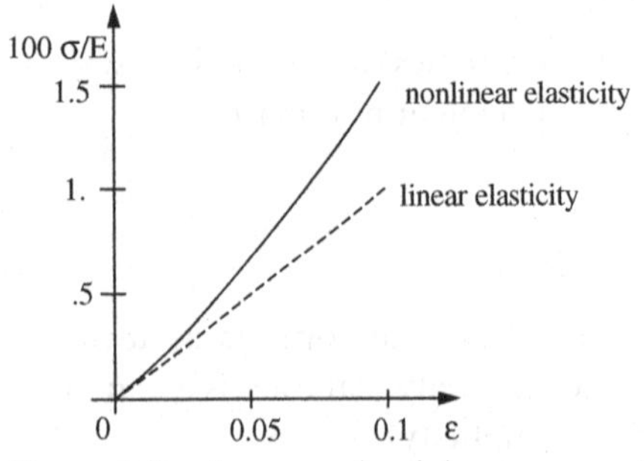


Figure 3.9. An example of the stress–strain curve for a nonlinear elastic material.

from the solution of the linear problem. Then we can write in this special case

$$R_m(\mathbf{U}) = \Delta R_m(\mathbf{U}) = \frac{1}{2}a\int_0^L \varepsilon^2 \frac{\mathrm{d}N_m}{\mathrm{d}x}\mathrm{d}x. \tag{3.77}$$

The local residual vectors are easy to express if we understand that the strain $\varepsilon$ is constant inside an element. The stiffness matrix must be evaluated by differentiation of Eq. (3.76), which immediately gives

$$K_{mn}(\mathbf{U}) \equiv \frac{\partial R_m}{\partial U_n} = \int_0^L E(1+a\varepsilon)\frac{\mathrm{d}N_m}{\mathrm{d}x}\frac{\mathrm{d}N_n}{\mathrm{d}x}\mathrm{d}x. \tag{3.78}$$

With the previous values of the displacement, computed for the linear elastic problem, the strain $\varepsilon^e$, the local contributions of the residual vector $\Delta r^e$, and that of the stiffness matrix $k^e$ can be obtained for each element number $e$:

- Element 1:

$$\varepsilon^1 = 0.155\times10^{-2},\ [\Delta r^1] = 0.01201\times10^{-2}E\begin{bmatrix}-1\\1\end{bmatrix},\ [k^1] = 1.155E\begin{bmatrix}1 & -1\\-1 & 1\end{bmatrix},$$

- Element 2:

$$\varepsilon^2 = 0.135\times10^{-2},\ [\Delta r^2] = 0.00911\times10^{-2}E\begin{bmatrix}-1\\1\end{bmatrix},\ [k^2] = 1.135E\begin{bmatrix}1 & -1\\-1 & 1\end{bmatrix},$$

- Element 3:

$$\varepsilon^3 = 0.095\times10^{-2},\ [\Delta r^3] = 0.00451\times10^{-2}E\begin{bmatrix}-1\\1\end{bmatrix},\ [k^3] = 1.095E\begin{bmatrix}1 & -1\\-1 & 1\end{bmatrix},$$

- Element 4:

$$\varepsilon^4 = 0.035\times10^{-2},\ [\Delta r^4] = 0.00061\times10^{-2}E\begin{bmatrix}-1\\1\end{bmatrix},\ [k^4] = 1.035E\begin{bmatrix}1 & -1\\-1 & 1\end{bmatrix}.$$

The assembly of the residual vector and of the stiffness matrix is performed exactly in the same way as for the linear case. After the boundary condition on the left-hand side is taken into account, it leads to the global vector and matrix:

$$[K] = \begin{bmatrix}1 & 0 & 0 & 0 & 0\\1.155 & 2.29 & -1.135 & 0 & 0\\0 & -1.135 & 2.23 & -1.095 & 0\\0 & 0 & -1.095 & 2.13 & -1.035\\0 & 0 & 0 & -1.035 & 1.035\end{bmatrix},\quad [R] = \begin{bmatrix}0\\-0.00290\\-0.00460\\-0.00390\\-0.00061\end{bmatrix}.$$

After Gaussian triangulation, they are transformed into

$$[K'] = \begin{bmatrix}1 & 0 & 0 & 0 & 0\\0 & 2.29 & -1.135 & 0 & 0\\0 & 0 & 1.668 & -1.095 & 0\\0 & 0 & 0 & 2.13 & -1.035\\0 & 0 & 0 & 0 & 0.276\end{bmatrix},\quad [R'] = \begin{bmatrix}0\\-0.00290\\-0.00604\\-0.00786\\-0.00638\end{bmatrix}.$$

Then the increment of displacement $\Delta\mathrm{U}$, corresponding to the first Newton–Raphson iteration, is obtained by solving the upper triangular system:

$$[K'][\Delta U] = [R']. \tag{3.79}$$

The solution of Eq. (3.79) is easily obtained, thus giving the new expression of the displacement:

$$\begin{array}{ll} \Delta U_1 = 0 & U_1 = 0 \\ \Delta U_2 = -1.0 \times 10^{-4} & U_2 = 0.145 \times 10^{-2} \\ \Delta U_3 = -1.8 \times 10^{-4} & U_3 = 0.272 \times 10^{-2} \\ \Delta U_4 = -2.2 \times 10^{-4} & U_4 = 0.362 \times 10^{-2} \\ \Delta U_5 = -2.3 \times 10^{-4} & U_5 = 0.397 \times 10^{-2} \end{array}$$

After this first iteration we observe that the correction to the initial solution is approximately 6%.

## PROBLEMS

### A. Proficiency Problems

1. Determine the Cholesky decomposition of the matrix

$$\mathbf{A} = \begin{bmatrix} 4 & -2 & 0 & 0 \\ -2 & 5 & 6 & 0 \\ 0 & 6 & 10 & -2 \\ 0 & 0 & -2 & 13 \end{bmatrix}.$$

2. Compute the two-point Gaussian integration approximation of the integral

$$I(a) = \int_0^a \cos(x)\,\mathrm{d}x.$$

Compare it with the exact value of $I(a)$ for small values of the parameter $a$, by utilizing a third-order expansion of both expressions. Compare the exact values of $I(a)$ with its approximation for $a = k\pi/20$, with $k = 1, 2, \ldots, 10$ (a hand calculator is needed).

3. A linear function is defined on the two-dimensional rectangle $[a, b] \times [c, d]$ by

$$f = \alpha x_1 + \beta x_2 + \gamma.$$

Verify that the $1 \times 1$ Gaussian integration formula allows us to integrate $f$ exactly.

4. We consider the nonlinear set of equations given by

$$\mathbf{R}(\mathbf{V}) = \begin{bmatrix} v_1 v_2 - v_2 - v_1 + 1 \\ -2v_1 v_2 + v_2^2 + v_2 - 2 \end{bmatrix} = \begin{bmatrix} 0 \\ 0 \end{bmatrix}.$$

Perform (manually) one Newton–Raphson iteration from the initial value: $v_1 = v_2 = 0$. Compare the norm of the residual $\mathbf{R}$ before and after the first iteration. Find the subincrementation $\gamma = 1/2^k$ so that the norm of the residual is decreased after the iteration.

5. The stationary heat equation is defined in a simple one-dimensional domain as indicated in the following figure.

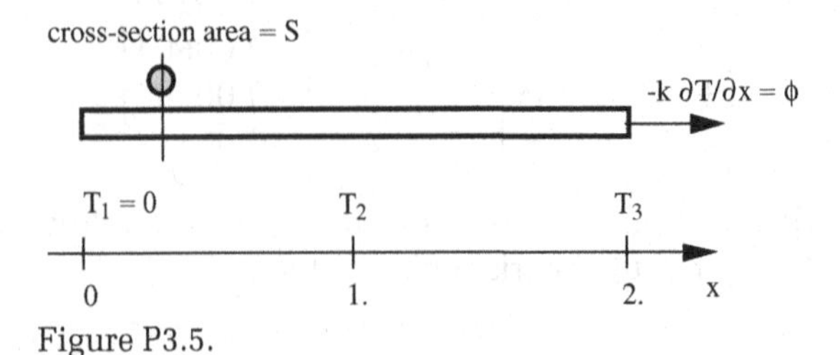


Figure P3.5.

The conductivity factor depends on the temperature $T$ according to

$$k = k_0 + k_1 T.$$

Utilizing the Galerkin method, write the discretized thermal equations, using the simple mesh outlined in the above figure with two linear elements. Indicate how it can be solved, and solve it approximately when $k_1 = -0.1k_0$.

## B. Depth Problems

6. The purpose of this problem is to show in a simple example that, in general, isoparametric elements do not give exactly the same interpolation as classical elements. A quadratic one-dimensional three-node element is considered, the nodes of which have coordinates

$$x_1 = -1, \quad x_2 = 0.3, \quad x_3 = 1.$$

   a. It is first considered as a classical element: determine directly the three shape functions $N_1(x)$, $N_2(x)$, and $N_3(x)$, and compute their value for $x_0 = 0$.
   b. The same element is now treated as an isoparametric element: define the mapping between the reference space (coordinate $\xi$) and the physical space (coordinate $x$). Compute the value of $\xi_0$ corresponding to $x_0$ and the numerical values of $N_1^0(\xi_0)$, $N_2^0(\xi_0)$, and $N_3^0(\xi_0)$, and compare it with the previous values of the ordinary shape functions $N_1(x_0)$, $N_2(x_0)$, and $N_3(x_0)$.

7. Find the coefficients of the one-dimensional three-point Gaussian integration formula by using the same method as in Exercise 3.3.

8. The Crout decomposition of a symmetric matrix $\mathbf{A}$ is written

$$\mathbf{A} = \mathbf{L}\mathbf{D}\mathbf{L}^T,$$

   where $\mathbf{L}$ is a lower triangular matrix with all diagonal terms equal to 1, and $\mathbf{D}$ is a diagonal matrix. Write an algorithm to determine the $\mathbf{L}$ and $\mathbf{D}$ matrices progressively. The advantage of the Crout decomposition is that the initial matrix $\mathbf{A}$ need not be positive, but only definite.

9. A nonlinear thermal problem is considered in which the conductibility parameter is a known function of the temperature; that is, we have $k = k(T)$. Show that the variational problem defined by the functional $I$ [Eq. (3.50)] *does not* correspond to the initial problem [Eqs. (3.46)–(3.49)].

## C. Numerical and Computational Problems

10. Write a FORTRAN code, using the Gaussian triangulation method for solving a banded linear system. Input data: the matrix $\mathbf{A}$, the dimension $n$ of $\mathbf{A}$, the half-bandwidth $l_b$, the right-hand-side vector $\mathbf{b}$. Output: the lower triangular matrix $\mathbf{L}$, the upper triangular matrix $\mathbf{U}$, the solution vector $\mathbf{x}$, and possibly an integer that indicates if the resolution was completed without an attempt to make a division by zero.

11. Write a FORTRAN code, using the Cholesky triangulation method for solving a symmetric banded linear system. Input data: the matrix $\mathbf{A}$, the dimension $n$ of $\mathbf{A}$, the half-bandwidth $l_b$, the right-hand-side vector $\mathbf{b}$. Output: the solution vector $\mathbf{X}$, and possibly an integer that indicates if the resolution was completed without a division by zero.

12. Write a FORTRAN code to solve a nonlinear system by the Newton–Raphson method. We suppose that the residual vector **R** and the matrix of its derivatives **H** are obtained by using other modules. The numerical solution of each linear system will be done with the Gaussian triangulation method (see Problem 10 or 11).
13. This is the same as Problem 12, except that the derivatives of the residual **R** are computed numerically by finite-difference approximation: compare the two methods.
14. Write a computer code that approximately solves Problem 5 for any discretization of the [1, 2] into $n$ intervals with linear one-dimensional elements.

CHAPTER FOUR

# Typical Finite Elements

The finite-element method is a powerful tool to stimulate the creativity of engineers, scientists, and applied mathematicians. Therefore it is not surprising that many elements have been presented in the literature and are available for a range of applications.

A finite element is defined not only by a geometric subdomain, or even the explicit mathematical expression of the shape functions. A finite element also includes the specification of whether it is an isoparametric element (as we shall essentially consider here) and the definition of the space integration formulas. In a broader sense we can also consider that the space discretization procedure for each variable, the time integration procedure when required, and the mathematical formulation of the physical problem also contribute to the FE definition.

Evaluation of the behavior of an element type will depend on the kind of problem that is considered, the required accuracy, and the time and effort allocated toward code development. In the realm of metal forming, the subject of this book, the choice of element formulation depends primarily on the following considerations.

1. Mode of deformation: the chosen element must approximate real material stiffness in all modes of deformation important to the physical problem being simulated. An element can be too stiff in some modes, producing "locking" for example, particularly when incompressibility constraints are considered; or it can be too compliant, producing "hourglass" and other spurious low-energy deformation modes.
2. Evolution: an element must respond reasonably to large deformation and it must be possible to determine appropriate strain or strain-rate hardening throughout the forming operation. That is, hardening laws must be integrated accurately.
3. Time efficiency: because metal-forming problems are very challenging in terms of computation time and storage, elements must be chosen that simplify the problem wherever possible without losing essential accuracy. For example, it is tempting but often impractical to rely on solid elements to solve thin-shell bending problems, or on shell–bending elements if membrane elements will suffice. Two-dimensional approaches are particularly efficient where feasible, taking advantage of symmetry or approximate symmetry.
4. Contact condition: contact and friction are crucial aspects in metal-deformation analysis. Special care must be taken to ensure that physical

contact conditions are enforced without introducing nonphysical effects or unacceptable numerical instability.

To these considerations, we add two other desirable properties that are becoming more valuable as research progresses.

1. Remeshing: an element should be compatible with an automatic meshing and remeshing procedure. This is desirable not only for non-steady-state processes in which a mesh may become too distorted for accuracy, but also for improving the accuracy of an initial mesh.
2. Error estimation: a posteriori error estimation allows a second analysis to be performed for the purpose of producing results of a specified accuracy or to generate meshes and other control parameters to optimize accuracy for a given size of computational resources.

The first two sections of this chapter are devoted to two-dimensional, axisymmetric, and three-dimensional elements, regardless of the specific problem to be addressed. Three illustrations are then discussed briefly: linear elasticity, time-dependent thermal problems, and incompressible materials.

## 4.1 Two-Dimensional Elements

### The Three-Node Triangular Element

This element was already defined in Section 2.3 (also see Exercise 2.8). This is the simplest two-dimensional element, and it can also be viewed as an isoparametric element as indicated in Fig. 4.1.

In the reference space with coordinates $(\xi_1, \xi_2)$, the shape functions can be defined easily; we can also introduce the three triangular coordinates $p$, $q$, and $r$ by

$$N_1^0 = p = \xi_1, \quad N_2^0 = q = \xi_2, \quad N_3^0 = r = 1 - \xi_1 - \xi_2. \tag{4.1}$$

This element can be used for linear (compressible) elasticity and thermal analysis. However, the plane-strain version of it cannot be used in the usual way for incompressible behavior, and it should be avoided for elastoplastic analysis when the incompressible plastic deformation is dominant.

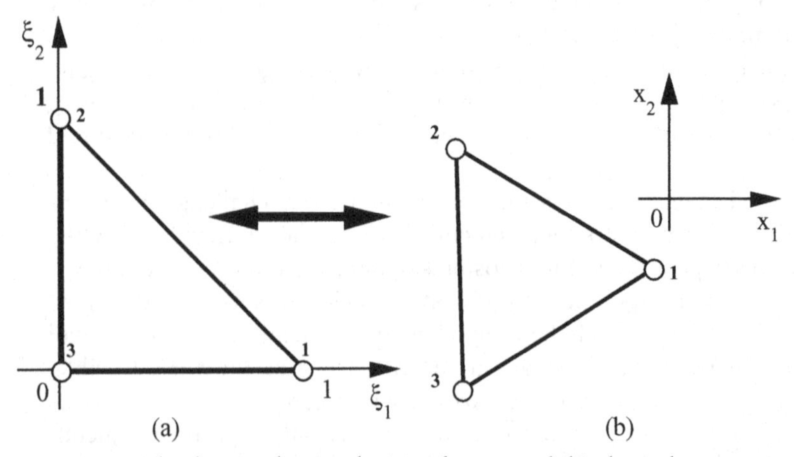


Figure 4.1. The three-node triangle: (a) reference and (b) physical spaces.

**Exercise 4.1: Show that the triangular coordinates *p*, *q*, and *r* represent the relative areas of the subtriangles defined in the following figure.**

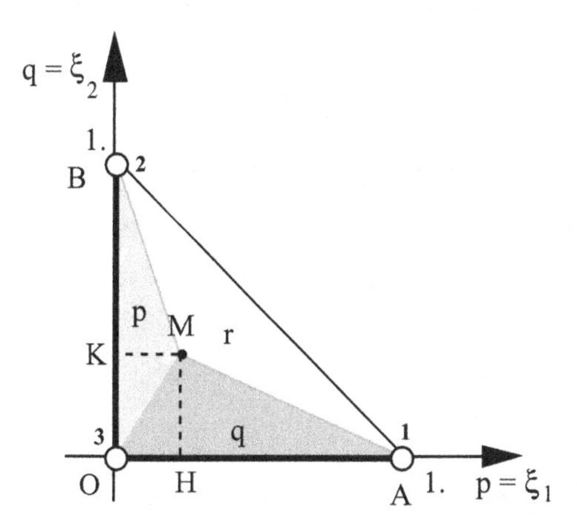


The surface of the reference element is given by

$$|S^0| = \frac{1}{2}OA, \quad OB = \frac{1}{2}. \tag{4.1–1}$$

The triangle $OAM$ has a basis with length $OA = 1$, and a height $MH = \xi_2 = q$, so that its relative area is

$$\frac{|S^q|}{|S^0|} = \frac{\frac{1}{2}OA\,MH}{\frac{1}{2}} = q. \tag{4.1–2}$$

The same calculation applies to triangle $OMB$ and gives

$$\frac{|S^p|}{|S^0|} = \frac{\frac{1}{2}OB\,MK}{\frac{1}{2}} = p. \tag{4.1–3}$$

Finally we observe that

$$r = 1 - p - q = \frac{|S^0| - |S^p| - |S^q|}{|S^0|} = \frac{|S^r|}{S^0}. \tag{4.1–4}$$

### The Two-Dimensional Mini-Element

This element is mainly used for modeling incompressible or near incompressible material flows. The general idea is to introduce an additional node at the centroid of the element and to divide the triangle into three subtriangles, denoted 1, 2, and 3 in black boxes in Fig. 4.2. The shape functions are the same as those defined for the three-node element in Eq. (4.1), plus the bubble function, $N_4^0$. The shape function $N_4^0$ is defined by the following conditions.

1. It takes a value of 1 at node 4, the centroid of the initial triangle.
2. It is equal to zero on the boundary of the triangle.
3. It is a linear function of the coordinates in each subtriangle 1, 2, and 3.

As the bubble shape function is equal to zero on the boundary of the element, it has no connection with degrees of freedom defined on neighboring elements. Therefore it is often possible to eliminate the internal node at the element level, in order to reduce the global number of unknowns.

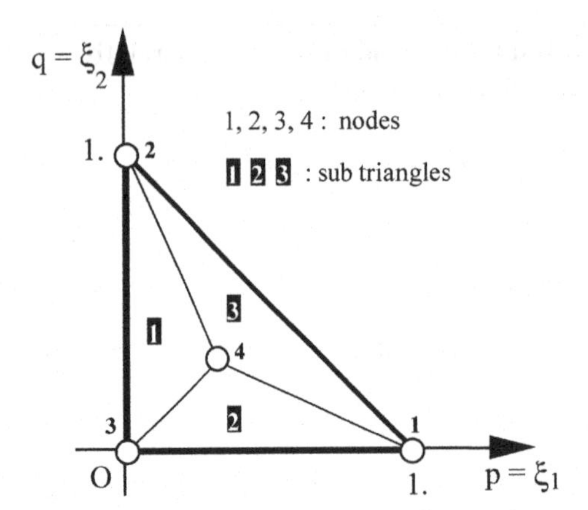


Figure 4.2. The two-dimensional mini-element in the reference space.

**The Six-Node Quadratic Triangular Element**

The reference element is pictured in Fig. 4.3(a), where the additional nodes, with respect to those of the previous linear element, are put at the middle of the edges; Fig. 4.3(b) shows the curved element in the physical space.

In the reference coordinate system, the shape functions are built as second-order polynomials of $\xi_1$ and $\xi_2$. These functions are better expressed in terms of the triangular coordinates $p$, $q$, and $r$, already defined by Eq. (4.1):

$$\begin{aligned} N_1^0 &= p(2p-1), & N_2^0 &= q(2q-1), \\ N_3^0 &= r(2r-1), & N_4^0 &= 4pq, \\ N_5^0 &= 4qr, & N_6^0 &= 4pr. \end{aligned} \tag{4.2}$$

---

**Exercise 4.2: Verify that the shape functions satisfy the usual relation**

$$\sum_{m=1}^{6} N_m^0(\xi_1, \xi_2) \equiv 1. \tag{4.2–1}$$

**In the reference coordinate system, calculate the integral**

$$S_m = \int_{\Omega_0} N_m^0(\xi_1, \xi_2)\, d\xi_1\, d\xi_2. \tag{4.2–2}$$

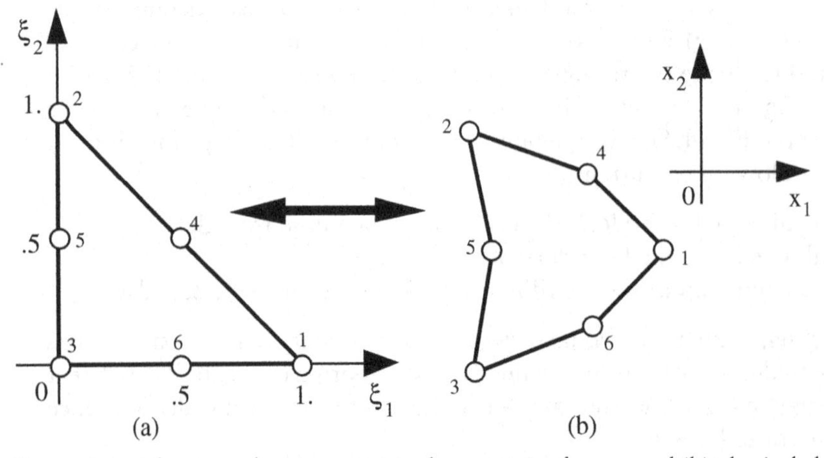


Figure 4.3. The six-node isoparametric element: (a) reference and (b) physical elements.

From Eq. (4.2) we have

$$\begin{aligned} &N_1^0 + N_2^0 + N_3^0 + N_4^0 + N_5^0 + N_6^0 \\ &\quad = p(2p-1) + q(2q-1) + r(2r-1) + 4pq + 4qr + 4pr \\ &\quad = p^2 + q^2 + r^2 + 2pq + 2pr + 2qr - (p+q+r) \\ &\quad = 2(p+q+r)^2 - (p+q+r) = 1. \end{aligned} \tag{4.2–3}$$

The last equality is obtained when one recalls that $p + q + r = 1$. We can distinguish two kinds of integrals, according to the $m$ index; if $m$ equals 1, 2, or 3, we have, for example,

$$S_1 = \int_0^1 \left[ \int_0^{1-\xi_2} \xi_1(2\xi_1 - 1)\, \mathrm{d}\xi_1 \right] \mathrm{d}\xi_2 = \int_0^1 \left[ \frac{2}{3}\xi_1^3 - \frac{1}{2}\xi_1^2 \right]_0^{1-\xi_2} \mathrm{d}\xi_2. \tag{4.2–4}$$

This is transformed according to

$$S_1 = \int_0^1 \left[ \frac{2}{3}(1-\xi_2)^3 - \frac{1}{2}(1-\xi_2)^2 \right] d\xi_2 = \left[ -\frac{2}{12}(1-\xi_2)^4 + \frac{1}{6}(1-\xi_2)^3 \right]_0^1 = 0. \tag{4.2–5}$$

Similarly, we can verify that $S_2 = S_3 = 0$. For the other values of $m$, we obtain

$$\begin{aligned} S_4 &= \int_0^1 \left( \int_0^{1-\xi_2} 4\xi_1\xi_2\, \mathrm{d}\xi_1 \right) \mathrm{d}\xi_2 = \int_0^1 \left[ 2\xi_1^2\xi_2 \right]_0^{1-\xi_2} \mathrm{d}\xi_2 \\ &= \int_0^1 2(1-\xi_2)^2\xi_2\, \mathrm{d}\xi_2 = 2\left( \frac{1}{4}\xi_2^4 - \frac{2}{3}\xi_2^3 + \frac{1}{2}\xi_2^2 \right)_0^1 = \frac{1}{6}. \end{aligned} \tag{4.2–6}$$

With the same approach, we can observe that $S_5 = S_6 = 1/6$.

---

### The Bubble Element

In a similar way to the linear triangular element, it is often useful for incompressible flow to introduce a seventh shape function, as a polynomial with an order of 3, which is equal to zero on the sides of the element, and to one at the centroid of the element. By its geometric definition, this shape function is pictured as a bubble function, with the expression

$$N_m^0 = 27pqr. \tag{4.3}$$

This element adds only internal degrees of freedom, which can be eliminated on most occasions.

Note: the complete basis set for third-order polynomials includes more than seven elements (i.e., 10), so we have here

$$\sum_{m=1}^{7} N_m^0 = 1 + 27pqr \neq 1.$$

### Numerical Integration in Triangular Elements

For triangles, special integration formulas were developed in order to obtain the best possible accuracy with a given number of integration points. The method is a straightforward generalization of Eq. (3.21) as an approximation of Eq. (3.20) to the two-dimensional reference element. The general formula for approximate integration of a given function $f$ of the triangular coordinates $p$, $q$, and $r$, with $n_Q$ integration points, can also be considered to be analogous to Eq. (3.29) in Chapter 3:

$$I(f) = \int_{\Omega^0} f(p,q,r)\,\mathrm{d}p\,\mathrm{d}q \cong \sum_{k=1}^{n_Q} h^k f(p^k, q^k, r^k). \tag{4.4}$$

Three cases will be examined in terms of the weighting factors $h^k$ and the triangular coordinates $p^k$, $q^k$, and $r^k$.

- one integration point:

$$h^1 = 0.5, \quad p^1 = q^1 = r^1 = \frac{1}{3}. \tag{4.5}$$

- three integration points:

$$h^1 = \frac{1}{6}, \quad p^1 = \frac{2}{3}, \quad q^1 = r^1 = \frac{1}{6}. \tag{4.6}$$

  The others are deduced by the two other permutations. With this scheme a polynomial with a degree of 1 is integrated exactly.
- six integration points:

$$h^1 = 0.054975871828, \quad p^1 = 0.816847572980, \quad q^1 = r^1 = 0.091576213510. \tag{4.7a}$$

The integration point numbers 2 and 3 are obtained from number 1 by permutations.

$$h^4 = 0.111690794839, \quad p^4 = 0.108103018168, \quad q^4 = r^4 = 0.445948490916. \tag{4.7b}$$

The integration point numbers 5 and 6 are obtained from number 4 by permutation. Here a polynomial with a degree of 3 is exactly integrated.

The approximate positions of integration points are shown in the reference element in Fig. 4.4.

### The Four-Node Quadrilateral Element

The reference element is the square $[-1, 1]^2$ with sides equal to 2. A one-to-one mapping between the reference system $(\xi_1, \xi_2)$ and the physical coordinate system $(x_1, x_2)$ is defined and allows for the transformation of the square into a general convex quadrilateral, as shown in Fig. 4.5.

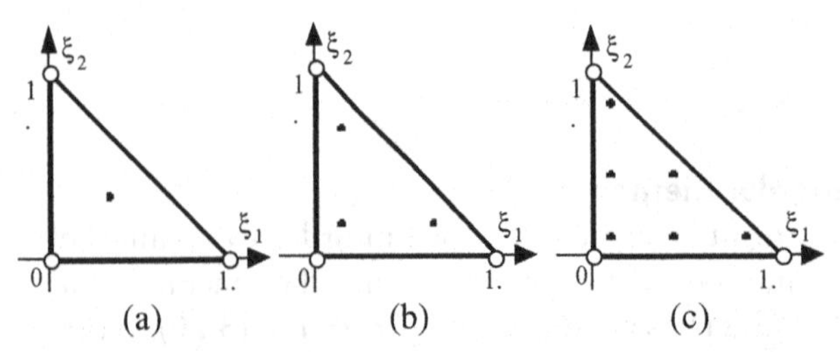


Figure 4.4. Integration points in the reference triangle: (a) one, (b) three, and (c) six integration points.

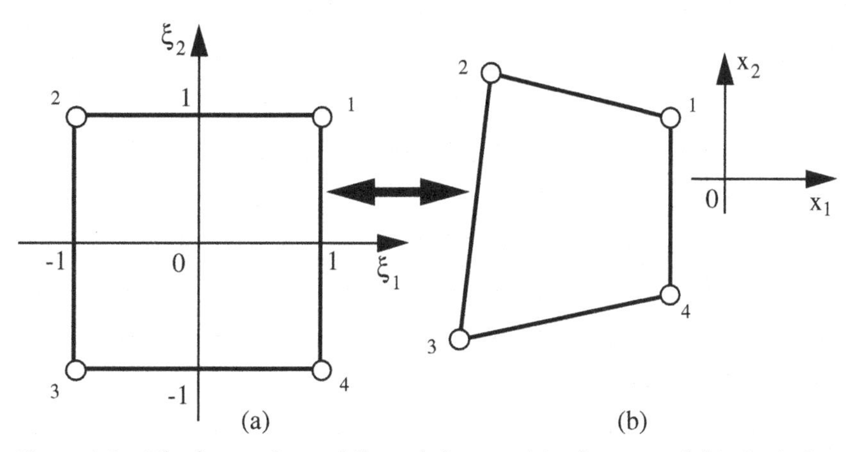


Figure 4.5. The four-node quadrilateral element: (a) reference and (b) physical spaces.

The shape functions in the reference space are expressed as the four products of the one-dimensional linear shape functions defined by Eq. (3.2) in Chapter 3:

$$\begin{aligned} N_1^0(\xi_1,\xi_2) &= \frac{(1+\xi_1)(1+\xi_2)}{4}, \quad N_2^0(\xi_1,\xi_2) = \frac{(1-\xi_1)(1+\xi_2)}{4} \\ N_3^0(\xi_1,\xi_2) &= \frac{(1-\xi_1)(1-\xi_2)}{4}, \quad N_4^0(\xi_1,\xi_2) = \frac{(1+\xi_1)(1-\xi_2)}{4} \end{aligned} \tag{4.8}$$

The four-node bilinear element (as it is linear in each direction) has been used for many applications, with generally four Gaussian integration points. It is also very popular for incompressible flow with a reduced integration formula, using only one Gaussian point for the penalty contribution. Another possibility for incompressible flow is to add an internal node and to divide the quadrilateral element into four triangles. However, the weakness of this element, compared with triangular elements, lies in the difficulty of developing a general code for remeshing complex domains without producing distorted elements, which may induce inaccurate numerical results.

### The Quadratic Nine-Node Quadrilateral Element

This element is also difficult to introduce in a general meshing code. However, it is rather accurate with a nine-point Gaussian integration scheme. With a four-point Gaussian reduced integration scheme for the divergence term, it can cope with incompressible flow. The reference element and the isoparametric element with possibly curved sides are represented in Fig. 4.6.

Again, the shape functions are easily written as the product of quadratic one-dimensional shape functions in $\xi_1$ and in $\xi_2$.

$$\begin{aligned} N_1 &= \frac{1}{4}\xi_1(1+\xi_1)\xi_2(1+\xi_2), \quad N_2 = \frac{-1}{4}\xi_1(1-\xi_1)\xi_2(1+\xi_2), \\ N_3 &= \frac{1}{4}\xi_1(1-\xi_1)\xi_2(1-\xi_2), \quad N_4 = \frac{-1}{4}\xi_1(1+\xi_1)\xi_2(1-\xi_2), \\ N_5 &= \frac{1}{2}(1-\xi_1^2)\xi_2(1+\xi_2), \quad N_6 = \frac{-1}{2}\xi_1(1-\xi_1)(1-\xi_2^2), \\ N_7 &= \frac{-1}{2}(1-\xi_1^2)\xi_2(1-\xi_2), \quad N_8 = \frac{1}{2}\xi_1(1+\xi_1)(1-\xi_2^2), \\ N_9 &= (1-\xi_1^2)(1-\xi_2^2). \end{aligned} \tag{4.9}$$

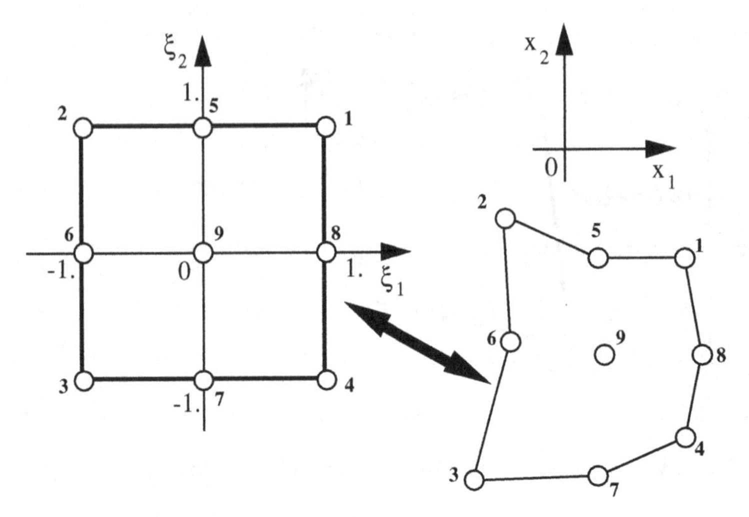


Figure 4.6. Nine-node quadrilateral elements.

## 4.2 Axisymmetric Elements

Many problems in practical forming processes are truly axisymmetric: cylinder or bar extrusion, wire drawing, and forging of axisymmetric parts. For those problems, it is more convenient to develop a special theory to take advantage of the symmetry. Another interesting case is the axisymmetric problem with torsion that is two dimensional in space and that corresponds to a three-component velocity formulation.

### The Classical Axisymmetric Problem

We shall consider the simple case of forging of a cylinder, Fig. 4.7(a), with barreling, Fig. 4.7(b), for which it is clear that the problem is invariant with respect to any rotation around the $z$ axis, for isotropic material behavior and axisymmetric boundary conditions. It is therefore clear that it is possible to consider the problem in the $(r, z)$ coordinate system [Fig. 4.7(c)]. For a thermal problem the temperature will depend only on $r$ and $z$; for a mechanical problem the velocity field will also depend on $r$ and $z$ only and have only two components, $v_r$ and $v_z$.

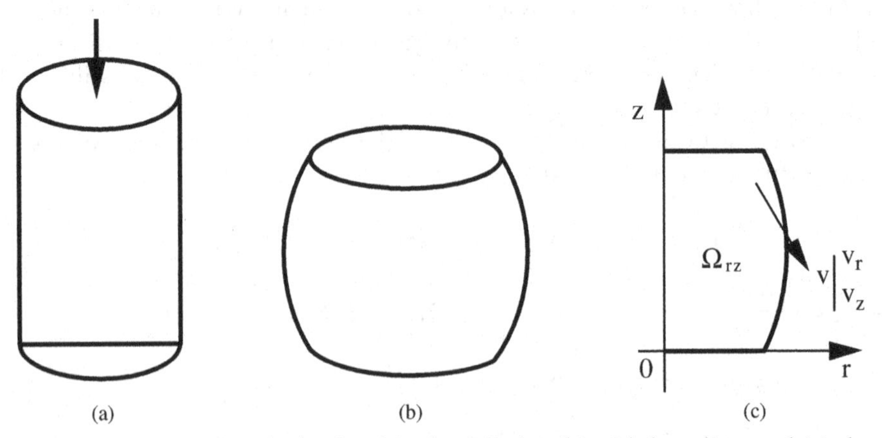


Figure 4.7. Axisymmetric forging (a) of a cylinder, (b) with barreling, and (c) the reference system.

For axisymmetric problems, the same elements as those described for a two-dimensional analysis, in Section 4.1, can be used by replacing $(x_1, x_2)$ with $(r, z)$. However, in the integral formulation, the integration must be extended to the volume of the part. For any function $f$ of $r$ and $z$, we have the following for an axisymmetric domain $\boldsymbol{\Omega}$:

$$\int_{\Omega} f(r,z)\,\mathrm{d}V = \int_0^{2\pi} \left( \int_{\Omega_{rz}} f(r,z) r\,\mathrm{d}r\,\mathrm{d}z \right) \mathrm{d}\theta = 2\pi \int_{\Omega_{rz}} f(r,z) r\,\mathrm{d}r\,\mathrm{d}z, \qquad (4.10)$$

in which $\Omega_{rz}$ represents half of a section of the part by a plane including the $Oz$ axis.

Remark: $2\pi$ in Eq. (4.10) is generally omitted in the final formula, which corresponds to the integral for a unit angle around the $Oz$ axis (expressed in radians).

For thermal analysis the temperature gradient is

$$\mathbf{grad}(T) = \begin{bmatrix} \partial T/\partial r \\ \partial T/\partial z \end{bmatrix}. \qquad (4.11)$$

Thus the transposition from the two-dimensional to the axisymmetric problem is limited to the introduction of an $r$ term in the surface integral.

For mechanical analysis, the situation is slightly more complicated, as the strain-rate tensor, which is derived from the velocity field, takes the form

$$\dot{\varepsilon} = \begin{bmatrix} \dfrac{\partial v_r}{\partial r} & 0 & \dfrac{1}{2}\left(\dfrac{\partial v_r}{\partial z} + \dfrac{\partial v_z}{\partial r}\right) \\ 0 & \dfrac{v_r}{r} & 0 \\ \dfrac{1}{2}\left(\dfrac{\partial v_r}{\partial z} + \dfrac{\partial v_z}{\partial r}\right) & 0 & \dfrac{\partial v_z}{\partial z} \end{bmatrix}. \qquad (4.12)$$

It is no longer possible to consider only the components on the $r$ and $z$ axes, as a new contribution appears on the $\theta$ axis (which is perpendicular to the plane of the section).

The stress tensor must have the same form:

$$\boldsymbol{\sigma} = \begin{bmatrix} \sigma_{rr} & 0 & \sigma_{rz} \\ 0 & \sigma_{\theta\theta} & 0 \\ \sigma_{rz} & 0 & \sigma_{zz} \end{bmatrix}. \qquad (4.13)$$

Finally, on the $Oz$ axis the boundary conditions are

$$v_r = 0, \qquad \sigma_{rz} = 0. \qquad (4.14)$$

### The Axisymmetric Problem with Torsion

We suppose that the workpiece remains axisymmetric during all the deformation processes, so that the geometry can be described again by half of a cross section containing the symmetry axis.

The velocity field is defined on the two-dimensional domain in the $(r, z)$ space, but it must have three components (see Fig. 4.8), including the component $v_\theta$, which will

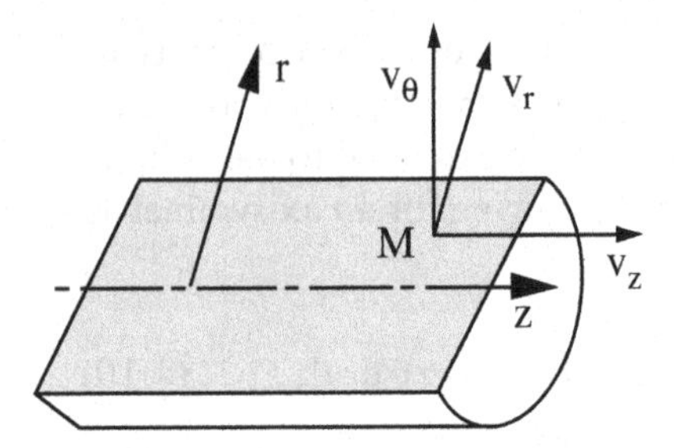


Figure 4.8. The axisymmetric geometrical problem with torsion.

represent the torsional effect:

$$\mathbf{v} = \begin{bmatrix} v_r(r,z) \\ v_\theta(r,z) \\ v_z(r,z) \end{bmatrix}. \tag{4.15}$$

The strain-rate tensor is more complicated than the axisymmetric one; it is shown to be[1]

$$\dot{\varepsilon} = \begin{bmatrix} \dfrac{\partial v_r}{\partial r} & \dfrac{1}{2}\left(\dfrac{\partial v_\theta}{\partial r} - \dfrac{v_\theta}{r}\right) & \dfrac{1}{2}\left(\dfrac{\partial v_r}{\partial z} + \dfrac{\partial v_z}{\partial r}\right) \\ \dfrac{1}{2}\left(\dfrac{\partial v_\theta}{\partial r} - \dfrac{v_\theta}{r}\right) & \dfrac{v_r}{r} & \dfrac{1}{2}\dfrac{\partial v_\theta}{\partial z} \\ \dfrac{1}{2}\left(\dfrac{\partial v_r}{\partial z} + \dfrac{\partial v_z}{\partial r}\right) & \dfrac{1}{2}\dfrac{\partial v_\theta}{\partial z} & \dfrac{\partial v_z}{\partial z} \end{bmatrix}. \tag{4.16}$$

Thus the stress tensor takes the general form

$$\sigma = \begin{bmatrix} \sigma_{rr} & \sigma_{r\theta} & \sigma_{rz} \\ \sigma_{r\theta} & \sigma_{\theta\theta} & \sigma_{\theta z} \\ \sigma_{rz} & \sigma_{\theta z} & \sigma_{zz} \end{bmatrix}. \tag{4.17}$$

The problem can be geometrically discretized with classical two-dimensional elements, but the unknown vector field has three components.

## 4.3 Three-Dimensional Elements

### The Four-Node Linear Tetrahedron

This element is the simple generalization of the triangle to the three-dimensional space. In Fig. 4.9 the four-node linear tetrahedron is drawn in the reference space.

The linear shape functions in the reference coordinates $(\xi_1, \xi_2, \xi_3)$ and the four tetrahedral coordinates $p, q, r$, and $s$ are defined by

$$N_1 = p = \xi_1, \quad N_2 = q = \xi_2, \quad N_3 = r = \xi_3, \quad N_4 = s = 1 - \xi_1 - \xi_2 - \xi_3. \tag{4.18}$$

This element is convenient for three-dimensional meshing and remeshing; however, it is not very accurate and it is not appropriate for incompressible materials.

[1] A. Moal, E. Massoni, and L.-L. Chenot, "A Finite Model for the Simulation of the Torsion and Torsion-Tension Tests," *Comp. Meth. Appl. Mech. Eng.* 103, 417–434 (1993).

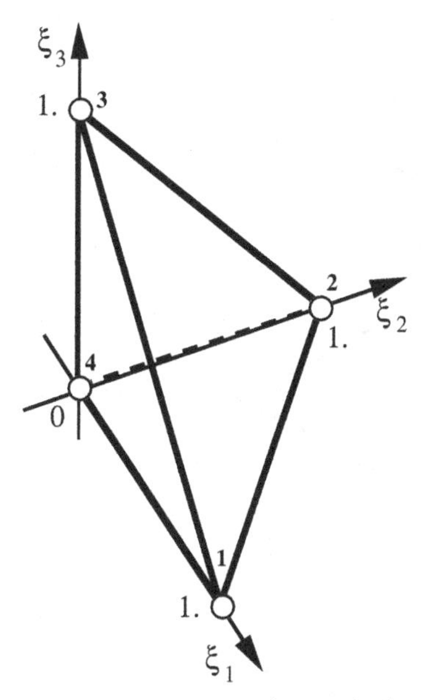


Figure 4.9. Four-node tetrahedron.

### The Three-Dimensional Mini-Element

The definition for the three-dimensional mini-element is similar to that of the two-dimensional one. As is shown in Fig. 4.10, a node is added at the centroid of the tetrahedron in order to define four subtetrahedra. The shape functions are the same as those of Eq. (4.18), to which we add the shape function that is equal to 1 on node 5, which is nil on the boundary of the tetrahedron and which is linearly interpolated on each subtetrahedron.

### The Ten-Node Quadratic Tetrahedral Element

The positions of the nodes and the node numbering are shown in Fig. 4.11(a) for the reference element and in Fig. 4.11(b) for the curved tetrahedron in the physical space.

The shape functions in the reference space are a generalization of those for the quadratic triangle:

$$\begin{aligned}
&N_1 = p(2p-1), \quad N_2 = q(2q-1), \quad N_3 = r(2r-1), \\
&N_4 = s(2s-1), \quad N_5 = 4pq, \quad N_6 = 4qr, \quad N_7 = 4rs, \\
&N_8 = 4ps, \quad N_9 = 4pr, \quad N_{10} = 4qs.
\end{aligned} \tag{4.19}$$

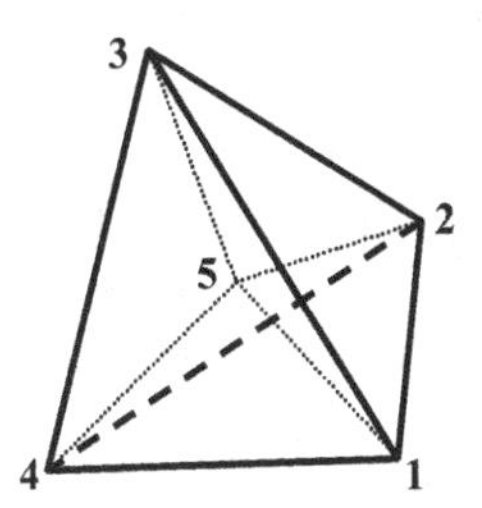


Figure 4.10. The three-dimensional mini-element.

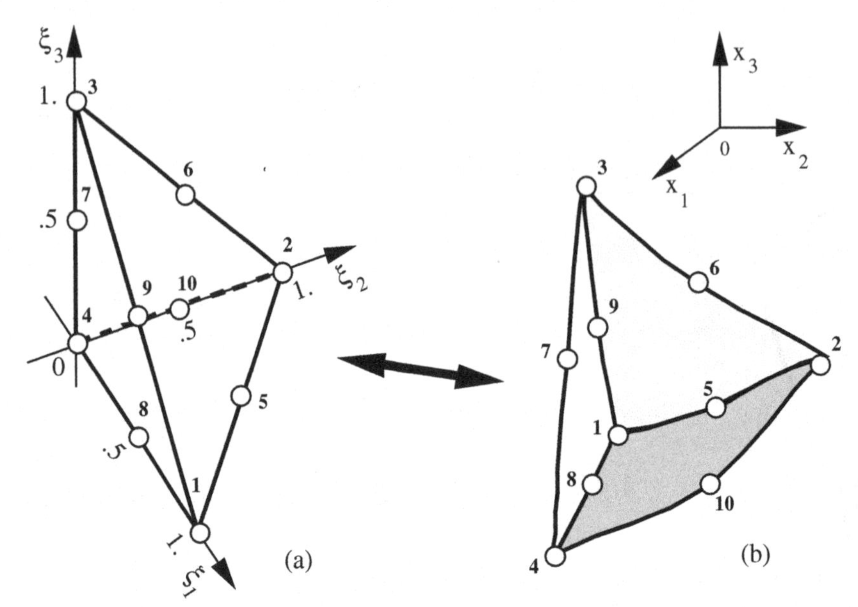


Figure 4.11. The ten-node quadratic isoparametric tetrahedron: (a) reference and (b) physical spaces.

**The Eight-Node Trilinear Element**

This element, which is often called the brick element, has been very popular for structural analyses as well as for the computation of incompressible deformation. In the reference coordinates, the domain of the element is the $[-1, 1]^3$ cube (see Fig. 4.12).

The eight shape functions are derived from the products of linear functions of the three reference coordinates $\xi_1, \xi_2$, and $\xi_3$.

$$
\begin{aligned}
N_1^0 &= \frac{1}{8}(1+\xi_1)(1+\xi_2)(1-\xi_3), & N_2^0 &= \frac{1}{8}(1-\xi_1)(1+\xi_2)(1-\xi_3),\\
N_3^0 &= \frac{1}{8}(1-\xi_1)(1-\xi_2)(1-\xi_3), & N_4^0 &= \frac{1}{8}(1+\xi_1)(1-\xi_2)(1-\xi_3),\\
N_5^0 &= \frac{1}{8}(1+\xi_1)(1+\xi_2)(1+\xi_3), & N_6^0 &= \frac{1}{8}(1-\xi_1)(1+\xi_2)(1+\xi_3),\\
N_7^0 &= \frac{1}{8}(1-\xi_1)(1-\xi_2)(1+\xi_3), & N_8^0 &= \frac{1}{8}(1+\xi_1)(1-\xi_2)(1+\xi_3).
\end{aligned}
\tag{4.20}
$$

**Membrane Elements**

A membrane is the idealization of a very thin part (like a sheet), for which we neglect variations of mechanical, thermal, and physical parameters through the thickness. A membrane is thus considered a geometric surface in three dimensions, with an additional parameter defined on this surface: the local thickness, which is considered as negligible with respect to the radii of curvature (see also footnote 2).

The classical derivation of the strain rate is not presented here, as the stress and strain-rate tensors must be expressed in a special reference system for which one reference vector is normal to the surface. This can be done with the introduction of

[2] A. Sawczuk, "On Plastic Shell Theories at Large Strains and Displacements," *Int. J. Mech. Sci.* 24, 231–244 (1982).

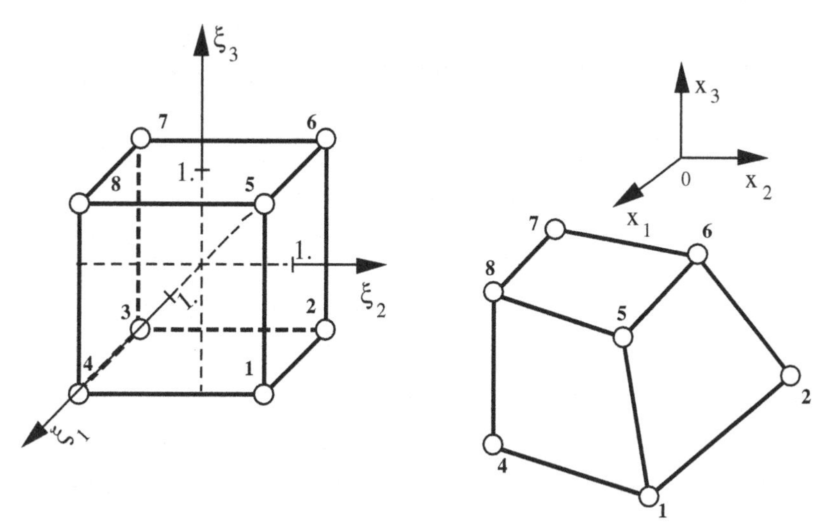


Figure 4.12. Eight-node trilinear element.

material coordinates $\theta^1, \theta^2$ describing the surface of the membrane. These coordinates can be chosen as the initial in-plane coordinates $x_1, x_2$ of the undeformed membrane, which is most often assumed to be planar initially (see Fig. 4.13).

---

**Exercise 4.3: For a membrane, show that partial differentiation with respect to the material coordinates $\theta^1, \theta^2$ allows us to define two vectors $\mathbf{g}_1$ and $\mathbf{g}_2$ that are tangential to the element surface in the physical space. Give the expression of the surface integral on the element, and of the normal unit vector $\mathbf{g}_3$ to the surface in the physical space.**

The geometric definition of the element surface in the physical space is given by:

$$\mathbf{x} = \mathbf{x}(\theta^1, \theta^2), \tag{4.3–1}$$

which can be differentiated with respect to $\xi_1$ and $\xi_2$ to give the two vectors

$$\mathbf{g}_i = \frac{\partial \mathbf{x}}{\partial \theta^i}, \qquad i = 1 \text{ or } 2. \tag{4.3–2}$$

These vectors are clearly tangential to the element surface. The vector product j is introduced:

$$\mathrm{j} = \mathbf{g}_1 \times \mathbf{g}_2, \tag{4.3–3}$$

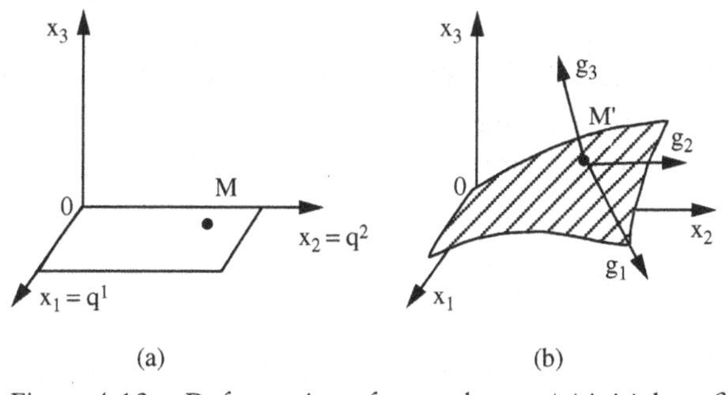


Figure 4.13. Deformation of a membrane: (a) initial configuration with the material coordinates and (b) deformed state with the local tangential reference system.

as is the surface Jacobian $j = \|\mathbf{j}\|$. We see immediately that any surface integral of a function $f$, over a surface domain $S$, can be computed by

$$\int_S f(\theta^1, \theta^2)\, j\, d\theta^1\, d\theta^2. \tag{4.3–4}$$

The vector product j is normal to both vectors $\mathbf{g}_1$ and $\mathbf{g}_2$ (if they are not collinear), so that it is also normal to the element surface; the unit normal is $\mathbf{g}_3$:

$$\mathbf{g}_3 = \mathbf{j}/|\mathbf{j}|. \tag{4.3–5}$$

---

The stress tensor is now defined by its contravariant components in the local system, by

$$\sigma = \sum_{ij} \sigma^{ij} \mathbf{g}_i \otimes \mathbf{g}_j. \tag{4.21}$$

The membrane approximation on the stress tensor can now be easily expressed by

$$\sigma^{i3} = \sigma^{3i} = 0, \qquad i = 1, 2, 3, \tag{4.22}$$

which means simply that the normal stress to the surface of the membrane is neglected, with the shear that also corresponds to the normal vector $\mathbf{g}_3$.

The new material reference system allows us to define the material derivatives of the velocity field by

$$v_{i/j} = \mathbf{g}_i \cdot (\partial \mathbf{v}/\partial \theta^j), \tag{4.23}$$

and the covariant components of the strain-rate tensor by

$$\dot{\varepsilon}_{ij} = \frac{1}{2}(v_{i/j} + v_{j/i}), \tag{4.24}$$

which will allow us to write

$$\dot{\varepsilon} = \sum_{ij} \dot{\varepsilon}_{ij} \mathbf{g}^i \otimes \mathbf{g}^j. \tag{4.25}$$

In Eq. (4.25), we utilize the reciprocal basis system $\mathbf{g}^1, \mathbf{g}^2, \mathbf{g}^3$, which is obtained from the material system by the following equalities:

$$\mathbf{g}^i \cdot \mathbf{g}_j = \delta_{ij} \tag{4.26}$$

($\delta_{ij}$ is the usual Kronecker symbol).

To describe this; any two-dimensional element can be used as an isoparametric element with a generalization of the concept described in Chapter 3, Section 3.2. If the three-node triangle is taken as an example, a one-to-one mapping is defined between the reference element $\mathbf{\Omega}_0$ in the local two-dimensional coordinates $(\xi_1, \xi_2)$ and the triangle $\mathbf{\Omega}_e$ in the physical three-dimensional space, as pictured in Fig. 4.14.

With the notation defined in Fig. 4.14, the mapping is written in the usual way for the element number $e$, in terms of the node coordinates $\mathbf{X}_m^e$:

$$\mathbf{x} = \sum_{m=1}^{3} \mathbf{X}_m^e N_m^0(\boldsymbol{\xi}), \tag{4.27}$$

which is also, with the component notations,

$$x_i = \sum_{m=1}^{3} X_{im}^e N_m^0(\xi_1, \xi_2) \quad \text{for } i = 1\text{–}3. \tag{4.28}$$

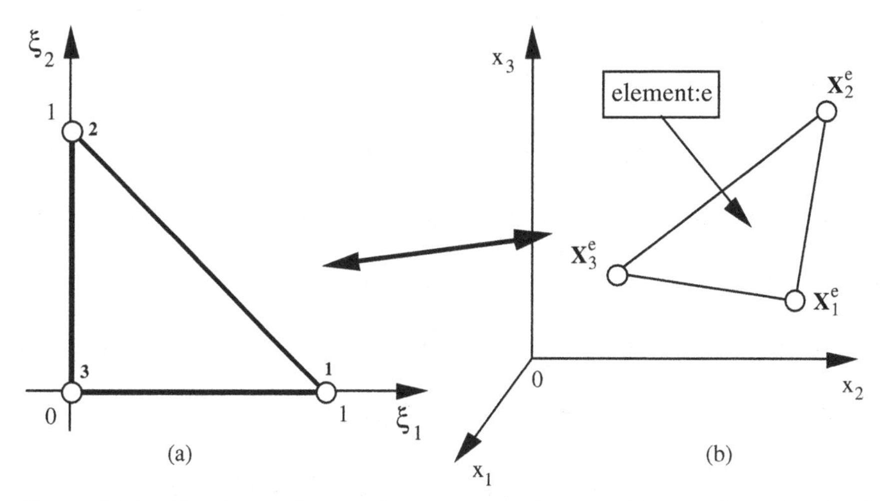


Figure 4.14. The linear three-node membrane element: (a) two-dimensional reference and (b) three-dimensional physical spaces.

The unknown functions are defined on the surface of the elements and are discretized in terms of the same shape functions $N_m^0$. For example, if the velocity field is the unknown, it is written in element number $e$ as

$$\mathbf{v} = \sum_{m=1}^{3} \mathbf{v}_m^e N_m^0(\boldsymbol{\xi}). \tag{4.29}$$

#### Thin-Shell Elements

In thin-shell theory, the in-plane tangential stresses and strains are still considered, but additional components are introduced to take into account the bending effects. The simplest way to introduce this effect is to use true three-dimensional elements with small thicknesses, as shown in Fig. 4.15. Many authors tested this strategy (see, e.g., footnotes 3 and 4). We observe that too much constraint is imposed when the

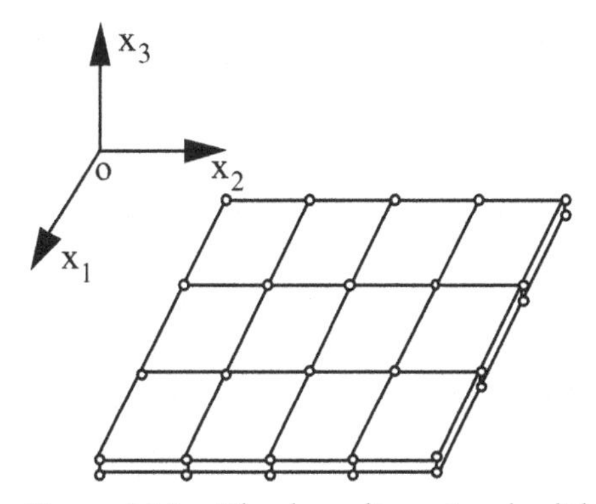


Figure 4.15. The three-dimensional solid-element approach for thin shells.

[3] T. Shimizu, E. Massoni, N. Soyris, and J.-L. Chenot, "A Modified 3-Dimensional Finite Element Model for Deep-Drawing analysis," in *Advances in Finite Deformation Problems in Material Processing and Structure*, N. Chandra et al., eds. (ASME, New York, 1991), AMD Vol. 125, pp. 113–118.

[4] M. Kawka and A. Makinouchi, "Finite Element Simulation of Sheet Metal Forming Processes by Simultaneous Use of Membrane, Shell and Solid Elements," in *Numerical Methods in Industrial Forming Processes*, J.-L. Chenot et al., eds. (A. A. Balkema, Rotterdam, 1992), pp. 491–496.

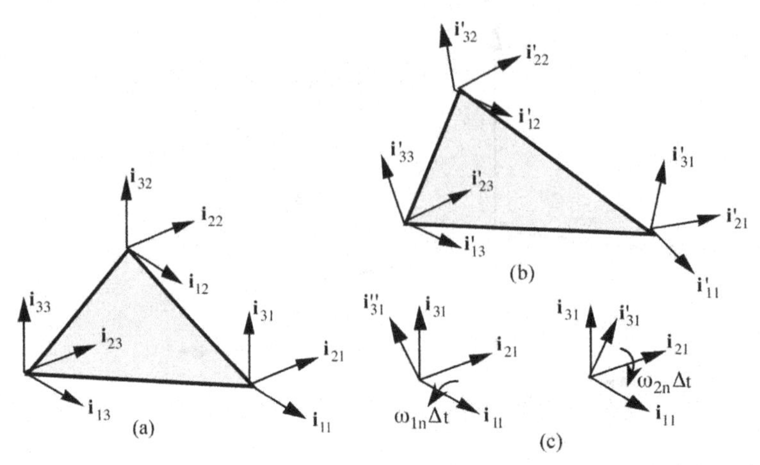


Figure 4.16. Kinematics in the thin-shell three-node element (a) before and (b) after an increment of deformation ($'$). (c) Rotation of the $\mathbf{i}_{3n}$ vector successively around $\mathbf{i}_{1n}$ with an angle of $\omega_{1n}$, $\Delta t$ and around $\mathbf{i}_{2n}$ with an angle of $\omega_{2n}\Delta t$.

thicknesses of the elements are small with respect to the other dimensions, thus leading to inaccurate results. This effect is referred to as locking,[5] and a remedy is to use reduced integration for the terms corresponding to bending.

A simplification can be introduced by considering a surface element that includes the membrane three-dimensional velocity vector, plus two additional parameters, $\omega_{1n}$ and $\omega_{2n}$, accounting for local rotation of the material unit vector $\mathbf{i}_{3n}$ at each node $n$. For a triangular element the total velocity vector is then given by

$$\mathbf{v} = \sum_{n=1}^{3} \mathbf{V}_n N_n(\xi_1, \xi_2) + \xi_3 \sum_{n=1}^{3} \frac{h_n}{2}(-\omega_{1n}\,\mathbf{i}_{2n} + \omega_{2n}\,\mathbf{i}_{1n}) N_n(\xi_1, \xi_2), \tag{4.30}$$

where $\xi_1$, and $\xi_2$ are the local coordinates for the surface description; $\xi_3$ is the coordinate that varies along the thickness. The kinematics in the element is pictured in Fig. 4.16.

## 4.4 Application to Linear Elasticity

Hooke's equations for an isotropic (or isotropic) linear elastic material were recalled in Section 1.5 and are written[6] as

$$\sigma = \lambda\theta\,\mathbf{I} + 2\mu\varepsilon = \mathbf{D}:\varepsilon, \tag{4.31}$$

or in indicial notation as

$$\sigma_{ij} = \lambda\left(\sum_k \varepsilon_{kk}\right)\delta_{ij} + 2\mu\varepsilon_{ij} = \sum_{kl} D_{ijkl} : \varepsilon_{kl}. \tag{4.32}$$

The unknown field is the displacement vector $\mathbf{u}$, defined in a geometric domain $\Omega$.

[5] This means that bending is nearly prevented for purely numerical (i.e., nonphysical) reasons.

[6] The elastic tensor is often denoted by $\mathbf{c}$, as mentioned in Chap. 1, but here we prefer the $\mathbf{D}$ notation, because it is often used in the FE computation of elastoplasticity.

On the boundary $\partial\Omega$ we impose two different conditions:

- a prescribed displacement on a part $\partial\Omega^u$ of $\partial\Omega$,

$$\mathbf{u} = \mathbf{u}^d, \tag{4.33a}$$

- a prescribed stress vector on the complementary part $\partial\Omega^s$ of $\partial\Omega$,

$$\boldsymbol{\sigma} \cdot \boldsymbol{n} = \mathbf{T}^d. \tag{4.33b}$$

An energy variational formulation[7] is used to find the solution of the elastic problem as the displacement field $\mathbf{u}$, satisfying the boundary condition Eq. (4.33a), and minimizing the functional

$$I(\mathbf{u}) = \frac{1}{2}\int_{\Omega} \varepsilon : \mathbf{D} : \varepsilon \, \mathrm{d}V - \int_{\partial\Omega^s} \mathbf{T}^d \cdot \mathbf{u} \, \mathrm{d}S. \tag{4.34}$$

The displacement field is discretized in terms of nodal displacement vectors $\mathbf{U}_n$, and global shape functions $N_n$, with the usual form:

$$\mathbf{u} = \sum_n \mathbf{U}_n N_n. \tag{4.35}$$

The discretized strain tensor is computed by using Eqs. (1.34) and (4.35):

$$\varepsilon_{ij} = \frac{1}{2}\sum_n \left( U_{in}\frac{\partial N_n}{\partial x_j} + U_{jn}\frac{\partial N_n}{\partial x_i} \right). \tag{4.36}$$

Because of its linear form, Eq. (4.33) can be cast into the more compact form[8]

$$\varepsilon = \mathbf{B} : \mathbf{U}, \tag{4.37a}$$

which is the symbolic translation of the component equality:

$$\varepsilon_{ij} = \sum_{k,n} B_{ijkn} U_{kn}. \tag{4.37b}$$

Remark: the $\mathbf{B}$ operator will also be used for velocity formulations for the relationship between the strain-rate tensor and the nodal velocity vectors.

Now the functional defined by Eq. (4.34) is discretized into

$$I(\mathbf{U}) = \frac{1}{2}\int_{\Omega} \mathbf{U}^T \cdot \mathbf{B}^T : \mathbf{D} : \mathbf{B} \cdot \mathbf{U} \, \mathrm{d}\mathcal{V} - \int_{\partial\Omega^s} \mathbf{T}^d \cdot \mathbf{N} \cdot \mathbf{U} \, \mathrm{d}\mathcal{S}, \tag{4.38}$$

which is a quadratic function of the components of the global displacement vector $\mathbf{U}$. The indicial form of this last equation is straightforward (even if it appears complicated):

$$I(\mathbf{U}) = \frac{1}{2}\int_{\Omega} \sum_{\substack{ijkn \\ \lambda\mu l\nu}} U_{kn} B_{ijkn} D_{ij\lambda\mu} B_{\lambda\mu\kappa\nu} U_{\kappa\nu} \, \mathrm{dV} - \int_{\partial\Omega^s} \sum_{kn} T_k^d U_{kn} \mathrm{N}_n \, \mathrm{dS}. \tag{4.39}$$

[7] R. H. Wagoner and J.-L. Chenot, *Fundamentals of Metal Forming* (Wiley, New York, 1997), pp. 181–183.

[8] Also see Exercise 5.1.

To find the approximate solution, we must minimize this functional with respect to any nonprescribed component $U_{kn}$ of the displacement vector $\mathbf{U}$. We thus obtain the following set of equations:

$$R_{kn}(\mathbf{U}) = \frac{\partial I(\mathbf{U})}{\partial U_{kn}} = \int_{\Omega} \sum_{ij\lambda\mu\kappa\nu} B_{ijkn} D_{ij\lambda\mu} B_{\lambda\mu\kappa\nu} U_{\kappa\nu} \, \mathrm{d}V - \int_{\partial\Omega^s} T_k^d N_n \, \mathrm{d}S = 0. \tag{4.40}$$

Equation (4.40) is usually recast into the compact symbolic equivalent:

$$\mathbf{R}(\mathbf{U}) = \frac{\partial I(\mathbf{U})}{\partial \mathbf{U}} = \int_{\Omega} \mathbf{B}^T : \mathbf{D} : \mathbf{B} \cdot \mathbf{U} \, \mathrm{d}\mathcal{V} - \int_{\partial\Omega^s} \mathbf{T}^d \cdot \mathbf{N} \, \mathrm{d}\mathcal{S} = 0. \tag{4.41}$$

Equation (4.40), or Eq. (4.41), with the appropriate displacement boundary conditions, is the final equation that must be explicitly calculated by assembly of the stiffness matrix and of the loading vector (as described for a simple case in Chapter 3) and solved numerically in terms of the components of the displacement vector $\mathbf{U}$.

## 4.5 The Time-Dependent Heat Problem

### The Classical Galerkin Formulation

The heat equation (for more details see Chapter 7) is written in the domain $\Omega$ of the part:

$$\rho c = \frac{\mathrm{d}T}{\mathrm{d}t} = \mathrm{div}(k \, \mathbf{grad} \, T) + \dot{q}_v \tag{4.42}$$

where

- $\rho$ is the material density,
- $c$ is the heat capacity,
- $\mathrm{d}T/\mathrm{d}t$ is the material derivative for a deformable body,
- $k$ is the thermal conductivity, and
- $\dot{q}_v$ is a heat source term, which is identified with the heat generation from plastic work.

On the part $\partial\Omega^\phi$ of the boundary $\partial\Omega$ a prescribed normal heat flux is given:

$$-k\frac{\partial T}{\partial \mathbf{n}} = \phi_f, \tag{4.43}$$

whereas the temperature itself can be imposed on the other part of the boundary.

The temperature field and its time derivative are discretized into finite-element form according to

$$T = \sum_n T_n N_n(\xi), \tag{4.44a}$$

$$\dot{T} = \sum_n \dot{T}_n N_n(\xi). \tag{4.44b}$$

The heat equation, Eq. (4.42), is multiplied by any shape function $N_m$ and is integrated over the domain $\Omega$:

$$\int_{\Omega} \left[ \rho c \frac{\mathrm{d}T}{\mathrm{d}t} - \mathrm{div}(k \, \mathbf{grad} \, T) - \dot{q}_v \right] N_m \, \mathrm{d}\mathcal{V} = 0. \tag{4.45}$$

The Green's theorem is used to transform the second contribution of the integral:

$$\int_{\Omega} -\mathrm{div}(k\,\mathbf{grad}\,T)N_m\,\mathrm{d}\mathcal{V}$$
$$= \int_{\Omega} k\,\mathbf{grad}\,T\cdot\mathbf{grad}\,N_m\,\mathrm{d}\mathcal{V} - \int_{\partial\Omega} k\,\mathbf{grad}\,T\cdot\mathbf{n}N_m\,\mathrm{d}\mathcal{S}. \tag{4.46}$$

With the help of Eq. (4.46), Eqs. (4.44a) and (4.44b) are substituted into Eq. (4.45) so that the Galerkin method gives the following semidiscretized[9] equation:

$$\mathbf{C}\cdot\frac{\mathrm{d}\mathbf{T}}{\mathrm{d}t} + \mathbf{H}\cdot\mathbf{T} + \mathbf{F} = 0, \tag{4.47}$$

where $\mathbf{C}$ is the heat capacity matrix, with

$$\mathrm{C}_{mn} = \int_{\Omega} \rho c N_m N_n\,\mathrm{d}\mathcal{V}, \tag{4.48a}$$

$\mathbf{H}$ is the heat conductivity matrix,

$$H_{mn} = \int_{\Omega} k\,\mathbf{grad}(N_m)\cdot\mathbf{grad}(N_n)\,\mathrm{d}\mathcal{V}, \tag{4.48b}$$

and $\mathbf{F}$ is the heat flux and volume source vector, which is given by

$$F_m = -\int_{\Omega} \dot{q}_v N_m\,\mathrm{d}\mathcal{V} + \int_{\partial\Omega^{\phi}} \phi_f N_m\,\mathrm{d}\mathcal{S}, \tag{4.48c}$$

where the boundary condition [Eq. (4.43)] was used.

### Time Integration and the Lumped Mass Matrix Formulation

Many numerical methods were tested for the time integration of Eq. (4.45). The Dupont scheme[10] seems to be an interesting compromise among cost, ease of implementation, and efficiency. The Dupont scheme belongs to a family of second-order schemes in which we introduce the following approximations in Eq. (4.45):

$$\mathbf{T} = a\mathbf{T}^{t-\Delta t} + \left(-\frac{3}{2} + 2a + g\right)\mathbf{T}^{t} + \left(-\frac{1}{2} + a + g\right)\mathbf{T}^{t+\Delta t}, \tag{4.49}$$

$$\frac{\mathrm{d}\mathbf{T}}{\mathrm{d}t} = (1-g)\frac{\mathbf{T}^{t} - \mathbf{T}^{t-\Delta t}}{\Delta t} + g\frac{\mathbf{T}^{t+\Delta t} - \mathbf{T}^{t}}{\Delta t}, \tag{4.50}$$

$$\mathbf{C} = \left(\frac{1}{2} - g\right)\mathbf{C}^{t+\Delta t} + \left(\frac{1}{2} + g\right)\mathbf{C}^{t}, \tag{4.51}$$

and similar expressions to that given by Eq. (4.51) for $\mathbf{H}$ and $\mathbf{F}$, with the values $a = 0.25$ and $g = 1$. When $\mathbf{T}_{t-\Delta t}$ and $\mathbf{T}_t$ are known, the unknown is $\mathbf{T}_{t+\Delta t}$ and the resulting equation is linear.

The finite-element solution may be physically unrealistic because of spatial temperature oscillations when a large value of heat flux is present with a relatively coarse

[9] The expression "semidiscretized" is used to indicate that at this stage only the space discretization is completed. The time discretization can then be performed with a different procedure. However, in many explicit time-discretization schemes, the time increment and the element size cannot be chosen independently.

[10] M. A. Hogge, "A Comparison of Two and Three Level Integration Schemes for Non-Linear Heat Conduction," in *Proceedings Numerical Methods in Heat Transfer* (Wiley, Chichester, 1981), p. 75.

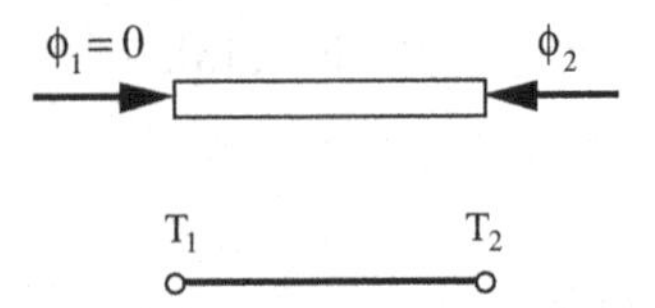


Figure 4.17. The one-linear-element example for heat transfer.

mesh, even if exact time integration is used. This can be illustrated by the very simple example in Fig. 4.17, with only one linear element; the solution is given in Exercise 4.4.

---

**Exercise 4.4: Using the Galerkin method, write the discretized equations for the problem in Fig. 4.17 and solve it. One linear element is used with unit length for simplicity, and the solution is approximated with only two nodal temperatures, $T_1$ and $T_2$, which are functions of time.**

The Galerkin method with finite-element discretization allows us to write a system of two ordinary differential equations:

$$\rho c \begin{bmatrix} 2/3 & 1/3 \\ 1/3 & 2/3 \end{bmatrix} \begin{bmatrix} \dot{T}_1 \\ \dot{T}_2 \end{bmatrix} + k \begin{bmatrix} 1 & -1 \\ -1 & 1 \end{bmatrix} \begin{bmatrix} T_1 \\ T_2 \end{bmatrix} = \begin{bmatrix} 0 \\ \phi_2 \end{bmatrix}. \tag{4.4–1}$$

This system is linear if all thermal parameters are considered constant, and it can be solved analytically by first solving for $\dot{T}_1$ and $\dot{T}_2$

$$\begin{bmatrix} \dot{T}_1 \\ \dot{T}_2 \end{bmatrix} = -\frac{k}{\rho c} \begin{bmatrix} 3 & -3 \\ -3 & 3 \end{bmatrix} \begin{bmatrix} T_1 \\ T_2 \end{bmatrix} + \begin{bmatrix} -\phi_2' \\ 2\phi_2' \end{bmatrix}, \tag{4.4–2}$$

where we put $\phi'2 = \phi_2/\rho c$. Equation (4.4–2) is easily derived from Eq. (4.4–1) by dividing both sides by $\rho c$ and multiplying them by the matrix

$$\begin{bmatrix} 2 & -1 \\ -1 & 2 \end{bmatrix},$$

which is the inverse of matrix

$$\begin{bmatrix} 2/3 & 1/3 \\ 1/3 & 2/3 \end{bmatrix}.$$

The eigenvalues of the matrix in Eq. (4.4–2) are computed, giving $6\sigma$ (where $\sigma = k/\rho c$) and 0. The orthogonal matrix of the eigenvectors is (see Section 2.5)[11]

$$\frac{1}{\sqrt{2}} \begin{bmatrix} 1 & -1 \\ 1 & 1 \end{bmatrix}. \tag{4.4–3}$$

We make the change of unknown functions

$$\begin{bmatrix} T_1' \\ T_2' \end{bmatrix} = \frac{1}{\sqrt{2}} \begin{bmatrix} 1 & -1 \\ 1 & 1 \end{bmatrix} \begin{bmatrix} T_1 \\ T_2 \end{bmatrix}, \tag{4.4–4}$$

and the same transformation for the time derivatives. That allows us to transform Eq. (4.4–2) into the simpler system:

$$\dot{T}_1' = -6\sigma T_1' - 3\frac{\phi_2'}{\sqrt{2}}, \quad \dot{T}_2' = \frac{\phi_2'}{\sqrt{2}}. \tag{4.4–5}$$

The solution of this is easily shown to be

$$T_1' = \frac{\phi_2'}{2\sqrt{2}\sigma}[\exp(-6\sigma t) - 1], \quad T_2' = \frac{\phi_2'}{\sqrt{2}}t. \tag{4.4–6}$$

After coming back to the initial unknown by inversion of Eq. (4.4–4), we obtain the final solution:

$$T_1 = \frac{\phi_2'}{4\sigma}[\exp(-6\sigma t) - 1] + \frac{\phi_2'}{2}t, \quad T_2 = -\frac{\phi_2'}{4\sigma}[\exp(-6\sigma t) - 1] + \frac{\phi_2'}{2}t, \tag{4.4–7}$$

[11] R. H. Wagoner and J.-L. Chenot, *Fundamentals of Metal Forming* (Wiley, New York, 1997).

which we rewrite as

$$T_1 = \frac{\phi_2}{12k}\{3[\exp(-6\sigma t) - 1] + 6\sigma t\}, \quad T_2 = \frac{\phi_2}{12k}\{-3[\exp(-6\sigma t) - 1] + 6\sigma t\}, \qquad (4.4\text{–}8)$$

or with the reduced time $t' = 6\sigma t$, as

$$T_1 = \frac{\phi_2}{12k}\{3[\exp(-t') - 1] + t'\}, \quad T_2 = \frac{\phi_2}{12k}\{3[\exp(-t') - 1] + t'\}. \qquad (4.4\text{–}9)$$

---

In Fig. 4.18 we plotted the reduced temperature, versus the reduced time, by assuming $6\sigma = 1$ and $\phi_2/12k = 1$. The actual curve can be obtained by giving the experimental value for $\sigma = k/\rho c$ and the heat flux $\phi_2$. We see that the temperature of node 1 shows a nonphysical evolution by first decreasing until $t = -6.59\sigma$ and reaching approximately $-10.8k/\phi_2$. Then it increases and finally, after approximately $18\sigma$, exhibits a reasonable behavior.

A classical remedy is to lump the consistent heat capacity matrix C. In the lumping transformation, the C matrix is replaced by a diagonal matrix $\mathbf{C}^L$, defined by

$$C_{ij}^L = \delta_{ij} \sum_k C_{ik}. \qquad (4.52)$$

---

**Exercise 4.5: For the same problem as the one analyzed in Exercise 4.4, compute the solution of the one-element heat flow with a lumped heat capacity matrix.**

After lumping, the heat capacity matrix in Eq. (4.4–1) is transformed into

$$\rho c \begin{bmatrix} 1 & 0 \\ 0 & 1 \end{bmatrix} \begin{bmatrix} \dot{T}_1 \\ \dot{T}_2 \end{bmatrix} + k \begin{bmatrix} 1 & -1 \\ -1 & 1 \end{bmatrix} \begin{bmatrix} T_1 \\ T_2 \end{bmatrix} = \begin{bmatrix} 0 \\ \phi_2 \end{bmatrix}, \qquad (4.5\text{–}1)$$

which immediately gives

$$\begin{bmatrix} \dot{T}_1 \\ \dot{T}_2 \end{bmatrix} = -\sigma \begin{bmatrix} 1 & -1 \\ -1 & 1 \end{bmatrix} \begin{bmatrix} T_1 \\ T_2 \end{bmatrix} + \begin{bmatrix} 0 \\ \phi_2' \end{bmatrix}. \qquad (4.5\text{–}2)$$

With the same method as in Exercise 4.4, we finally get

$$T_1 = \frac{\phi_2}{12k}\{3[\exp(-t'/3) - 1] + t'\}, \quad T_2 = \frac{\phi_2}{12k}\{-3[\exp(-t'/3) - 1] + t'\}. \qquad (4.5\text{–}3)$$

---

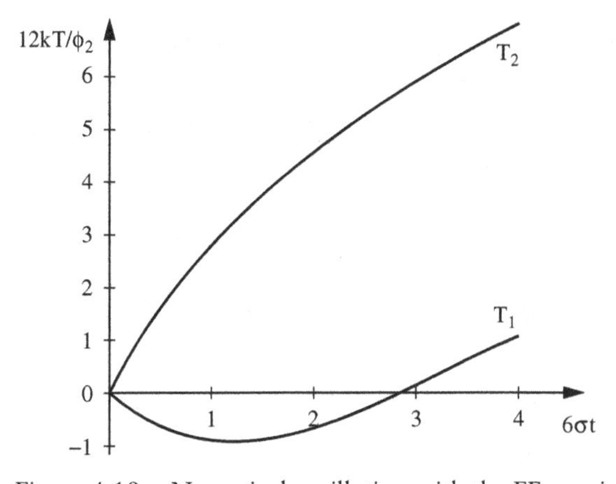


Figure 4.18. Numerical oscillation with the FE consistent formulation.

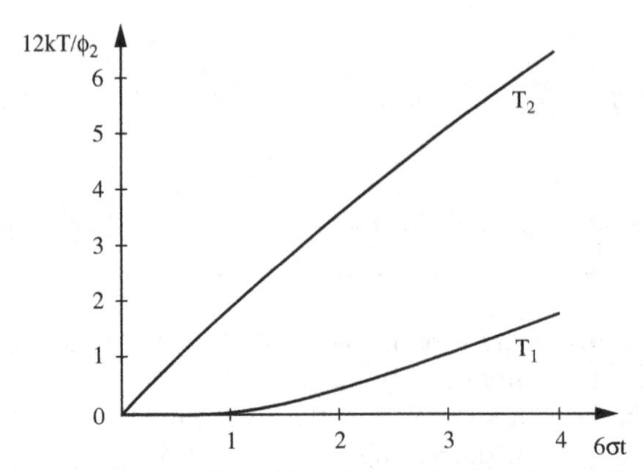


Figure 4.19. Effect of lumping on the one-element FE problem.

This method has the advantage of completely suppressing the oscillation of the solution that is due to spatial discretization, as is clear in Fig. 4.19. However, when the elements are refined, a loss of accuracy with respect to the consistent Galerkin formulation can occur.

### Lumping of Quadratic Elements

The one-dimensional quadratic element is considered at first as pictured in Fig. 4.20. If an attempt is made to compute directly the lumped heat capacity matrix, we observe that the resulting diagonal matrix has the form

$$\mathrm{C}^{eL} = \frac{\rho c a}{6}\begin{bmatrix} 1 & 0 & 0 \\ 0 & 4 & 0 \\ 0 & 0 & 1 \end{bmatrix}. \tag{4.53}$$

After assembly, we observe that the intermediate nodes (here node 2) will not be equivalent to the other nodes that correspond to the elements ends (here nodes 1 and 3). Even when the elements are identical (i.e., they can be deduced from one of them by a translation), the global heat capacity matrix will take the form

$$\mathrm{C} = \frac{\rho c a}{6}\begin{bmatrix} 1 & 0 & 0 & 0 & 0 & 0 & \cdots & 0 & 0 & 0 \\ 0 & 4 & 0 & 0 & 0 & 0 & \cdots & 0 & 0 & 0 \\ 0 & 0 & 2 & 0 & 0 & 0 & \cdots & 0 & 0 & 0 \\ 0 & 0 & 0 & 4 & 0 & 0 & \cdots & 0 & 0 & 0 \\ 0 & 0 & 0 & 0 & 2 & 0 & \cdots & 0 & 0 & 0 \\ 0 & 0 & 0 & 0 & 0 & 4 & \cdots & 0 & 0 & 0 \\ \cdots & \cdots & \cdots & \cdots & \cdots & \cdots & \cdots & \cdots & \cdots & \cdots \\ 0 & 0 & 0 & 0 & 0 & 0 & \cdots & 2 & 0 & 0 \\ 0 & 0 & 0 & 0 & 0 & 0 & \cdots & 0 & 4 & 0 \\ 0 & 0 & 0 & 0 & 0 & 0 & \cdots & 0 & 0 & 1 \end{bmatrix}.$$

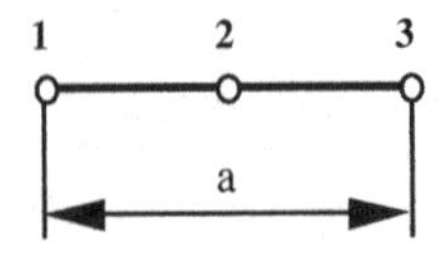


Figure 4.20. The quadratic one-dimensional element.

Then we conclude that the classical quadratic shape functions are not appropriate for the lumped formulation, and a change of basis function must be made. The principle for this change is to build new basis functions that will give equivalent terms for a regular mesh after assembly. In the one-dimensional case the extremity nodes can only be adjacent to two elements, so it is easy to convince ourselves that the new shape functions must be

$$N'_1 = N_1 + \frac{1}{8}N_2, \quad N'_2 = \frac{3}{4}N_2, \quad N'_3 = N_3 + \frac{1}{8}N_2. \tag{4.54}$$

We can verify that the lumped local heat capacity matrix is now

$$\mathbf{C}^{eL} = \frac{\rho c a}{6}\begin{bmatrix} 1.5 & 0 & 0 \\ 0 & 3 & 0 \\ 0 & 0 & 1.5 \end{bmatrix}. \tag{4.55}$$

Obviously, for identical elements, the global heat capacity matrix will have equal terms for any inner node number.

The situation is worse for the two-dimensional quadratic six-node triangular element, where the element lumped matrix has null diagonal terms for each node corresponding to an apex of the triangle. For the two-dimensional problem, in the most regular mesh, a node corresponds to six triangles if it is an apex, and to two triangles if it is at midside. The new shape functions are then[12]

$$N'_1 = N_1 + \frac{1}{8}(N_2 + N_6), \quad N'_2 = \frac{3}{4}N_2. \tag{4.56}$$

The apex terms of the three-dimensional ten-node quadratic tetrahedron are negative. However, the three-dimensional case is more complicated, as space cannot be filled with regular tetrahedrons. The mean value of the number of neighbors can be used, and this gives

$$N'_1 = N_1 + 0.13(N_2 + N_6 + N_7), \quad N'_2 = 0.74N_2. \tag{4.57}$$

It appears that the coefficients are close to the one- and two-dimensional cases.

## PROBLEMS

### A. Proficiency Problems

1. A triangle is defined in the $Ox_1x_2$ reference system by its nodal coordinates

   $M_1(2, 2)$, $M_2(-2, 3)$, $M_3(-2, -1)$,

   as shown in the figure below. Write the equations defining the mapping between the reference system and the physical space. Use these equations to transform the shape functions in order to express it in terms of the physical coordinates $x_1$ and $x_2$.

[12] J.-L. Chenot, Y. Tronel, and N. Soyris, "Finite Element Calculation of Thermo-Coupled Large Deformation in Hot Forging, in *Advanced Computational Methods in Heat Transfer* II, L. C. Wrobel, C. A. Brebbia, and A. J. Nowak, eds. (Computational Mechanics Pub., Southampton, co-published with Elsevier Applied Science, London, 1992), pp. 493–511.

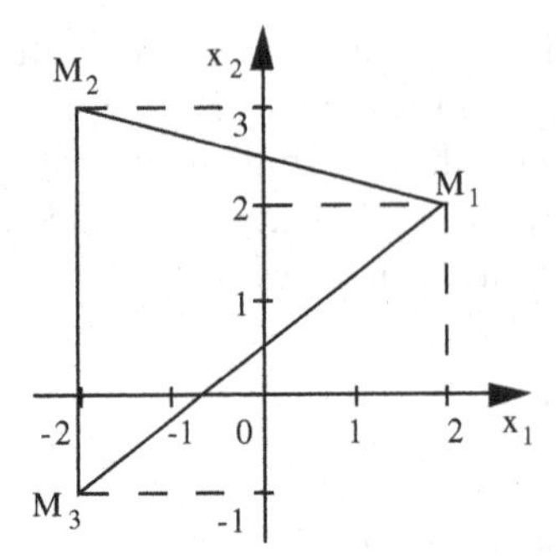

Figure P4.1.

2. For the triangle in the above figure, compute the Jacobian matrix for the transformation between the reference space and the physical space. Using the one-point integration formula for triangles, calculate the area of the triangle. Verify that the area is computed exactly in this case.
3. In the following figure, a quadrilateral is considered with the following nodal coordinates:

$$M_1(1,2),\ M_2(-2,1),\ M_3(-1,-1),\ M_4(2,-2).$$

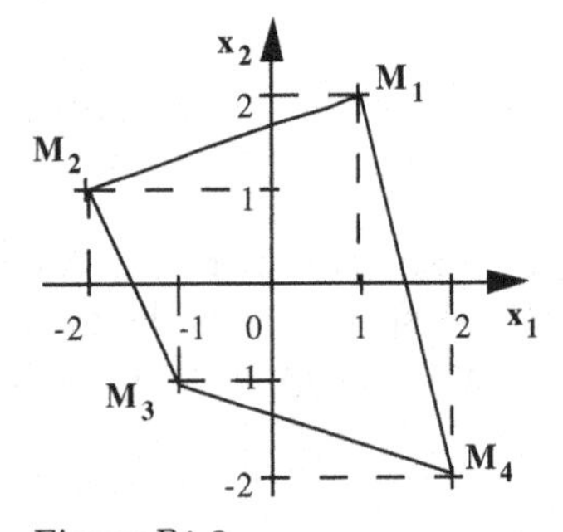

Figure P4.3.

Define the one-to-one mapping between the reference space and the physical space. Compute the Jacobian determinant $J$ as a function of the local coordinates $\xi_1$ and $\xi_2$. Find the values $-1 < \xi_1 < 1$ and $-1 < \xi_2 < 1$ for which $J$ is optimum.

4. A quadratic triangular element is defined as pictured in the figure below.

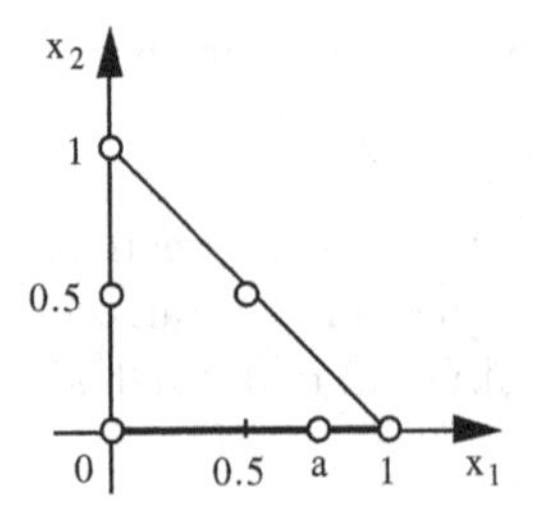

Figure P4.4.

Note that only one node is moved from the position it has in the reference element. Compute the Jacobian determinant of the mapping between the reference space and the physical one. Determine the extremum values the $a$ parameter can be given in order that the Jacobian determinant remains positive.

5. A two-dimensional velocity problem is discretized into four-node square elements with side lengths equal to 2. Express the incompressibility condition at the center of the element; show that generally the incompressibility condition is not fulfilled elsewhere. Verify that if the zero velocity flux through the element boundary is imposed, the same equation as before must be written.
6. Compute the unit normal vector $\mathbf{n}$ to the face containing nodes 1, 2, and 3 of a tetrahedron equal to the reference tetrahedron.
7. A simple two-dimensional elastic problem is defined on a trapezoidal sample in the following figure; the Lamé coefficients of the material are denoted by $\lambda$ and $\mu$.

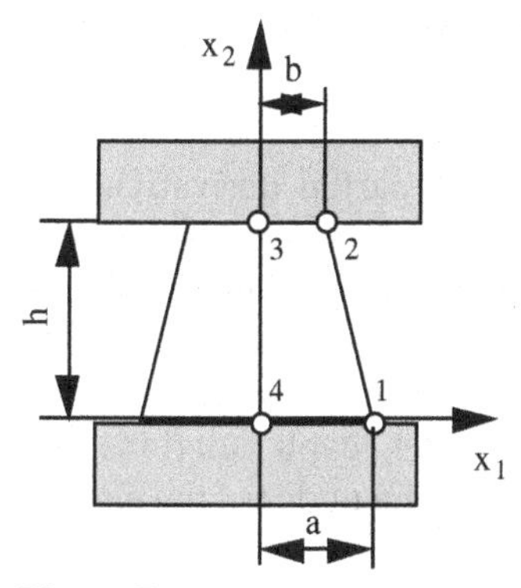


Figure P4.7.

The sample is upset by the upper surface with a small vertical displacement equal to $\Delta h$. Because of the symmetry of the process, only one-half of the part need be discretized. If one bilinear quadrilateral element with nodes 1, 2, 3, and 4 is used, show that only the horizontal displacements $U_1$ and $U_2$ of nodes 1 and 2 respectively are unknown. If a zero friction stress is assumed on the lower and upper surfaces, write the discretized energy functional, using one Gaussian integration point. Approximately solve the problem by minimizing the functional with respect to the unknown displacements.

## B. Depth Problems

8. The Green's equation is considered with the special form

$$\int_{\Omega} X_1^2 \, \mathrm{d}S = \int_{\partial\Omega} \frac{X_1^3}{3} n_1 \, \mathrm{d}l,$$

where $\Omega$ is the reference triangle and $n_1$ the $ox_1$ component of the normal to the boundary $\partial\Omega$ of $\Omega$. Use the one-point integration formula for triangles to compute the left-hand side, and the one-point Gaussian integration formula for each side of the triangle at the left-hand side. Compare both sides of the Green's equation and explain the difference; propose a remedy to find the exact value of both sides.

9. Generalize Problem 4 to the quadratic tetrahedral element.

## C. Numerical and Computational Problems

10. Write a FORTRAN module for the computation of the value and the partial derivatives of each shape function of the reference element for
    - a one-dimensional linear element
    - a two-dimensional linear triangular element
    - a three-dimensional linear tetrahedral element

    Input data: the dimension of the physical space (1, 2, or 3), and the local coordinate(s).
11. This is the same problem as Problem 10, with the quadratic shape functions.
12. Write a general FORTRAN module for the interpolation of either a function or a vector field and its partial derivatives with respect to the local coordinates inside an element.

    Input data: the dimension of the vector space (1 for a scalar function), the number of nodes of the element, the numerical values of the shape function, and its derivatives for each node of the element.
13. Write a FORTRAN subroutine for the calculation of the partial derivatives of a scalar or a vector function with respect to the *physical coordinates*. The scalar or vector function is interpolated within an isoparametric element. The module corresponding to Problem 12 should be used.

    Input data: dimension of the vector function, dimension of the physical space, number of nodes of the element, partial derivatives of the shape functions with respect to the local coordinates, and global coordinates of the nodes of the element in the physical space.
14. Write a computer code that generates a random element:
    - a linear or a quadratic triangle in two-dimensions
    - a linear or a quadratic tetrahedron in three-dimensions

    Using Problems 10–13, compute the Jacobian determinant $J$ of the mapping between the local and the physical spaces at each node. Check that $J$ remains strictly positive at each node, and if this condition is not fulfilled then generate a new element.
15. Write a FORTRAN module that generates a random geometric point inside a given element defined in the physical space. The case of two (triangle) or three (tetrahedron) dimensions will be analyzed separately. Input data: space dimension, a code defining the element type, number of nodes, and node coordinates. It is suggested to use the constraints on the local coordinates before calculating the physical coordinates, and modules elaborated in Problem 13.
16. Consider a scalar function defined in a physical space (of dimension 2 or 3) by a polynomial of degree 1 or 2. Write a computer code that generates random elements in the physical space, and random geometrical points in the elements. For each point of a given element, compute the partial derivatives of the function with respect to the physical coordinates, either directly from the polynomial expression, or by using the isoparametric finite-element discretization of the function, defined from the analytical nodal values. Compare the results. Before coding, write the general flow chart of the program and the names of the subroutines that were written for the previous problems and that are necessary here.

CHAPTER FIVE

# Classification of Finite-Element Formulations

The finite-element formulation of a given physical problem depends primarily on the nature of the problem and, to some extent, on several numerical choices the scientist decides to make. In this chapter we shall review five different topics, linked to plastic forming, which have a prominent influence on the way the problem will be solved.

First, the explicit and implicit formulations originally corresponded to different problems, distinguished by the deformation rate level, but they are now used for a much larger range of processes.

Second, historically the first attempts to model metal-forming processes by the finite-element method were based on the flow formulation, or the rigid-plastic material model, in which elastic deformation is neglected. Despite an additional cost and a more complex treatment, the elastoplastic approach is more and more often applied to forming problems in engineering.

Third, the Eulerian formulation, when applicable, is very efficient to treat steady-state processes, whereas the updated Lagrangian formulation is the most popular method for the other cases involving large strain.

Fourth, the displacement or velocity approaches were mostly used in the past for practical applications, as they are more economical. However, mixed methods provide an additional flexibility for the finite-element problem formulations, which can result in more satisfactory solutions, both from the mathematical and from the numerical point of view.

Fifth, the problem of time integration of constitutive equations has received considerable attention during the past 20 years, and some progress is still necessary to achieve complete accuracy for any real process.

The following notation must be introduced in order to simplify the following formulas.

**B** is defined in terms of the shape functions (see Exercise 5.1). This operator allows us to express the discretized small strain (or strain-rate) tensor as a function of the displacement (or velocity) vector; we have

$$\Delta\varepsilon = \mathbf{B} \cdot \Delta\mathrm{U} \quad \text{or } \dot{\varepsilon} = \mathbf{B} \cdot \mathrm{V}. \tag{5.1a}$$

Note that we have used the dot (or scalar) product notation, with the consequence that the order of the operator **B** and of the vector of the nodal unknowns can be exchanged. The same remark is applicable to formula (5.1b).

$\mathbf{B}_n$ is a nodal value of the $\mathbf{B}$ operator, such that

$$\Delta\varepsilon = \sum_n \mathbf{B}_n \cdot \Delta\mathbf{U}_n \quad \text{or } \dot{\varepsilon} = \sum_n \mathbf{B}_n \cdot \mathbf{V}_n. \tag{5.1b}$$

$B_{ijkn}$ is the component notation of the $\mathbf{B}$ operator, with

$$\Delta\varepsilon_{ij} = \sum_{kn} B_{ijkn}\Delta U_{kn} \quad \text{or } \dot{\varepsilon}_{ij} = \sum_{kn} B_{ijkn} V_{kn}. \tag{5.1c}$$

$\mathbf{N}$ is a vector of the shape functions: it is defined by as many components as the number of degrees of freedom. With this global notation, the discretized displacement can be written with the compact form

$$\mathbf{u} = \mathbf{N} \cdot \mathbf{U} \tag{5.1d}$$

or the velocity field

$$\mathbf{v} = \mathbf{N} \cdot \mathbf{V}. \tag{5.1e}$$

$N_n$ is a shape function with number $n$ (i.e., component number $n$ of $\mathbf{N}$), such that

$$N_n(\mathbf{X}_m) = \delta_{nm}.$$

Other notation is as follows, where

$\mathbf{U}$ is a vector of the components of the nodal displacement, or a global displacement vector,
$\mathbf{U}_n$ is a displacement vector at node $n$ at time $t$, with components $\mathbf{U}_{in}$ with $i = 1, 2, (3)$,
$\mathbf{V}$ is a vector of the components of the nodal velocity vectors, or the global velocity vector,
$\mathbf{V}^t$ is the global velocity vector at time $t$,
$\mathbf{V}^t_n$ is the velocity vector at node $n$ at time $t$, with components $V^t_{in}$ with $i = 1, 2, (3)$,
$\mathbf{X}$ is the vector of all the components of the nodal position vectors, or the global coordinate vector,
$\mathbf{X}^t$ is the global coordinate vector at time $t$,
$\mathbf{X}^t_n$ is the position vector of node $n$ at time $t$, with components $X^t_{in}$ with $i = 1, 2, (3)$, and
$\mathbf{\Omega}_t$ is the domain of interest at time $t$.

---

**Exercise 5.1: Determine the explicit form for the components of the B operator, from its definition and the fundamental interpolation form of the velocity in terms of the shape functions.**

From Eq. (5.1e), the interpolated velocity field is

$$v_k = \sum_n V_{kn} N_n. \tag{5.1–1}$$

The components of the strain-rate tensor follow from its definition and Eq. (5.1–1):

$$\dot{\varepsilon}_{ij} = \frac{1}{2}\left(\frac{\partial v_i}{\partial x_j} + \frac{\partial v_j}{\partial x_i}\right) = \frac{1}{2}\sum_n \left(V_{in}\frac{\partial N_n}{\partial x_j} + V_{jn}\frac{\partial N_n}{\partial x_i}\right). \tag{5.1–2}$$

Equation (5.1–2) must be transformed to allow the identification of the **B** operator:

$$\dot{\varepsilon}_{ij} = \frac{1}{2} \sum_{kn} V_{kn} \left( \frac{\partial N_n}{\partial x_j} \delta_{ki} + \frac{\partial N_n}{\partial x_i} \delta_{kj} \right). \tag{5.1–3}$$

It is easy to convince oneself that Eqs. (5.1–2) and (5.1–3) are equivalent, if $\delta$ is the Kronecker symbol. When Eqs. (5.1c) and (5.1–3) are compared, we conclude that

$$B_{ijkn} = \frac{1}{2} \left( \frac{\partial N_n}{\partial x_j} \delta_{ki} + \frac{\partial N_n}{\partial x_i} \delta_{kj} \right). \tag{5.1–4}$$

---

## 5.1 Implicit and Explicit Formulations

From the physical point of view, the role played by kinetic energy is clear. We can distinguish accordingly three types of processes.

1. In the quasi-static problems, the kinematic energy is insignificant (say less than 1%, or even 0.1%) of the total energy. Most of the forming processes fall into this category, as can be shown by determining the order of magnitude of each contribution.
2. In the high-strain-rate phenomena, or purely dynamic processes, the kinetic energy is overwhelmingly dominant, and the constitutive equation of the material can be completely neglected, or considered as a small perturbation. This is the case of processes with a very high energetic impact, such as shape charges.
3. In the intermediate, or dynamic situation, both kinetic energy and reversible or irreversible deformation energy have a similar order of magnitude. This case is well represented for example by the elastic vibration of structures, or by moderate velocity impacts such as crashes, or high-velocity forming processes, or by the processing of semisolid or liquid metal in casting.

Formerly, the numerical approaches used to solve a process were clearly separated, according to whether the problem corresponded to type 1 or type 2, and the associated formulations took full advantage of the physical situation. Today there appears to be a tendency to solve quasi-static problems with methods designed for dynamic analysis, and we can observe the inverse trend to adapt quasi-static implicit methods to problems of intermediate rates or higher. For more details on the problem, refer to footnotes 1–5.

The general method is well documented in the literature: here we shall restrict ourselves to summarizing the main points in order to make a clear distinction between the different situations that can be encountered. The important vector fields are the

[1] T. Belytschko and C. S. Tsay, "Explicit Algorithms for Non-Linear Dynamics of Shells," *Comp. Meth. Appl. Mech. Eng.* 43, 251–276 (1984).

[2] K. Schweizerhof and J.-O. Hallquist, "Explicit Integration Schemes and Contact Algorithms for Thin Sheet Metal Forming," in *FE-Simulation of 3-D Sheet Metal Forming Processes in Automotive Industry* (VDI Berichte 894, VDI Verlag, Dusseldorf, 1991), pp. 405–440.

[3] J. C. Nagtegaal and L. M. Taylor, "Comparison of Implicit and Explicit Finite Element Methods for Analysis of Sheet Forming Problems," in *FE-Simulation of 3-D Sheet Metal Forming Processes in Automotive Industry* (VDI Berichte 894, VDI Verlag, Dusseldorf, 1991), pp. 705–726.

[4] K. J. Bathe, *Finite Element Procedures* (Prentice-Hall, Englewood Cliffs, NJ, 1995).

[5] J.-L. Chenot and F. Bay, "An Overview of Numerical Modelling Technique," *J. Mater. Process. Technol.* 80–81, pp. 8–15 (1998).

position vector $\mathbf{x}$, the velocity $\mathbf{v}$, and the acceleration $\gamma$. They are discretized in the usual way in terms of the shape functions $N_n$, according to

$$\mathbf{x} = \sum_n \mathbf{X}_n N_n \quad \text{or } \mathbf{x} = \mathbf{N} \cdot \mathbf{X}, \tag{5.2a}$$

$$\mathbf{v} = \sum_n \mathbf{V}_n N_n \quad \text{or } \mathbf{v} = \mathbf{N} \cdot \mathbf{V}, \tag{5.2b}$$

$$\gamma = \sum_n \mathbf{\Gamma}_n N_n \quad \text{or } \gamma = \mathbf{N} \cdot \mathbf{\Gamma}, \tag{5.2c}$$

where $\mathbf{X}_n$, $\mathbf{V}_n$, and $\mathbf{\Gamma}_n$ are respectively the nodal position vector, the nodal velocity vector, and the nodal acceleration vector for node number $n$, and $\mathbf{X}, \mathbf{V}, \Gamma$ are the global vectors for the same variables. In complex material behavior we have a general constitutive equation linking the stress tensor to other state variables, such as elastic strain or irreversible strain (or strain rate) tensors, internal state variables, temperature, or even the time derivative of the stress tensor itself. For simplicity, we shall make the assumption that the material is purely viscous, and therefore that the stress tensor is simply a function of the velocity field by means of the strain-rate tensor:

$$\sigma = \sigma[\dot{\varepsilon}(\mathbf{v})] = \sigma(\mathbf{v}).$$

According to Eq. (5.1b) the discretized strain-rate tensor is written in terms of the $\mathbf{B}$ operator. The virtual velocity field $\mathbf{v}^*$ is also discretized with the same shape functions:

$$\mathbf{v}^* = \sum_n \mathbf{V}_n^* N_n, \tag{5.3a}$$

so that the discretized virtual strain rate can be written as

$$\dot{\varepsilon}^* = \sum_n \mathbf{B}_n \mathbf{V}_n^*. \tag{5.3b}$$

With this notation, the general virtual work principle[6] can be written for any discretized virtual velocity field. Then we write the corresponding equation for each independent $\mathbf{V}^*$: we can choose successively any vector with zero for each component except one. For any $m$ index, the resulting set of equations is expressed as

$$\int_\Omega \rho\gamma N_m \, dV - \int_\Omega \rho \mathbf{g} N_m \, dV + \int_\Omega \sigma(\mathbf{v}) : \mathbf{B}_m \, dV - \int_{\partial\Omega^s} \mathbf{T}^d N_m \, dS = 0. \tag{5.4}$$

This equation must be written for any node number $m$, and for any component, except for those corresponding to prescribed velocities. When acceleration is not neglected, the first term of Eq. (5.4) is transformed according to

$$\int_\Omega \rho\gamma N_m \, dV = \int_\Omega \rho \left[ \sum_n \mathbf{\Gamma}_n N_n \right] N_m \, dV = (\mathbf{M} \cdot \mathbf{\Gamma})_m \tag{5.5a}$$

On the right-hand side of Eq. (5.5a) we have the component number $m$ of the product of the mass matrix $\mathbf{M}$ and of the vector of nodal acceleration $\Gamma$. The mass matrix is

[6] R. H. Wagoner and J.-L. Chenot, *Fundamentals of Metal Forming* (Wiley, New York, 1997), p. 176.

built according to

$$M_{minj} = \delta_{ij} \int_{\Omega} \rho N_m N_n \, dV, \tag{5.5b}$$

where $mi$ (resp. $nj$) are the indices for the component $i$ (resp. $j$) corresponding to node number $m$ (resp. $n$). Now we can write the other terms of Eq. (5.4), that is, those that do not include the acceleration, with the notation $\mathbf{R}(\mathbf{V}, \mathbf{X})$, in order to transform the discretized form of the virtual work into a more compact form:

$$\mathbf{M} \cdot \mathbf{\Gamma} - \mathbf{R}(\mathbf{V}, \mathbf{X}) = 0. \tag{5.5c}$$

**Implicit Formulations**

In most metal-forming processes the influence of inertia forces can be neglected, as they contribute to less than 1% of the energy variation of the system (a forging example is analyzed in Problem 5.7). With this approximation, Eq. (5.5c) is written at time $t$ as

$$\mathbf{R}(\mathbf{V}^t, \mathbf{X}^t) = 0. \tag{5.6}$$

Knowing the domain $\Omega_t$, and the global coordinate vector $\mathbf{X}^t$, at time $t$, we have to solve Eq. (5.6) in order to compute the global velocity vector $\mathbf{V}^t$. Then the nodal update can be performed according to the Euler scheme:

$$\mathbf{X}^{t+\Delta t} = \mathbf{X}^t + \Delta t \mathbf{V}^t. \tag{5.7}$$

This simple general scheme is called implicit, as it is impossible to compute the velocity $\mathbf{V}^t$ with a simple expression. A global linear or nonlinear equation must be solved, which involves all the velocity components.

A "more implicit" scheme can be preferred, in which the update is performed with the velocity corresponding to the end of the time increment:

$$\mathbf{X}^{t+\Delta t} = \mathbf{X}^t + \Delta t \mathbf{V}^{t+\Delta t}. \tag{5.8}$$

In this case, we have generally a more complicated system, which is also written at the end of the increment, and must be solved in terms of the velocity $\mathbf{V}^{t+\Delta t}$:

$$\mathbf{R}(\mathbf{X}^{t+\Delta t}, \mathbf{V}^{t+\Delta t}) = \mathbf{R}(\mathbf{X}^t + \Delta t \mathbf{V}^{t+\Delta t}, \mathbf{V}^{t+\Delta t}) = 0. \tag{5.9}$$

When acceleration is taken into account,[7] Eq. (5.5c) written at time $t$ becomes

$$\mathbf{M}^t \cdot \mathbf{\Gamma}^t - \mathbf{R}(\mathbf{V}^t, \mathbf{X}^t) = 0. \tag{5.10}$$

If we suppose that the coordinate and velocity vectors are known at time $t$, the global acceleration vector $\mathbf{\Gamma}^t$ can be calculated by solving a linear system with matrix $\mathbf{M}^t = \mathbf{M}(\mathbf{X}^t)$. The update can be performed according to

$$\begin{aligned} \mathbf{V}^{t+\Delta t} &= \mathbf{V}^t + \Delta t \mathbf{\Gamma}^t, \\ \mathbf{X}^{t+\Delta t} &= \mathbf{X}^t + \Delta t \mathbf{V}^t + \frac{1}{2} \Delta t^2 \mathbf{\Gamma}^t. \end{aligned} \tag{5.11}$$

[7] M. P. Miles, L. Fourment, and J.-L. Chenot, "Inertia Effects in Finite-Element Simulation of Metal Forming Processes," *J. Mater. Process. Technol.* 45, 19–24 (1994).

**Explicit Formulations**

If in Eq. (5.10) the mass matrix $\mathbf{M}^t$ is replaced by a diagonal matrix $\mathbf{D}^{Mt}$, then calculation of the acceleration global vector is straightforward, as it requires only inversion of a diagonal matrix:

$$\mathbf{\Gamma}^t = (\mathbf{D}^{Mt})^{-1}\mathbf{R}(\mathbf{X}^t, \mathbf{V}^t). \tag{5.12}$$

The method is called explicit because no linear or nonlinear system has to be solved. The transformation that is used to transform the mass matrix is called lumping (which is similar to the lumping of the thermal discretized equation in Section 4.5). It is an approximation in which the components on a line of the matrix are summed and put on the diagonal:

$$D_{ij}^{Mt} = \delta_{ij} \sum_j M_{ij}^t. \tag{5.13}$$

Another approach clearly originates from the development of the finite-difference method. We consider first the more straightforward discretization of the equilibrium equation, which for a nodal expression mainly requires the calculation of $\mathrm{div}(\sigma) = 0$. This can be done schematically in two steps.

1. Within each element the strain-rate tensor is evaluated from the nodal values of the velocity, using the interpolation functions within each element with the help of Eq. (5.1–2), and the stress tensor is then obtained with the constitutive equation. At this stage the stress field is discontinuous at element interfaces and requires a first smoothing procedure to define the stress tensor at each node. Without entering into details of the various procedures that were proposed, we find that one of the simplest examples of smoothing consists of the evaluation of any stress tensor at a given node as the arithmetic average of the contributions of each element containing this node (see Fig. 5.1).
2. Now the divergence of the stress field can be calculated again within each element, from the previously obtained nodal values of the stress tensor, and smoothed at the nodal level by a similar procedure. This simple approach shows that, if at time $t$ the variables $\mathbf{X}^t$ and $\mathbf{V}^t$ are known, then the acceleration

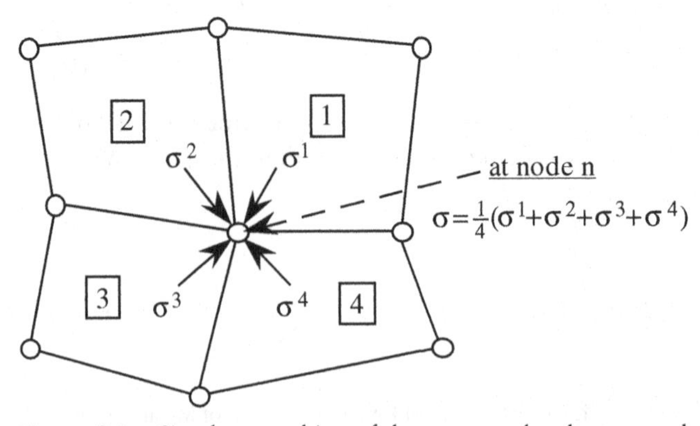


Figure 5.1. Simple smoothing of the stress at the element nodes.

vector can be deduced with the equilibrium equation by

$$\mathbf{\Gamma}^t = \frac{1}{\rho}\{\mathrm{div}[\sigma(\mathbf{V}^t)] + \mathbf{g}\}. \tag{5.14}$$

At time $t + \Delta t$ we can obtain the new variables by expressions quite similar to those of Eqs. (5.10a) and (5.10b), and the integration scheme can proceed further with the same steps.

The major differences with respect to the previous implicit approach are as follows.

First, these schemes are conditionally stable; that is, the time step must satisfy a Courant type of condition in order that, during a time increment, elastic waves will not cross more than one element. For an elastoplastic material with density $\rho$ and Young modulus $E$, the time increment $\Delta t$ and the mesh size $h$ must verify

$$\Delta t < \frac{h}{\sqrt{E/\rho}}.$$

We see that for steel processing the time step should be less than $10^{-6}$ s. For a forming process lasting between 0.1 and 10 s, the number of time increments would be prohibitive. It is therefore necessary to increase the velocity of the tools artificially for metal forming (by a factor of 100–1000), or equivalently to increase the material density. However, this artificial change of the physical parameters can perturb the solution, as shown, for example, in footnote 8.

Second, possible elastic oscillations of the solution could be exaggerated by the explicit time-integration scheme, if no appropriate numerical damping is added,[9] but again they can affect the accuracy of the result.

Third, the second method, which requires two levels of numerical differentiation and subsequent smoothing, can lead to important numerical errors. This is especially crucial when we try to analyze quasi-static processes, in which the acceleration term is small and can be even smaller than the numerical errors on the calculation of div($\sigma$).

The third drawback can be at least partially overcome by a modification of the FEM (which is often referred to as the finite-volume method, or FVM). A control volume $C_n$ is defined around any node $n$, as suggested in Fig. 5.2 for a two-dimensional mesh composed of triangles and quadrilaterals.

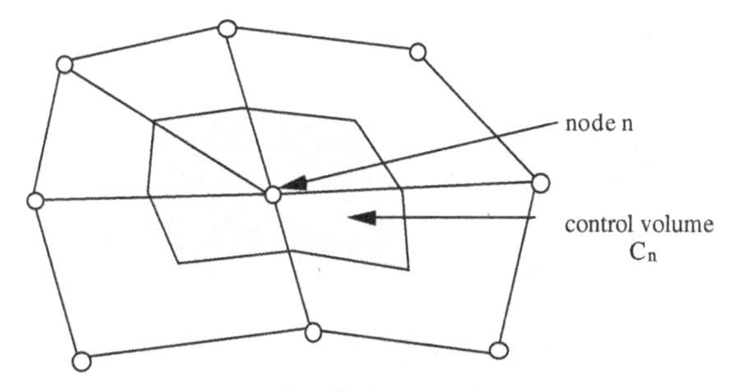


Figure 5.2. Control volume.

[8] S. P. Wang, S. Choudhry, and T. B. Wertheimer, "Comparison Between the Static Implicit and Dynamic Explicit Methods for FEM Simulation of Sheet Forming Processes," in *Simulation of Materials Processing: Theory, Methods and Applications*, J. Huétink and F. Baaijens, eds. (Balkema, Rotterdam, 1998), pp. 245–250.

[9] Physical damping can be efficient in the sense that viscosity softens also physical oscillations.

The dynamic equation is integrated in an approximate way in the control domain $C_n$ with volume $\mathrm{vol}_n$ and boundary $\partial C_n$:

$$\rho \mathrm{vol}_n \boldsymbol{\Gamma}_n + \int_{C_n} \rho \mathbf{g} \, \mathrm{d}V - \int_{\partial C_n} \boldsymbol{\sigma} \cdot \mathbf{n} \, \mathrm{d}S. \tag{5.15}$$

This method is largely employed in fluid flow, and more specifically in die filling for casting. For incompressible materials (solid or liquid), the pressure field $p$ may produce large oscillations of the solution. A compromise, which is often used, is to solve the pressure field by a partially implicit formulation, which leads to a linear system involving the nodal values of the pressure only.

## 5.2 Rigid-Plastic or Elastoplastic Approximation

In many problems involving large or very large deformation, it is possible to consider two levels of approximation, according to the desired degree of accuracy, and also the kind of results that are requested (e.g., residual stresses), or the need for further calculations (e.g., springback effect).

The first one is the rigid-plastic approximation or, for rate-dependent plasticity, rigid-viscoplastic approximation, in which the elasticity effects are completely neglected. This idealization, also referred to as the flow formulation, is based on the experimental observation that elastic strains remain small, generally 0.2–0.5%, as compared with irreversible strains, which are often greater than 10%. There is no crucial problem for the usual viscoplastic constitutive laws: when the strain rate is vanishingly small, the stress tends also to zero. This is not the case for plastic behavior, for which we have seen[10] that only the direction of the strain-rate tensor is relevant. For example, if we recall the flow rule associated with the Von Mises criterion, we have

$$\mathbf{s} = \frac{2}{3} \frac{\sigma_0}{\dot{\bar{\varepsilon}}} \dot{\boldsymbol{\varepsilon}}. \tag{5.16}$$

Furthermore, it is clear that if $\dot{\varepsilon}$ tends to zero, the deviatoric stress tensor takes a zero-over-zero indeterminate form. For this drawback to be obviated, different kinds of regularization can be introduced. One of the most convenient methods replaces Eq. (5.16) by

$$\mathbf{s} = \frac{2}{3} \frac{\sigma_0}{\sqrt{\dot{\bar{\varepsilon}}^2 + \dot{\bar{\varepsilon}}_0^2}} \dot{\boldsymbol{\varepsilon}}, \tag{5.17}$$

which imposes that $\mathbf{s}$ tends to zero when $\dot{\varepsilon}$ does. Provided $\dot{\bar{\varepsilon}}_0$ is small enough, the difference between Eqs. (5.16) and (5.17) is small, except when $\dot{\bar{\varepsilon}} < \dot{\bar{\varepsilon}}_0$, for which $\mathbf{s}$ is small itself. This regularization can also be introduced for viscoplastic materials in order to avoid an infinite derivative of the equivalent stress with respect to $\dot{\bar{\varepsilon}}$, when $\dot{\bar{\varepsilon}}$ tends to zero (and a similar one can also be used for the friction law).

The elastoplastic approximation is more realistic, as it takes into account the elastic contribution to the total strain rate:

$$\dot{\boldsymbol{\varepsilon}} = \dot{\boldsymbol{\varepsilon}}^e + \dot{\boldsymbol{\varepsilon}}^p, \tag{5.18}$$

[10] R. H. Wagoner and J.-L. Chenot, *Fundamentals of Metal Forming* (Wiley, New York, 1997), Chaps. 5 and 7.

where the upper index is $e$ for the elastic contribution and $p$ for the plastic (or the viscoplastic) one.

This approach is necessary when we analyze processes in which the elastic effects play a significant role; for example, springback effects after deep drawing, residual stresses after cold forming, skin-pass at the end of sheet rolling. This formulation is more complicated, even when the small strain approximation is used, as the stress tensor is not an explicit function of the strain increment. Moreover, if large rotations occur in the process, the small strain formulation, derived from the material derivative of the stress tensor, is no longer appropriate because it violates the mechanical principle of objectivity. A new form of stress tensor differentiation must be introduced to satisfy objectivity; for isotropic solid materials, the Jauman derivative is the more often used.[11] Finally, despite the fact that in theory no indeterminate form of the stress tensor can arise for elastoplastic material, a regularization in the form suggested by Eq. (5.17) can be utilized to improve the numerical behavior, when both elastic and plastic strain-rate tensors are small.

From the following it is clear that, however more approximate, the rigid-plastic or purely viscoplastic approaches are simpler to implement in computer codes, and often they are more economical with respect to computer time, as the constitutive law is explicit and therefore requires no iteration at the element level. Often much larger steps can be solved in a stable manner, and the equilibrium equation can in some formulations be solved directly, rather than solving for balance update forces.

Another advantage of the first formulation is clear when stationary processes are considered. With the flow formulation the equations of a steady-state process are easy to write and to solve (iteratively), only one time. When elastoplastic or elastoviscoplastic processes are investigated, the solution of the stationary state is not easy and very few references are available in the literature for a direct procedure: most works use an incremental approach to reach the steady state as a limit case, but here the nonlinear equations are to be solved a rather large number of times, that is, for each increment.

## 5.3 Incremental, Rate, and Flow Formulations

This section is often viewed as connected to the previous one, linking elastoplastic (or elastic viscoplastic) constitutive equations to incremental formulation, and rigid-plastic or purely viscoplastic behavior to rate or flow formulation. In fact this linkage is only a historical artifact, as shown in this section.

### Incremental Approach

The incremental approach was the first attempt to model small elastic deformation of a solid by the finite-element method. The general procedure was outlined in Section 4.4 [Eqs. (4.31)–(4.41)]. The final system given by Eq. (4.40) or Eq. (4.41) is linear, the numerical solution of which is (at least in theory) very straightforward. By this procedure, the FE solution of the elastic problem is readily obtained by performing only one displacement increment. When the elastic problem is not linear, either because

[11] The Jauman derivative is defined by the formula $\mathrm{d}_J\sigma/\mathrm{d}t = \mathrm{d}\sigma/\mathrm{d}t + \omega\sigma - \sigma\omega$, where the material derivative of the stress tensor is used in the right-hand side of the above equation, and $\omega$ is the rate of rotation antisymmetrical tensor, the components of which are $\omega_{ij} = \frac{1}{2}(\partial v_i/\partial x_j - \partial v_j/\partial x_i)$.

of the material or geometry, the solution of the resulting nonlinear system requires an iterative procedure (as in Section 3.6 for a one-dimensional rod).

However, for time-dependent plastic or viscoplastic problems, because of the constitutive equation, it is no longer possible to solve the problem with only one increment. The total displacement $\mathbf{u}$ has to be decomposed into small increments $\Delta\mathbf{u}$, in order to follow the history of each material point. Moreover, because of the nonlinear behavior, the problem is linearized between the present value $\mathbf{u}$ of the displacement before an increment, and the incremented one $\mathbf{u}' = \mathbf{u} + \Delta\mathbf{u}$. The procedure is then always schematically the same:

1. An approximation of the constitutive equation is written in terms of the increment of strain derived from $\Delta\mathbf{u}$.
2. The virtual work equation is written for the new configuration and the new stress state.
3. The resulting equations are linearized and solved iteratively by the Newton–Raphson method (or a similar scheme).

Only the first step will be illustrated for the elastic perfectly plastic material, by a simple incremental form of the elastoplastic behavior. As a first approximation of Eq. (5.18), the total strain increment is decomposed into an elastic strain increment plus a plastic strain increment:

$$\Delta\varepsilon = \Delta\varepsilon^e + \Delta\varepsilon^p. \tag{5.19}$$

Hooke's law[12] is written for the elastic contribution, giving

$$\Delta\sigma = \lambda\Delta\theta^e\mathbf{1} + 2\mu\Delta\varepsilon^e = \mathbf{D}:\Delta\varepsilon^e. \tag{5.20}$$

If the yield function is denoted by $f$, the plastic condition at the beginning of the increment is $f(\sigma) \leq 0$, and at the end of the increment it is $f(\sigma + \Delta\sigma) \leq 0$. Three different cases must be investigated according to the value of the yield criterion at the beginning of the increment, and at the end, with a first guess where the stress increment is calculated as if the total strain increment were completely elastic. We put

$$\Delta\sigma^e = \lambda\Delta\theta\mathbf{1} + 2\mu\Delta\varepsilon. \tag{5.21}$$

Case 1: This is the case of pure elastic loading:

$$f(\sigma) \leq 0, \qquad f(\sigma + \Delta\sigma^e) \leq 0.$$

In this case our guess is realistic: the increment must be considered as entirely elastic, and we conclude that

$$\Delta\varepsilon^e = \Delta\varepsilon, \qquad \Delta\sigma^e = \Delta\sigma, \qquad \Delta\varepsilon^p = 0.$$

Case 2: This case corresponds to an elastoplastic loading:

$$f(\sigma) = 0, \qquad f(\sigma + \Delta\sigma^e) > 0.$$

A discretized form of the flow rule must be introduced; we choose here

$$\Delta\varepsilon^p = \Delta\lambda_p\frac{\partial f}{\partial\sigma}(\sigma), \qquad \Delta\lambda_p > 0, \tag{5.22a}$$

[12] Here we have denoted by $\mathbf{D}$ the four-rank elasticity tensor.

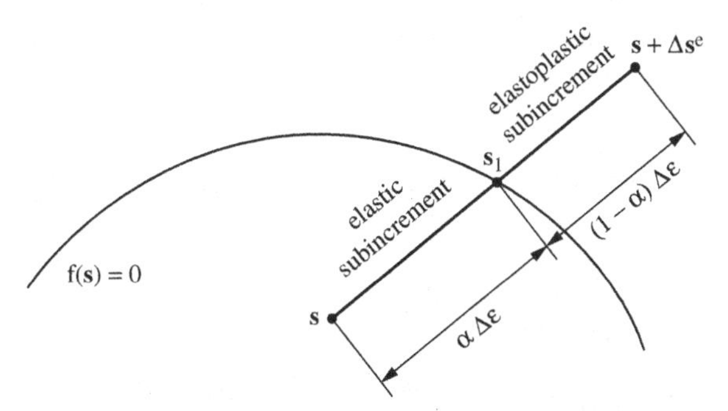


Figure 5.3. Elastic to elastoplastic transition during an increment (in the deviatoric stress tensor space).

with

$$f(\sigma + \Delta\sigma) = 0. \tag{5.22b}$$

This is the explicit form of the flow rule. An alternative form of Eq. (5.22a) is given by the implicit form[13]

$$\Delta\varepsilon^p = \Delta\lambda_p \frac{\partial f}{\partial \sigma}(\sigma + \Delta\sigma), \qquad \Delta\lambda_p > 0. \tag{5.23}$$

Case 3: We have to analyze a last situation, which corresponds to the elastic to elastoplastic transition. It is defined by

$$f(\sigma) < 0, \qquad f(\sigma + \Delta\sigma^e) > 0.$$

We shall admit that the first part of the increment corresponds to a purely elastic behavior; that is, we put

$$f(\sigma + \alpha\Delta\sigma^e) = 0. \tag{5.24}$$

Here $\alpha$ is the smallest root of Eq. (5.24) satisfying $0 < \alpha < 1$, Eq. (5.24) corresponds to a first elastic subincrement $\Delta\varepsilon_1 = \alpha\Delta\varepsilon$, and a second subincrement, $\Delta\varepsilon_2 = (1-\alpha)\Delta\varepsilon$, is elastoplastic and is therefore formally identical to Case 2. This is illustrated in Fig. 5.3.

The implicit scheme is generally preferred, as it always gives a unique solution for convex yield criteria and exhibits a more stable numerical behavior. Both schemes can be viewed as limit cases of the generalized trapezoidal rule, which is potentially more accurate and will be presented in Section 5.6.

---

**Exercise 5.2: For the von Mises yield criterion and the associated flow rule, give a graphical interpretation of Eq. (5.23) in the space of the deviatoric stress tensor (if we assume $2\mu = 1$). Examine the explicit and the implicit forms.**

In the deviatoric stress space, the von Mises yield function is a hypersphere with radius $\sqrt{\frac{2}{3}}\sigma_0$, if $\sigma_0$ is the uniaxial yield stress. The explicit case is illustrated in the left-hand figure below. If $\Delta\mathbf{e}^e$ is the deviatoric strain increment, with our hypothesis Hooke's law reduces to $\Delta\mathbf{s} = \Delta\mathbf{e}^e$ so that

$$\Delta e = \Delta e^e + \Delta e^p = \Delta s + \Delta\lambda_p \frac{\partial f}{\partial \mathbf{s}}(\mathbf{s}), \tag{5.2–1}$$

[13] This is equivalent to the "radial return" method.

which indicates that the deviatoric strain increment is decomposed into two "vectors," one being on the normal to the yield surface.

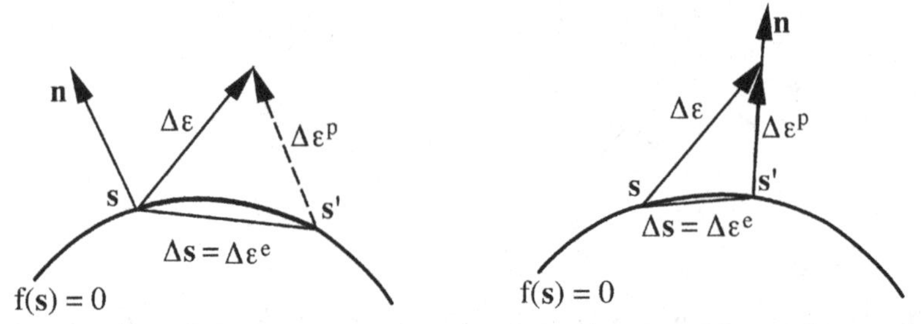


The implicit case follows immediately in the right-hand figure above, where the normal at the yield surface corresponds here to the end of the increment.

---

More complex constitutive equations can be analyzed and solved with an incremental formulation, but the major limitation of this approach lies in its inability to allow a more accurate integration scheme than the first order with respect to the time increment; this is especially true for the time evolution of the domain.

**The Flow Formulation**

Strictly speaking, the flow formulation is devoted to steady-state processes, when elasticity is neglected, and when the stress tensor is a function of the velocity field only, or more precisely of the strain-rate tensor: $\boldsymbol{\sigma} = \boldsymbol{\sigma}(\dot{\varepsilon})$. The virtual work principle[14] can be applied easily and gives, for any virtual velocity field $\mathbf{v}^*$,

$$\int_{\Omega} \boldsymbol{\sigma}(\dot{\varepsilon}) : \dot{\varepsilon}^* \, \mathrm{d}V - \int_{\partial\Omega^p} \mathbf{T}^d \cdot \mathbf{v}^* \, \mathrm{d}S = 0. \tag{5.25}$$

After discretization of the unknown velocity field, and of the virtual velocity, we finally get a nonlinear set of equations, the unknown of which is the vector of the nodal velocities $\mathbf{V}$:

$$\int_{\Omega} \boldsymbol{\sigma}(\dot{\varepsilon}) : \mathbf{B} \, \mathrm{d}V - \int_{\partial\Omega^p} \mathbf{T}^d \cdot \mathbf{N} \, \mathrm{d}S = 0. \tag{5.26a}$$

It is generally solved iteratively with the Newton–Raphson method. When the stress tensor derives from the strain-rate tensor through a viscoplastic potential $\varphi$,[15] by minimizing the discrete functional we obtain an expression similar to Eq. (5.26a):

$$\int_{\Omega} \frac{\partial \varphi}{\partial \dot{\varepsilon}}(\dot{\varepsilon}) : \mathbf{B} \, \mathrm{d}V - \int_{\partial\Omega^p} \mathbf{T}^d \cdot \mathbf{N} \, \mathrm{d}S = 0. \tag{5.26b}$$

In this case the tangential matrix used in the Newton–Raphson method is symmetric and comes from Eq. (5.26b) by differentiation with respect to $\mathbf{V}$:

$$\mathbf{K} = \int_{\Omega} \left( \frac{\partial^2 \varphi}{\partial \dot{\varepsilon}^2}(\dot{\varepsilon}) : \mathbf{B} \right) : \mathbf{B} \, \mathrm{d}V. \tag{5.27}$$

[14] R. H. Wagoner and J.-L. Chenot, *Fundamentals of Metal Forming* (Wiley, New York, 1997), Chap. 5, Sect. 4.

[15] R. H. Wagoner and J.-L. Chenot, *Fundamentals of Metal Forming* (Wiley, New York, 1997), Chap. 5, Sect. 7.

This method can be extended to non-steady-state problems, and most authors have done it with the most straightforward integration scheme of the material coordinates. At time $t$, $\Omega_t$ represents the domain, $\mathbf{X}^t$ the nodal position vectors, and $\mathbf{V}^t$ the nodal velocities. We suppose that $\boldsymbol{\Omega}_t$ and $\mathbf{X}^t$ are known; then the numerical resolution of Eq. (5.26a), or Eq. (5.26b), allows us to determine $\mathbf{V}^t$. The nodal position vectors are updated by the one-step forward Euler scheme:

$$\mathbf{X}^{t+\Delta t} = \mathbf{X}^t + \Delta t \mathbf{V}^t, \tag{5.28}$$

such that the new discretized domain $\boldsymbol{\Omega}_{t+\Delta t}$ is known, and the procedure can be repeated until the end of the process.

If the flow formulation is analyzed in more detail, we see immediately that, for nonstationary problems, it can be considered as a special case of the incremental formulation. To prove this assumption we have only to remark that Eq. (5.28) is

$$\Delta \mathbf{U}^t = \Delta t \mathbf{V}^t \tag{5.29}$$

and observe that the above method can be applied with the displacement vector as the major unknown, instead of the velocity. The major difference is the absence of elasticity, which eliminates the need for a local resolution of the constitutive law. However, both methods can in turn be seen as peculiar cases of a more general approach, which will now be surveyed, although it is not common for solid mechanics analysis.

### More General Velocity Formulations

Let us first consider briefly a quasi-static problem in which inertia can be neglected. Here the assumption will be made that at time $t$, the domain $\Omega_t$, the vector of nodal positions $\mathbf{X}^t$, and the vector of nodal velocities $\mathbf{V}^t$ are known. The unknowns will be the same variables at time $t + \Delta t$, that is, $\Omega_{t+\Delta t}$, $\mathbf{X}^{t+\Delta t}$, and $\mathbf{V}^{t+\Delta t}$. It is possible to interpolate the velocity linearly by putting

$$\mathbf{V}^{t+\tau} = \mathbf{V}^t + \frac{\tau}{\Delta t}(\mathbf{V}^{t+\Delta t} - \mathbf{V}^t) \tag{5.30}$$

for any time $\tau$ lying in the interval $[0, \Delta t]$. During the same time interval, the trajectories of the nodes can be found by integration with respect to time $\tau$ of Eq. (5.30), which is the definition of the velocity:

$$\frac{\mathrm{d}\mathbf{X}^{t+\tau}}{\mathrm{d}\tau} = \mathbf{V}^{t+\tau}, \tag{5.31}$$

which immediately gives

$$\mathbf{X}^{t+\tau} = \mathbf{X}^t + \tau \mathbf{V}^t + \frac{\tau^2}{2\Delta t}(\mathbf{V}^{t+\Delta t} - \mathbf{V}^t). \tag{5.32}$$

Then, by putting $\tau = \Delta t$, we find the value of the vector of coordinates at the end of the increment is obviously

$$\mathbf{X}^{t+\Delta t} = \mathbf{X}^t + \frac{1}{2}\Delta t(\mathbf{V}^{t+\Delta t} + \mathbf{V}^t), \tag{5.33}$$

which is known as the trapezoidal rule for the time integration of the nodal positions, and which is asymptotically more accurate[16] than the scheme of Eq. (5.28). Without

[16] That means that the error caused by time discretization tends to zero faster when the time increment tends to zero.

entering into details for elastoplastic or elastic viscoplastic materials,[17] we remark that obviously the strain-rate tensor at the end of the increment depends only on $\mathbf{X}^{t+\Delta t}$ and $\mathbf{V}^{t+\Delta t}$. This can be verified by calculating the strain-rate tensor in the usual way:

$$\dot{\varepsilon}^{t+\Delta t} = \frac{1}{2}\left[\frac{\partial \mathbf{v}^{t+\Delta t}}{\partial \mathbf{x}^{t+\Delta t}} + \left(\frac{\partial \mathbf{v}^{t+\Delta t}}{\partial \mathbf{x}^{t+\Delta t}}\right)^T\right]. \tag{5.34}$$

We recall that the continuous vector functions $\mathbf{x}^{t+\Delta t}$ and $\mathbf{v}^{t+\Delta t}$ are interpolated linearly in terms of the corresponding nodal values and the FE shape functions, so that with Eqs. (5.33) and (5.34) we can put

$$\dot{\varepsilon}^{t+\Delta t} = \dot{\varepsilon}(\mathbf{X}^{t+\Delta t}, \mathbf{V}^{t+\Delta t}) = \dot{\varepsilon}(\mathbf{V}^{t+\Delta t}).$$

If we keep in mind the hypothesis that the stress tensor can be expressed with the strain-rate tensor only, that is, $\boldsymbol{\sigma}^{t+\Delta t} = \boldsymbol{\sigma}(\dot{\varepsilon}^{t+\Delta t})$, we come readily to the conclusion that the virtual work equation can be written formally on the configuration at the end of the increment:

$$\int_{\Omega_{t+\Delta t}} \boldsymbol{\sigma}(\mathbf{V}^{t+\Delta t}) : \dot{\varepsilon}^* \, dV - \int_{\partial\Omega^s_{t+\Delta t}} \mathbf{T}^{(t+\Delta t)d} \cdot \mathbf{v}^* \, dS = 0, \tag{5.35}$$

provided the virtual velocity $\mathbf{v}^*$ is also defined on the domain $\Omega_{t+\Delta t}$. It is evident that Eq. (5.35) is far more complicated than the form given by Eq. (5.25) written at time $t$. Nevertheless, if, with the help of Eq. (5.33), we choose $\mathbf{V}^{t+\Delta t}$ as the principal unknown, we see that the domain $\Omega_{t+\Delta t}$ depends itself on this unknown and Eq. (5.35) is nonlinear, but has to be solved with respect to $\mathbf{V}^{t+\Delta t}$ only. If we come back to Eq. (5.20), the nodal acceleration can also be calculated on the time interval by

$$\boldsymbol{\Gamma}^{t+\tau} = \frac{d\mathbf{V}^{t+\tau}}{d\tau} = \frac{1}{\Delta t}(\mathbf{V}^{t+\Delta t} - \mathbf{V}^t). \tag{5.36}$$

We observe that the acceleration vector is constant on the interval $[t, t + \Delta t]$ with this scheme. Also, when $\boldsymbol{\Gamma}^{t+\tau}$ is substituted in Eqs. (5.30) and (5.32), we obtain exactly the same expressions as those given respectively in Eqs. (5.10a) and (5.10b). In addition, the virtual work equation with the inertia term [Eq. (5.6)] can be written alternately on the domain $\Omega_{t+\Delta t}$ at the end of the increment.

## 5.4 Lagrangian Versus Eulerian Schemes

The more "natural" formulation is probably the Lagrangian one, in which the memory of the initial positions of the material points are kept during the whole deformation process. However, for stationary processes, in which the domain of interest can be considered as fixed, the Eulerian description is more convenient, mostly for problems in which the constitutive equation includes weak or negligible memory effects.

### The Lagrangian Formulations in the Finite-Element Method

We must distinguish between the total Lagrangian method, in which all of the problem is defined on the initial configuration $\Omega_0$ corresponding to the time origin

[17] See C. Bohatier and J.-L. Chenot, "Finite Element Formulation for Non Steady State Large Viscoplastic Deformation," *Int. J. Numer. Meth. Eng.* 21, 1697–1708 (1985).

$t = 0$; and the updated Lagrangian method, for which the reference configuration during a time increment $[t, t + \Delta t]$ is the domain at the beginning of the increment, $\Omega_t$.

The total Lagrangian method is not the easiest one to utilize in plasticity or in viscoplasticity, because these constitutive laws depend on each increment of the deformation; in other words, they are path dependent. Furthermore, there is no need to refer the current increment to a reference state far removed, and any attempt to do so carries significant mathematical complexity in the kinematics of large deformation. The total Lagrangian method seems more appropriate for large displacements and large deformations of purely elastic materials (such as rubber). In this case, because of the reversible nature of the problem, the path need not be known precisely, and the deformation process is best described physically with only the initial and the final configuration. As we are not concerned here by this kind of material, we shall focus ourselves on the updated Lagrangian method, for which several strategies can be used.

Let $\Omega_t$ be the configuration at the beginning of the increment and $\Omega_{t+\Delta t}$ the configuration at the end of the increment. We suppose that both domains are meshed and that any node $n$ of the second one with coordinates $\mathbf{X}^{t+\Delta t}$ is linked to the corresponding first one $\mathbf{X}^t$ by the displacement formula:

$$\mathbf{X}_n^{t+\Delta t} = \mathbf{X}_n^t + \Delta \mathbf{U}_n^t. \tag{5.37}$$

The strain increment $\Delta\varepsilon$ can be calculated with the spatial derivatives of the displacement increment $\Delta\mathbf{U}_n^t$, and the strain-rate tensor is approximated by

$$\dot{\varepsilon}^t = \frac{1}{\Delta t} \Delta \varepsilon^t. \tag{5.38}$$

The stress tensor at the end the increment $\boldsymbol{\sigma}^{t+\Delta t}$ can be evaluated with the constitutive equation, and the equilibrium equation will be satisfied in the weak sense on the configuration at the end of the increment:

$$\int_{\Omega_{t+\Delta t}} \boldsymbol{\sigma}^{t+\Delta t} : \dot{\varepsilon}^* \, \mathrm{d}V - \int_{\partial\Omega^s_{t+\Delta t}} \mathbf{T}^{(t+\Delta t)d} \cdot \mathbf{v}^* \, \mathrm{d}S = 0. \tag{5.39}$$

This formulation allows us to take into account large displacements during the increment. However, if the domain variation during the increment can be considered as small enough, Eq. (5.39) is often replaced by the approximation

$$\int_{\Omega_t} \boldsymbol{\sigma}^{t+\Delta t} : \dot{\varepsilon}^* \, \mathrm{d}V - \int_{\partial\Omega^s_t} \mathbf{T}^{(t+\Delta t)d} \cdot \mathbf{v}^* \, \mathrm{d}S = 0, \tag{5.40}$$

which is more easy to calculate as the domain of integration is known at time $t$. When additional field variables are added to the material description, several situations can be found.

First, the evolution of the variable is governed by a simple differential equation, and no coupling with the deformation process is introduced. An example is represented by the equivalent strain, when the material exhibits no strain hardening: the equation of evolution is

$$\frac{\mathrm{d}\bar{\varepsilon}^t}{\mathrm{d}t} = \dot{\bar{\varepsilon}}^t, \tag{5.41}$$

where the time derivative is a material derivative. Equation (5.41) can be discretized according to an explicit first-order scheme:

$$\Delta\bar{\varepsilon}^t = \dot{\bar{\varepsilon}}^t \Delta t. \tag{5.42}$$

It is important to note that the updated value of the equivalent strain,

$$\bar{\varepsilon}^{t+\Delta t} = \bar{\varepsilon}^t + \Delta\bar{\varepsilon}^t,$$

is given to the updated position of the node. This procedure is perfectly consistent with the material derivative concept, for which precisely the material point is followed during its trajectory.

Second, the evolution of a variable, which is again ruled by an ordinary differential equation but which involves coupling with the material flow. This is the case for the equivalent strain when strain hardening is introduced. To make the things simpler, we shall suppose that the stress tensor depends only on the displacement field and on the equivalent strain:

$$\boldsymbol{\sigma}^{t+\Delta t} = \boldsymbol{\sigma}(\Delta\boldsymbol{\varepsilon}^t, \bar{\varepsilon}^{t+\Delta t}). \tag{5.43}$$

If Eq. (5.43) is substituted into Eq. (5.39), and Eq. (5.42) is taken into account, we have coupled equations that must be solved simultaneously. In practice, for relatively small time increments, it is possible to solve these equations alternately; first Eq. (5.39) is solved with a constant value of $\boldsymbol{\varepsilon}^t$, and then the equivalent strain is incremented by Eq. (5.42). The process can be iterated for a given increment of time, but generally this is not necessary.

Third, for thermal coupling the heat equation is itself governed by a partial differential equation, and we shall see in Chapter 7 that for irreversible processes a source term must be added to take into account heat dissipated by deformation. Moreover, the constitutive equation of the material depends on temperature. When the coupling is not too strong, a simultaneous solution of the equilibrium equation and of the thermal equation can be replaced by an alternate scheme (see footnote 18 for applications to forging, and footnotes 19 and 20 for applications to sheet forming).

### The Euler Description

When the problem is stationary in a fixed domain $\Omega$, we can use the Euler description in which the velocity field defined in $\Omega$ depends only on space coordinates $\mathbf{v}(\mathbf{x})$. This velocity field is discretized in the usual way and the functional of the problem (if any) is minimized on the fixed domain $\Omega$. However, there are two cases of practical interest in which the Euler formulation induces difficulties for solving the complete problem. The calculation of a free surface will be examined first. We suppose that the

[18] J.-L. Chenot, "Three Dimensional Finite Element Modelling of the Forging Process," in *Computational Plasticity. Models, Software and Applications*, D. R. J. Owen et al., eds. (Pineridge Press, Swansea, 1989), pp. 793–816.

[19] R. H. Wagoner, Yong H. Kim, and Y. T. Keum, "3-D Sheet Forming Analysis Including the Effects of Strain Hardening, Rate Sensitivity, Anisotropy, Friction, Heat Generation, and Transfer," Advanced Technology of Plasticity, Japan Society of Technology of Plasticity, 1990, pp. 1751–1756.

[20] Y. H. Kim and R. H. Wagoner, "A 3-D Finite Element Method for Non-Isothermal Sheet Forming Processes," *Int. J. Mech. Sci.* **33**, 911–926 (1991).

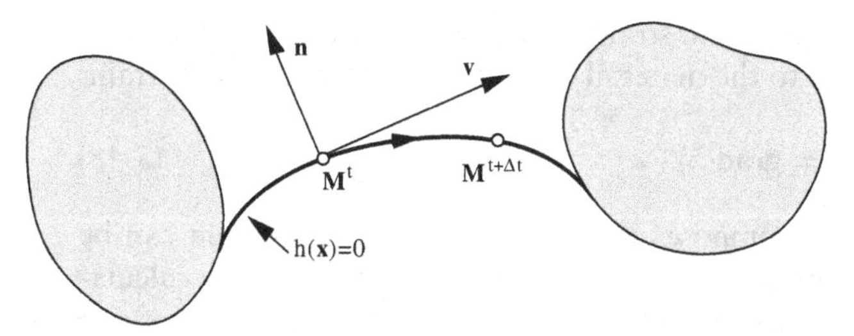


Figure 5.4. The free-surface kinematic condition.

free surface $\partial\Omega^F$ is an invariant part of the boundary $\partial\Omega$. If the external pressure can be neglected, the stress condition on the free surface will be written as

$$\mathrm{T} = \boldsymbol{\sigma} \cdot \mathbf{n} = 0, \tag{5.44}$$

where **n** is the normal to the free surface. The second condition is better expressed if we suppose that the equation of the free surface can be written as

$$h(\mathbf{x}) = 0, \tag{5.45}$$

which means that it does not depend explicitly on time. If **x** is a material point position vector, we have

$$\frac{\mathrm{d}\mathbf{x}}{\mathrm{d}t} = \mathbf{v},$$

and by time differentiation of Eq. (5.42) we see that

$$\frac{\partial h}{\partial \mathbf{x}}(\mathbf{x}) \cdot \mathbf{v} = 0 = \mathbf{n} \cdot \mathbf{v}, \tag{5.46}$$

where we have recalled that the vector $\partial h/\partial \mathbf{x}$ is proportional to the normal **n**. Moreover, Eq. (5.46) indicates that for a steady-state process, if a material point is on the free surface at any time, it remains on it until it reaches an obstacle (see Fig. 5.4). This is also the proof to the assumption that for a steady state the free surface is invariant.

From the following we deduce that the free surface can be determined by calculating the trajectories of the material points. This can be done approximately by writing the discretized equations as

$$\frac{\Delta x_1}{v_1} = \frac{\Delta x_2}{v_2} = \frac{\Delta x_3}{v_3} = \Delta t, \tag{5.47}$$

which is obviously another form of Eq. (5.28) used for transient problems. The numerical process is schematically illustrated in Fig. 5.5 for one material point with $M^0$ as the initial position.

The second problem concerns other physical parameters that represent a kind of memory of the material such as equivalent strain, and temperature. The simplest

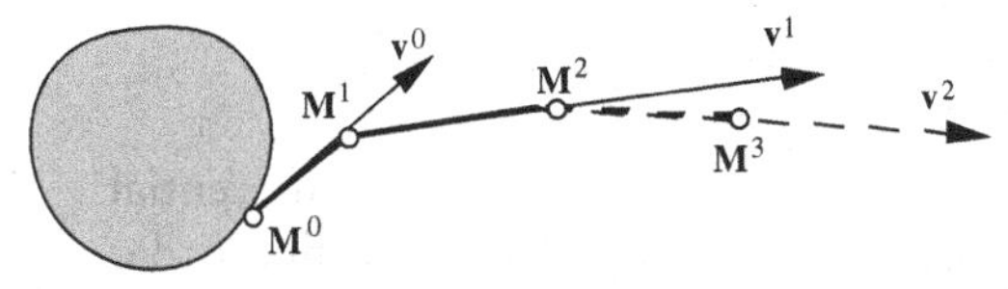


Figure 5.5. Numerical calculation of a trajectory on the free surface.

example will be outlined with the equivalent strain $\bar{\varepsilon}$. The basic equation is obtained by equating the equivalent strain rate to the material derivative of the equivalent strain:

$$\dot{\bar{\varepsilon}} = \frac{\mathrm{d}\bar{\varepsilon}}{\mathrm{d}t} = \frac{\partial \bar{\varepsilon}}{\partial t} + \mathbf{grad}(\bar{\varepsilon}) \cdot \mathbf{v} = \mathbf{grad}(\bar{\varepsilon}) \cdot \mathbf{v}, \tag{5.48}$$

where the assumed stationary state imposes to cancel $\partial\bar{\varepsilon}/\partial t$. This problem can be treated globally by a Galerkin approach; when $\dot{\bar{\varepsilon}}$ is known from the velocity calculation, $\bar{\varepsilon}$ can be deduced from the set of equations (one equation for each value of $n$)

$$\int_{\Omega} \mathbf{grad}(\bar{\varepsilon}) \cdot \mathbf{v} N_n \, \mathrm{d}V - \int_{\Omega} \dot{\bar{\varepsilon}} N_n \, \mathrm{d}V = 0, \tag{5.49}$$

and appropriate boundary conditions.

A simpler solution can be constructed when the trajectories are known (at least approximately). Along a trajectory, Eqs. (5.47) and (5.48) may be used to write

$$\dot{\bar{\varepsilon}}(\mathbf{x})\Delta t \cong \bar{\varepsilon}(\mathbf{x} + \Delta\mathbf{x}) - \bar{\varepsilon}(\mathbf{x}) \cong \mathbf{grad}(\bar{\varepsilon}) \cdot \Delta\mathbf{x}, \tag{5.50}$$

so that, if $\bar{\varepsilon}$ is known at a material point $M_0$ of the trajectory, this equation shows that it can be calculated progressively for any discretized point on the same trajectory, following $M_0$ in the flow.

#### An Alternative Formulation

In the previous subsections we presented two methods:

- the Lagrange method, in which the nodes of the mesh are moved as if they were material points, that is, with the kinematic displacements;
- the Euler representation, for which the nodes are fixed in space.

Both methods can be considered limit cases of a more general theory: the arbitrary Lagrange–Euler (ALE) method. In the ALE method we introduce a velocity of the mesh $\mathbf{v}_M$, which is allowed to be "arbitrary" with respect to the kinematic velocity field $\mathbf{v}$, except that it must preserve the boundary. We then have on $\partial\Omega$

$$(\mathbf{v} - \mathbf{v}_M) \cdot \mathbf{n} = 0, \tag{5.51}$$

which indicates that a boundary node must have a relative velocity that is tangential to the boundary. For any memory function, such as the equivalent strain $\bar{\varepsilon}$, the incremental update of the nodal values must be revised. For example, we shall change Eq. (5.42) to

$$\Delta\bar{\varepsilon}^t = \dot{\bar{\varepsilon}}^t \Delta t + \mathbf{grad}(\bar{\varepsilon}^t) \cdot (\mathbf{v}^t - \mathbf{v}_M^t)\Delta t. \tag{5.52}$$

The additional term in Eq. (5.52) is null if $\mathbf{v}^t = \mathbf{v}_M^t$, that is, for the Lagrangian formulation, and it corresponds to Eq. (5.50) if $\mathbf{v}_M^t = 0$, which is the Euler scheme.

Remark: numerical treatment of Eq. (5.52) can be done at the element level when $\bar{\varepsilon}$ is only defined at integration points.

## 5.5 Mixed Methods

Here we consider a physical problem that can be described by a partial differential equation with an order of 2, with second-order space derivatives. The first-order derivatives, or a function of these derivatives, generally possess a physical meaning.

This is obviously true for the mechanical and thermal problems we have in mind:

1. In thermal analysis the heat flux is proportional to the temperature gradient, or is derived from it by a linear transformation.
2. In continuum mechanics the stress tensor is a function of the displacement gradient, or of the velocity gradient.

These physical variables can be introduced as additional, or secondary unknowns to the primary unknowns. To formulate the mathematical problem properly, we must impose two kinds of physical conditions: a conservation law, and a constitutive law linking primary and secondary unknowns. In practice these two conditions can be specified either with the Galerkin method or by building a functional when it is possible. A special mention is deserved of the incompressible flow, to which a lot of work has been devoted.

### Galerkin Formulation of the Mixed Method

To illustrate the main point, we again consider the unsteady heat equation (stating energy conservation), with the simple form

$$\rho c\dot{T} = \mathrm{div}[k\,\mathbf{grad}(T)]. \tag{5.53}$$

The heat flux vector $\phi$, with components $(\phi_1, \phi_2, \phi_3)$, is now considered an unknown linked to the temperature gradient by the constitutive equation:

$$\phi = -k\,\mathbf{grad}(T). \tag{5.54}$$

This new relation allows us to transform the heat equation into

$$\rho c\dot{T} = -\mathrm{div}(\phi). \tag{5.55}$$

Both unknowns $T$ and $\phi$ can be discretized in terms of shape functions $N_n$ and $Q_q$, which need not be the same:

$$T = \sum_n T_n N_n, \tag{5.56a}$$

$$\phi = \sum_q \Phi_q Q_q. \tag{5.56b}$$

The time derivative $\dot{T}$ is also discretized, in a similar way as $T$ in Eq. (5.56a).

To obtain the explicit form of the discretized equations, we shall proceed as follows. Equation (5.55) is multiplied by any shape function $N_{n'}$, the discretized expansions of $\dot{T}$ and $\phi$ in Eqs. (5.56a) and (5.56b) are substituted, and the resulting expression is integrated over the domain $\Omega$, to give the first set of Galerkin equations:

$$\int_\Omega pc\left(\sum_n \dot{T}_n N_n\right) N_{n'}\,\mathrm{d}V + \int_\Omega \mathrm{div}\left(\sum_q \Phi_q Q_q\right) N_{n'}\,\mathrm{d}V = 0. \tag{5.57a}$$

A similar transformation is applied to Eq. (5.54), after having multiplied it by any $Q_{q'}$:

$$\int_\Omega k\,\mathbf{grad}\left(\sum_n T_n N_n\right) Q_{q'}\,\mathrm{d}V + \int_\Omega \left(\sum_q \Phi_q Q_q\right) Q_{q'}\,\mathrm{d}V = 0. \tag{5.57b}$$

**Exercise 5.3: Put Eqs. (5.57a) and (5.57b) in matrix form. Show that, if the shape functions $N_n$ and $Q_q$ are identical and a simple change of unknown is made, then the matrices are symmetric.**

The heat capacity matrix C was already introduced in Eq. (4.46). A matrix **M** is defined with

$$M_{iqi'q'} = \delta_{ij} \int_\Omega Q_q Q_{q'} \, \mathrm{d}V. \tag{5.3–1}$$

In Eq. (5.57a), the divergence operator is linear and will be represented by the matrix D, the components of which are

$$D_{n'iq} = \int_\Omega N_{n'} \frac{\partial Q_q}{\partial x_i} \, \mathrm{d}V \tag{5.3–2}$$

and the gradient operator in Eq. (5.57b) by the matrix **G** so that

$$G_{iq'n} = \int_\Omega Q_{q'} \frac{\partial N_n}{\partial x_i} \, \mathrm{d}V. \tag{5.3–3}$$

With this notation, the global system involving the nodal temperature array $T$, its time derivative $\dot{T}$, and the heat flux array $\mathbf{\Phi}$ (which is not necessarily composed of the nodal values) can be written with the following matrix form:

$$\begin{bmatrix} \mathbf{C} & \mathbf{0} \\ \mathbf{0} & \mathbf{0} \end{bmatrix} \begin{bmatrix} \dot{\mathbf{T}} \\ \mathbf{0} \end{bmatrix} + \begin{bmatrix} \mathbf{0} & \mathbf{D} \\ \mathrm{k}\mathbf{G} & \mathbf{M} \end{bmatrix} \begin{bmatrix} \mathbf{T} \\ \mathbf{\Phi} \end{bmatrix} = \begin{bmatrix} \mathbf{0} \\ \mathbf{0} \end{bmatrix}. \tag{5.3–4}$$

To this equation, appropriate boundary conditions must be imposed on $T$ and $\dot{T}$. When the shape functions $N$ and $Q$ are identical, Eqs. (5.3–2) and (5.3–3) show that **G** is the transpose of **D**, that is, $\mathbf{G} = \mathbf{D}^T$. We have now simply to remark that if Eq. (5.54) is replaced by $\phi' = -\mathbf{grad}(T)$, then the **D** matrix is replaced by $k\mathbf{D}$ in Eq. (5.3–4), so that the second "super" matrix is also symmetric.

### Functional Formulation of the Mixed Problem

Considering a linear elastic problem defined in a domain $\Omega$, we find the unknowns will be the displacement field **u** and the stress tensor field $\boldsymbol{\sigma}$. A functional $J$ of these two variables is introduced:

$$J(\mathbf{u}, \boldsymbol{\sigma}) = \int_\Omega \boldsymbol{\sigma} : \boldsymbol{\varepsilon}(\mathbf{u}) \, \mathrm{d}V - \frac{1}{2} \int_\Omega \boldsymbol{\sigma} : \mathbf{D}^{-1} : \boldsymbol{\sigma} \, \mathrm{d}V - \int_{\partial\Omega^s} \mathbf{T}^d \cdot \mathbf{u} \, \mathrm{d}S, \tag{5.58}$$

where $\boldsymbol{\varepsilon}(\mathbf{u})$ stands for the (small) strain tensor associated with **u**, and $\mathbf{D}^{-1}$ is the inverse[21] of the fourth-rank elastic tensor **D** defined in Eq. (5.20), which allows us to express the constitutive equation by $\boldsymbol{\varepsilon} = \mathbf{D}^{-1} : \boldsymbol{\sigma}$.

**Exercise 5.4: With a generalization of the formal variational calculation outlined,[22] show that Eq. (5.58) corresponds to the elastic problem.**

According to the method given in footnote 23, we put by definition

$$F = \boldsymbol{\sigma} : \boldsymbol{\varepsilon}(\mathbf{u}) - \frac{1}{2} \boldsymbol{\sigma} : \mathbf{D}^{-1} : \boldsymbol{\sigma}, \quad f = \mathbf{T}^d \cdot \mathbf{u}, \tag{5.4–1}$$

[21] The fourth-rank compliance tensor $\mathbf{D}^{-1}$ is also denoted by **S**.

[22] R. H. Wagoner and J.-L. Chenot, *Fundamentals of Metal Forming* (Wiley, New York, 1997), Chap. 5, Sect. 5.

[23] R. H. Wagoner and J.-L. Chenot, *Fundamentals of Metal Forming* (Wiley, New York, 1997), Problem 5.10.

so that we get by differentiation with respect to each component of the stress tensor

$$\frac{\partial \mathbf{F}}{\partial \sigma} = \varepsilon(\mathbf{u}) - \mathbf{D}^{-1} : \sigma = 0, \tag{5.4–2}$$

which is the constitutive equation. For the displacement, we remark first that only its derivatives are present in $F$; then for each component $u_i$,

$$\sum_j \frac{\partial}{\partial x_j}\left(\frac{\partial F}{\partial u_{i,j}}\right) = \sum_j \frac{\partial \sigma_{ij}}{\partial x_j} = 0, \qquad \text{where } u_{i,j} = \frac{\partial u_j}{\partial x_j}, \tag{5.4–3}$$

which is the equilibrium equation. Here the boundary condition is obtained by equating

$$\sum_j \frac{\partial F}{\partial u_{i,j}} n_j = \sum_j \sigma_{ij} n_j, \quad \frac{\partial f}{\partial u_i} = T_i^d. \tag{5.4–4}$$

On the part of the boundary where the displacement is not prescribed, this last relation is equivalent to the stress vector condition:

$$\sigma \cdot \mathbf{n} = \mathbf{T} = \mathbf{T}^d. \tag{5.4–5}$$

We observe that the stress boundary condition is verified with the functional form of Eq. (5.58), but the displacement must be imposed in addition:

$$\mathbf{u} = \mathbf{u}^d \text{ on } \partial\Omega^u. \tag{5.4–6}$$

---

Now the finite-element discretization procedure can proceed in the usual way by introducing the nodal displacement array $\mathbf{U}$ and the nodal stress array $\Sigma$:

$$\mathbf{u} = \mathbf{N} \cdot \mathbf{U} = \sum_n \mathbf{U}_n N_n,$$

$$\sigma = \mathbf{Q} \cdot \Sigma = \sum_q \Sigma_q Q_q.$$

The discrete form of Eq. (5.58) is given by substituting the previous FE expressions, giving

$$J(\mathbf{U}, \Sigma) = \sum_{ijqn} \left(\int_\Omega Q_q \frac{\partial N_n}{\partial x_j}\, \mathrm{d}V\right) \Sigma_{ijq} \mathbf{U}_{in}$$
$$-\frac{1}{2} \sum_{ijqi'j'q'} \left(\int_\Omega Q_q D^{-1}_{iji'j'} Q_{q'}\, \mathrm{d}V\right) \Sigma_{ijq} \Sigma_{i'j'q'} - \sum_{in} \left(\int_{\partial\Omega^s} T_i^d \cdot N_n\, \mathrm{d}S\right) U_{in}. \tag{5.59}$$

To obtain the solution of the elastic problem with the mixed formulation, we must equate to zero the derivative of the discretized functional [Eq. (5.59)], with respect to any component of the nodal unknowns:

$$\forall i, n \frac{\partial J(\mathbf{U}, \Sigma)}{\partial U_{in}} = 0, \quad \forall k, j, q \frac{\partial J(u, \Sigma)}{\partial \Sigma_{kjq}} = 0. \tag{5.60}$$

After this calculation we get a symmetric linear system, which can be denoted symbolically:

$$\begin{bmatrix} 0 & \mathrm{C} \\ \mathrm{C}^T & \mathrm{E} \end{bmatrix} \begin{bmatrix} \mathrm{U} \\ \Sigma \end{bmatrix} = \begin{bmatrix} \mathrm{S} \\ 0 \end{bmatrix}. \tag{5.61}$$

The **C** and **E** matrices are readily obtained by a simple identification of Eqs. (5.60) and (5.61):

$$C_{inkjq} = \delta_{ik} \int_\Omega Q_q \frac{\partial N_n}{\partial x_j} \, \mathrm{d}V,$$

$$E_{ijqi'j'q'} = \int_\Omega Q_q D^{-1}_{iji'j'} Q_{q'} \, \mathrm{d}V.$$

The load vector **S** comes from the surface integral of Eq. (5.59) and is calculated in the usual way.

---

**Exercise 5.5: Show that the functional defined by**

$$\mathbf{K}(\mathbf{v}, \sigma) = \int_\Omega \sigma : \dot{\varepsilon}(\mathbf{v}) \, \mathrm{d}V - \int_\Omega \varphi'(\sigma) \, \mathrm{d}V - \int_{\partial\Omega^s} \mathbf{T}^d \cdot \mathbf{v} \, \mathrm{d}S \qquad (5.5\text{–}1)$$

**corresponds to a mixed formulation of the viscoplastic problem, provided $\varphi'$ is the complementary viscoplastic potential, so that**

$$\dot{\varepsilon} = \frac{\partial \varphi'}{\partial \sigma}(\sigma).$$

As in Exercise 5.4 the volume integrand is put equal to $F$:

$$F(\mathbf{v}, \sigma) = \sigma : \dot{\varepsilon}(\mathbf{v}) - \varphi'(\sigma). \qquad (5.5\text{–}2)$$

With this form, we observe that differentiating with respect to the stress tensor, and equating to 0, is equivalent to the constitutive equation[24]

$$\frac{\partial F(\mathbf{v}, \sigma)}{\partial \sigma} = 0 = \dot{\varepsilon}(\mathbf{v}) - \frac{\partial \varphi'(\sigma)}{\partial \sigma}. \qquad (5.5\text{–}3)$$

The end of the proof proceeds exactly as in Exercise 5.4, if the function $J$ is replaced by $K$, and the displacement **u** by the velocity **v**. The incompressible case is analyzed in the next subsection.

---

### The Mixed Method for Incompressible Flow

Special attention is devoted here to the enforcement of the incompressibility condition by the mixed method for viscoplastic materials, for which a viscoplastic potential $\varphi$ can be defined (see footnote 25). The deviatoric stress tensor is obtained by

$$\mathrm{s} = \frac{\partial \varphi(\dot{\varepsilon})}{\partial \dot{\varepsilon}}, \quad \text{with } tr(\dot{\varepsilon}) = \mathrm{div}(\mathbf{v}) = 0. \qquad (5.62)$$

This issue has been addressed in a large number of publications, both from the theoretical point of view and from the computational, or engineering aspect; only a brief introduction can be given here. Velocity field **v** and hydrostatic pressure $p$ variables are introduced as unknowns, and the functional is

$$K_1(\mathbf{v}, p) = \int_\Omega \varphi(\dot{\varepsilon}) \, \mathrm{d}V - \int_\Omega p \, \mathrm{div}(\mathbf{v}) \, \mathrm{d}V - \int_{\partial\Omega^s} \mathbf{T}^d \cdot \mathbf{v} \, \mathrm{d}S. \qquad (5.63)$$

The derivation is only summarized here, as the procedure is analogous to that used in the other cases. With the help of (Eq. 5.62), the partial differential equation of the

[24] R. H. Wagoner and J.-L. Chenot, *Fundamentals of Metal Forming* (Wiley, New York, 1997).
[25] R. H. Wagoner and J.-L. Chenot, *Fundamentals of Metal Forming* (Wiley, New York, 1997), Section 5.7 and Eqs. 5.106, 5.109.

problem is easily seen to be

$$\sum_j \frac{\partial}{\partial x_j}\left(\frac{\partial \varphi(\dot{\varepsilon})}{\partial \dot{\varepsilon}_{ij}} - p\delta_{ij}\right) = \mathrm{div}(\mathbf{s} - p\mathbf{1}) = \mathrm{div}(\boldsymbol{\sigma}) = 0, \tag{5.64}$$

which is the equilibrium equation. By differentiation with respect to $p$ of the functional in Eq. (5.63), the incompressibility condition is obtained. It is important to note here that the velocity field need not to be a priori incompressible, it appears as a result of the calculation. Finally, the boundary stress condition is also obtained as in the previous examples.

The velocity field is discretized according to Eq. (5.2b), whereas the pressure field is generally expressed in terms of different shape functions:

$$p = \sum_q P_q Q_q \quad \text{or } p = \mathbf{P}\cdot\mathbf{Q}. \tag{5.65}$$

However, the shape functions $N_n$ for $\mathbf{v}$ and $Q_q$ for $p$ cannot be chosen arbitrarily.[26] The discretized equations are obtained by substitution of the velocity and pressure fields into Eq. (5.63) and differentiation with respect to the velocity components:

$$\int_\Omega \mathbf{s}:\mathbf{B}\,\mathrm{d}V - \int_\Omega p\,\mathrm{trace}(\mathbf{B})\,\mathrm{d}V - \int_{\partial\Omega^s} \mathbf{T}^d\cdot\mathbf{N}\,\mathrm{d}S = 0, \tag{5.66a}$$

and with respect to the pressure components:

$$\int_\Omega \mathrm{div}(\mathbf{v})\mathbf{Q}\,\mathrm{d}V = 0. \tag{5.66b}$$

The indicial forms of Eqs. (5.66a) and (5.66b) are respectively:

$$\forall n, k \int_\Omega \sum_{ij} S_{ij} B_{ijnk}\,\mathrm{d}V - \int_\Omega p \sum_i B_{iink}\,\mathrm{d}V - \int_{\partial\Omega^s} T_k^d N_n\,\mathrm{d}S = 0, \tag{5.67a}$$

$$\forall q \int_\Omega \mathrm{div}(\mathbf{v}) Q_q\,\mathrm{d}V = 0. \tag{5.67b}$$

These equations can be summarized in symbolic form as follows:

$$\mathbf{R}(\mathbf{X}, \mathbf{V}, \mathbf{P}) = 0. \tag{5.68}$$

These equations can be solved simultaneously with respect to all the unknowns: the velocity vector $\mathbf{V}$ and the pressure vector $\mathbf{P}$, by a classical Newton–Raphson iterative method.

## 5.6 Material Integration Schemes

The problem can be stated as follows: given the stress at the beginning of the increment and the strain increment, find the stress increment, or at least a good approximation of it, using the material constitutive equation. Even though the total process is generally decomposed into rather small increments, for nonlinear plastic or viscoplastic

[26] The choice of "good" shape functions involves mathematical developments that are outside the scope of this book. Some examples of choices will be given in the applications chapters.

problems, some care must be taken to transform the rate form of the material behavior into an acceptable incremental law. We shall first consider a rather general case in which the constitutive equation takes the form

$$\dot{\sigma}(t) = \mathrm{S}[\sigma(t), \dot{\varepsilon}(t), \bar{\varepsilon}(t)]. \tag{5.69}$$

The exact incremental form, which cannot be evaluated analytically in general, can be written as

$$\begin{aligned}\sigma(t+\Delta t) &= \sigma(t) + \int_t^{t+\Delta t} \dot{\sigma}(\tau)\,\mathrm{d}\tau \\ &= \sigma(t) + \int_0^{\Delta t} \mathrm{S}[\sigma(t+\theta), \dot{\varepsilon}(t+\theta), \bar{\varepsilon}(t+\theta)]\,\mathrm{d}\theta.\end{aligned} \tag{5.70}$$

**Explicit Integration Scheme**

The simplest approximation is obtained by considering that all the variables are approximately constant during the time increment, and equal to their initial value. With this hypothesis, Eq. (5.70) immediately gives

$$\sigma(t+\Delta t) = \sigma(t) + \mathrm{S}[\sigma(t), \dot{\varepsilon}(t), \bar{\varepsilon}(t)]\Delta t. \tag{5.71}$$

This material integration formula is convenient for coding, but it may lack stability, as is often the case for explicit schemes.

**Implicit Integration Scheme**

The basic hypothesis of this scheme is analogous to the previous one, except that the variables are given their value at the end of the increment; thus Eq. (5.70) may be rewritten as

$$\sigma(t+\Delta t) = \sigma(t) + \mathrm{S}\{\sigma(t+\Delta t), \dot{\varepsilon}[(t+\Delta t), \bar{\varepsilon}(t+\Delta t)]\}\Delta t. \tag{5.72}$$

If the strain-rate tensor is assumed be constant during the time increment, and equal to $\Delta\varepsilon/\Delta t$, the stress tensor $\sigma(t+\Delta t)$ at the end of increment cannot be obtained immediately, as it appear at both sides of Eq. (5.72) and then necessitates the resolution of a nonlinear system. However, although computationally more complicated, the implicit integration scheme is widely preferred when numerical instability can occur.

**The Generalized Trapezoidal Scheme**

A semi-implicit, two-level scheme can be proposed as a weighted compromise between the two previous ones:

$$\begin{aligned}\sigma(t+\Delta t) = \sigma(t) &+ (1-\eta)\mathrm{S}[\sigma(t), \dot{\varepsilon}(t), \bar{\varepsilon}(t)]\Delta t \\ &+ \eta\mathrm{S}[\sigma(t+\Delta t), \dot{\varepsilon}(t+\Delta t), \bar{\varepsilon}(t+\Delta t)]\Delta t,\end{aligned} \tag{5.73}$$

where $\eta$ is a numerical parameter ranging from 0 to 1, and the best accuracy (for small increments) is generally obtained for $\eta = 0.5$. This integration scheme involves the mechanical parameters at the beginning and at the end of the increment, and it is again implicit with respect to the stress tensor. An alternative form of this scheme can be written as

$$\sigma(t+\Delta t) = \sigma(t) + \mathrm{S}\{\sigma(t+\eta\Delta t), \dot{\varepsilon}(t+\eta\Delta t), \bar{\varepsilon}(t+\eta\Delta t)\}\Delta t. \tag{5.74}$$

Even more precise schemes can be proposed by using several integration subintervals in the interval $[t, t+\Delta t]$; these schemes are referred as substepping methods and will not be discussed further.

---

**Exercise 5.6: For the simple case of an elastic viscoplastic material obeying the Norton–Hoff law with no strain hardening, give the forms of the material integration scheme. Examine the case when a displacement formulation is used. The viscoplastic strain-rate tensor will be expressed by**

$$\dot{\varepsilon}^p = \frac{\partial \varphi'}{\partial \sigma} = \frac{1}{2K}\left(\frac{\bar{\sigma}}{\sqrt{3}K}\right)^{\frac{1}{m}-1} \mathbf{s}. \tag{5.6–1}$$

The additive decomposition of the strain-rate tensor is recalled [from Eq. (5.18)]:

$$\dot{\varepsilon} = \dot{\varepsilon}^e + \dot{\varepsilon}^p,$$

where the elastic contribution must obey

$$\dot{\sigma} = \lambda_e \dot{\theta}^e \mathbf{1} + 2\mu \dot{\varepsilon}^e = \mathbf{D}_e : \dot{\varepsilon}^e. \tag{5.6–2}$$

The viscoplastic part is given by Eq. (5.6); thus we can write

$$\dot{\sigma} = \mathbf{D}_e : \left[\dot{\varepsilon} - \frac{1}{2K}\left(\frac{\bar{\sigma}}{\sqrt{3}K}\right)^{\frac{1}{m}-1} \mathbf{s}\right] \equiv \mathbf{S}(\dot{\varepsilon}, \sigma). \tag{5.6–3}$$

The explicit material integration scheme is derived from Eq. (5.67) (here no dependence on the equivalent strain is assumed):

$$\sigma(t+\Delta t) = \sigma(t) + \left\{\lambda_e \dot{\theta}^e(t)\mathbf{1} + 2\mu_e\left[\dot{\varepsilon}(t) - \frac{1}{2K}\left(\frac{\bar{\sigma}(t)}{\sqrt{3}K}\right)^{\frac{1}{m}-1} \mathbf{s}(t)\right]\right\}\Delta t, \tag{5.6–4}$$

where we used the following relations:

$$\dot{\theta}^p = tr(\dot{\varepsilon}^p) = 0, \quad \dot{\theta} = tr(\dot{\varepsilon}) = tr(\dot{\varepsilon}^e) = \dot{\theta}^e. \tag{5.6–5}$$

In the same way the implicit material integration scheme is

$$\sigma(t+\Delta t) = \sigma(t) + \left\{\lambda_e \dot{\theta}^e(t+\Delta t)\mathbf{1} + 2\mu_e\left[\dot{\varepsilon}(t+\Delta t) - \frac{1}{2K}\left(\frac{\bar{\sigma}(t+\Delta t)}{\sqrt{3}K}\right)^{\frac{1}{m}-1} \mathbf{s}(t+\Delta t)\right]\right\}\Delta t, \tag{5.6–6}$$

and the semi-implicit scheme will be written as

$$\sigma(t+\Delta t) = \sigma(t) + \frac{1}{2}\lambda_e[\dot{\theta}^e(t) + \dot{\theta}^e(t+\Delta t)]\mathbf{1}\Delta t + \mu_e\left(\dot{\varepsilon}(t) + \dot{\varepsilon}(t+\Delta t) - \frac{1}{2K}\left\{\left[\frac{\bar{\sigma}(t)}{\sqrt{3}K}\right]^{\frac{1}{m}-1} \mathbf{s}(t) + \left[\frac{\bar{\sigma}(t+\Delta t)}{\sqrt{3}K}\right]^{\frac{1}{m}-1} \mathbf{s}(t+\Delta t)\right\}\right)\Delta t. \tag{5.6–7}$$

---

These integration schemes are consistent and satisfactory when the constitutive law can be expressed with a single equation of the general form given by Eq. (5.69). This is not the case for elastoplasticity, which deserves special attention. The main differences are clear when it is recalled that as soon as the yield criterion is reached the purely elastic deformation turns into an elastoplastic regime, and that the yield criterion must not be violated. The previous analysis is thus still valid for the purely elastic increments. When plasticity is reached, the change of constitutive equation must be taken into account as well as the yield criterion constraint. We shall first consider the case of a purely elastoplastic increment, in order to have a law of the form given by Eq. (5.69). However, if Eq. (5.71) is applied without any change, we obtain

$$\sigma(t+\Delta t) = \sigma(t) + \lambda_e \dot{\theta}^e(t+\Delta t)\mathbf{1}\Delta t + 2\mu_e\left\{\dot{\varepsilon}(t+\Delta t) - \lambda_p(t+\Delta t)\frac{\partial f}{\partial \sigma}[\sigma(t+\Delta t), \bar{\varepsilon}(t+\Delta t)]\right\}\Delta t. \tag{5.75}$$

With this form, the yield criterion will be generally violated at the end of the increment. This drawback can be avoided by determining the plastic multiplier $\dot{\lambda}_p$ in such a way that

$$f[\sigma(t+\Delta t), \bar{\varepsilon}(t+\Delta t)] = 0. \tag{5.76}$$

This method can also be used in conjunction with the schemes given by Eq. (5.70), (5.72), or (5.73).[27] Finally, if during the increment there is a transition point between a purely elastic and an elastoplastic deformation, then the time increment is decomposed into two subincrements. In the same way as in Section 5.3, Case 3, the first subincrement will correspond to the elastic process, and the second one to the elastoplastic deformation; the previous procedures will be applied separately to each subincrement.

## PROBLEMS

### A. Proficiency Problems

1. Calculate the numerical values of the 24 components of the **B** operator for a plane strain problem and a unit linear triangle.
2. A higher-order integration scheme can be constructed from a linear interpolation of the acceleration field. To obtain the explicit time integration formulas, the nodal acceleration vector is interpolated in the interval $\tau \in [0, \Delta t]$ according to

$$\mathbf{\Gamma}^{t+\tau} = \mathbf{\Gamma}^t + \frac{\tau}{\Delta t}(\mathbf{\Gamma}^{t+\Delta t} - \mathbf{\Gamma}^t).$$

   Calculate the velocity and the nodal coordinate vectors at the end of the increment, as functions of their initial values at time $t$ and of the acceleration vectors at time $t$ and $t+\Delta t$.
3. A cylindrical sample of initial length $l = l_0$ and initial cross-sectional area $S = S_0$ is subjected to a uniaxial tension along its axis with a constant force $F$. The material of the sample is incompressible and viscoplastic according to the Norton–Hoff law $\sigma = \sigma_0 \dot{\varepsilon}^n$. Show that the evolution law for the length can be written:

$$\dot{\varepsilon} = \dot{l}/l = (\sigma : \sigma_0)^{1/m} = \left(\frac{Fl}{\sigma_0 l_0 S_0}\right)^{1/m}.$$

   Calculate the exact analytical solution of this differential equation in $l$. We assume that $l_0 = 0.1$ m, $S_0 = 10^{-4}$ m$^2$, $\sigma_0 = 100$ MPa s$^{-1}$, $F = 2 \times 10^3$ N, $m = 1/3$. Solve

[27] Note here that when the yield criterion is quadratic, Eqs. (5.69) and (5.70) will give the same result.

numerically the problem with a one-step explicit Euler scheme with $\Delta t = 3\,\text{s}$ for $0 < t < 30\,\text{s}$. Compare the analytical solution with the numerical one. Again compute a numerical solution for the same value of the time step for a one-step implicit Euler scheme, and a two-level semi-implicit scheme.

4. Consider an elastic perfectly plastic material with a yield stress $\sigma_0$ and Lamé coefficients $\lambda = 100\sigma_0$ and $\mu = 70\sigma_0$. The initial stress state $\boldsymbol{\sigma}$ and the strain increment $\Delta\boldsymbol{\varepsilon}$ are given by

$$\boldsymbol{\sigma} = \begin{bmatrix} 0.8\sigma_0 & \sqrt{3}/5\sigma_0 & 0 \\ \sqrt{3}/5\sigma_0 & 0.8\sigma_0 & 0 \\ 0 & 0 & 0 \end{bmatrix}, \quad \Delta\boldsymbol{\varepsilon} = \begin{bmatrix} 0.01 & 0 & 0 \\ 0 & 0 & 0 \\ 0 & 0 & 0.01 \end{bmatrix}.$$

Verify that the initial state corresponds to a plastic state and that the increment of deformation cannot be purely elastic. Compute the final stress and the elastic and plastic strain increments, using either the explicit or the implicit scheme. Discuss the results.

Perform the same calculation for the same initial stress, but for a strain increment given by

$$\Delta\boldsymbol{\varepsilon} = \begin{bmatrix} 0 & 0 & 1 \\ 0 & 0 & 0 \\ 0 & 0 & 0 \end{bmatrix}.$$

Perform the same calculation for the same strain increment, but for an initial stress given by

$$\boldsymbol{\sigma} = \begin{bmatrix} 0.8\sigma_0 & 0.3\sigma_0 & 0 \\ 0.3\sigma_0 & 0.8\sigma_0 & 0 \\ 0 & 0 & 0 \end{bmatrix}.$$

5. A velocity field is defined for $x_1 > 0$ and $x_2 > 0$ by the expressions for its components:

$$v_1 = \frac{1}{2}x_1^2, \quad v_2 = -x_1 x_2.$$

Determine first the analytical representation of the free-surface line starting at point $A$ with coordinates (1, 1) as pictured in the figure below, and compute the $x_2$ coordinate of point $B$ with coordinate $x_1 = 2$.

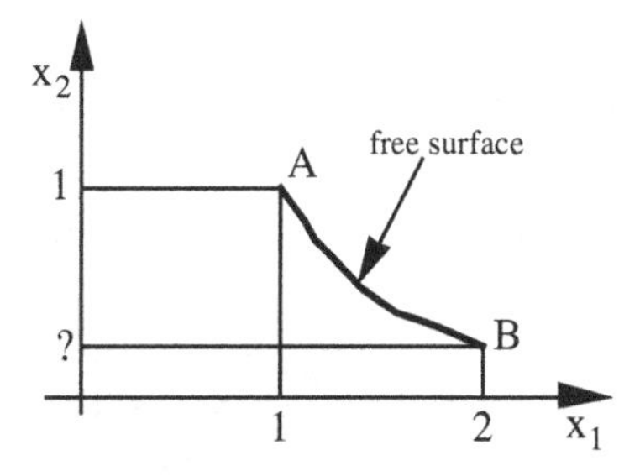


Figure P5.5.

Compute the free surface $AB$ numerically by using the scheme outlined in Section 5.4 (Eq. 5.47) with a constant increment $\Delta x_1 = 0.1$. Compare the $x_2$ coordinate of point $B$ for the analytical solution and for the numerical solution. Explain the result.

## B. Depth Problems

6. Again compute a numerical solution for Problem 1, with the same value of the time step for a one-step implicit Euler scheme, and a two-level semi-implicit scheme. To achieve this result a (simple) numerical method must be used to solve the implicit equations.

7. Compute the plastic energy for plane strain upsetting of an initially square workpiece, for a total reduction of height of 50% (see the figure below).

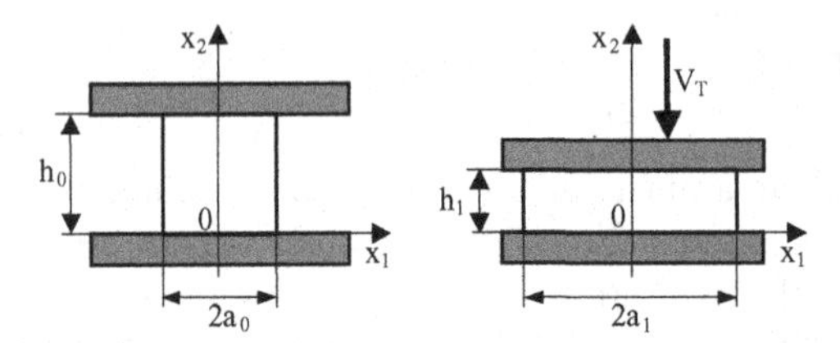


Figure P5.7.

A velocity field corresponding to a homogeneous strain-rate tensor will be assumed at each time step. Compute the corresponding change in kinetic energy, and compare it with the plastic energy. Assume that the initial side of the square workpiece is 0.1 m, the plastic yield stress is $\sigma_0 = 10^9$ Pa, the density of the material is 7600 Kg/m$^3$, and the velocity of the upper die is $V_T = 1$ m/s. In this case, show that the ratio between the change of kinetic energy and the plastic energy is less than $10^{-4}$.

8. A one-dimensional sample is subjected to a traction with a constant strain rate $\dot{\varepsilon}$. The constitutive equation of the material is elastic viscoplastic (Norton) with a rate sensitivity index of $m = 0.5$. Write the differential equation satisfied by the stress $\sigma$ and solve it for a zero initial stress. Compare the result with the purely viscoplastic material.

9. A one-dimensional viscoelastic sample, with a Young's modulus of $E$ and a viscosity of $\eta$, is subjected to a constant tension $F$ to one side and clamped at the other side. A mass $M$ is tied to the end of the sample; the mass of the latter will be neglected (see figure below).

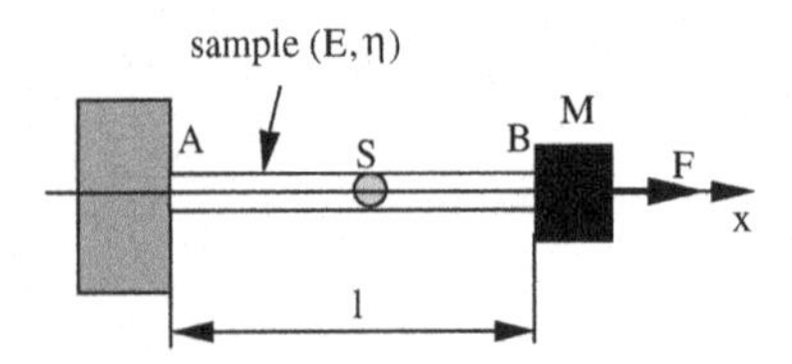


Figure P5.9.

The sample is supposed to be uniformly deformed along the $x$ axis. Write the dynamic equations for the point $B$ at the right-hand side of the sample. Show that there is one purely dynamic equation relating the second derivative of $l$, and a constitutive equation involving the stress in the sample (assumed to be constant along the $x$ axis) and the time derivative of the stress. Propose a numerical algorithm to integrate the differential equations in terms of a time increment $\Delta t$, where for a given length $l$ and its time derivative, the stress is updated, and when the stress is known the acceleration of point $B$ is computed so that $l$ and $\dot{l}$ are updated.

## C. Numerical and Computational Problems

10. Write a FORTRAN code that approximately solves the *elastoplastic* incremental equations for a given strain increment, using the three different schemes presented in Section 5.6.

11. Write a FORTRAN subroutine that approximately solves the *elastic viscoplastic* incremental equations for a given strain increment, using the three different schemes presented in Section 5.6.

12. Write a FORTRAN code to solve numerically Problem 9. The physical parameters defining the problem will be given the following values: $l_0 = 0.1, S_0 = 10^{-4}, M = 10^4, F = 10^4, E = 10^9, h = 10^{10}$. Compare the results for $t = 0$ to 1 with $\Delta t = 0.1$, 0.025, and 0.01.

CHAPTER SIX

# Auxiliary Equations: Contact, Friction, and Incompressibility

In this chapter we gathered the three main problems one faces in metal-forming simulation once the element and the constitutive formulation are chosen:

1. The contact problem is still among the most challenging ones, as it introduces a very stiff constraint, which results in a discontinuity of the velocity field at the onset of the contact.
2. The appropriate introduction of the friction law is also crucial in forming processes, as the material flow can be more sensitive to a small variation of the friction parameters than to a change of rheology.
3. The introduction in the finite-element formulation of the incompressibility requirement for dense materials, when elasticity is neglected. This topic is not straightforward from the engineering point of view, so it has motivated very theoretical works in the field of functional analysis; we shall not review these here as it is far beyond the scope of this book.

## 6.1 The Contact Problem

The contact problem with tools is probably the most challenging issue in modeling metal-forming processes. It is intrinsically a difficult mathematical problem, as it corresponds to a time discontinuity of the velocity when contact is established and, even in the stationary case, it involves unilateral conditions. Moreover if contact is not taken into account properly, or if the approximation is too crude, the final results of the computation are greatly affected and can be completely wrong. For simplicity we shall restrict ourselves here to rigid tools. The main methods that were tested in the literature were contact elements,[1] nodal approaches,[2,3] penalty formulations,[4–6] Lagrange multiplier methods,[7,8] and the CFS method.[9–11]

[1] A. Heege and P. Alart, "A Frictional Contact Element for Strongly Curved Contact Problems," *Int. J. Num. Meth. Eng.* 39, 165–184 (1996).

[2] J.-L. Chenot, "Finite Element Calculation of Unilateral Contact with Friction in Non Steady-State Processes," in *Numerical Techniques for Engineering Analysis and Design*, G. N. Pande and J. Middleton, eds. (Nijhoff, Dordrecht, 1987), T1, pp. 1–10.

[3] L. Fourment, K. Mocellin, and J.-L. Chenot, "An implicit contact algorithm for the 3D simulation of the forging process," in *Computational Plasticity – Fundamental and Applications, Proceedings of the Fifth International Conference*, D. R. J. Owen et al., eds. (Pineridge Press, Swansea, 1997), pp. 873–877.

[4] P. Papadopoulos, R. E. Jones, and J. M. Solberg, "A Novel F. E. Formulation for Frictionless Contact Problems," *Int. J. Num. Meth. Eng.* 38, 2603–2617 (1995).

### The Contact Problem Expressed in Terms of Velocity

Let us suppose that the tool in contact with the part (for which the domain is $\Omega$) can be represented by a function $g$ of space coordinates $\mathbf{x}$, and of time $t$ if it moves, such that

$$g(\mathbf{x}, t) = 0 \tag{6.1}$$

corresponding to the tool surface, and

$$g(\mathbf{x}, t) < 0 \tag{6.2}$$

is obeyed for any point in the interior of the tool. We know that the outer normal to the tool is then defined by

$$\mathbf{n} = \frac{1}{|\partial g/\partial \mathbf{x}|}\frac{\partial g}{\partial \mathbf{x}}. \tag{6.3}$$

At a given time $t$, we suppose that the tool velocity is $\mathbf{v}^{to}$ and the material velocity is $\mathbf{v}$ for a material point with coordinates $\mathbf{x}$ on the tool surface. The rate condition for the material point to glide on the tool surface can be written[12] as

$$(\mathbf{v} - \mathbf{v}^{to}) \cdot \mathbf{n} = \Delta\mathbf{v} \cdot \mathbf{n} = 0. \tag{6.4}$$

---

**Exercise 6.1: Give the proof of the tool-part contact condition expressed by Eq. (6.4).**

At time $t$ a material point with coordinates $\mathbf{x}'(t)$ is on the tool surface if

$$g[\mathbf{x}'(t), t] = 0. \tag{6.1–1}$$

At time $t + \mathrm{d}t$ the same material point on the tool surface has moved such that the new coordinate vector $\mathbf{x}'(t + \mathrm{d}t)$ will satisfy

$$g[\mathbf{x}'(t + \mathrm{d}t), t + \mathrm{d}t] = 0, \tag{6.1–2}$$

which is transformed with a first-order Taylor expansion and gives

$$0 = g[\mathbf{x}'(t + \mathrm{d}t), t + \mathrm{d}t] \cong g[\mathbf{x}'(t), t] + \frac{\partial g}{\partial \mathbf{x}}[\mathbf{x}'(t), t]\frac{\mathrm{d}\mathbf{x}'}{\mathrm{d}t}\mathrm{d}t + \frac{\partial g}{\partial t}[\mathbf{x}'(t), t]\,\mathrm{d}t. \tag{6.1–3}$$

[5] P. R. Dawson, D. E. Boyce, G. M. Eggert, and A. J. Baudoin, "A consistent penalty method for contact between a deforming viscoplastic workpiece and a rigid tool," *Int. J. Num. Meth. Eng.* 38, 3969–3987 (1995).

[6] G. Zavarise and R. L. Taylor, "A force method for contact problems with large penetration," in *Computational Plasticity – Fundamental and Applications, Proceedings of the Fifth International Conference*, D. R. J. Owen et al., eds. (Pineridge Press, Swansea, 1997), pp. 280–285.

[7] P. Wriggers, "Finite Element Algorithms for Contact Problems," *Archi. Comput. Meth. Eng.* 2, 4, 1–49 (1995).

[8] D. Pantuso and K. J. Bathe, "Finite element analysis of thermo-elastoplastic solids in contact," in *Computational Plasticity – Fundamental and Applications, Proceedings of the Fifth International Conference*, D. R. J. Owen et al., eds. (Pineridge Press, Swansea, 1997), pp. 72–87.

[9] M. J. Saran and R. H. Wagoner, "A Consistent Implicit Formulation for Nonlinear Finite Element Modeling with Contact and Friction, Part 1 – Theory," *ASME Trans. J. Appl. Mech.* 58, 499–506 (1991).

[10] M. J. Saran and R. H. Wagoner, "A Consistent Implicit Formulation for Nonlinear Finite Element Modeling with Contact and Friction, Part 2 – Numerical Verification and Results," *ASME Trans. J. Appl. Mech.* 58, 507–512 (1991).

[11] D. Zhou and R. H. Wagoner, *An N-CFS Algorithm for Sheet Forming FE Modelling, Metal Forming Process Simulation in Industry* (Druckerei GmbH, Monchengladbach, Germany, 1994), Vol. II, pp. 57–74.

[12] It can also be viewed as an infinitesimal distance of gliding during the time interval $\mathrm{d}t$ (see also Exercise 6.1).

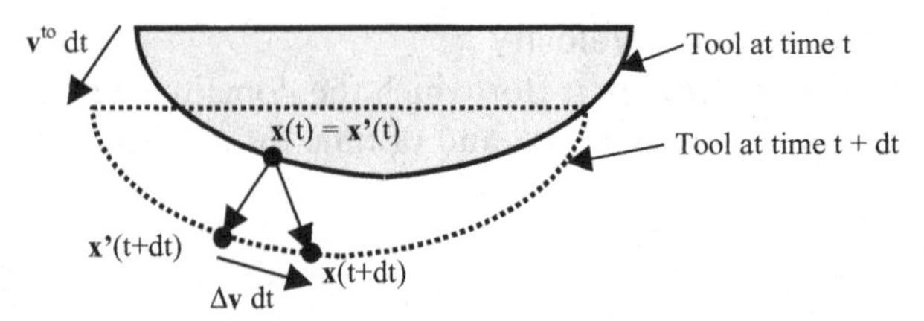


Figure 6.1. Analysis of contact.

By using Eq. (6.1) for the material point $\mathbf{x}'(t)$, the definition of the velocity, and dividing by d$t$, we transform Eq. (6.1–3) into

$$\frac{\partial g}{\partial \mathbf{x}}[\mathbf{x}'(t), t]\mathbf{v}^{to} + \frac{\partial g}{\partial t}[\mathbf{x}'(t), t] \cong 0. \tag{6.1–4}$$

If a material point of the part, with coordinates $\mathbf{x}(t)$, is moving on the tool surface, a similar derivation can be used to derive the following:

$$\frac{\partial g}{\partial \mathbf{x}}[\mathbf{x}(t), t]\mathbf{v} + \frac{\partial g}{\partial t}[\mathbf{x}(t), t] \cong 0. \tag{6.1–5}$$

Now we suppose that the geometric points have the same location at time $t$, that is, $\mathbf{x}(t) = \mathbf{x}(t')$, and we subtract Eq. (6.1–4) from Eq. (6.1–5), getting

$$\frac{\partial g}{\partial \mathbf{x}}[\mathbf{x}(t), t]\,(\mathbf{v} - \mathbf{v}^{to}) = 0. \tag{6.1–6}$$

This is equivalent to Eq. (6.4) if we recall the definition of the normal to the tool surface given by Eq. (6.3); obviously we must also put

$$\Delta\mathbf{v} = \mathbf{v} - \mathbf{v}^{to}. \tag{6.1–7}$$

The previous analysis is summarized in Fig. 6.1. For general curved surfaces, it is approximate when finite increments of time are considered, and it is exact for stationary states when only velocities are considered.

---

Equation (6.4) is not sufficient to describe all the possible physical situations that can occur in reality. In fact, on one hand we cannot allow for the material points of the part to penetrate in the tool; on the other hand, we must allow it to lose contact. That is mathematically expressed by (see Fig. 6.2)

$$\Delta\mathbf{v} \cdot \mathbf{n} \geq 0. \tag{6.5}$$

For a material point in contact with the tool, we need a physical criterion to decide whether Eq. (6.4) or Eq. (6.5) must be imposed. If we consider that adhesion between the tool and the part is negligible, that is, no surface traction is possible in the contact

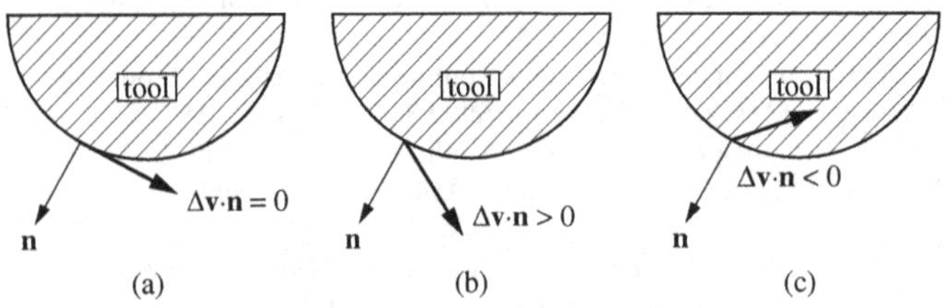


Figure 6.2. Contact conditions: (a) gliding contact, (b) loss of contact, and (c) nonphysical evolution.

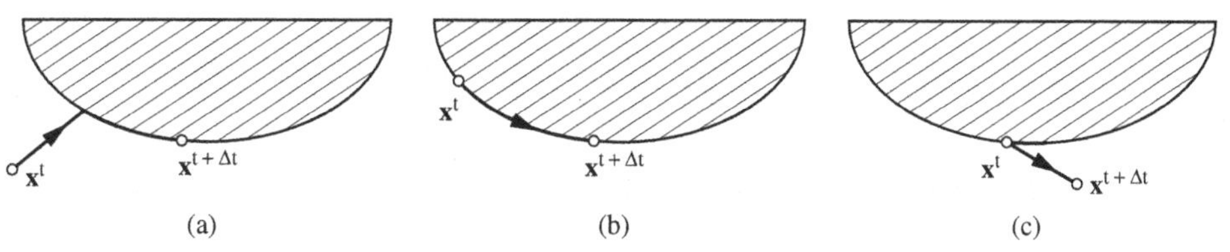


Figure 6.3. Incremental contact (the tool is supposed fixed here): (a) contact establishment, (b) gliding contact, and (c) loss of contact.

zone, we finally obtain the complete set of contact equations:

$$\text{if } \sigma_n = (\boldsymbol{\sigma} \cdot \mathbf{n}) \cdot \mathbf{n} < 0 \text{ then } \Delta\mathbf{v} \cdot \mathbf{n} = 0, \tag{6.5a}$$

$$\text{if } \sigma_n = (\boldsymbol{\sigma} \cdot \mathbf{n}) \cdot \mathbf{n} = 0 \text{ then } \Delta\mathbf{v} \cdot \mathbf{n} \geq 0, \tag{6.5b}$$

$$\sigma_n = (\boldsymbol{\sigma} \cdot \mathbf{n}) \cdot \mathbf{n} > 0 \text{ is impossible.} \tag{6.5c}$$

All the previous equations are valid for infinitesimal displacements, and they apply without change to stationary processes that can be described with the velocity field only, using an Eulerian reference system.

### The Incremental Contact Equations

When finite time intervals are considered, approximations must be made, which are often chosen in order to achieve a compromise between accuracy and programming effort. We shall only consider here a velocity formulation with a one-step time-integration method.[13] With these hypotheses we have:

$$\mathbf{x}^{t+\Delta t} = \mathbf{x}^t + \Delta t \mathbf{v}^t. \tag{6.6}$$

It can be immediately transformed to a displacement formulation by putting $\Delta\mathbf{u}^t = \Delta t \mathbf{v}^t$. Starting at time $t$ with a material point $\mathbf{x}^t$, we have three cases to examine (see Fig. 6.3).

Case 1: The material point is on the free surface at the beginning of the increment and comes into contact during the increment. The discrete equations are

$$g(\mathbf{x}^t, t) > 0 \quad \text{and } g(\mathbf{x}^{t+\Delta t}, t + \Delta t) = 0. \tag{6.7}$$

Case 2: The material point is already in contact at the beginning of the increment, and the normal stress is negative at the beginning and at the end of the time increment, so that the material point must remain in contact at the end of the increment:

$$\sigma_n^t < 0 \quad \text{and } g(\mathbf{x}^t, t) = 0, \tag{6.8a}$$

$$\sigma_n^{t+\Delta t} < 0 \quad \text{and } g(\mathbf{x}^{t+\Delta t}, t + \Delta t) = 0. \tag{6.8b}$$

Case 3: The material point is in contact at the start of the increment, but the normal stress is null so that the material point is allowed to leave the tool surface:

$$\left[\sigma_n^t \geq 0 \text{ and } g(\mathbf{x}^t, t) = 0\right]; \quad \text{then } g(\mathbf{x}^{t+\Delta t}, t + \Delta t) \geq 0. \tag{6.9}$$

[13] More accurate results can be obtained with two-step methods, in conjunction with Eq. (6.4).

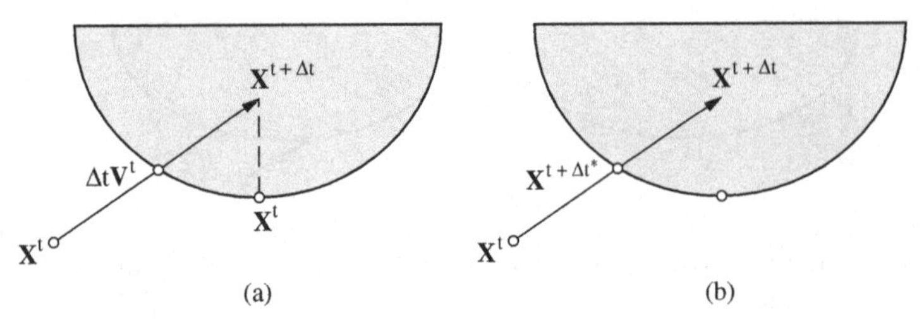


Figure 6.4. Noniterative contact algorithms: (a) projection on the tool surface and (b) reduction of the time step.

### The Nodal Contact Algorithms

When a finite-element discretization is built, the previous constraints cannot be imposed on every geometric point of the discrete boundary in contact with the tools, when its shape is complicated. The contact conditions can instead be imposed either at the nodal points of the mesh, or in an average sense as outlined later.

We shall consider a nodal point in contact at time $t$, with coordinates $\mathbf{X}_n^t$: in the following a nodal coordinate vector will be denoted more simply by $\mathbf{X}^t$ as the node number $n$ is not essential here. The simplest algorithm for cases 1 and 2 consists of calculating $\mathbf{X}^{t+\Delta t}$ and projecting it on the tool surface into $\mathbf{X}'$, so that the length $|\mathbf{X}^{t+\Delta t} - \mathbf{X}'|$ is minimum; see Fig. 6.4(a). This method is very simple and does not introduce a significant increase in computational time. However, the projection method does not satisfy the mechanical equations. Therefore this additional approximation can produce errors, for example on the volume conservation, if the number of contact events is important.

To improve accuracy, a more efficient method (for case 1 only) consists of reducing the time increment $\Delta t$ into $\Delta t^*$ so that we have exactly

$$g(\mathbf{X}^{t+\Delta t^*}, t + \Delta t^*) = 0. \tag{6.10}$$

The value of $\Delta t^*$ depends on the integration scheme that is selected. For the simplest one, the one-step explicit scheme we use here, Eq. (6.10), gives

$$g(\mathbf{X}^t + \Delta t^* \mathbf{V}^t, t + \Delta t^*) = 0. \tag{6.11}$$

The computation of the appropriate time increment $\Delta t^*$ must be done for any node coming into contact, and the smaller $\Delta t^*$ must be selected, so that only one node will come into contact during one time step; see Fig. 6.4(b).

Another scheme can be built, which is slightly more complicated, but which allows us to keep the same value of the time increment and to satisfy exactly the contact condition at the end of the increment. For this purpose the iterative scheme, which is used to compute the velocity field $\mathbf{V}^{(k)}$, is modified at each iteration $k$. Keeping the one-step explicit scheme for simplicity, we put at iteration number $k$

$$\mathbf{X}^{(k)} = \mathbf{X} + \mathbf{V}^{(k)} \Delta t. \tag{6.12}$$

The $t$ superscript is omitted here to condense the notation. The velocity vector is modified into a new velocity vector $\mathbf{V}'^{(k)}$, which is defined in such a way that

$$\mathbf{X}'^{(k)} = \mathbf{X} + \mathbf{V}'^{(k)} \Delta t \tag{6.13}$$

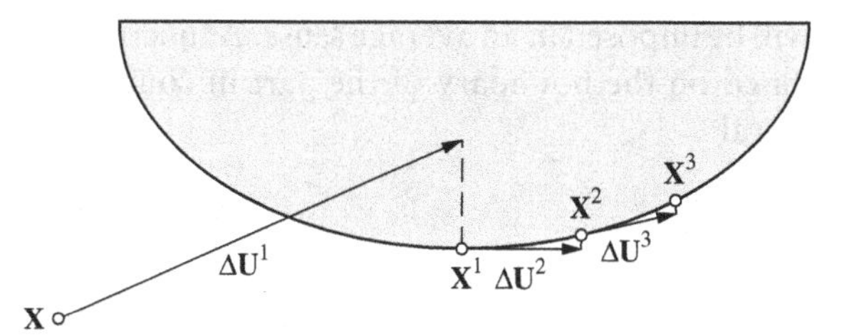


Figure 6.5. Iterative implicit contact algorithm.

is obtained by projection of $\mathbf{X}^{(k)}$ on the tool surface. Then the iterations can proceed with the new boundary condition[14]:

$$\left[\mathbf{V}^{(k+1)} - \mathbf{V}'^{(k)}\right]\mathbf{n}' = 0 \tag{6.14}$$

if $\mathbf{n}'$ is the normal corresponding to $\mathbf{X}'^{(k)}$. The evolution of the displacement $\Delta\mathbf{U}^{(k)} = \mathbf{V}^{(k)}\Delta t$ is schematically pictured for the first iterations in Fig. 6.5.

### The Penalty Method

This method can take various forms depending on whether the penalty constraint is imposed on the velocity term or incrementally, at node level or in average with an integral form. We shall restrict ourselves to the nodal incremental form; for more details the interested reader is referred to Ref. 15.

In the first approach a nodal point with coordinates $\mathbf{X}_l^t$, which is at a distance $d_l^t$ of the tool surface at time $t$, will be penalized if it tends to penetrate the tool. For that we consider the normal relative displacement along the outside normal to the tool surface $\mathbf{n}_l^t$, during the time increment $\Delta t$:

$$\Delta\mathbf{u}_l^t \cdot \mathbf{n}_l^t = \left(\mathbf{V}_l^t - \mathbf{v}_{\text{tool}}\right)\mathbf{n}_l^t \Delta t. \tag{6.15}$$

Penetration of the tool will occur if this normal displacement is negative and its absolute value is greater than $d_l^t$. With the introduction of the Mac Cauley bracket[16] and $I_c$, the set of indices of nodes that can be in contact with the tool, during a time increment, the penalty contribution to the functional will be

$$\sum_{l \in I_c} \rho \left\langle \left(\mathbf{V}_l^t - \mathbf{v}_{\text{tool}}\right) \mathbf{n}_l^t \left(\Delta t - d_l^t\right)\right\rangle^2. \tag{6.16}$$

After differentiation with respect to the velocity components, we obtain the following additional contribution in the nonlinear equations of the problem:

$$\sum_{l \in I_c} \rho \left\langle \left(\mathbf{V}_l^t - \mathbf{v}_{\text{tool}}\right) \mathbf{n}_l^t \left(\Delta t - d_l^t\right)\right\rangle \mathbf{n}_l^t. \tag{6.17}$$

### The Lagrange Multiplier Method

Instead of imposing contact conditions at nodes, we find it possible to treat the incremental problem of contact expressed by Eq. (6.7) with the Lagrange multiplier

[14] This new condition corresponds to the best linear approximation of the tool surface, i.e., by the local tangential plane.

[15] L. Fourment, K. Mocellin, and J.-L. Chenot, "An Implicit Contact Algorithm for the 3-D Simulation of the Forging Process," in *Computational Plasticity – Fundamental and Applications, Proceeding of the Fifth International Conference*, D. R. J. Owen et al., eds. (Pineridge Press, Swansea, 1997), pp. 873–879.

[16] The Mac Cauley bracket is defined by $\langle x\rangle = x$ if $x \geq 0$, and $\langle x\rangle = 0$ if $x < 0$.

technique, so that the contact condition will be imposed in an average sense. A function $\lambda^c$, called the Lagrange multiplier, is defined on the boundary of the part in contact with the tools, and we introduce the integral

$$\Phi_c = \int_{\partial\Omega^f} \lambda^c g(\mathbf{x}^{t+\Delta t}, t + \Delta t)\, \mathrm{d}S, \tag{6.18}$$

which is obviously equal to zero when the contact condition is satisfied at the end of the increment. We suppose that the mechanical problem can be formulated as the minimization of a functional $\Phi$ of the velocity field $\mathbf{v}$, subject to the contact condition as a constraint, and the displacement increment is calculated by the one-step explicit scheme.[17] To take into account the contact condition, we can introduce a new functional according to

$$\Phi_L(\mathbf{v}^t, \lambda^c) = \Phi(\mathbf{v}^t) + \int_{\partial\Omega^f} \lambda^c g(\mathbf{x}^t + \Delta t\mathbf{v}^t, t + \Delta t)\, \mathrm{d}S. \tag{6.19}$$

The solution of the problem is then given by simultaneously solving the two equations:

$$R_L(\mathbf{v}^t, \lambda^c) = \frac{\partial\Phi(\mathbf{v}^t)}{\partial\mathbf{v}} + \int_{\partial\Omega^f} \lambda^c \frac{\partial g(\mathbf{x}^t + \Delta t\mathbf{v}^t, t + \Delta t)}{\partial\mathbf{v}} \Delta t\, \mathrm{d}S = 0, \tag{6.20a}$$

$$\forall\lambda^* \int_{\partial\Omega^f} \lambda^* g(\mathbf{x}^t + \Delta t\mathbf{v}^t, t + \Delta t)\, \mathrm{d}S = 0. \tag{6.20b}$$

Here $\lambda^c$ can be interpreted as the reaction force, which must be negative. If we take the simple example of a two-dimensional problem, which is computed with bilinear elements, a discontinuous piecewise constant function can be chosen for $\lambda^c$. Equation (6.20b) is then replaced by

$$\forall e \int_{\partial\Omega^f \cap \partial\Omega^e} g(\mathbf{x}^t + \Delta t\mathbf{v}^t, t + \Delta t)\, \mathrm{d}S = 0, \tag{6.21}$$

which means that the average value of the $g$ function is equal to zero on any element boundary in contact with the tool.

When the formulation cannot be deduced from a functional form, the same method applies simply by replacing, for example, $\partial\Phi/\partial\mathbf{v}(\mathbf{v}^t)$ by the vector $\mathbf{r}(\Delta\mathbf{u})$, where $\mathbf{r}$ is the equation obtained from the virtual work principle, with the displacement field $\Delta\mathbf{u}$ as unknown.

The drawback of this method is that it imposes the additional unknown function $\lambda^c$, which is defined by element, on the boundary in contact.

## 6.2 Friction

We shall suppose that, in the general case, the friction shear stress is expressed by a relation of the form

$$\tau = -\alpha(\Delta\mathbf{v}, \sigma_n)\frac{\Delta\mathbf{v}}{|\Delta\mathbf{v}|}, \tag{6.22}$$

[17] More sophisticated integration schemes, involving the velocity field at different time increments, could of course be used with this formulation.

where

$\Delta\mathbf{v}$ is the tangential velocity difference between the tool and the part, and
$\alpha$ is a function of $\Delta\mathbf{v}$ and of the normal stress $\sigma_n$, which must be positive when contact occurs (and null if there is no contact).

In general, $\boldsymbol{\tau}$ is merely introduced in the expression of the virtual work principle[18] as a prescribed component of the stress vector on the boundary $\partial\Omega^f$ in contact with the tools. Another part $\partial\Omega^T$ of the boundary may correspond to an external stress vector load $\mathbf{T}^d$. For any virtual velocity field $\mathbf{v}^*$, the resulting equation will be written as

$$\int_\Omega \boldsymbol{\sigma} : \dot{\boldsymbol{\varepsilon}}^* \, \mathrm{d}V - \int_{\partial\Omega^T} \mathbf{T}^d \cdot \mathbf{v}^* \, \mathrm{d}S - \int_{\partial\Omega^f} \alpha(\Delta\mathbf{v}, \sigma_n) \frac{\Delta\mathbf{v}}{|\Delta\mathbf{v}|} \cdot \mathbf{v}^* \, \mathrm{d}S = 0. \tag{6.23}$$

The most difficult issue here is to introduce the value of $\sigma_n$, which requires in fact the global solution of the problem. Indeed, the stress tensor field is generally obtained after discretization and numerical resolution of the finite-element problem by post processing.

Three methods can be used:

a. When a Lagrange multiplier method is used the $\lambda^c$ parameter is substituted for $\sigma_n$.
b. When an iterative scheme is utilized to compute the velocity (or displacement field), the normal stress introduced in the friction law corresponds to the previous iteration.
c. An even simpler algorithm makes use of the normal stress computed from the previous increment.

It is worthy to remark that, when a stationary process is computed directly, only methods a or b can be used. A particular case, which is very convenient for practical applications, is when the friction shear stress does not depend on the normal stress:

$$\boldsymbol{\tau} = -\alpha(\Delta\mathbf{v}) \frac{\Delta\mathbf{v}}{|\Delta\mathbf{v}|}. \tag{6.24a}$$

For isotropic friction Eq. (6.24a) is again simplified; we must have

$$\boldsymbol{\tau} = -\alpha(|\Delta\mathbf{v}|) \frac{\Delta\mathbf{v}}{|\Delta\mathbf{v}|}, \tag{6.24b}$$

and the last term in Eq. (6.23) will be

$$\int_{\partial\Omega^c} \alpha(|\Delta\mathbf{v}|) \frac{\Delta\mathbf{v}}{|\Delta\mathbf{v}|} \cdot \mathbf{v}^* \, \mathrm{d}S. \tag{6.25}$$

---

**Exercise 6.2: Show that in Eq. (6.24b) $\boldsymbol{\tau}$ can be derived from a friction potential $\varphi_f$.**

We introduce the scalar function

$$\varphi_f(|\Delta\mathbf{v}|) = -\int_0^{|\Delta\mathbf{v}|} \alpha(u) \, \mathrm{d}u, \tag{6.2–1}$$

[18] R. H. Wagoner and J.-L. Chenot, *Fundamentals of Metal Forming* (Wiley, New York, 1997), Eq. 5.72, p. 176.

and we compute successively

$$\frac{\partial \varphi_f(|\Delta \mathbf{v}|)}{\partial \Delta v_i} = -\alpha(|\Delta \mathbf{v}|)\frac{\partial |\Delta \mathbf{v}|}{\partial \Delta v_i} = -\alpha(|\Delta \mathbf{v}|)\frac{\Delta v_i}{|\Delta \mathbf{v}|}. \tag{6.2–2}$$

Using Eq. (6.24b), by obvious identification we see that

$$\tau_i = \frac{\partial \varphi_f(|\Delta \mathbf{v}|)}{\partial \Delta v_i}. \tag{6.2–3}$$

---

The viscoplastic (Norton-like) friction law can be written as

$$\boldsymbol{\tau} = -\alpha_f |\Delta \mathbf{v}|^p \frac{\Delta \mathbf{v}}{|\Delta \mathbf{v}|}, \tag{6.26}$$

which, according to the result of Exercise 6.2, is easily shown to derive from the friction viscoplastic potential:

$$\varphi_f(|\Delta \mathbf{v}|) = -\frac{\alpha_f}{p+1}|\Delta \mathbf{v}|^{p+1}. \tag{6.27}$$

We can then conclude that the complete functional associated with a viscoplastic material with a viscoplastic friction law must be written as

$$\Phi(\mathbf{v}) = \int_\Omega \frac{K}{m+1}(\sqrt{3}\dot{\bar{\varepsilon}})^{m+1}\,\mathrm{d}V + \int_{\partial\Omega_c} \frac{\alpha_f}{p+1}|\Delta \mathbf{v}|^{p+1}\,\mathrm{d}S. \tag{6.28}$$

Note: additional terms can arise from incompressibility if the penalty method is used, and possibly from prescribed external forces.

The limit case is the Tresca law: it corresponds to $p = 0$ and can give rise to numerical problems. A regularization of the friction law is then necessary; for example,

$$\varphi_f(|\Delta \mathbf{v}|) = -\alpha_f\left(|\Delta \mathbf{v}|^2 + \Delta v_0^2\right)^{1/2}, \tag{6.29}$$

where $\Delta v_0$ is a numerical parameter, small enough with respect to the mean value of $|\Delta \mathbf{v}|$. The "regularized" friction law is now derived with the help of Eq. (6.2–3):

$$\boldsymbol{\tau} = -\alpha_f \frac{\Delta \mathbf{v}}{\left(|\Delta \mathbf{v}|^2 + \Delta v_0^2\right)^{1/2}}. \tag{6.30}$$

In Fig. 6.6 the regularized friction law and the initial one are compared.

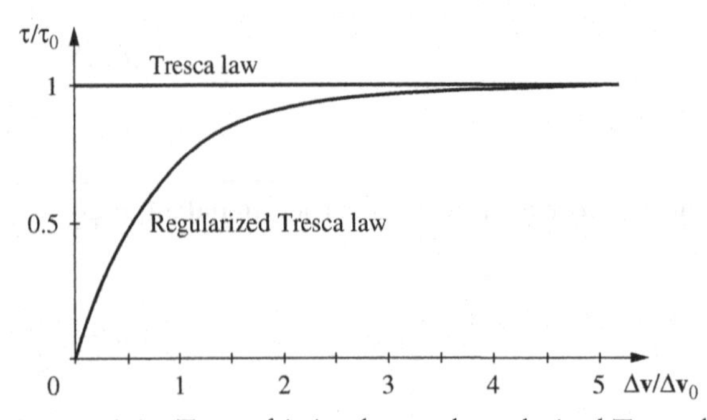


Figure 6.6. Tresca friction law and regularized Tresca law.

## 6.3 Incompressibility

We must distinguish two different situations, which are taken into account with different approaches: the steady-state processes and the incremental or transient processes.

### Incompressibility for Steady-State Flow

We suppose here that the problem is expressed in terms of the velocity field; from Ref. 19 the incompressibility condition comes from a constant density and is written as

$$\mathrm{div}(\mathbf{v}) = 0. \tag{6.31}$$

For viscous or viscoplastic flow, the incompressibility condition must be imposed as a mathematical constraint on the velocity field. This condition can be achieved classically by the Lagrange multiplier method or by the penalty method.

### The Lagrange Multiplier Method

We suppose that the velocity field is found by the minimization of the functional $\Phi$, subject to the constraint given by Eq. (6.31). In the same way as in Section 6.1 for the contact constraint, the variational problem with incompressibility is equivalent to a new variational problem with a functional of $\mathbf{v}$ and of a scalar function $\lambda$, defined by

$$\Psi(\mathbf{v}, \lambda) = \Phi(\mathbf{v}) + \int_\Omega \lambda \, \mathrm{div}(\mathbf{v}) \, \mathrm{d}V. \tag{6.32}$$

The formal variation of $\Psi$ gives two equations:

$$\left[\frac{\partial \Psi(\mathbf{v}, \lambda)}{\partial \mathbf{v}}, v^*\right] = \int_\Omega \frac{\partial \varphi}{\partial \dot{\varepsilon}} \cdot \dot{\varepsilon}^* \, \mathrm{d}V + \int_\Omega \lambda \, \mathrm{div}(\mathbf{v}^*) \, \mathrm{d}V - \int_{\partial\Omega^T} \mathbf{T}^d \cdot \mathbf{v}^* \, \mathrm{d}S = 0, \tag{6.33a}$$

$$\left[\frac{\partial \Psi(\mathbf{v}, \lambda)}{\partial \lambda}, \lambda^*\right] = \int_\Omega \lambda^* \, \mathrm{div}(\mathbf{v}) \, \mathrm{d}V = 0, \tag{6.33b}$$

for any variation $\mathbf{v}^*$ of $\mathbf{v}$ and $\lambda^*$ of $\lambda$.[20] The scalar function $\lambda$ is the Lagrange multiplier, which can be identified with the opposite of the pressure, that is, $\lambda = -p$.

This is easily seen by mere identification when the virtual work principle is first written:

$$\int_\Omega \sigma \cdot \dot{\varepsilon}^* \, \mathrm{d}V - \int_{\partial\Omega^T} \mathbf{T}^d \cdot \mathbf{v}^* \, \mathrm{d}S = 0. \tag{6.34}$$

With the help of the definition of the deviatoric stress tensor $\mathbf{s}$, Eq. (6.38) can be transformed into

$$\int_\Omega \mathbf{s} \cdot \dot{\varepsilon}^* \, \mathrm{d}V - \int_\Omega p \cdot \mathrm{div}(\mathbf{v}^*) \, \mathrm{d}V - \int_{\partial\Omega^T} \mathbf{T}^d \cdot \mathbf{v}^* \, \mathrm{d}S = 0, \tag{6.35}$$

[19] R. H. Wagoner and J.-L. Chenot, *Fundamentals of Metal Forming* (Wiley, New York, 1997), Eq. 5.42, p. 168.

[20] In fact, $\mathbf{v}^*$ and $\lambda^*$ are null where $\mathbf{v}$ and $\lambda$ are prescribed.

which is analogous to Eq. (6.33a) if we put $p = -\lambda$. Now Eq. (6.33b) imposes the incompressibility condition in the weak sense.

The functions $\mathbf{v}$ and $p$ can be discretized with a standard finite-element approximation respectively in terms of the nodal velocity vectors $\mathbf{V}_n$ and the nodal pressures $P_m$:

$$\mathbf{v} = \sum_n \mathbf{V}_n N_n, \tag{6.36a}$$

$$p = \sum_m P_m M_m. \tag{6.36b}$$

However, the shape functions $N_n$ and $M_m$ must verify compatibility conditions in order that the approximation is consistent. The $M_m$ shape functions can be discontinuous across element boundaries as no derivative of $p$ appears in the functional given by Eq. (6.35). For example, in a two-dimensional plane-strain problem, we can consider the linear triangle or the bilinear quadrilateral element for the velocity field, and a constant pressure on each element. The first element is not appropriate as the incompressibility condition is imposed too strongly and causes locking: the three-node triangle is too stiff for an incompressible flow (see Exercise 6.3). The four-node quadrilateral element is not entirely satisfactory from the theoretical point of view (which will not be discussed here). Nevertheless, this latter element was extensively used in applications, as it provides good results in most occasions, at least with regard to the velocity field.

### The Lagrange Multiplier Method with the Mini-Element

The mixed formulation with the mini-element (described in Section 4.1) is becoming more popular because of its reliability and cost effectiveness. The velocity field will be expressed by

$$\mathbf{v} = \mathbf{v}^u + \mathbf{v}^b = \sum_n \mathbf{V}_n^u N_n + \sum_e \mathbf{V}_e^b N_e^b, \tag{6.37}$$

where $N_n$ is the global shape function corresponding to the linear triangular element, and $N_e^b$ is the additional "bubble" shape function, defined in the element number $e$, according to Section 4.1. The pressure field is simply discretized by the linear triangular elements:

$$p = \sum_n P_n N_n. \tag{6.38}$$

For a Newtonian linear behavior, we shall write the integral formulation Eqs. (6.35) and (6.33b) by introducing the usual triangular element discretization, denoted by $\mathbf{v}^u$, $\dot{\varepsilon}^u$, $\mathbf{s}^u$, and the bubble contribution $\mathbf{v}^b$, $\dot{\varepsilon}^b$, $\mathbf{s}^b$:

$$\int_\Omega (\mathbf{s}^u + \mathbf{s}^b) : (\dot{\varepsilon}^{u^*} + \dot{\varepsilon}^{b^*})\, \mathrm{d}V - \int_\Omega p \,\mathrm{div}(\mathbf{v}^{u^*} + \mathbf{v}^{*^b})\, \mathrm{d}V - \int_{\partial\Omega^T} \mathbf{T}^d\, (\mathbf{v}^{u^*} + \mathbf{v}^{*^b})\, \mathrm{d}S = 0, \tag{6.39a}$$

$$-\int_\Omega p^* \,\mathrm{div}(\mathbf{v}^u + \mathbf{v}^b)\, \mathrm{d}V = 0. \tag{6.39b}$$

Separating the different contributions, we rewrite these equations as

$$\int_\Omega \mathbf{s}^u : \dot{\varepsilon}^{u^*}\,\mathrm{d}V + \int_\Omega \mathbf{s}^b : \dot{\varepsilon}^{u^*}\,\mathrm{d}V - \int_\Omega p\,\mathrm{div}(\mathbf{v}^{u^*})\,\mathrm{d}V - \int_{\partial\Omega^T} \mathbf{T}^d \cdot \mathbf{v}^{u^*}\,\mathrm{d}S = 0, \tag{6.40a}$$

$$\int_\Omega \mathbf{s} : \dot{\varepsilon}^{b^*}\,\mathrm{d}V + \int_\Omega \mathbf{s}^b : \dot{\varepsilon}^{b^*}\,\mathrm{d}V - \int_\Omega p\,\mathrm{div}(\mathbf{v}^{*^b})\,\mathrm{d}V = 0, \tag{6.40b}$$

$$-\int_\Omega p^*\,\mathrm{div}(\mathbf{v}^u)\,\mathrm{d}V - \int_\Omega p^*\,\mathrm{div}(\mathbf{v}^b)\,\mathrm{d}V = 0. \tag{6.40c}$$

Equations (6.40a)–(6.40c) are now put in the symbolic linear form, with a symmetric matrix, according to

$$\begin{bmatrix} \mathbf{K}_{uu} & \mathbf{K}_{ub} & \mathbf{K}_{up} \\ \mathbf{K}_{bu} & \mathbf{K}_{bb} & \mathbf{K}_{bp} \\ \mathbf{K}_{pu} & \mathbf{K}_{pb} & 0 \end{bmatrix} \begin{bmatrix} \mathbf{V}^u \\ \mathbf{V}^b \\ \mathbf{P} \end{bmatrix} = \begin{bmatrix} \mathbf{S}^u \\ 0 \\ 0 \end{bmatrix}. \tag{6.41}$$

The unknown $\mathbf{V}^b$ can be eliminated at the element level; that is, if we consider that Eq. (6.41) is now written for the degrees of freedom of one element only, we can write

$$\mathbf{V}^b = -\mathbf{K}_{bb}^{-1}(\mathbf{K}_{bu} \cdot \mathbf{V}^u + \mathbf{K}_{bp} \cdot \mathbf{P}), \tag{6.42}$$

so that the system of Eq. (6.41) can be simplified into

$$\begin{bmatrix} \mathbf{K}'_{uu} & \mathbf{K}'_{up} \\ \mathbf{K}'_{pu} & \mathbf{K}'_{pp} \end{bmatrix} \begin{bmatrix} \mathbf{V}^u \\ \mathbf{P} \end{bmatrix} \doteq \begin{bmatrix} \mathbf{S}^u \\ 0 \end{bmatrix}. \tag{6.43}$$

The matrices in Eq. (6.43) are computed with the help of Eqs. (6.41) and (6.42) by mere identification:

$$\begin{aligned} \mathbf{K}'_{uu} &= \mathbf{K}_{uu} - \mathbf{K}_{ub} \cdot \mathbf{K}_{bb}^{-1} \cdot \mathbf{K}_{bu}, \\ \mathbf{K}'_{up} &= \mathbf{K}'_{pu} = \mathbf{K}_{up} - \mathbf{K}_{ub} \cdot \mathbf{K}_{bb}^{-1} \cdot \mathbf{K}_{bp}, \\ \mathbf{K}'_{pp} &= -\mathbf{K}_{pb} \cdot \mathbf{K}_{bb}^{-1} \cdot \mathbf{K}'_{bp}. \end{aligned} \tag{6.44}$$

The situation is slightly more complicated with nonlinear constitutive equations, but the methodology is basically the same. In addition, some simplifications can be introduced to reduce the volume of computation: they take advantage of the fact that the bubble contribution is smaller than the usual velocity field.

With the use of the three-dimensional mini-element, the method can also be extended to three-dimensional computation.

---

**Exercise 6.3: A Newtonian (or viscoplastic) flow is considered in a square domain that is meshed with $n \times n$ nodes. Show that a triangular mesh with a constant pressure is not acceptable, while quadrilateral elements can be considered as a first approach (see figure below).**

**Consider a mixed velocity and pressure formulation where the two-dimensional mini-element described in Section 4.1 is used with the mesh (a) of the figure below, where we suppose that the internal velocity is eliminated. Verify that this discretization is acceptable.**

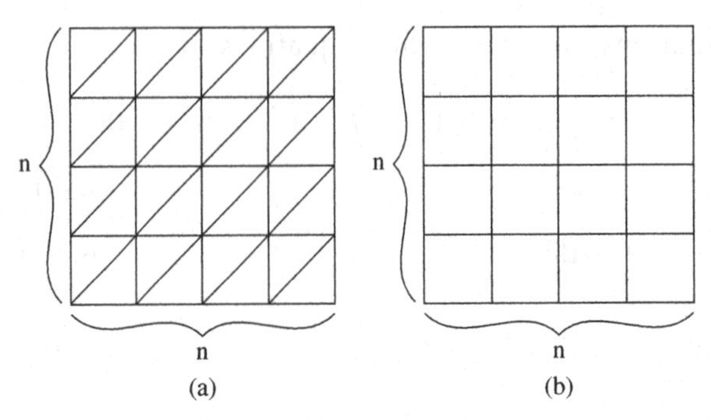


From both (a) and (b) figures above, it is easy to observe that the total number of velocity field components is $2n^2$. Equation (6.33b) shows that with a pressure field constant by linear triangular element we have as many incompressibility constraints as we have elements, that is, $2(n-1)^2$ in figure (a) above. If we compare these two numbers we note that, when $n$ is rather large, the number of constraints is approximately equal to the number of degrees of freedom, so that there remain very few unknowns to take into account the constitutive equations.

In the right-hand-side figure above, when the square is meshed with bilinear quadrilaterals, the total number of degrees of freedom is the same. But with a pressure field constant by element, the number of constraints is equal to $(n-1)^2$, so that this number is approximately half of the number of degrees of freedom.

In the mixed formulation we introduce initially $2n^2 + 4(n-1)^2$ components for the velocity field, and $n^2$ components for the pressure field. The $4(n-1)^2$ internal components are eliminated at the element level. Finally, we have $2n^2$ components for the velocity field and $n^2$ discrete equations for the incompressibility.

### The Penalty Method

This method for enforcing incompressibility is approximate; we introduce a large enough positive number $\kappa$, the penalty factor, such that the modified functional $\Phi_\kappa$ is defined by

$$\Phi_\kappa(\mathbf{v}) = \Phi(\mathbf{v}) + \frac{1}{2}\int_\Omega \kappa[\mathrm{div}(\mathbf{v})]^2\,\mathrm{d}V. \tag{6.45}$$

Formal variation of Eq. (6.45) with respect to $\mathbf{v}$ gives the following equation, for any $\mathbf{v}^*$:

$$\left[\frac{\partial\Phi_\kappa(\mathbf{v})}{\partial\mathbf{v}}, \mathbf{v}^*\right] = \int_\Omega \frac{\partial\varphi}{\partial\dot{\varepsilon}} : \dot{\varepsilon}^*\,\mathrm{d}V + \int_\Omega \kappa\,\mathrm{div}(\mathbf{v})\mathrm{div}(\mathbf{v}^*)\,\mathrm{d}V - \int_{\partial\Omega^T} \mathbf{T}^d\cdot\mathbf{v}^*\,\mathrm{d}S = 0. \tag{6.46}$$

This is equivalent to the virtual work principle if we use a viscoplastic constitutive equation with compressibility governed by

$$\mathrm{p} = -\kappa\,\mathrm{div}(\mathbf{v}). \tag{6.47}$$

If $\kappa$ is large enough, it is clear by inspection of Eq. (6.45) that the divergence of the velocity field will be small when the minimum of the functional is reached. If we suppose that $\kappa$ tends to infinity, the discrete form of Eq. (6.45) will impose exact incompressibility at the integration points for the second term. This shows that the number of integration points for the penalty term must be generally less than for the other contribution; otherwise we should also observe locking: this is called a *reduced integration scheme.* With a similar approach as in Exercise 6.3, we can easily see that three-node triangles again are not appropriate for plane-strain incompressible flows, and that in this case, four-node bilinear quadrilaterals require only one integration point per element for the penalty term.

In fact, for practical computational reasons, we cannot choose $\kappa$ arbitrarily large, and a reasonable compromise is, for example,[21]

$$10^5 K \leq k \leq 10^7 K. \tag{6.48}$$

for a Norton–Hoff viscoplastic material with consistency $K$.

The advantage of the penalty method is that it allows formulation of the problem in terms of the velocity field alone. Therefore the total number of degrees of freedom is smaller than in the mixed, or Lagrange, multiplier method for which we must add the pressure unknowns. The drawback of the penalty method results from the condition that gives linear systems with a relatively bad conditioning so that iterative methods cannot be used.

---

**Exercise 6.4: Consider an initially square domain with side equal to 2, which is deformed in plane strain by two flat tools moving symmetrically with velocity equal to 1 (see the figure below). Discretize a quarter of the problem by considering one square element with side 1. Introduce nodal velocity vectors that satisfy symmetry conditions and boundary conditions and take the values $v^a$ and $v^b$ for the horizontal components on the right side. Express the velocity field in the element and the strain-rate tensor. Write the integral equations for a purely Newtonian material:**

$$\mathbf{s} = 2\eta\dot{\varepsilon}.$$

**The friction law is linear with the form**

$$\tau = -\alpha_f \eta \Delta \mathbf{v}.$$

**Compare the penalty formulation with coefficient $\rho\eta$, and the mixed formulation with a constant pressure in the element.**

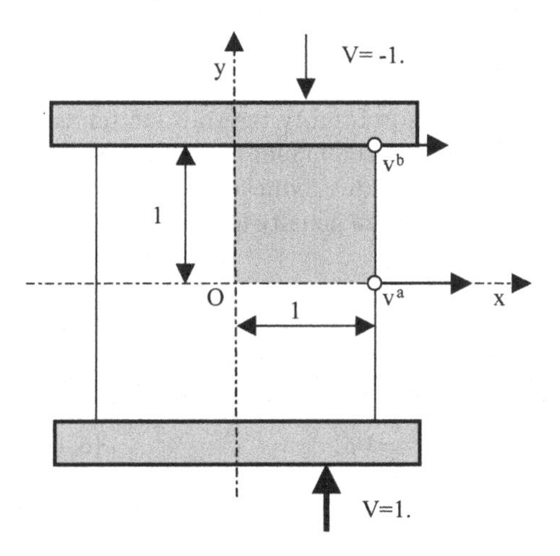


[21] When the computation is done in double-precision floating-point arithmetic, with a relative precision of $10^{-14}$ to $10^{-15}$.

The discretized velocity field that respects the symmetry and boundary conditions has the form

$$v_1 = x(1-y)v^a + xyv^b,$$
$$v_2 = -y, \tag{6.4–1}$$

and a similar expression for the virtual velocity field:

$$v_1^* = x(1-y)v^{a*} + xyv^{b*},$$
$$v_2^* = 0.$$

The strain-rate tensor $\dot{\varepsilon}$ is then

$$\begin{bmatrix} (1-y)v^a + yv^b & \frac{x}{2}(v^b - v^a) \\ \frac{x}{2}(v^b - v^a) & -1 \end{bmatrix}. \tag{6.4–2}$$

With the penalty method, the virtual work equations are obtained in the usual way by putting

$$\int_\Omega 2\eta\dot{\varepsilon} : \dot{\varepsilon}^* \, dV + \rho_p\eta \int_\Omega \mathrm{div}(\mathbf{v})\mathrm{div}(\mathbf{v}^*)\, dV - \int_{\partial\Omega^c} \tau \cdot \mathbf{v}^* \, dS = 0. \tag{6.4–3}$$

We obtain the first equation by putting $v^{a*} = 1$ and $v^{b*} = 0$:

$$R^a = 0 = 2\eta\dot{\varepsilon} \int_0^1 \int_0^1 \left\{ [(1-y)v^a + yv^b](1-y) + 2\frac{1}{2}x(v^b - v^a)\frac{-x}{2} \right\} dx\,dy$$
$$+ \rho\eta \int_0^1 \int_0^1 [(1-y)v^a + yv^b - 1](1-y)\, dx\,dy. \tag{6.4–4}$$

Taking into account the friction term, we obtain the second equation in a similar way with $v^{a*} = 0$ and $v^{b*} = 1$:

$$R^b = 0 = 2\eta \int_0^1 \int_0^1 \left\{ [(1-y)v^a + yv^b]y + 2\frac{1}{2}x(v^b - v^a)\frac{x}{2} \right\} dx\,dy$$
$$+ \rho\eta \int_0^1 \int_0^1 [(1-y)v^a + yv^b - 1]y\,dx\,dy + \alpha_f\eta \int_0^1 v^b x^2\, dx. \tag{6.4–5}$$

When all the integrals in Eqs. (6.4–4) and (6.4–5) are calculated exactly, we obtain the linear system:

$$(6+2\rho)v^a + \rho v^b = 3\rho, \quad \rho v^a + (6+2\rho+2\alpha_f)v^b = 3\rho. \tag{6.4–6}$$

We observe that when the penalty coefficient $\rho$ tends to infinity the solution tends to $v^a = 1$ and $v^b = 1$, which is nonphysical as the friction coefficient plays no role. This is the reason why the penalty term must be integrated with a reduced scheme.

Here we take a one-integration-point scheme so that the penalty term in Eqs. (6.4–4) and (6.4–5) becomes:

$$\rho\left(\frac{v^a + v^b}{2} - 1\right). \tag{6.4–7}$$

Using Eq. (6.4–7) for the penalty contribution, we find the linear system becomes

$$(6+3\rho)v^a + 3\rho v^b = 6\rho, \quad 3\rho v^a + (6+3\rho+2\alpha_f)v^b = 6\rho. \tag{6.4–8}$$

The solution of this is readily obtained:

$$v^a = \frac{2\rho(3+\alpha_f)}{\rho(6+\alpha_f)+6+2\alpha_f},$$
$$v^b = \frac{6\rho}{\rho(6+\alpha_f)+6+2\alpha_f}. \tag{6.4–9}$$

Now we see that when $\rho$ tends to infinity the solution still depends on the friction coefficient:

$$v^a = \frac{2(3+\alpha_f)}{6+\alpha_f} = 1 + \frac{\alpha_f}{6+\alpha_f},$$

$$v^b = \frac{8}{8+3\alpha_f} = 1 - \frac{\alpha_f}{6+\alpha_f}. \quad (6.4\text{–}10)$$

In the mixed formulation with a constant pressure in the element, the penalty term is replaced by the integral,

$$-\int_\Omega p \operatorname{div}(\mathbf{v}^*)\,dV = -p\left(\frac{1}{2}v^{a*} + \frac{1}{2}v^{b*}\right), \quad (6.4\text{–}11)$$

and the integral form of the incompressibility is now

$$-\int_\Omega p^* \operatorname{div}(\mathbf{v})\,dV = -p^*\left(\frac{1}{2}v^a + \frac{1}{2}v^b - 1\right) = 0. \quad (6.4\text{–}12)$$

The new linear system to be solved is $3 \times 3$:

$$6v^a - 3\frac{p}{\eta} = 0,$$

$$(6+2\alpha_f)v^b - 3\frac{p}{\eta} = 0,$$

$$v^a + v^b - 2 = 0. \quad (6.4\text{–}13)$$

The solution of this is identical to Eqs. (6.4–10) for the nodal velocities, and the pressure is

$$p = 2\eta\left(1 + \frac{\alpha_f}{6+\alpha_f}\right). \quad (6.4\text{–}14)$$

---

### Incompressibility for Nonstationary Processes

If time integration was exact, the condition expressed by Eq. (6.47) would ensure volume conservation at each time step. To show that this is not the case, we can consider a very simple upsetting example, as discussed in Exercise 6.4.

---

Exercise 6.5: **Consider an initially square domain in plane strain with a uniform deformation at each time step (see figure below). Write the velocity field at each intermediate configuration for an upsetting velocity equal to $h_0$, and compute the final volume with an explicit one-step time-integration scheme, for a 50% total reduction of height.**

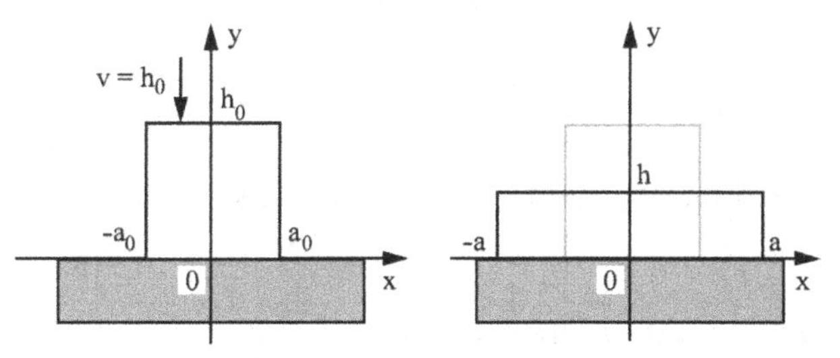


For the configuration corresponding to the right-hand figure above, taking into account the unit tool velocity, the incompressible velocity field can be written as

$$v_x = \frac{h_0}{h}x, \quad v_y = -\frac{h_0}{h}y. \quad (6.5\text{–}1)$$

We observe that the velocity of the right-hand side of the part is

$$\frac{da}{dt} = \frac{h_0}{h}a. \tag{6.5–2}$$

In this case the solution can be computed exactly, but in practical examples a time-integration scheme must be used, so that we can compare both results. With time steps equal to $\Delta t = 1/2n$, after $k$ increments the height of the part will be exactly

$$h_k = h_0\left(1 - \frac{k}{2n}\right). \tag{6.5–3}$$

The explicit integration scheme allows the following to be written:

$$a_{k+1} = a_k + \frac{da_k}{dt}\Delta t = a_k + \frac{a_k}{1-(k/2n)}\frac{1}{2n}. \tag{6.5–4}$$

This gives

$$a_{k+1} = a_k\frac{2n-k+1}{2n-k}, \tag{6.5–5}$$

which, by recurrence, is obviously

$$a_{k+1} = \frac{2n+1}{2n-k}. \tag{6.5–6}$$

The final half-width will be

$$a_n = \frac{2n+1}{n+1}. \tag{6.5–7}$$

The error in surface conservation is

$$\frac{\Delta S}{S_0} = \frac{a_0h_0 - a_nh_n}{a_0h_0} = 1 - \frac{2n+1}{2n+2} = \frac{1}{2n+2}. \tag{6.5–8}$$

For $n = 1$ we have $\Delta S/S_0 = 25\%$; $n = 10$ corresponds to $\Delta S/S_0 = 4\%$.

---

One way to improve the volume conservation is to select a more accurate time-integration procedure; this can be achieved, for example, with the two-level semi-implicit scheme discussed in Section 5.1. Another method is to enforce volume conservation during each finite time increment. For any subdomain $\omega_t$ at time $t$, we want to require that

$$\int_{\omega_{t+\Delta t}} dV = \int_{\omega_t} dV. \tag{6.49}$$

A domain change is made in the first integral, in order to come back to the subdomain at time $t$, giving

$$\int_{\omega_t} J\, dV = \int_{\omega_{t+\Delta t}} dV, \tag{6.50}$$

where $J$ is the Jacobian of the transformation from $t$ to $t + \Delta t$ with expression

$$J = \det\left(\mathrm{F}_t^{t+\Delta t}\right). \tag{6.51}$$

Here $\mathrm{F}_t^{t+\Delta t}$ is the deformation gradient with components

$$\mathrm{F}_{ijt}^{t+\Delta t} = \frac{\partial x_i^{t+\Delta t}}{\partial x_j^t}. \tag{6.52}$$

If Eq. (6.47) must hold for any subdomain $\omega_t$, we come to the general incremental conservation formula:

$$J - 1 = \det(\mathbf{F}_t^{t+\Delta t}) - 1 = 0. \tag{6.53}$$

Now if the new material coordinates are expressed in terms of the velocity field $\mathbf{v}^t$, and the explicit scheme given by Eq. (6.6) is used, the deformation gradient is then written as

$$\mathbf{F}_t^{t+\Delta t} = \frac{\partial \mathbf{x}^{t+\Delta t}}{\partial \mathbf{x}^t} = \mathbf{I} + \Delta t \frac{\partial \mathbf{v}^t}{\partial \mathbf{x}^t}, \tag{6.54}$$

and the condition of Eq. (6.53) is expressed by

$$\det\left(\mathbf{I} + \Delta t \frac{\partial \mathbf{v}^t}{\partial \mathbf{x}^t}\right) - 1 = 0. \tag{6.55}$$

In two dimensions, this latter expression is easily shown to give (after division by $\Delta t$)

$$\operatorname{div}(\mathbf{v}^t) + \Delta t \det\left(\frac{\partial \mathbf{v}^t}{\partial \mathbf{x}^t}\right) = 0, \tag{6.56}$$

and a slightly more complicated expression in three dimensions. The penalty approach can be used with the incremental expression of Eq. (6.56), so that the penalty term becomes

$$\frac{1}{2}\int_\Omega \kappa \left[\operatorname{div}(\mathbf{v}^t) + \Delta t \det\left(\frac{\partial \mathbf{v}^t}{\partial \mathbf{x}^t}\right)\right]^2 \mathrm{d}V = 0. \tag{6.57}$$

However, this expression is more complicated than the classical penalty formulation, as it consists of quadratic and higher-order terms. Another approach is to introduce the relative density $\rho_r$ and to allow for short fluctuations around the theoretical value of 1. If the relative density at time $t$ is not exactly equal to 1 and we want to obtain 1 at time $t + \Delta t$, we should write

$$\int_{\omega_t} \rho_r^t \,\mathrm{d}V = \int_{\omega_{t+\Delta t}} \rho_r^{t+\Delta t} \,\mathrm{d}V = \int_{\omega_{t+\Delta t}} \mathrm{d}V. \tag{6.58}$$

With the same approach as previously used, we conclude that, instead of Eq. (6.56), we must impose

$$\det\left(\mathrm{I} + \Delta t \frac{\partial \mathbf{v}^t}{\partial \mathbf{x}^t}\right) - \rho_r^t = 0. \tag{6.59}$$

If the time increment is small enough, we can use only the first term and transform Eq. (6.56) into

$$\operatorname{div}(\mathbf{v}^t) - \frac{\rho_r^t - 1}{\Delta t} = 0. \tag{6.60}$$

In fact, as a result of the approximation introduced in Eq. (6.60), the condition $\rho_r^{t+\Delta t} = 1$ will not be satisfied exactly, and the next density must be computed exactly by

$$\rho_r^{t+\Delta t} = \frac{\rho_r^t}{\det[\mathbf{I} + \Delta t(\partial \mathbf{v}^t / \partial \mathbf{x}^t)]}, \tag{6.61}$$

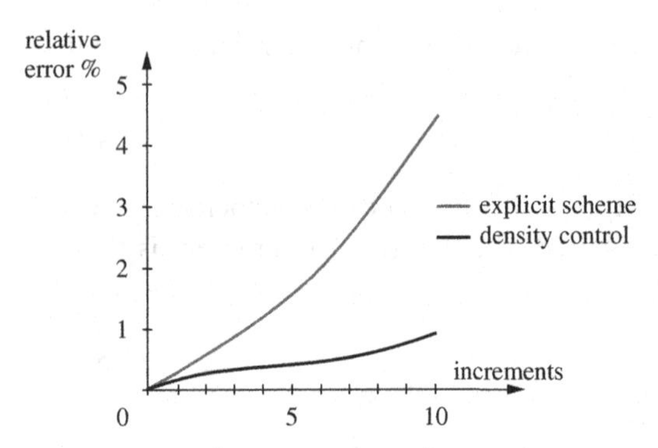


Figure 6.7. Relative error in surface evolution.

so that the process can be repeated at the next time step. With this scheme the volume conservation is approximately guaranteed at each time step.

If the same calculation as in Exercise 6.4 is repeated with the approximate density control, we obtain a much better surface conservation,[22] as shown in Fig. 6.7. The result above can be improved by the control of the time step so that the maximum strain increment remains constant.

## PROBLEMS

### A. Proficiency Problems

1. The equation of a planar tool is given the usual form:

$$g(\mathbf{x}) = ax_1 + bx_2 + cx_3 - h = 0,$$

with the condition

$$a^2 + b^2 + c^2 = 1.$$

Show that the unit normal to the plane has the components $(a, b, c)$. We suppose that the normal is oriented toward the outside of the tool. Verify that for each point interior to the tool we have $g(\mathbf{x}) < 0$. For any interior point, determine its orthogonal projection on the surface of the tool given by the previous equation.

2. Show that with linear elements it is equivalent to impose a zero velocity divergence in each element with a one-integration-point scheme, or a zero velocity flux through the surface of the elements.
Note: the flux $\phi$ of the velocity field $\mathbf{v}$, through the surface $\Gamma$ with normal $\mathbf{n}$, is defined by

$$\phi = \int_\Gamma \mathbf{v} \cdot \mathbf{n} \, dS.$$

3. For a mesh of the same structure as shown in the left-hand figure of Exercise 6.3 with quadratic triangles, show that a piecewise constant pressure field is acceptable in regard to the number of constraints on the velocity field.

[22] Which could be kept as constant if the time step was calculated in order that the deformation increment would be constant at each time step.

## B. Depth Problem

4. Verify that a second-order implicit scheme, with the classical incompressibility formulation, allows having the element density constant up to the second order.

## C. Numerical and Computational Problem

5. The surface of a tool can be defined approximately by a set of adjacent triangles. Write a FORTRAN subroutine that determines the minimum distance between a point $M$ and a triangle $ABC$. Input data: vector coordinates $\mathbf{x}_B$ of point $M$, vectors coordinates $\mathbf{x}_A, \mathbf{x}_B, \mathbf{x}_C$ of the apex of the triangle; output data: minimum distance $d$.

CHAPTER SEVEN

# Thermomechanical Principles

This chapter is devoted to the analysis of the interactions between thermal phenomena and deformation processes. In hot forming, the effect of temperature is especially important, as in general the tools are used at relatively low temperature[1] and thus produce a highly heterogeneous temperature field. For the workpiece itself, there is always a competition between the cooling effect that is due to the contact with tools and radiation plus convection on free surfaces on one hand, and heating by plastic deformating in the core of the part or by the friction dissipation on the surface in contact on the other hand. In fact, it may happen for forming of complex parts that the final result is cooling in some regions and heating in other regions. For a complete analysis of a forming sequence, different situations must be considered:

1. The preform can be heated in an oven and then transferred to the forming machine: in this case the strain and stress evolution can be neglected as a first approach, and a purely thermal analysis may be sufficient.
2. During the forming process, large plastic or elastoplastic deformation (respectively viscoplastic or elastic viscoplastic) is imposed on the workpiece, while the local slipping distance at the tool–part interface can also be important. A coupled thermal and mechanical analysis with large strain theory is therefore necessary.
3. After forming, the part is cooled down until it reaches room temperature. During this step, the material undergoes generally small deformation (but possibly with noticeable displacements that can alter the final geometry, while the stress state is completely changed). The prediction of the residual stresses after forming necessitates following accurately the temperature and stress and strain history at any point of the part during cooling. Moreover, phase changes in the solid state may occur during the cooling process. The situation is the same for possible further heat treatments.

Even cold-forming processes produce plastic work, at high stress level, which is mainly transformed into heat and can raise the temperature of the workpiece as much as 100–200 °C in one sequence.[2]

[1] This is so that they are hard enough to avoid excessive deformation.

[2] See Problem 6 at the end of this chapter.

In this chapter, the classical heat equation is established for essentially undeformable bodies. Various kinds of boundary conditions relevant in metal forming are reviewed. Then a more complex section (Section 7.2) is devoted to an introduction to the thermodynamics of continuum mechanics (it can be skipped at first reading). In the last section, aspects of mechanical and thermal coupling are discussed and analyzed from the computational point of view.

## 7.1 The Elementary Heat Equations

The equations for heat propagation can be analyzed similarly to the mechanical equation: a general law of heat conservation is first built, then the material response to heat flux is experimentally determined – which is the analog to the constitutive equation, and then boundary conditions are examined that can be linear (conduction) or nonlinear (radiation).

### Elementary Introduction to the Definition of Heat

The elementary definition of heat was given with a reference material: liquid water. It is the input thermal energy that produces a 1 °C increase of temperature for the unit mass of liquid water (without phase change); that is,

$$Q = \rho c V \Delta T, \tag{7.1}$$

where $Q$ is the heat energy in calories (cal), $\rho$ is the density in g cm$^{-3}$, $V$ is the volume of the sample in cm$^3$, $c$ is the specific heat that is taken to 1 for liquid water, and $\Delta t$ is the temperature increase in °C. This is shown to be convertible (theoretically) into mechanical energy,[3] so that the same unit can be used for thermal and mechanical energy, namely the joule (J). The usual SI set of units is then used and $\rho$ is given in kg/m$^3$, $c$ in J/kg/°C, and $V$ in m$^3$.

### Definition of Heat Flux

Heat flux is described by a vector field $\mathbf{q}$, which is defined accordingly: $q = |\mathbf{q}|$ is the amount of heat energy that crosses the unit surface per unit time; $\mathbf{q}$ is parallel to the normal $\mathbf{n}$ to the surface and corresponds to the same direction if the transfer of energy in the $\mathbf{n}$ direction is positive; that is,

$$\mathbf{q} = q\mathbf{n}; \tag{7.2}$$

otherwise it has the opposite direction.

---

**Exercise 7.1: We consider a flat slab of thickness $a$ and of square section with side $b$. Compute the mean heat flux when the heat energy goes along the $x_1$ direction, normal to the shadowed surface (see figure below), and increases the temperature of the slab of $\Delta T$, in a time $\Delta t$. The material has a specific heat $c$ and a density equal to $\rho$.**

[3] To achieve this transformation, a reversible process should be necessary, which is not possible in practical situations.

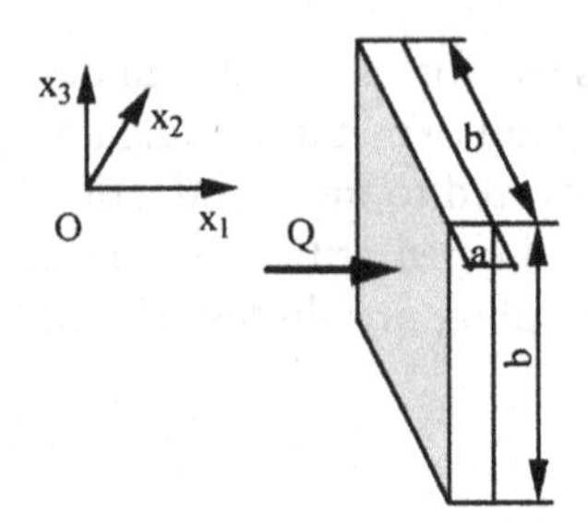


The heat capacity of the slab is

$$Q = \rho cab^2 \Delta T. \tag{7.1–1}$$

The mean power is then

$$\dot{Q} = \frac{Q}{\Delta t} = \rho cab^2 \frac{\Delta T}{\Delta t}, \tag{7.1–2}$$

so that the heat per unit time and per unit area is written

$$q = \frac{\dot{Q}}{b^2} = \rho ca \frac{\Delta T}{\Delta t}. \tag{7.1–3}$$

The heat vector comes immediately:

$$\mathbf{q} \leftrightarrow [q] = \begin{bmatrix} \rho ca(\Delta T/\Delta t) \\ 0 \\ 0 \end{bmatrix}. \tag{7.1–4}$$

---

**The Conduction Law**

We suppose that the slab depicted in Exercise 7.1 has its first face normal to the $x$ axis (shadowed) with a uniform temperature $T_1$ and the second face with temperature $T_2$. For steady state, the heat per unit time is assumed to be proportional to the temperature difference $\Delta T = T_2 - T_1$; that is, we put[4]

$$\dot{Q} = -k_1 \Delta T \qquad \text{with } k_1 > 0. \tag{7.3}$$

If the material is homogeneous, we can also assume that for any thickness a:

$$\dot{Q} = -k_2 \frac{\Delta T}{a} \qquad \text{with } k_2 > 0. \tag{7.4}$$

Finally, the homogeneity also allows us to suppose that for any area $b^2$ of the slab we have the same relationship:

$$\dot{Q} = -kb^2 \frac{\Delta T}{a} \qquad \text{with } k > 0. \tag{7.5}$$

Each side of Eq. (7.5) is divided by $b^2$ and we get

$$q = -k \frac{\Delta T}{a}. \tag{7.6}$$

If we suppose now that the thickness tends to zero,[5] the finite quotient in Eq. (7.6) tends to the derivative of $T$ with respect to $x_1$, and the one-dimensional *Fourier* law

[4] The negative sign comes from the experimental (common) evidence that heat always flows in the direction of decreasing temperature.

[5] We can also consider that around a given $x_1$ coordinate, a "subslab" of decreasing thickness d$a$ is introduced, and we let d$a$ tend to zero.

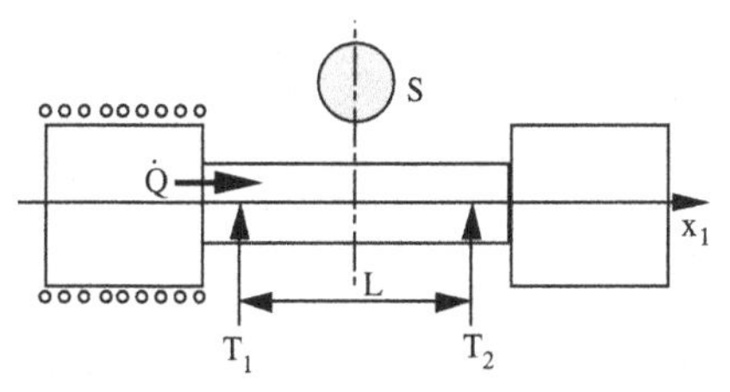


Figure 7.1. Principle of experimental measurement of the thermal conductivity parameter $k$.

is obtained:

$$q = -k\frac{dT}{dx_1}, \tag{7.7}$$

where $k$ is the material *thermal conductivity*, which is given in J m/s/°C (in SI units). For isotropic materials, which possess identical properties for any orientation by definition, the three-dimensional Fourier law can be written as

$$\mathbf{q} = -k\,\mathbf{grad}(T). \tag{7.8}$$

In the general case, the (scalar) heat conductivity factor $k$ can be a function of the position and of the temperature, that is, $k = k(\mathbf{x}, T)$.

For anisotropic materials the Fourier law is generalized according to

$$\mathbf{q} = -\mathbf{C}\cdot\mathbf{grad}(T), \tag{7.9}$$

where $\mathbf{C}$ is a symmetrical positive definite $3 \times 3$ matrix in three dimensions.

The basic principle for heat conductivity experimental measurement is summarized as follows (see Fig. 7.1):

1. The left-hand side of the sample is heated by an electric heater, the power of which is measured, while the right-hand side of the sample is cooled.
2. The experimental device is designed in order that most of the heat flows along the $Ox_1$ direction, perpendicular to the section with area $S$.
3. The temperatures $T_1$ and $T_2$ are measured at precise locations separated by a distance $L$.
4. Steady state is reached, by checking that the temperatures $T_1$ and $T_2$ remain constant.

The experimental thermal conductivity is then approximated by

$$k = -\frac{\dot{Q}L}{S(T_2 - T_1)}. \tag{7.10}$$

**The Elementary Heat Equation**

The elementary heat equation can be obtained by writing the conservation of heat in an elementary cube, with sides equal to $\Delta x$ and its faces perpendicular to the coordinate axes, as pictured in Fig. 7.2.

Through face number 1 of the cube, the inward normal of which has components (1, 0, 0), and during the time increment $\Delta t$, the heat entering the sample is

$$\Delta Q_1 = q_1(x_1, x_2, x_3)\Delta x^2 \Delta t. \tag{7.11a}$$

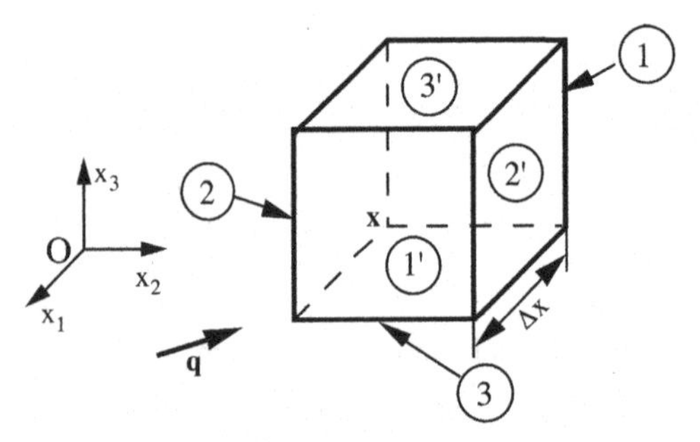


Figure 7.2. Heat conservation in an elementary cube.

For the opposite face number 1′, with inward normal components (−1, 0, 0), the corresponding heat will be

$$\Delta Q_{1'} = -q_1(x_1 + \Delta x, x_2, x_3)\Delta x^2 \Delta t. \tag{7.11b}$$

Analogously, the heat entering faces 2, 2′, 3, 3′ are respectively written as

$$\Delta Q_2 = q_2(x_1, x_2, x_3)\Delta x^2 \Delta t, \tag{7.11c}$$

$$\Delta Q_{2'} = -q_2(x_1, x_2 + \Delta x, x_3)\Delta x^2 \Delta t, \tag{7.11d}$$

$$\Delta Q_3 = q_3(x_1, x_2, x_3)\Delta x^2 \Delta t, \tag{7.11e}$$

$$\Delta Q_{3'} = -q_3(x_1, x_2, x_3 + \Delta x)\Delta x^2 \Delta t. \tag{7.11f}$$

We must also take into account the variation of temperature $\Delta T$ of the sample during $\Delta t$; by definition it corresponds to the heat

$$\Delta Q_c = \rho c \Delta T \Delta x^3. \tag{7.11g}$$

Possibly a last contribution must be introduced to deal with the internal source of heat (chemical, plastic deformation, electric dissipation, etc.):

$$\Delta Q_i = \dot{w} \Delta x^3 \Delta t. \tag{7.11h}$$

If the contributions to heat are summed up, we conclude that

$$\Delta Q_c = \Delta Q_1 + \Delta Q_{1'} + \Delta Q_2 + \Delta Q_{2'} + \Delta Q_3 + \Delta Q_{3'} + \Delta Q_i. \tag{7.12}$$

We divide both sides of Eq. (7.12) by $\Delta x^3 \Delta t$ and we remark that

$$\frac{-q_1(x_1 + \Delta x, x_2, x_3) + q_1(x_1, x_2, x_3)}{\Delta x} \cong -\frac{\partial q_1}{\partial x_1}(x_1, x_2, x_3)\Delta x, \tag{7.13}$$

and similar expressions for coordinates $x_2$ and $x_3$.

Introducing Eq. (7.13), and the similar expressions for the other coordinates, into Eq. (7.12), we get

$$\rho c \frac{\partial T}{\partial t} = -\sum_i \frac{\partial q_i}{\partial x_i} + \dot{w}. \tag{7.14}$$

With the help of the Fourier law, Eq. (7.14) is transformed into the *heat equation*:

$$\rho c \frac{\partial T}{\partial t} = \mathrm{div}[k\, \mathbf{grad}(T)] + \dot{w}, \tag{7.15}$$

or the component form:

$$\rho c \frac{\partial T}{\partial t} = -\sum_i \frac{\partial}{\partial x_i}\left(k \frac{\partial T}{\partial x_i}\right) + \dot{w}. \tag{7.16}$$

### Boundary Conditions

When a part of the boundary is in contact with another body or a fluid, the heat transfer is generally approximated by

$$q_n = \alpha(T - T_{\text{ext}}), \tag{7.17}$$

where $q_n$ is the heat flux normal to the interface between the material and the other body with temperature $T_{\text{ext}}$. When the Fourier law is used, Eq. (7.8), this condition is rewritten as

$$-k\frac{\partial T}{\partial n} = -k\,\mathbf{grad}(T)\cdot\mathbf{n} = \alpha(T - T_{\text{ext}}). \tag{7.18}$$

For free surfaces (or contact with transparent bodies), the radiation condition is used when the temperature is high enough, expressed as follows:

$$-k\frac{\partial T}{\partial n} = -k\,\mathbf{grad}(T)\cdot\mathbf{n} = \varepsilon_r\sigma_r\left(T^4 - T_{\text{ext}}^4\right), \tag{7.19}$$

where

$\varepsilon_r$ is the emissivity parameter that depends on the characteristics of the surface,
$\sigma_r$ is the Stefan constant, and
$T_{\text{ext}}$ is the outside temperature.

### The Steady-State Heat Equation

When the time-dependent term is dropped in Eq. (7.15), the steady-state heat equation is obtained:

$$\text{div}[k\,\mathbf{grad}(T)] + \dot{w} = 0. \tag{7.20}$$

It is worthy to note that this equation is valid only for undeformed bodies, or is a good approximation when the displacements can be considered small. We shall see in the next section that the material derivative must be substituted for the partial derivative with respect to time in the elementary heat equation. We must then consider an additional term for bodies undergoing large deformation, which is the steady-state contribution of the material derivative, and which gives rise to

$$\rho c\,\mathbf{grad}(T)\cdot\mathbf{v} = \text{div}[k\,\mathbf{grad}(T)] + \dot{w}. \tag{7.21}$$

The simplest form of the steady-state heat equation corresponds to the case when the body is not deformable, the heat conductivity parameter is constant, and the source term is nil:

$$\text{div}[\mathbf{grad}(T)] = \Delta T = 0. \tag{7.22}$$

Here the $\Delta$ operator is the Laplacian, the component equivalent form being

$$\sum_i \frac{\partial^2 T}{\partial x_i^2} = 0. \tag{7.23}$$

Some useful experimental values of the main thermal parameters are given in Table 7.1.

**Table 7.1. Thermal Data**

| Material | Density (kg/m$^3$) | Specific Heat (J/kg/°C) | Thermal Conductivity (W/m/°C) |
|---|---|---|---|
| Steel (0.9% C) | | | |
| 20 °C | $7.8 \times 10^3$ | $4.6 \times 10^2$ | 46 |
| 900 °C | | $6.3 \times 10^2$ | 29 |
| Aluminum | | | |
| 20 °C | $2.7 \times 10^3$ | $0.9 \times 10^3$ | $2.4 \times 10^2$ |
| 600 °C | | $1.2 \times 10^3$ | $2.5 \times 10^2$ |
| Copper | | | |
| 20 °C | $8.95 \times 10^3$ | $3.8 \times 10^2$ | $3.9 \times 10^2$ |
| 900 °C | | $5.2 \times 10^2$ | $3.4 \times 10^2$ |

## 7.2 Thermodynamic Principles for Continuous Media

### Energy Conservation: The First Thermodynamic Principle

The energy conservation is first considered in the elementary form of the balance of various contributions. But the major result of this approach, the heat equation for deformable bodies, requires the introduction of the entropy function and will be analyzed in the next subsection.

The classical form of the first thermodynamic principle for a closed system is expressed as

$$\frac{dE}{dt} + \frac{dK}{dt} = \frac{dW}{dt} + \frac{dQ}{dt}, \tag{7.24}$$

where

$E$ is the internal energy,
$K$ is the kinetic energy,
$W$ is the work of external and internal forces, and
$Q$ is the input of heat.

The first principle can be written for any subdomain $\omega$ of a deformed body: Eq. (7.24) is then written with a $\omega$ index for the different contributions. If we define the density of internal energy $e$, we can write

$$\frac{dE_\omega}{dt} = \frac{d}{dt}\int_\omega \rho e \, dV = \int_\omega \rho \frac{de}{dt} dV, \tag{7.25}$$

where use has been made of the equation expressing the time derivative of an integral containing the density $\rho$.[6] Similarly, the rate of kinetic energy is

$$\frac{dK_\omega}{dt} = \frac{d}{dt}\int_\omega \frac{1}{2}\rho \mathrm{v}^2 dV = \int_\omega \rho \mathrm{v} \cdot \gamma \, dV, \tag{7.26}$$

where $\gamma$ is the acceleration: $\gamma = d\mathrm{v}/dt$.

[6] The general form of the equation is $(d/dt)\int_{\omega_t} \rho f dV = \int_{\omega_t} \rho (df/dt) dV$. This is from R. H. Wagoner and J.-L. Chenot, *Fundamentals of Metal Forming* (Wiley, New York, 1997), p. 170, Eq. 5.45.

The rate of work of external and internal forces is written as

$$\frac{dW_\omega}{dt} = \int_\omega \rho \mathbf{g} \cdot \mathbf{v} dV + \int_{\partial\omega} \mathbf{T} \cdot \mathbf{v} dS, \tag{7.27}$$

in which $\mathbf{T} = \boldsymbol{\sigma} \cdot \mathbf{n}$ is the stress vector and $\mathbf{n}$ is the outward normal. This equation is transformed with the application of the Green's theorem to the divergence of the stress tensor[7] to give

$$\frac{dW_\omega}{dt} = \int_\omega \rho \mathbf{g} \cdot \mathbf{v} dV + \int_\omega \boldsymbol{\sigma} : \mathbf{grad}(\mathbf{v}) dV + \int_\omega \mathrm{div}(\boldsymbol{\sigma}) \cdot \mathbf{v} dV. \tag{7.28}$$

We recall that $\boldsymbol{\sigma}$ is symmetric and the definition of the strain tensor is

$$\dot{\varepsilon} = \frac{1}{2}[\mathbf{grad}(\mathbf{v}) + \mathbf{grad}(\mathbf{v})^T], \tag{7.29}$$

so that we have the equality

$$\boldsymbol{\sigma} : \mathbf{grad}(\mathbf{v}) = \boldsymbol{\sigma} : \dot{\varepsilon}. \tag{7.30}$$

Now Eq. (7.28) can be modified with Eqs. (7.30) and (1.48) to give

$$\frac{dW_\omega}{dt} = \int_\omega \rho \mathbf{v} \cdot \boldsymbol{\gamma} dV + \int_\omega \boldsymbol{\sigma} : \dot{\varepsilon} dV. \tag{7.31}$$

Finally, the heat contribution is obtained by expressing the total heat flux input through the surface of $\omega$:

$$\frac{dQ_\omega}{dt} = -\int_{\partial\omega} \mathbf{q} \cdot \mathbf{n} dS, \tag{7.32}$$

where $\mathbf{q}$ is the heat flux vector on the surface $\partial\omega$, and we assume no internal heat is produced.[8] An equivalent form is obtained with the help of the divergence theorem, Eq. (5.8)[9]:

$$\frac{dQ_\omega}{dt} = -\int_\omega \mathrm{div}(\mathbf{q}) dV. \tag{7.33}$$

Now the different terms appearing in the first principle are combined, that is, Eqs. (7.25), (7.26), (7.31), and (7.33). We remark that the kinetic term appears in both sides, and can thus be simplified, so that we have

$$\int_\omega \rho \frac{de}{dt} dV = \int_\omega \boldsymbol{\sigma} : \dot{\varepsilon} dV - \int_\omega \mathrm{div}(\mathbf{q}) dV. \tag{7.34}$$

This equality between integrals must be satisfied for any subdomain $\omega$, and thus it allows statement of the final local form:

$$\rho \frac{de}{dt} = \boldsymbol{\sigma} : \dot{\varepsilon} - \mathrm{div}(\mathbf{q}). \tag{7.35}$$

[7] The equation is written here as $\int_\omega \mathrm{div}(\boldsymbol{\sigma}) \cdot \mathbf{v} dV = -\int_\omega \boldsymbol{\sigma} : \mathbf{grad}(\mathbf{v}) dV - \int_{\partial\omega} (\boldsymbol{\sigma} \cdot \mathbf{n}) \cdot \mathbf{v} dV$. It is Eq. 5.72 with $\Omega = \omega$, $\delta \mathbf{u} = \mathbf{v}$, and $\mathbf{T} = \boldsymbol{\sigma} \cdot \mathbf{n}$ of R. H. Wagoner and J.-L. Chenot, *Fundamentals of Metal Forming* (Wiley, New York, 1997), p. 176.

[8] If this is not the case, an additional term must be introduced in the left-hand side of Eq. (7.24).

[9] R. H. Wagoner and J.-L. Chenot, *Fundamentals of Metal Forming* (Wiley, New York, 1997).

### The Second Thermodynamic Principle and the Clausius–Duhem Inequality

Let us examine the elementary form of the second thermodynamic principle. A general closed system is considered, which is first supposed at a homogeneous temperature in the whole domain, but is allowed to vary with time. The second fundamental law of thermodynamics assumes that there exists a state function $S$, the entropy of the system, which verifies

$$\frac{\mathrm{d}S}{\mathrm{d}t} \geq \frac{1}{T}\frac{\mathrm{d}Q}{\mathrm{d}t}, \tag{7.36}$$

where the heat input $Q$ has the same definition as in Eq. (7.24).

It is important to note that the inequality in Eq. (7.36) becomes strict when the process is irreversible, and an equality when the process is reversible. Using this remark, we find it possible to decompose the entropy function into two contributions: $S^r$, the reversible part, and $S^i$, the irreversible one, and rewrite Eq. (7.36) in a nondifferential form as

$$S = S^r + S^i,$$

with

$$\frac{\mathrm{d}S^r}{\mathrm{d}t} = \frac{1}{T}\frac{\mathrm{d}Q}{\mathrm{d}t}, \quad \text{and} \quad \frac{\mathrm{d}S^i}{\mathrm{d}t} \geq 0. \tag{7.37}$$

Now the generalization of Eq. (7.36) to continuum mechanics proceeds in the same way as previously for the first principle: a local density of entropy $s$ is defined so that, for any subdomain $\omega$, the corresponding entropy will be expressed as

$$S_\omega = \int_\omega \rho s \,\mathrm{d}V. \tag{7.38}$$

Taking into account this definition, we can rewrite Eq. (7.36)[10]:

$$\frac{\mathrm{d}}{\mathrm{d}t}\int_\omega \rho s \mathrm{d}V \geq \int_{\partial\omega} -\frac{\mathbf{q}}{T}\cdot \mathbf{n}\,\mathrm{d}S, \tag{7.39}$$

where the entropy flux through the boundary is defined by taking Eq. (7.32) into account, and with the introduction of the local temperature $T$.[11] As in the previous subsection, Eq. (7.39) is easily transformed and gives

$$\int_\omega \rho\frac{\mathrm{d}s}{\mathrm{d}t}\mathrm{d}V + \int_\omega \mathrm{div}\left(\frac{\mathbf{q}}{T}\right)\mathrm{d}S \geq 0. \tag{7.40}$$

Using the definition of the divergence operator, we see immediately that

$$\mathrm{div}\left(\frac{\mathbf{q}}{T}\right) = \frac{\mathrm{div}(\mathbf{q})}{T} - \frac{\mathbf{q}\cdot\mathbf{grad}(T)}{T^2}. \tag{7.41}$$

Now if we recall that Eq. (7.40) must be satisfied for any subdomain $\omega$, and make use of Eq. (7.41), we obtain

$$\rho\frac{\mathrm{d}s}{\mathrm{d}t} + \frac{\mathrm{div}(\mathbf{q})}{T} - \frac{\mathbf{q}\cdot\mathbf{grad}(T)}{T^2} \geq 0, \tag{7.42}$$

[10] In order that the surface element will not be confused with the differential of the entropy function, it is denoted here by d$S$ (with italic) in the remaining of the section.

[11] It is easy to convince oneself that any other definition – for example, the mean temperature in the subdomain – cannot be consistent with a local treatment.

which is the simplest form of the Clausius–Duhem relation. Another form can be obtained if Eq. (7.35) is subtracted from Eq. (7.42) multiplied by $T$:

$$\rho\left(T\frac{ds}{dt}-\frac{de}{dt}\right)+\boldsymbol{\sigma}:\dot{\boldsymbol{\varepsilon}}-\frac{\mathbf{q}\cdot\mathbf{grad}(T)}{T}\geq 0. \tag{7.43}$$

Following many authors, we shall prefer to use the free energy state function $\Psi$, which is defined by

$$\Psi=e-Ts, \tag{7.44}$$

and which gives immediately

$$\frac{d\Psi}{dt}=\frac{de}{dt}-\frac{dT}{dt}s-T\frac{ds}{dt}. \tag{7.45}$$

Equation (7.45) allows the transformation of the second form of the Clausius–Duhem inequality, Eq. (7.43), by substituting $e$ into $\psi$ to obtain

$$-\rho\left(\frac{d\Psi}{dt}+s\frac{dT}{dt}\right)+\boldsymbol{\sigma}:\dot{\boldsymbol{\varepsilon}}-\frac{\mathbf{q}\cdot\mathbf{grad}(T)}{T}\geq 0. \tag{7.46}$$

---

**Exercise 7.2: Express the third form of the Clausius–Duhem equation when the process is reversible, with no heat exchange, and with an approximate constant and homogeneous temperature.**

The first condition allows writing Eq. (7.46) with the equality sign. The second condition corresponds to no heat flux, that is, $\mathbf{q}=0$. The third condition indicates that the temperature in the domain is constant in space and time so that

$$\mathbf{grad}(T)=0,\quad \text{and}\ \frac{dT}{dt}=0. \tag{7.2–1}$$

Taking into account these three conclusions leads us to rewrite Eq. (7.46) as

$$-\rho\frac{d\Psi}{dt}+\boldsymbol{\sigma}:\dot{\boldsymbol{\varepsilon}}=0. \tag{7.2–2}$$

---

### State Variables and State Functions

The system under consideration is defined locally by

- kinematic variables: the six independent components $\varepsilon_{ij}$ of the (small) strain tensor $\boldsymbol{\varepsilon}$,
- temperature $T$.

In contrast, state functions help to follow the evolution of the system. We have already introduced

- the internal energy $e$,
- the entropy $s$,
- the free energy $\Psi$.

They correspond to different descriptions of the system but, as state functions, they share the following properties:

- They are functions of the state variables and possibly of the internal variables.
- They are independent of the path between two states described by values of the state variables and internal variables.

Neglecting the influence of the internal variables for the moment, we suppose that the only state variables are the (small) strains $\varepsilon_{ij}$ and $T$. With our hypotheses we can thus write for the free energy state function:

$$\frac{d\Psi}{dt} = \frac{\partial \Psi}{\partial \varepsilon} : \frac{d\varepsilon}{dt} + \frac{\partial \Psi}{\partial T}\frac{dT}{dt}. \tag{7.47}$$

As we assumed that the strain tensor remains small, we are allowed to write $\dot{\varepsilon} = (d\varepsilon/dt)$. We also put $\dot{T} = (dT/dt)$ and Eq. (7.47) becomes

$$\frac{d\Psi}{dt} = \frac{\partial \Psi}{\partial \varepsilon} : \dot{\varepsilon} + \frac{\partial \Psi}{\partial T}\dot{T}. \tag{7.48}$$

Similar expressions can also be deduced for $s$ and $e$.

If Eq. (7.48) is put into the expression of the Clausius–Duhem inequality, Eq. (7.46), we get

$$\left(\sigma - \rho\frac{\partial \Psi}{\partial \varepsilon}\right) : \dot{\varepsilon} - \rho\left(\frac{\partial \Psi}{\partial T} + s\right)\dot{T} - \frac{\mathbf{q}\cdot\mathbf{grad}(T)}{T} \geq 0. \tag{7.49}$$

We suppose that a transformation occurs without deformation, that is, with $\dot{\varepsilon} = 0$ for any time, and with a homogeneous temperature field for which $\mathbf{grad}(T) = 0$. The remaining term must therefore satisfy

$$-\rho\left(\frac{\partial \Psi}{\partial T} + s\right)\dot{T} \geq 0.$$

As $\dot{T}$ can be either positive or negative, we conclude that[12]

$$s = -\frac{\partial \Psi}{\partial T}. \tag{7.50}$$

As $\Psi$ and $s$ are state functions, the equality given by Eq. (7.50) must be independent of the path, and therefore holds for general transformations. The same argument can be used for deformed bodies with uniform distribution of temperature, for which Eq. (7.49) reduces to

$$\left(\sigma - \rho\frac{\partial \Psi}{\partial \varepsilon}\right) : \dot{\varepsilon} \geq 0.$$

Considering that this inequality must be true for any strain-rate tensor $\dot{\varepsilon}$, we have necessarily

$$\sigma = \rho\frac{\partial \Psi}{\partial \varepsilon}. \tag{7.51}$$

Finally, taking into account Eqs. (7.50) and (7.51), we see that

$$\frac{\mathbf{q}\cdot\mathbf{grad}(T)}{T} \leq 0. \tag{7.52}$$

This last condition can be fulfilled if we put[13]

$$\mathbf{q} = -k\,\mathbf{grad}(T), \tag{7.53}$$

where the conductivity factor $k$ is positive and may depend on temperature.

[12] This proof is obviously approximate, as it is difficult to produce temperature change, in a real material, together with a homogeneous temperature distribution.

[13] More general forms for the heat flux can be used; for example, if the material is thermally anisotropic, one can introduce the linear relationship of Eq. (7.9). We note here that C is a definite and positive matrix, for the Clausius–Duhem inequality to be satisfied.

If the definition of the free energy function $\Psi$, Eq. (7.45), is used to transform the expression of the first principle, Eq. (7.35), the following form is derived:

$$\rho\left(\frac{\mathrm{d}\Psi}{\mathrm{d}t} + T\frac{\mathrm{d}s}{\mathrm{d}t} + s\frac{\mathrm{d}T}{\mathrm{d}t}\right) - \boldsymbol{\sigma} : \dot{\boldsymbol{\varepsilon}} + \mathrm{div}(\mathbf{q}) = 0. \tag{7.54}$$

Equations (7.45), (7.50), and (7.51) are used to simplify this expression, which therefore becomes

$$\rho T\frac{\mathrm{d}s}{\mathrm{d}t} + \mathrm{div}(\mathbf{q}) = 0. \tag{7.55}$$

The time derivative of the entropy state function is developed in a similar way as in Eq. (7.45) for the free energy:

$$\frac{\mathrm{d}s}{\mathrm{d}t} = \frac{\partial s}{\partial \boldsymbol{\varepsilon}} : \dot{\boldsymbol{\varepsilon}} + \frac{\partial s}{\partial T}\dot{T}. \tag{7.56}$$

This last equation allows us to rewrite Eq. (7.55) as

$$\rho T\frac{\partial s}{\partial T}\dot{T} + \rho T\frac{\partial s}{\partial \boldsymbol{\varepsilon}} : \dot{\boldsymbol{\varepsilon}} + \mathrm{div}(\mathbf{q}) = 0. \tag{7.57}$$

One remarks that, first by differentiation of Eq. (7.50) with respect to $\boldsymbol{\varepsilon}$, and second with the help of Eq. (7.51), it is possible to write

$$\frac{\partial s}{\partial \boldsymbol{\varepsilon}} = \frac{\partial}{\partial \boldsymbol{\varepsilon}}\left(-\frac{\partial \Psi}{\partial T}\right) = -\frac{\partial}{\partial T}\left(\frac{\partial \Psi}{\partial \boldsymbol{\varepsilon}}\right) = -\frac{1}{\rho}\frac{\partial \boldsymbol{\sigma}}{\partial T}. \tag{7.58}$$

The specific heat is defined by

$$c = T\frac{\partial s}{\partial T}. \tag{7.59}$$

The *heat equation for deformed bodies* is finally derived from Eq. (7.57) with the help of Eqs. (7.53), (7.58), and (7.59):

$$\rho c\dot{T} = \rho c\frac{\mathrm{d}T}{\mathrm{d}t} = \mathrm{div}[k\,\mathbf{grad}(T)] + T\frac{\partial \boldsymbol{\sigma}}{\partial T} : \dot{\boldsymbol{\varepsilon}}. \tag{7.60a}$$

It is often observed experimentally that, in practical situations, the last term on the right-hand side of Eq. (7.60a) can be neglected, leading to the well-known simplified form:

$$\rho c\dot{T} = \rho c\frac{\mathrm{d}T}{\mathrm{d}t} = \mathrm{div}[k\,\mathbf{grad}(T)]. \tag{7.60b}$$

### Internal State Variables

Internal state variables are introduced to describe the changes of the state functions with respect to physical phenomena that are difficult to measure directly. They will be denoted by $\boldsymbol{\alpha}$, with components $\alpha_i$, and they can represent the equivalent plastic strains from a reference configuration, or the phase change proportion, the density of dislocations, or any physical variable describing the internal state of the material. If the free energy function $\Psi$ is now considered a function of the (small) strain $\boldsymbol{\varepsilon}$, the internal variables $\boldsymbol{\alpha}$, and the temperature $T$, Eq. (7.48) must be modified into

$$\frac{\mathrm{d}\Psi}{\mathrm{d}t} = \frac{\partial \Psi}{\partial \boldsymbol{\varepsilon}} : \dot{\boldsymbol{\varepsilon}} + \frac{\partial \Psi}{\partial \boldsymbol{\alpha}} \cdot \dot{\boldsymbol{\alpha}} + \frac{\partial \Psi}{\partial T}\dot{T}. \tag{7.61}$$

In view of Eq. (7.61), the thermodynamic forces $\mathbf{A}$, associated with the internal variables $\boldsymbol{\alpha}$, can be defined formally, by analogy with Eq. (7.51):

$$\mathbf{A} = \rho\frac{\partial \Psi}{\partial \boldsymbol{\alpha}}.$$

This definition of the thermodynamic forces is introduced into Eq. (7.61), and the resulting expression is put into Eq. (7.45) (which gives the basic relation between free energy, internal energy, and entropy); then we get

$$\frac{de}{dt} = \boldsymbol{\sigma} : \dot{\boldsymbol{\varepsilon}} + \mathbf{A} \cdot \dot{\boldsymbol{\alpha}} + T\frac{ds}{dt}. \tag{7.62}$$

We use the same approach as in the previous subsection, where the influence of the internal variables was neglected. Starting again from Eq. (7.54), and recalling that the first term of the left-hand side is the time derivative of the internal energy, we find that Eq. (7.62) shows that Eq. (7.55) must be rewritten with one more term:

$$\rho T\frac{ds}{dt} + \mathbf{A} \cdot \dot{\boldsymbol{\alpha}} + \operatorname{div}(\mathbf{q}) = 0. \tag{7.63}$$

Now the rate of entropy is developed in terms of the time derivatives of the state and internal variables. The analog of Eq. (7.56) includes an additional contribution, involving the internal variables $\boldsymbol{\alpha}$ and its time derivative. For this to be written in a proper way, the following relation is introduced:

$$\frac{\partial s}{\partial \boldsymbol{\alpha}} \cdot \dot{\boldsymbol{\alpha}} = \frac{\partial}{\partial \boldsymbol{\alpha}}\left(-\frac{\partial \Psi}{\partial T}\right) \cdot \dot{\boldsymbol{\alpha}} = -\frac{\partial}{\partial T}\left(\frac{\partial \Psi}{\partial \boldsymbol{\alpha}}\right) \cdot \dot{\boldsymbol{\alpha}} = -\frac{1}{\rho}\frac{\partial \mathbf{A}}{\partial T} \cdot \dot{\boldsymbol{\alpha}}. \tag{7.64}$$

The heat equation is finally obtained by developing Eq. (7.63) and taking into account the definition of the heat capacity, Eq. (7.59), and Eqs. (7.58) and (7.64):

$$\rho c\dot{T} = \rho c\frac{dT}{dt} = \operatorname{div}[k\,\mathbf{grad}(T)] - \mathbf{A} \cdot \dot{\boldsymbol{\alpha}} + T\left(\frac{\partial \boldsymbol{\sigma}}{\partial T} : \dot{\boldsymbol{\varepsilon}} + T\frac{\partial \mathbf{A}}{\partial T} \cdot \dot{\boldsymbol{\alpha}}\right). \tag{7.65}$$

With the same hypothesis as in the previous subsection, the second thermodynamic principle is deduced from Eq. (7.46) with Eqs. (7.50), (7.51), and the definition of the thermodynamic forces associated with internal variables:

$$\mathbf{A} \cdot \dot{\boldsymbol{\alpha}} - \frac{k[\mathbf{grad}(T)]^2}{T} \geq 0. \tag{7.66}$$

## 7.3 Thermoelasticity

The thermodynamic approach of Section 7.3 can be illustrated in a relatively simple way: if an appropriate free energy function is introduced, the classical Hooke law for linear elastic material can be deduced. In addition, the thermal behavior is also obtained. When the influence of temperature is neglected, an energy formulation of the problem can be given, where the solution is the minimum of the energy functional.

### Mechanical and Thermal Behavior of a Linear Elastic Material

The free energy is approximated by a quadratic function of the strain tensor $\boldsymbol{\varepsilon}$, and of the difference $T - T_0$ between the temperature of the material and a reference temperature $T_0$. It is written

$$\rho\Psi = \rho\Psi_0 - \rho s_0(T - T_0) + \frac{1}{2}\lambda\theta^2 + \mu\boldsymbol{\varepsilon} : \boldsymbol{\varepsilon} - (3\lambda + 2\mu)\alpha\theta(T - T_0)$$
$$- \frac{\rho c}{2T_0}(T - T_0)^2, \tag{7.67}$$

where

$\theta = \mathrm{tr}(\varepsilon) = \sum_i \varepsilon_{ii}$, and
$\lambda, \mu$ are the usual Lamé coefficients.

If both $\varepsilon = 0$ and $T = T_0$, the free energy function is equal to the reference free energy $\rho\Psi_0$. The thermomechanical relationship is obtained by applying Eq. (7.51):

$$\sigma = \rho\frac{\partial \Psi}{\partial \varepsilon} = [\lambda\theta - (3\lambda + 2\mu)\alpha(T - T_0)]\mathbf{1} + 2\mu\varepsilon, \tag{7.68a}$$

which is the conventional notation for the component expression:

$$\sigma_{ij} = \rho\frac{\partial \Psi}{\partial \varepsilon_{ij}} = [\lambda\theta - (3\lambda + 2\mu)\alpha(T - T_0)]1_{ij} + 2\mu\varepsilon_{ij}, \tag{7.68b}$$

where $\mathbf{1}$ is the unit tensor, and $1_{ij} = \delta_{ij}$ is the Kronecker symbol.

From Eq. (7.68a), or Eq. (7.68b), we see that the stress state corresponds to the superposition of a purely elastic body obeying Hooke's law and of an additional pressure originating from an isotropic dilatation with the $\alpha$ parameter.[14] The entropy state function is derived with Eq. (7.50), from the expression of the free energy [Eq. (7.67)]:

$$\rho s = -\rho\frac{\partial \Psi}{\partial T} = \rho s_0 + (3\lambda + 2\mu)\alpha\theta + \frac{\rho c}{T_0}(T - T_0). \tag{7.69}$$

For an undeformed body with temperature equal to $T_0$, the entropy is equal to the reference value $s_0$. The internal energy $e$ is easy to deduce from the expressions of $\Psi$, $s$, and Eq. (7.44):

$$\rho e = \rho\Psi + \rho T s = \rho\Psi_0 + \rho s_0 + \frac{1}{2}\lambda\theta^2 + \mu\varepsilon : \varepsilon + (3\lambda + 2\mu)\alpha\theta T_0 + \frac{\rho c}{2T}(T^2 - T_0^2). \tag{7.70}$$

---

**Exercise 7.3: Write the heat equation for a thermoelastic solid with the free energy given by Eq. (7.67).**

In this case, in which no internal state variables were introduced, the heat equation is given by Eq. (7.60a), for which we have to calculate

$$\frac{\partial \sigma}{\partial T} = -(3\lambda + 2\mu)\alpha\theta\mathbf{1}. \tag{7.3–1}$$

When this expression is introduced into Eq. (7.60a) we obtain

$$\rho c\frac{\mathrm{d}T}{\mathrm{d}t} = \mathrm{div}[k\,\mathbf{grad}(T)] - (3\lambda + 2\mu)T\alpha\mathbf{1} : \dot{\varepsilon}. \tag{7.3–2}$$

If we remark that

$$\mathbf{1} : \dot{\varepsilon} = \mathrm{tr}(\dot{\varepsilon}) = \dot{\theta}, \tag{7.3–3}$$

we see that the heat equation is finally

$$\rho c\frac{\mathrm{d}T}{\mathrm{d}t} = \mathrm{div}[k\,\mathbf{grad}(T)] - (3\lambda + 2\mu)T\alpha\dot{\theta}. \tag{7.3–4}$$

On the right-hand side of Eq. (7.3–4), the second term corresponds to some extent to the work produced by the pressure forces.

---

[14] The $\alpha$ parameter is *not* an internal variable here.

### Energy Conjugate Variables

The free energy function appears as a potential of the strain tensor from which the stress tensor can be derived by Eq. (7.51). A complementary potential $\Psi'$ can be defined by the following relationship:

$$\rho\Psi'(\boldsymbol{\sigma}) = \boldsymbol{\sigma} : \boldsymbol{\varepsilon} - \rho\Psi(\boldsymbol{\varepsilon}), \tag{7.71}$$

which can be used to obtain the strain tensor by a similar derivation:

$$\boldsymbol{\varepsilon} = \rho\frac{\partial\Psi'}{\partial\boldsymbol{\sigma}}(\boldsymbol{\sigma}). \tag{7.72}$$

It is easy to show that Eqs. (7.51) and (7.72) are equivalent. By differentiation of Eq. (7.71) divided by $\rho$, we obtain for any $\delta\boldsymbol{\varepsilon}$ and $\delta\boldsymbol{\sigma}$:

$$\frac{\partial\Psi'}{\partial\boldsymbol{\sigma}} : \delta\boldsymbol{\sigma} = \delta\left(\frac{\boldsymbol{\sigma}}{\rho}\right) : \boldsymbol{\varepsilon} + \frac{\boldsymbol{\sigma}}{\rho} : \delta\boldsymbol{\varepsilon} - \frac{\partial\Psi}{\partial\boldsymbol{\varepsilon}} : \delta\boldsymbol{\varepsilon}.$$

If we assume that the evolution of the density $\rho$ can be neglected, the following equation is obtained:

$$\frac{\partial\Psi'}{\partial\boldsymbol{\sigma}} : \delta\boldsymbol{\sigma} = \delta\boldsymbol{\sigma} : \boldsymbol{\varepsilon} + \boldsymbol{\sigma} : \delta\boldsymbol{\varepsilon} - \frac{\partial\Psi}{\partial\boldsymbol{\varepsilon}} : \delta\boldsymbol{\varepsilon}. \tag{7.73}$$

If the expression of $\boldsymbol{\sigma}$ given by Eq. (7.51) is introduced into Eq. (7.73), and keeping in mind that $\delta\boldsymbol{\sigma}$ can be given any value, we come to the conclusion that Eq. (7.72) holds.

---

**Exercise 7.4: Give the explicit form of the complementary potential for a purely elastic potential.**

From the expression of $\Psi$ in Eq. (7.67), and of $\boldsymbol{\sigma}$ with $T = T_0 = 0$ in Eq. (7.68a), we have

$$\rho\Psi'(\boldsymbol{\sigma}) = (\lambda\theta\mathbf{1} + 2\mu\boldsymbol{\varepsilon}) : \boldsymbol{\varepsilon} - \frac{1}{2}\lambda\theta^2 - \mu\boldsymbol{\varepsilon} : \boldsymbol{\varepsilon},$$

which gives

$$\rho\Psi'(\boldsymbol{\sigma}) = \frac{1}{2}\lambda\theta^2 - \mu\boldsymbol{\varepsilon} : \boldsymbol{\varepsilon}, \tag{7.4–1}$$

where we remark that $\Psi = \Psi'$. We have to substitute $\boldsymbol{\sigma}$ for $\boldsymbol{\varepsilon}$ in Eq. (7.4–1) by inverting Eq. (7.68a). We first compute the trace of the tensors:

$$\sum_i \sigma_{ii} = -3p = 3\lambda\theta + 2\mu\theta, \tag{7.4–2}$$

which allows expression of the dilatation $\theta$ in term of the pressure $p$:

$$\theta = -\frac{1}{\lambda + \frac{2}{3}\mu}p.$$

Introducing this relationship in the constitutive equation leads to

$$\boldsymbol{\varepsilon} = \frac{1}{2\mu}\left(\boldsymbol{\sigma} + \frac{\lambda}{\lambda + 2\mu/3}p\mathbf{1}\right), \tag{7.4–3}$$

so that with Eqs. (7.4–2) and (7.4–3), Eq. (7.4–1) is now

$$\rho\Psi'(\boldsymbol{\sigma}) = \frac{1}{2}\frac{\lambda}{\lambda + \frac{2}{3}\mu}p^2 + \frac{1}{4\mu}\left[\boldsymbol{\sigma} : \boldsymbol{\sigma} + \frac{2\lambda}{\lambda + \frac{2}{3}\mu}p\,\mathrm{tr}(\boldsymbol{\sigma}) + \frac{3\lambda^2}{(\lambda + \frac{2}{3}\mu)^2}p^2\right],$$

or in the more compact form,

$$\rho\Psi'(\boldsymbol{\sigma}) = \frac{1}{4\mu}\boldsymbol{\sigma} : \boldsymbol{\sigma} - \frac{3\lambda}{4(\lambda + \frac{2}{3}\mu)\mu}p^2. \tag{7.4–4}$$

---

## 7.4 Elastoviscoplasticity and Elastoplasticity

Different methods can be developed to adapt the previous approach to irreversible processes:

- The irreversible deformation is considered as a tensor internal state variable.[15]
- The thermodynamic approach can also be modified to incorporate the irreversible deformation: this approach will be followed in this section.

We suppose first that the strain tensor $\varepsilon$ remains small, and we decompose it into an elastic contribution $\varepsilon^e$, and an irreversible part $\varepsilon^p$, according to

$$\varepsilon = \varepsilon^e + \varepsilon^p, \tag{7.74a}$$

or with the indicial notation:

$$\varepsilon_{ij} = \varepsilon^e_{ij} + \varepsilon^p_{ij}. \tag{7.74b}$$

The free energy function will depend on the elastic strain and temperature only; that is to say, we have $\Psi(\varepsilon^e, T)$. With Eq. (7.74a) the Clausius–Duhem inequality, Eq. (7.46), becomes

$$-\rho\left(\frac{d\Psi}{dt} + s\frac{dT}{dt}\right) + \boldsymbol{\sigma} : \dot{\varepsilon}^e + \boldsymbol{\sigma} : \dot{\varepsilon}^p - \frac{\mathbf{q}\cdot\mathbf{grad}(T)}{T} \geq 0, \tag{7.75}$$

so that the analog of Eq. (7.50) will be

$$\left(\boldsymbol{\sigma} - \rho\frac{\partial\Psi}{\partial\varepsilon^e}\right) : \dot{\varepsilon}^e + \boldsymbol{\sigma} : \dot{\varepsilon}^p - \rho\left(\frac{\partial\Psi}{\partial T} + s\right)\frac{dT}{dt} - \frac{\mathbf{q}\cdot\mathbf{grad}(T)}{T} \geq 0, \tag{7.76}$$

where we used the expansion of Eq. (7.50) with $\dot{\varepsilon}^e$ instead of $\varepsilon$. If the thermoelastic contributions verify separately the equations of Section 7.3, the first and third terms of Eq. (7.76) vanish, giving

$$\boldsymbol{\sigma} : \dot{\varepsilon}^p - \frac{\mathbf{q}\cdot\mathbf{grad}(T)}{T} \geq 0. \tag{7.77}$$

As Eq. (7.77) must hold for any condition, we can assume a special deformation process with a homogeneous temperature field, with the consequence that, in this case, the general equation reduces to

$$\boldsymbol{\sigma} : \dot{\varepsilon}^p \geq 0. \tag{7.78}$$

Finally, according to our previous derivations, we admit that for more general deformations paths, for which the plastic contribution to the deformation does not remain small, but the elastic stain tensor can be considered as small, then Eq. (7.78) still holds, while Eq. (7.74a) is replaced by

$$\dot{\varepsilon} = \dot{\varepsilon}^e + \dot{\varepsilon}^p, \tag{7.79a}$$

which means

$$\dot{\varepsilon}_{ij} = \dot{\varepsilon}^e_{ij} + \dot{\varepsilon}^p_{ij}. \tag{7.79b}$$

[15] See H. Ziegler, *An Introduction to Thermodynamics* (North-Holland, Amsterdam, 1983).

### Viscoplasticity

To satisfy Eq. (7.78), one can introduce a viscoplastic potential $\varphi$, function of the irreversible strain rate $\dot{\varepsilon}^p$.[16]

A complementary potential $\varphi'$ can also be defined so that

$$\dot{\varepsilon}^p = \frac{\partial \varphi'(\boldsymbol{\sigma})}{\partial \boldsymbol{\sigma}} \qquad \text{or } \dot{\varepsilon}^p_{ij} = \frac{\partial \varphi'(\boldsymbol{\sigma})}{\partial \sigma_{ij}}. \tag{7.80}$$

It is obtained by putting

$$\varphi'(\boldsymbol{\sigma}) = \boldsymbol{\sigma} : \dot{\varepsilon}^p - \varphi(\dot{\varepsilon}^p). \tag{7.81}$$

---

**Exercise 7.5: Give the expression of the complementary potential $\varphi'$ for the Norton viscoplastic potential.**

The equivalent stress $\bar{\sigma}$ is defined by

$$\bar{\sigma}^2 = \frac{3}{2}\mathbf{s} : \mathbf{s}, \tag{7.5–1}$$

which is immediately transformed with the help of Eq. (5.111),[17] giving

$$\bar{\sigma}^2 = \frac{3}{2} 4K^2(\sqrt{3}\dot{\bar{\varepsilon}}^p)^{2m-2}\dot{\varepsilon}^p : \dot{\varepsilon}^p \tag{7.5–2}$$

or

$$\bar{\sigma} = \sqrt{3}K(\sqrt{3}\dot{\bar{\varepsilon}}^p)^m. \tag{7.5–3}$$

Now Eq. (7.81), which defines the complementary potential, can be calculated with Eq. (7.5–3). For an incompressible material $\varphi'(\boldsymbol{\sigma})$ depends in fact on the deviatoric stress tensor only, and will therefore be denoted by $\varphi'(s)$:

$$\varphi'(\mathbf{s}) = 2K(\sqrt{3}\dot{\bar{\varepsilon}}^p)^{m-1}\dot{\varepsilon}^p : \dot{\varepsilon}^p - \frac{K}{m+1}(\sqrt{3}\dot{\bar{\varepsilon}}^p)^{m+1}, \tag{7.5–4}$$

which is also

$$\varphi'(\mathbf{s}) = \frac{m}{m+1}K(\sqrt{3}\dot{\bar{\varepsilon}}^p)^{m+1}. \tag{7.5–5}$$

Equation (7.5–3) allows the final form to be written:

$$\varphi'(\mathbf{s}) = \frac{m}{m+1}K\left(\frac{\bar{\sigma}}{\sqrt{3}K}\right)^{\frac{m+1}{m}}. \tag{7.5–6}$$

Equation (7.80) can now be used to deduce the stress – strain-rate relationship:

$$\dot{\varepsilon}^p = \frac{1}{2K}\left(\frac{\bar{\sigma}}{\sqrt{3}K}\right)^{\frac{1}{m}-1}\mathbf{s}. \tag{7.5–7}$$

---

### Elastoplasticity

For plastic deformation the stress tensor is independent of the irreversible strain-rate tensor $\dot{\varepsilon}^p$. This means that if we use again the previous formalism, Eq. (7.80) shows that the potential $\varphi$ must be homogeneous of degree one with respect to $\dot{\varepsilon}^p$.[18] But as the strain-rate tensor corresponding to a given stress tensor (or deviatoric stress

[16] R. H. Wagoner and J.-L. Chenot, *Fundamentals of Metal Forming* (Wiley, New York, 1997), Chap. 8, Sec. 5.8.

[17] R. H. Wagoner and J.-L. Chenot, *Fundamentals of Metal Forming* (Wiley, New York, 1997).

[18] In order that, after differentiation to get the stress tensor, the result will be of degree zero, i.e., invariant when the plastic strain-rate tensor is multiplied by any scalar.

tensor) is not unique (it can be multiplied by any positive number), it is not possible to build a complementary potential in the same way as in Eq. (7.80). It must therefore be replaced by the flow rule concept:

$$\dot{\varepsilon}^p = \dot{\lambda}_p \frac{\partial \phi}{\partial \sigma}(\sigma) \qquad \text{with } \dot{\lambda}_p \geq 0. \tag{7.82}$$

Here $\phi$ is again a positive convex function, as it can be proved mathematically from Eq. (7.78). Moreover, from the experimental approach it is observed that plastic deformation cannot occur until the stress state reaches a limit defined by a yield criterion $f$, so that

if $f(\sigma) < 0$, only elastic deformation occurs;
if $f(\sigma) = 0$ and $(\partial\phi/\partial\sigma)(\sigma) : \sigma \geq 0$, we have plastic deformation;
and $f(\sigma) > 0$ is meaningless.

More often the yield function $f$ is used also in Eq. (7.82) instead of $\phi$, and the flow rule is termed *associated* in this case.

#### Elastic Viscoplastic Behavior

In complex metal-forming situations, when the temperature may vary over a wide range, the mechanical behavior must also vary between an elastoplastic behavior and an elastic viscoplastic behavior. The Perzyna-like constitutive equation[19] is a convenient way to achieve this goal. In the case of a power law, the corresponding isotropic viscoplastic potential is written in the form

$$\varphi'(\sigma) = \frac{m}{m+1} K \left\langle \frac{\bar{\sigma} - R}{K} \right\rangle^{(m+1)/m}, \tag{7.83}$$

where

the effective (or equivalent) stress $\bar{\sigma}$ is defined in the usual way by Eq. (7.5–1);
the consistency is a function of the equivalent strain and of the temperature, $K(\bar{\varepsilon}, T)$;
the yield stress $R$ is also a function of the same parameters, $R(\bar{\varepsilon}, T)$; and
the Mac Cauley bracket is defined by $\langle x \rangle = 0$ if $x \leq 0$, $\langle x \rangle = x$ if $x > 0$.

Now the irreversible strain rate is written:

$$\dot{\varepsilon}^p = \frac{3}{2\bar{\sigma}} \left\langle \frac{\bar{\sigma} - R}{K} \right\rangle^{\frac{1}{m}} \mathbf{s}. \tag{7.84}$$

#### The Heat Equation for Plastic or Viscoplastic Materials

The thermodynamic approach outlined in Section 7.2 for the heat equation has to be slightly adapted. With the strain-rate decomposition hypothesis, Eq. (7.54) becomes

$$\rho \left( \frac{d\Psi}{dt} + T \frac{ds}{dt} + s \frac{dT}{dt} \right) - \sigma : \dot{\varepsilon}^e - \sigma : \dot{\varepsilon}^p + \operatorname{div}(\mathbf{q}) = 0. \tag{7.85}$$

[19] P. Perzyna, "The Constitutive Equations for Rate Sensitive Plastic Materials," *Quart. Appl. Math.* 20, 321–332 (1963).

With the same procedure as described previously, that is, with the assumption that the stress is derived from the free energy as a function of the elastic strain, and with Eqs. (7.50) and (7.59), we obtain a new form of the heat equation:

$$\rho c \frac{\mathrm{d}T}{\mathrm{d}t} = \mathrm{div}[k\,\mathbf{grad}(T)] + \boldsymbol{\sigma} : \dot{\boldsymbol{\varepsilon}}^p + T\frac{\partial \boldsymbol{\sigma}}{\partial T} : \dot{\boldsymbol{\varepsilon}}^e. \tag{7.86a}$$

The additional term $\dot{w} = \boldsymbol{\sigma} : \dot{\boldsymbol{\varepsilon}}^p$ represents the rate of plastic or viscoplastic work, which is supposed to be entirely converted into heat here. The third contribution to the heat equation on the right-hand side of Eq. (7.86a) is often neglected on an experimental basis, giving the simpler form

$$\rho c \frac{\mathrm{d}T}{\mathrm{d}t} = \mathrm{div}[k\,\mathbf{grad}(T)] + \boldsymbol{\sigma} : \dot{\boldsymbol{\varepsilon}}^p. \tag{7.86b}$$

The evolution of some of the internal parameters can be viewed as the storage of a fraction of plastic work, so that a fraction of $\dot{w}$ is *not* converted into heat. It can be shown that, when internal parameters are taken into account, the new form of the heat equation will be

$$\rho c \dot{T} = \mathrm{div}[k\,\mathbf{grad}(T)] + \boldsymbol{\sigma} : \dot{\boldsymbol{\varepsilon}}^p - \mathbf{A} \cdot \dot{\boldsymbol{\alpha}} + T\left(\frac{\partial \boldsymbol{\sigma}}{\partial T} : \dot{\boldsymbol{\varepsilon}}^e + \frac{\partial \mathbf{A}}{\partial T} \cdot \dot{\boldsymbol{\alpha}}\right), \tag{7.87}$$

which is clearly the analog of Eq. (7.65) for plastic or viscoplastic materials.

## 7.5 Thermomechanical Coupling

### Introduction of the Material Dilatation

If the material is assumed elastic viscoplastic, or elastoplastic, we assume that Eq. (7.68a), or Eq. (7.68b), still holds for the elastic contribution:

$$\boldsymbol{\sigma} = [\lambda\theta^e - (3\lambda + 2\mu)\alpha_d(T - T_0)]\mathbf{1} + 2\mu\boldsymbol{\varepsilon}^e. \tag{7.88}$$

The coefficient $\alpha_d$, where the subscript "$d$" was added here to avoid confusion with the internal parameters, corresponds to linear dilatation. Equation (7.88) can be differentiated with respect to time to give the rate form[20,21]

$$\dot{\boldsymbol{\sigma}} = [\lambda\dot{\theta}^e - (3\lambda + 2\mu)\alpha_d\dot{T}]\mathbf{1} + 2\mu\dot{\boldsymbol{\varepsilon}}^e. \tag{7.89}$$

The trace operator is applied to both sides of Eq. (7.89) to give

$$\dot{p} = -\left(\lambda + \frac{2}{3}\mu\right)(\dot{\theta}^e - 3\alpha_d\dot{T}). \tag{7.90}$$

However, if the material is rigid plastic or rigid viscoplastic, Eq. (7.90) cannot be used. It is replaced by the density evolution that is due to dilatation with the parameter $\alpha_d$ for one-dimensional deformation[22]:

$$\mathrm{div}(\mathbf{v}) - 3\alpha_d\dot{T} = 0. \tag{7.91}$$

When the process is stationary, Eq. (7.91) is rewritten as

$$\mathrm{div}(\mathbf{v}) = 3\alpha_d\dot{T} = 3\alpha_d\mathbf{v} \cdot \mathbf{grad}(T). \tag{7.92}$$

[20] The Jauman derivative must be used instead of the material derivative if large rotations can occur. It is defined by the equation $\mathrm{d}_J\boldsymbol{\sigma}/\mathrm{d}t = \dot{\boldsymbol{\sigma}} - \boldsymbol{\omega}\boldsymbol{\sigma} + \boldsymbol{\sigma}\boldsymbol{\omega}$.

[21] When the parameters do not depend on temperature; otherwise, they must be differentiated with respect to temperature.

[22] Here no approximation is made on the derivative provided that $\dot{T}$ is the material derivative.

### The Heat Equation When Plastic or Viscoplastic Deformation Occurs

In the more general case, Eq. (7.87) is used where the internal parameters can be introduced to model different kinds of phenomenona related to changes of the microstructure of the material: evolution of the dislocation density, phase changes, and so on.

For *elastic plastic* (or viscoplastic) deformation without phase change, a first approximation can be introduced. We shall assume that the effect of the evolution of the internal vector parameter $\boldsymbol{\alpha}$ describing the microstructure of the material is taken into account approximately by simply making a constant fraction $r$ of the viscoplastic work transformed into heat. The remaining $(1-r)$ contribution is stored in the material to model metallurgical evolution at the grain or dislocation level. Generally the $r$ parameter is taken in the range 0.9–1.

Taking into account these two approximations, and using an elastic viscoplastic constitutive equation, we can then rewrite the heat equation:

$$\rho c\dot{T} = \text{div}[k\,\mathbf{grad}(T)] + r\boldsymbol{\sigma} : \dot{\boldsymbol{\varepsilon}}^p + T\frac{\partial\boldsymbol{\sigma}}{\partial T} : \dot{\boldsymbol{\varepsilon}}^e. \tag{7.93}$$

In the *purely plastic* (or viscoplastic) behavior, the elastic strain-rate tensor can be neglected in most forming problems[23]:

$$\rho c\dot{T} = \text{div}[k\,\mathbf{grad}(T)] + r\boldsymbol{\sigma} : \dot{\boldsymbol{\varepsilon}}^p. \tag{7.94}$$

For the well-known Norton–Hoff viscoplastic material, Eq. (7.94) may be written as

$$\rho c\frac{\mathrm{d}T}{\mathrm{d}t} = \text{div}[k\,\mathbf{grad}(T)] + rK(\sqrt{3}\dot{\bar{\varepsilon}})^{m+1}. \tag{7.95}$$

### Thermal Dependency of the Constitutive Equations and Friction Law

When the temperature range is significant, most of the mechanical or thermal parameters exhibit at least noticeable evolutions. The first approach is to use a linear interpolation: that can been done for the Lamé elastic parameters and for the thermal parameters. For the viscoplastic behavior, an exponential form is preferred as it corresponds to the classical activation energy law. The viscoplastic consistency $K$ is allowed to vary according to

$$K(T) = K_1 \exp(\beta/T). \tag{7.96}$$

The rate sensitivity index $m$ is most often taken as a constant, but when a temperature variation can be measured experimentally, a linear variation is used:

$$m(T) = m_0 + m_1 T. \tag{7.97}$$

If a viscoplastic friction law is utilized and written

$$\tau(T) = -\alpha_f K(T)|\Delta\mathbf{v}|^{p-1}\Delta\mathbf{v}, \tag{7.98}$$

then the thermal evolution of the friction is taken into account thanks to the consistency function $K(T)$. The Coulomb friction law can also be temperature dependent with an exponential form.

[23] This is not the case when the plastic contribution to the strain-rate tensor is of the same order of magnitude as the elastic one; and more specifically for cyclic loading, when the strain-rate tensor exhibits large variations.

### Thermal Influence on the Boundary Conditions

The boundary conditions given by Eqs. (7.18) and (7.19) depend on the surface temperature. They can represent, for example, the free surface condition, which is composed of the sum of a convective term and a radiative term. The contact zone with the tool must incorporate a conductive term, plus a surface heat dissipation that is due to friction. The first approximation is to assume that this last contribution is shared between tool and part according to

$$\phi_f = \frac{b}{b + b_{\text{to}}} \alpha_f K(T) |\Delta \mathbf{v}|^{p+1}, \tag{7.99}$$

where $b$ and $b_{\text{to}}$ are the effusivity of the material of the part and of the tool, respectively.

The material effusivity is defined from the conductivity parameter by

$$b = \sqrt{\rho c k}. \tag{7.100}$$

### Finite-Element Formulation for the Coupled Thermal and Mechanical Problem

The mechanical equation generally expressed as the virtual work principle is written and the unknown field[24] is discretized with the finite-element method. We note that the dilatation term introduces the time derivative of the temperature, while the temperature itself is present in different places as a result of the thermal variation of constitutive parameters and of the boundary conditions. As usual the arrays $\mathbf{V}$, $\mathbf{X}$, $\dot{\mathbf{T}}$, and $\mathbf{T}$ denote the set of nodal values of the velocity, nodal coordinates, time derivatives of the temperature, and temperature, respectively. The set of mechanical equations is then written symbolically:

$$\mathbf{R}(\mathbf{V}, \mathbf{X}, \dot{\mathbf{T}}, \mathbf{T}) = 0, \tag{7.101}$$

with the usual relation

$$\mathbf{V} = \frac{\mathrm{d}\mathbf{X}}{\mathrm{d}t}, \tag{7.102}$$

and with the definition of $\dot{\mathbf{T}}$:

$$\dot{\mathbf{T}} = \frac{\mathrm{d}\mathbf{T}}{\mathrm{d}t}. \tag{7.103}$$

Similarly, the heat equation, Eqs. (7.93), (7.94), or (7.95), is written in integral form according to Section 4.6, and it is discretized into finite elements. It is clear that the heat dissipation of plastic or viscoplastic work (and possibly of elastic work) will introduce a term that contains the velocity. The analog of Eq. (7.101) for the heat equation is then

$$\mathbf{S}(\mathbf{V}, \mathbf{X}, \dot{\mathbf{T}}, \mathbf{T}) = 0. \tag{7.104}$$

We then have a system of ordinary differential equations, given in an implicit form by the nonlinear equations Eqs. (7.101) and (7.104), and the derivatives in Eqs. (7.102) and (7.103).

[24] We shall consider the velocity field here, but it can also be the displacement field, or one of them plus the pressure field in mixed formulation, etc.

**Resolution Procedures of the Discretized Problem**

The most satisfactory procedure is to solve the complete system simultaneously with respect to $\mathbf{V}$ and $\dot{\mathbf{T}}$. The Newton–Raphson method can be utilized. Starting from an initial value of $\mathbf{V}$ and $\dot{\mathbf{T}}$, any iteration is formally the same; new values of these parameters are computed from the increments $\Delta\mathbf{V}$ and $\Delta\dot{\mathbf{T}}$ by solving the usual system:

$$\begin{bmatrix} \partial\mathbf{R}/\partial\mathbf{V} & \partial\mathbf{R}/\partial\dot{\mathbf{T}} \\ \partial\mathbf{S}/\partial\mathbf{V} & \partial\mathbf{S}/\partial\dot{\mathbf{T}} \end{bmatrix} \begin{bmatrix} \Delta\mathbf{V} \\ \Delta\dot{\mathbf{T}} \end{bmatrix} = \begin{bmatrix} -\mathbf{R}(\mathbf{V}, \mathbf{X}, \dot{\mathbf{T}}, \mathbf{T}) \\ -\mathbf{S}(\mathbf{V}, \mathbf{X}, \dot{\mathbf{T}}, \mathbf{T}) \end{bmatrix}. \tag{7.105}$$

It is worthy to note that two matrices were already computed:

$$\mathbf{H} = \frac{\partial\mathbf{R}}{\partial\mathbf{V}} \tag{7.106}$$

is the usual matrix derivative for the purely mechanical problem, while

$$\mathbf{C} = \frac{\partial\mathbf{S}}{\partial\dot{\mathbf{T}}} \tag{7.107}$$

is the heat capacity matrix for the thermal problem alone. The other matrices in Eq. (7.105) express the coupling condition, where

$\partial\mathbf{R}/\partial\mathbf{V}$ is the thermal coupling in the mechanical equation, and
$\partial\mathbf{S}/\partial\mathbf{V}$ represents the coupling of mechanics on the temperature distribution.

The potentially more consistent and accurate scheme is to solve simultaneously Eqs. (7.101) and (7.104). This can be done with the Newton–Raphson method, Eq. (7.105), on all the nodal unknowns, but this significantly increases the CPU time and is used very seldom if ever in the literature. This can also be done with an alternative method, which consists of solving the mechanical problem alone, then putting the new velocity field in the heat equation, and solving for the time derivative of the temperature; then the velocity field is updated, taking into account the new value of $\dot{T}$, and so on. This method converges rapidly when the coupling terms are smaller than the main matrices.

The time-integration scheme can be chosen independently in this case: the one-step explicit scheme is often kept for the mechanical equation while more sophisticated methods are necessary for the heat equation like the Dupont scheme, which was briefly described in Section 4.6.

With a one-step explicit formulation for the velocity, it is consistent to use an approximate coupling scheme; that is, the coupling is not performed within each time increment so that the flow chart of the code is as follows. Starting from the domain $\Omega_t$ at time $t$, and temperatures $\mathbf{T}^t$, and time derivative of the temperature $\dot{\mathbf{T}}^{t-\Delta t}$:

1. Calculate $\mathbf{V}^t$ with consistency $K^t$, and $\dot{\mathbf{T}}^{t-\Delta t}$.
2. Update the domain to get $\Omega_{t+\Delta t}$.
3. Compute the time derivative $\dot{\mathbf{T}}^t$ of temperature field on the domain $\Omega_t$, using the plastic work dissipation calculated with $\mathbf{V}^t$.
4. Update the new temperature distribution $\mathbf{T}^{t+\Delta t}$ on the domain $\Omega_{t+\Delta t}$; then go to step 1, for the next time increment.

## PROBLEMS

### A. Proficiency Problems

1a. A cylinder with radius $R$ and height $h$ is heated by a radial heat flux $q$ perpendicular to its axis. In cylindrical coordinates $(r, q, z)$ we assume

$$\mathbf{q} = \begin{bmatrix} q_r \\ q_\theta \\ q_z \end{bmatrix} = \begin{bmatrix} q \\ 0 \\ 0 \end{bmatrix}.$$

Compute the time derivative of the mean temperature of the cylinder, the material of which has a density $\rho$ and a specific heat $c$.

1b. Perform the same calculation if the vector heat flux is defined for $\mathbf{x}_1 \geq 0$ in Cartesian coordinates by

$$\mathbf{q} = \begin{bmatrix} q_1 \\ q_2 \\ q_3 \end{bmatrix} = \begin{bmatrix} q \\ 0 \\ 0 \end{bmatrix}.$$

and is null for $x_1 < 0$ (the axis of the cylinder being the $Ox_3$ axis).

2. A cylindrical sample of radius $R = 1$ cm and length $L = 10$ cm is electrically heated at one end by a current of intensity $I = 10$ A and of voltage $U = 20$ V (see Fig. 7.1). At the other end the sample is cooled. The temperature in the sample is recorded at two points close to the axis and separated by a distance of 10 cm: the records of temperature evolution is given in the figure below. We suppose that the energy loss on the lateral surface of the sample is negligible. Compute the thermal conductivity of the material, and discuss the validity and the precision of the result.

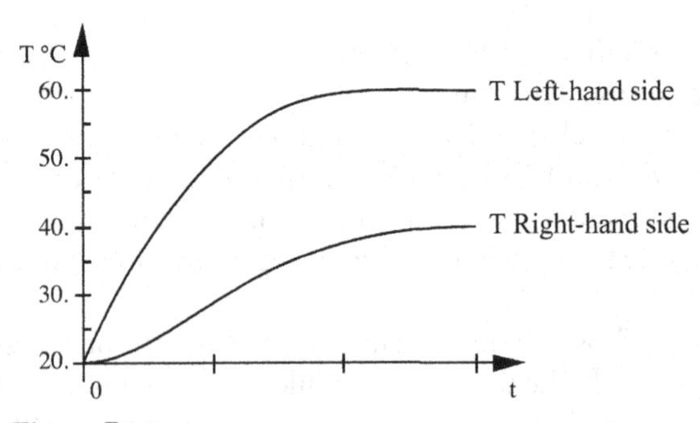


Figure P7.2.

3. We consider two bodies separated by a thin layer (of thickness $h$) of a material with a large conductivity parameter $k$ (see figure below).

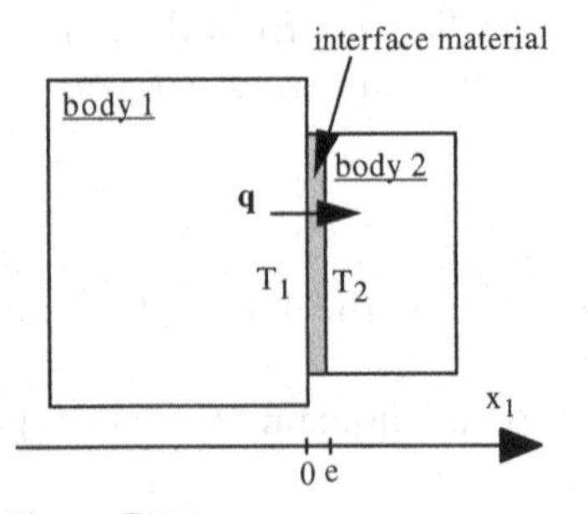


Figure P7.3.

Show that, if we assume that the thickness $e$ of the intermediate material is negligible, the normal heat flux between the two bodies can be expressed in terms of the contact resistance $R$ by

$$q = -\frac{1}{R}(T_2 - T_1).$$

4. A material is heated so that its temperature is increased $\Delta T$. The one-dimensional dilatation constant is $\alpha_d$; show that the deformation of the body caused by heating only is

$$\Delta \varepsilon = \alpha_d \Delta T \mathbf{1}.$$

Prove that the density change $\Delta\rho$ during the dilatation is related to the volume change $\Delta V$ by

$$\frac{\Delta\rho}{\rho} = -\frac{\Delta V}{V},$$

and deduce the final relation for incompressible materials:

$$\Delta\theta = 3\alpha_d \Delta T.$$

5. The deformation of a viscoplastic body is described by a velocity field with components

$$\mathbf{v} = \begin{bmatrix} v_1 \\ v_2 \\ v_3 \end{bmatrix} = \begin{bmatrix} ax_1^2 \\ ax_1x_2 \\ -3ax_1x_3 \end{bmatrix},$$

and the temperature field is

$$T = b\left(x_1^3 + x_2^3 + x_3^3\right).$$

Compute the rate of internal energy.

6. Evaluate the temperature raise during a homogeneous adiabatic deformation corresponding to an equivalent strain $\bar{\varepsilon}$. Compute the numerical value for a steel with yield stress $\sigma_0 = 500$ MPa and for $\bar{\varepsilon} = 1$, neglecting strain hardening.

## B. Depth Problems

7. A two-dimensional steady-state thermal problem is considered in the rectangular domain with sides $a$ and $b$, as pictured in the figure below.

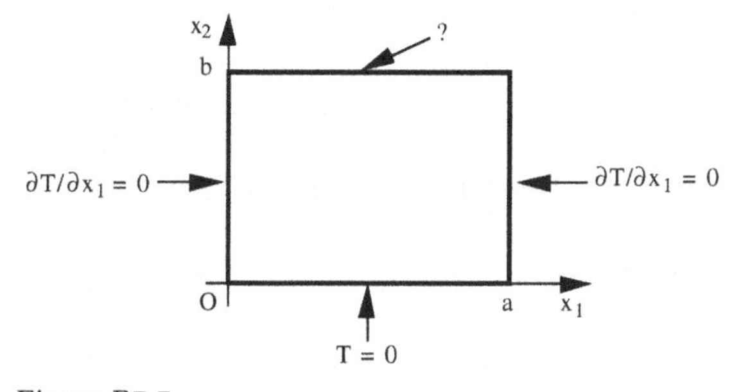


Figure P7.7.

The thermal conductivity is constant and the boundary conditions are

$$\text{for } x_1 = 0, \text{ or } x_1 = a, \quad -k\frac{\partial T}{\partial x_1} = 0;$$

$$\text{for } x_2 = 0, \quad T = 0.$$

An analytical solution is searched for with the form

$$T = f(x_1)g(x_2).$$

Show that the $f$ and $g$ functions must satisfy ordinary differential equations, and that $f$ is a solution that depends only on two parameters (one integer and one real parameter) when the boundary conditions are taken into account. Deduce the general form of $g$ and of the solution $T$. Find the temperature distribution for this solution on the fourth boundary $x_2 = b$.

8. With the same kind of methodology, solve the transient heat equation by putting

$$T = f(x_1)g(x_2)h(t).$$

9. Using boundary conditions given by Eqs. (7.17)–(7.19), write the semidiscretized integral equation corresponding to the heat equation, using the same method as in Chap. 4, Section 4.5. Show that the heat conductivity matrix **H** and the load vector **F** are modified.

10. The one-dimensional Sellars and Tegart viscoplastic law is given by

$$\dot{\varepsilon} = A[\sinh(\alpha\sigma)]^n,$$

where $A$, $\alpha$, and $n$ are experimental parameters that depend on the material ($A$ depends also on the temperature). Show for small values of the stress $s$ that this law is a good approximation of the Norton law. Express the form of the viscoplastic complementary potential and of the potential corresponding to the Sellars and Tegart law for incompressible material. Generalize the stress–strain-rate constitutive equation to three-dimensional problems.

11a. Compute the integral form of the coupling terms of Eq. (7.101) for a Norton–Hoff viscoplastic material with an exponential variation of the consistency, Eq. (7.96), the linear rate sensitivity index, Eq. (7.97), a friction law, given by Eq. (7.98), and a dilatation equation with a constant coefficient, Eq. (7.91).

11b. Compute the coupling terms for the thermal equation with the viscoplastic dissipation, Eq (7.95), transformed into an integral form.

## C. Numerical and Computational Problems

12. Numerically solve Problem 6 by a finite discretization of the rectangular domain into triangles, with the same boundary conditions. Observe the evolution of the error between the analytical solution and the finite-element approximation, when the mesh is uniformly refined.

13. Numerically solve Problem 7 by a finite discretization of the rectangular domain into triangles, with the same boundary conditions. The initial temperature distribution at time $t = 0$ will be the same as that of the analytical solution. Observe the evolution of the error between the analytical solution and the finite-element approximation, when the mesh is uniformly refined, or when the time increment is refined.

CHAPTER EIGHT

# Sheet-Metal Formability Tests

In this chapter,[1] sheet-metal *formability tests*, such as the tensile test, the plane-strain test, and the in-plane stretching test are analyzed. The FEM and experimental methods were used in order to demonstrate how two-dimensional in-plane simulation can help interpret and develop such tests, as well as to understand the nature of material behavior and governing mechanics.

A successful sheet-metal forming process can convert an initially flat sheet into a useful part of the desired shape. The major failures that may be encountered are splitting, wrinkling, and shape distortion. The deformed part can be considered unusable if any one of these failures occurs. Formability tests may be used to assess the capacity of a sheet to be deformed into a useful part.

Since sheet-metal forming operations are diverse in type, extent, and rate, many formability tests have been proposed. No single one can provide an accurate indication of the formability for all situations. Formability tests can be divided into two types: intrinsic and simulative. The intrinsic tests measure the basic material properties under certain stress–strain states, for example, the uniaxial tensile test and the plane-strain tensile test. Simulative tests subject the material to deformation that closely resembles a particular press-forming operation. A simulative test can provide limited and specific information that may be sensitive to factors other than the material properties, such as the thickness, surface condition, lubrication, and geometry and type of tooling.

## 8.1 Tensile Test

### Invariance of Strain Distribution to Material Strength and Deformation Rate[2]

A two-dimensional finite-element simulation of the tensile test will be presented in this section for a case in which the work-hardening and strain-rate sensitivity can be written as follows:

$$\sigma = k\varepsilon^n\dot{\varepsilon}^m. \tag{8.1}$$

[1] Chapter 8 was drafted by Dr. Dajun Zhou, Department of Material Science and Engineering, The Ohio State University. It is based on a series of papers by R. H. Wagoner and co-workers as referenced throughout.

[2] K. Chung and R. H. Wagoner, "Invariance of Neck Formation to Material Strength and Strain Rate for Power-Law Materials," *Metal Trans. A.* **17A**, 1631 (1986).
K. Chung and R. H. Wagoner, "Invariance of Plastic Strains with Respect to Imposed Rate at Boundary," *Metals and Materials* (Vol. 4, No. 1), 25–31 (1998).

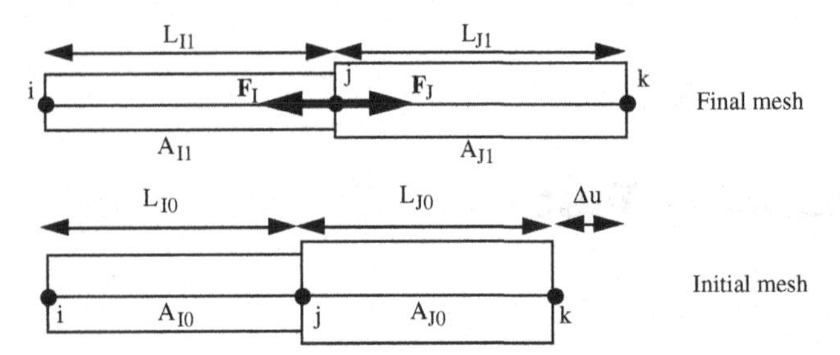


Figure 8.1. A simple one-dimensional problem with three nodes and two elements.

If this constitutive law is obeyed, then the plastic strain distributions for problems with displacement boundary conditions (i.e., no nonzero forces are specified) are independent of material strength constant, $k$, and strain rate, $\dot{\varepsilon}$. Since the strain rate is determined by imposed deformation velocity, the plastic strain distribution may be shown to be independent of imposed deformation velocity as long as Eq. (8.1) is obeyed.

In order to illustrate this concept in the simplest way, we will consider a non-uniform tensile bar of a strain-hardening and strain-rate-sensitive material. A procedure for analyzing this problem in one dimension, using three nodes and two elements, is outlined below.

1. The FEM mesh and dimensions are shown in Fig 8.1, including all geometrical and boundary information:
   a. The element length and cross-section area for element $I$ and $J$ and initial and final configuration;
   b. The boundary condition – fixed at node $i$ and prescribed displacement, $\Delta u$, at node $k$.
2. The mesh is deformed over a time interval of $\Delta t$. After $\Delta t$, nodes $j$ and $k$ are found in new positions and both elements have new lengths and strain increments. The force equilibrium at node $j$ is simply expressed as

$$F_I = F_J$$
$$\sigma_I A_I = \sigma_J A_J, \tag{8.2}$$

   where

$$F_I = k(\varepsilon_I + \Delta\varepsilon_I)^n \dot{\varepsilon}_I^m A_{I0} \exp(-\varepsilon_I - \Delta\varepsilon_I) \tag{8.3}$$

   and

$$F_J = k(\varepsilon_J + \Delta\varepsilon_J)^n \dot{\varepsilon}_J^m A_{J0} \exp(-\varepsilon_J - \Delta\varepsilon_J). \tag{8.4}$$

   The strain rates are dependent on incremental strain and the time interval as follows:

$$\dot{\varepsilon}_I = \Delta\varepsilon_I/\Delta t, \quad \text{and } \dot{\varepsilon}_J = \Delta\varepsilon_J/\Delta t. \tag{8.5}$$

   Substituting Eqs. (8.3) and (8.4) into (8.2) reveals that the strength constant, $k$, cannot effect the equilibrium. Any constant multiplying both sides of Eq. (8.2) also cannot change the status for equilibrium. Thus, the strains are invariant with respect to material strength.
3. What happens when the time interval for the same incremental strain changes to a shorter one, say from $\Delta t$ to $\Delta t^*$? Obviously, substituting a new time interval into Eq. (8.5) will change the strain rates in each element. Replacing

these new strain rates in Eqs. (8.3) and (8.4) will certainly change the nodal force (as does a change of strength coefficient). But, as the forces in Eqs. (8.3) and (8.4) change simultaneously with respect to their original value by a factor of $(\Delta t/\Delta t^*)^m$, the equilibrium will not be affected. Thus the strain distribution is invariant to imposed deformation velocity.

For the case of multiaxial deformation, in order to show the invariance of strain distribution, we have to use the concepts of effective strain increment and effective stress. In addition, one more assumption has to be made: the homogeneity of the employed yield function. Because of the homogenous nature of the governing equations, multiplying the strain rate or stress by a constant will have a multiplicative effect on the effective strain rate or effective stress. Mathematically, this idea can be expressed as follows:

$$\dot{\bar{\varepsilon}} = f(\alpha\dot{\varepsilon}_{ij}) = \alpha f(\dot{\varepsilon}_{ij}), \tag{8.6}$$

or

$$\bar{\sigma} = f(\alpha^m \sigma_{ij}) = \alpha^m f(\sigma_{ij}), \tag{8.7}$$

where $\alpha = (\Delta t/\Delta t^*)$, for example.

Following the same procedure, we find it easy to prove that an unchanged strain distribution and the associated stress distribution (which will differ by a constant factor) satisfy the equilibrium condition for each node when only a strength constant or a strain-rate change occurs.

The invariance of the strain distribution will not hold in the following situations.

1. In nonisothermal condition cases created by deformation-induced heating, the overall strain distribution will depend on heat generation and heat flow, which depend on rate and strength.
2. If the strain rate is too large (e.g., $>10^2$) inertial effects can become significant. A quasi-static mechanical equilibrium analysis, such as that in Eq. (8.2), should be replaced by a dynamic equilibrium analysis.

Chung and Wagoner[3] extended the development above to problems involving contact and friction. As long as Coulomb friction holds, the same strength and rate invariance properties hold.

In industry practice, changing a forming process from a slow-moving machine to a fast one or vice versa may be observed to affect the strain distribution inside the deforming workpiece. The invariance principle discussed above indicates that the strain-rate sensitivity of the material is not likely to be responsible for those differences, as has commonly been supposed. Instead, one should look to other effects for explanation: varying material rate sensitivity with respect to strain rate, lubrication viscosity effects, and deformation-induced heating. At very high rates, inertia may also be important.

### Effect of Deformation Heating

When metal is plastically deformed, most of the strain energy absorbed (usually ~90–95%) is converted into heat and the temperature of the metal increases.

[3] K. Chung and R. H. Wagoner, "Invariance of Neck Formation to Material Strength and Strain Rate for Power-Law Materials," *Metal. Trans. A.* **17A**, 1631 (1986).

Temperature rises of 100 °C have been reported[4] during tensile testing of sheet metal under normal test conditions with a strain rate comparable to that encountered in press forming. Since the temperature increment is caused by the work of deformation, the temperature distribution is related to the strain distribution. The temperature rise is maximum in the neck region, where the local strain is the highest; as a general rule, a higher temperature will result in a reduced flow stress during deformation. A temperature gradient will preferentially soften the material in the neck region and enhance the existing strain gradient, so that it has a detrimental effect on the strain localization.

In order to include the temperature effect in a constitutive law, Eq. (8.1) was modified as follows[5]:

$$\sigma = k(\varepsilon + \varepsilon_0)^n(\dot{\varepsilon}/\dot{\varepsilon}_0)^m(1 - \beta\Delta T). \tag{8.8}$$

For IF (interstitial free) steel, the parameters were found to be $k = 566$ MPa, $n = 0.219$, $m = 0.018$, $\dot{\varepsilon}_0 = 0.002$, $\varepsilon_0 = -0.014$, and $\beta = 0.0011$ (1/°C). A linear reduction of flow stress with temperature increment is assumed in the above equation for simplicity.

The first law of thermodynamics states the conservation of energy for thermodynamic systems (for more details, see Chapter 7). Assuming a complete conversion of plastic deformation energy to heat and no heat flow (local adiabatic condition), the temperature increment in a sample can be expressed as an integration:

$$dq = dw \rightarrow \int_{298}^{T} \rho C_p \, dT = \int_0^{\varepsilon} \sigma \, d\varepsilon, \tag{8.9}$$

where $\rho$ and $C_p$ are the density and heat capacity at constant pressure, respectively. For IF steel, $\rho = 7.85$ g cm$^{-3}$, and $C_p = 0.464$ J g$^{-1}$ °C$^{-1}$.

---

**Exercise 8.1: An element of IF sheet steel, tensile test sample was deformed to a uniform strain of 0.2 in 1 s. Calculate the temperature rise and the final flow stress from Eqs. (8.1) and (8.9); assume $\varepsilon_0 = 0$ and $m = 0$.**

The temperature rise is

$$\int_{298}^{T} \rho C_p \, dT = \int_0^{0.2} k\varepsilon^n \, d\varepsilon,$$

$$\rho C_p \Delta T = \frac{k}{1.219} 0.2^{1.219},$$

$$\Delta T = \frac{k}{1.219 \rho C_p} 0.2^{1.219},$$

$$= \frac{566 \times 1{,}000{,}000 \text{ N m}^{-2}}{7.85 \times 0.001(\text{g/mm}^3) \times 0.464(\text{J/g °C}) \times 1.219} \times 0.1406,$$

$$= \frac{566 \times 10^6 \text{ m}^{-3}}{0.00444(1/10^{-9}\text{m}^3)(1/°\text{C})} \times 0.1406 = 17.9 \text{ °C},$$

[4] R. A. Ayres, "Thermal Gradients, Strain Rate, and Ductility in Sheet Steel Tensile Specimens," *Metal. Trans. A*. 16A, 37 (1985).

[5] M. R. Lin and R. H. Wagoner, "Effect of Temperature, Strain, and Strain Rate on the Tensile Flow Stress of I. F. Steel and Stainless Steel Type 310," *Scripta Metal.* 20, 143–148 (1986).

and the final flow stress is

$$\sigma = 566 \times 0.2^{0.219}(1 - 0.0011\,\Delta T) = 390 \text{ MPa}.$$

Compared with that of the isothermal case, the flow stress is reduced 2%.

---

The effect of deformation heating becomes more important in actual stamping operations, where higher forming speed is involved. The physical problem is a coupled and interactive one. In order to model this complicated effect of heat generation, heat flow, and thermal softening during a nonisothermal, non-uniform deformation, a large displacement, large-strain finite-element code for sheet-metal stamping operations is used. We will study an example of deformation-induced heating in tensile testing by a two-dimensional FEM next.

The governing equations include mechanical equilibrium, energy conservation, and transient heat flow. The mechanical equilibrium equation in this two-dimensional tensile test model neglects inertial forces, body forces, and the elastic response.

The energy conservation equation has a local form in a controlled domain. The heat causing a temperature change in the domain is a combination of three parts, written as three terms on the right of the following equation:

- the heat conducted from and to adjacent domains by the temperature gradient;
- the heat generated inside the domain by the plastic deformation work; and
- the heat lost from the outer surface of this domain to its environment by conduction and convection according to the temperature difference, $(T - T_\infty)$:

$$\rho C_p \frac{\partial T}{\partial t} = \nabla(k \nabla T) + \dot{q} - 2\frac{h}{\Delta}(T - T_\infty). \tag{8.10}$$

Boundary conditions are expressed at two ends of the tensile specimen, $\partial S_1$,

$$T = \hat{T} \quad \text{on } \partial S_1, \tag{8.11}$$

and on the specimen surface, $\partial S_2$,

$$k\frac{\partial T}{\partial \mathbf{n}} + \frac{h}{\Delta}(T - T_\infty) = -\hat{q} \quad \text{on } \partial S_2, \tag{8.12}$$

where $k$ is heat conductivity, $\dot{q}$ is the heat generation rate by plastic deformation, $h$ is the heat convection coefficient, $\Delta$ is the current sheet thickness, $\hat{T}$ and $\hat{q}$ are the prescribed boundary temperature and heat flux over the boundaries $\partial S_1$ and $\partial S_2$, respectively, and $\mathbf{n}$ is the outward normal vector on $\partial S_2$.

The average heat generation rate, $\dot{q}$, during a small time interval $\Delta t$ is defined by

$$\dot{q} = \eta \sigma \dot{\varepsilon}, \tag{8.13}$$

where $\dot{\varepsilon}$ is the effective strain rate, and $\eta$ is the fraction of strain energy converted to heat; $\eta$ is usually assumed to be 0.9 from experimental data.[6]

A stepwise decoupled approach is used to analyze the current problem,[7] since our work has shown that changes in the temperature distribution are only weakly

[6] M. R. Lin and R. H. Wagoner, "Effect of Temperature, Strain, and Strain Rate on the Tensile Flow Stress of I. F. Steel and Stainless Steel Type 310," *Scripta Metal.* 20, 143–148 (1986).

[7] Y. H. Kim and R. H. Wagoner, "An Analytical Investigation of Deformation-Induced Heating in Tensile Testing," *Int. J. Mech. Sci.* 29, 179–194 (1987).

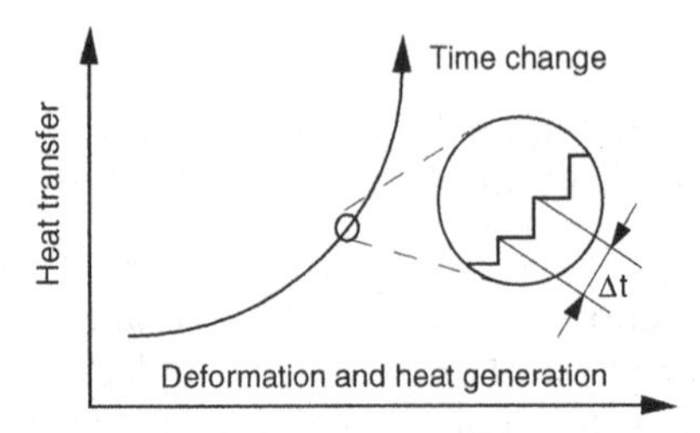


Figure 8.2. A schematic view of the stepwise decoupled approach to solving a coupled thermoplasticity problem.

coupled to the mechanical properties [by means of $\beta$ in Eq. (8.8)]. Also, for the time step required for solution of mechanical equilibrium, there is little benefit in accuracy to iterating for a simultaneous solution of thermal and mechanical equation. The principle of this approach is to separate the problem of heat generation, conduction, and material transport into two stepwise independent parts. At each time step, the heat generation and material transport are regarded as occurring instantaneously, followed immediately by the heat transfer, which occurs as in a stationary medium. Figure 8.2 shows schematically the concept of this stepwise decoupled approach. As long as a sufficiently small time step is used, this approach is suitable for solving coupled thermoplasticity problems.

A process flow chart is presented in Fig. 8.3, illustrating the computation procedure of the finite-element simulation of this two-dimensional tensile test problem.[8] The plastic deformation loop was carried out iteratively in order to solve the nonlinear mechanical equilibrium using the Newton–Raphson method. The temperature change calculation is straightforward and no iteration is necessary by virtue of its linear nature, when the transfer takes place in a stationary medium and when the thermal properties $k$, $h$, $\rho$, and $C_p$ do not depend on temperature, time, stress, or strain.

An ASTM E-8 standard sheet tensile specimen geometry and corresponding finite-element mesh (reduced to one quarter by symmetry) are shown in Fig. 8.4. A 0.6% taper over a distance of 25 mm to the specimen center is incorporated to ensure the repeatability of tensile test results. No force boundary conditions are specified, except a traction-free condition on the free surface, the $Y^+$ boundary. For a displacement-controlled uniaxial tensile test to be simulated at uniform cross-head speed, the displacement-boundary conditions are specified at the boundary nodes as follows: the $X^+$ boundary has uniform velocity in the $X$ direction where no movement is allowed in the $Y$ direction. The $X^-$ boundary, by symmetry, has no movement in the $X$ direction and no external force in the $Y$ direction. Four thermal boundary conditions were used in this FEM study separately, namely, isothermal, adiabatic, and transfer characteristics of immersion in water or air.

Temperature data were collected experimentally from multiple thermocouples capacitance-discharge welded on the specimen.[9]

The temperature distributions for various rates when tested in air are given in Fig. 8.5. The simulations agree well with the measurements. A temperature gradient is established by the heat conduction from the center to the ends. The gradient is not

[8] K. S. Raghavan and R. H. Wagoner, "Analysis of Nonisothermal Tensile Tests Using Measured Temperature Distributions," *Int. J. Plast.* 3, 33–49 (1987).

[9] M. R. Lin and R. H. Wagoner, "Effect of Temperature, Strain, and Strain Rate on the Tensile Flow Stress of I. F. Steel and Stainless Steel Type 310," *Scripta Metal.* 20, 143–148 (1986).

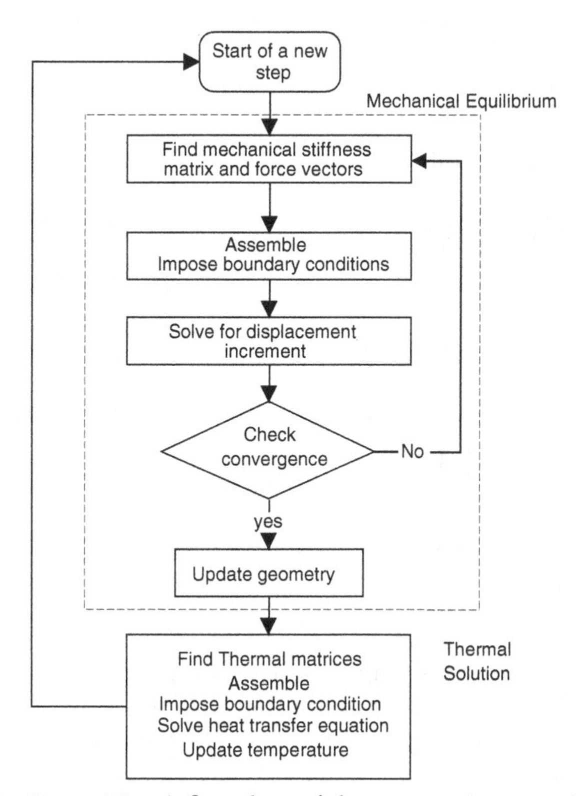


Figure 8.3. A flow chart of the computation procedure in the finite-element simulation of a nonisothermal tensile test problem.

significant before necking. After necking occurs, however, a sharp temperature rise occurs corresponding to strain localization. At a strain rate of $10^{-1}$ $s^{-1}$, the predicted temperature rise can be as high as 75 °C; for relatively low strain rates, the temperature variation is less severe because longer times allow for significant homogenization by heat conduction and transfer to the environment.

The FEM-predicted effect of heat transfer conditions on uniform and total elongation is shown in Fig. 8.6.[10] The numerical "failure elongation" depends on the choice

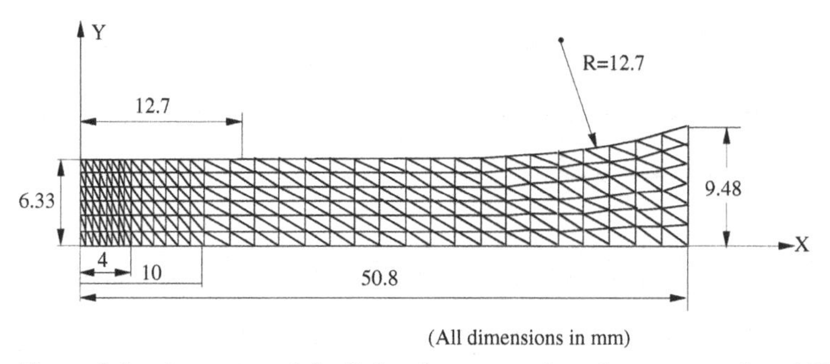


Figure 8.4. A quarter of the finite-element mesh and geometry of an ASTM E-8 standard sheet tensile specimen.

[10] Y. H. Kim and R. H. Wagoner, "An Analytical Investigation of Deformation-Induced Heating in Tensile Testing," *Int. J. Mech. Sci.* 29, 179–194 (1987).

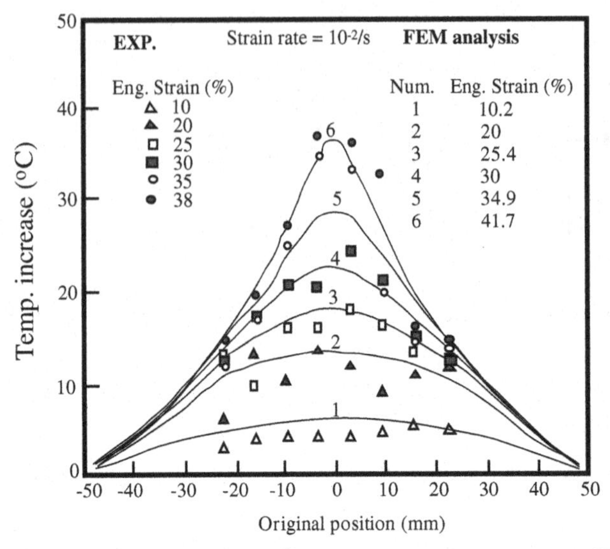


Figure 8.5. Comparison of temperature measurements and FEM simulation of sheet tensile tests.

of failure criterion, which was taken to be a true strain at the center of the specimen of 0.7. This value was set to agree with experimental observation and may depend on the choice of mesh and other simulation parameters. As shown in Fig. 8.6, both uniform elongation and total elongation are nearly constant for isothermal and adiabatic[11] cases for various rates. For the tests in air and in water, the lines lie between the two

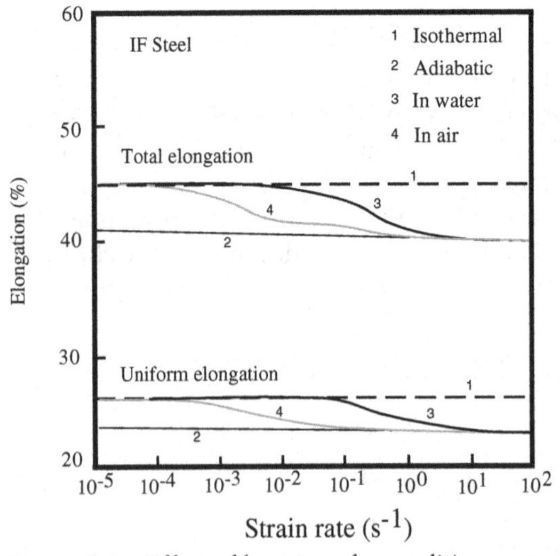


Figure 8.6. Effect of heat transfer conditions on uniform and total tensile elongation from the FEM.[12]

[11] Adiabatic here means locally adiabatic; i.e., that no heat transfer takes place to and from each material element.

[12] Y. H. Kim and R. H. Wagoner, "An Analytical Investigation of Deformation-Induced Heating in Tensile Testing," *Int. J. Mech. Sci.* 29, 179–194 (1987).

extreme cases; that is, they approach the isothermal case for relatively low strain rates and the adiabatic case for relatively high rates.

### Work-Hardening and Rate Sensitivity Effects

Work-hardening and strain-rate sensitivity are key material behaviors for sheet-metal forming operations. Closed-form analyses of the relationship between the uniform elongation in a uniaxial tension test and material properties are limited to one-dimensional prenecking analysis, such as the classical analysis by Considere.[13] Experiments are limited to attainable combinations of $n$ and $m$ values in real material, and results may be observed by variation of other material parameters. The FEM can be used to study these effects, with several advantages.

1. The effect of $n$ and $m$ can be examined fully and separately from each other, while other properties can be isolated.
2. The uniform strain range and the postuniform (necking) region can be examined. Some forming operations exploit both ranges.
3. Two- or three-dimensional capability allows better accuracy and understanding of the strain localization process, compared with one-dimensional simplified analyses, such as that presented elsewhere,[14] where no lateral stress is considered.

A rate sensitive, Hollomon type of material hardening law of the form

$$\sigma = k(\varepsilon + \varepsilon_0)^n(\dot{\varepsilon}/\dot{\varepsilon}_0)^m \tag{8.14}$$

was used, which is an isothermal version of Eq. (8.8) used in the nonisothermal analysis. $K$ and $\dot{\varepsilon}_0$ are chosen arbitrarily, since they do not affect strains. (See the invariance property that was discussed in the beginning of this chapter.) A von Mises yield function is used to define the effective stress and effective strain. To explore the effect of various values of $n$ and $m$, five cases of $n$ values (from $n = 0.0$ to $n = 0.5$), and eight cases of $m$ values (from $m = -0.002$ to $m = 0.3$) were used for the simulation.

The mesh and boundary condition is similar to those used in the nonisothermal study, as shown in Fig. 8.4. The only difference lies in the 1% taper within a 50.8-mm gage length. A failure criterion was adopted of the form that $\dot{\varepsilon}_C/\dot{\varepsilon}_{AV} = 5$, where $\dot{\varepsilon}_C$ is the strain rate in the center element and $\dot{\varepsilon}_{AV}$ is the strain rate average over the specimen length.

The variation of total elongation value, $e_f$, with respect to $n$ and $m$ values, is plotted in Figs. 8.7 and 8.8, respectively.[15] Figure 8.7 shows that the total elongation increases almost linearly with $n$ in the region of small $m$, and the rate of increase is larger at higher $m$ values. Figure 8.8 shows that $e_f$ also increases as $m$ increases in an accelerated manner with a higher than linear order. There is a strong interaction between $n$ and $m$ in determining the total elongation. This synergy becomes more important for large values of $n$ and $m$.

[13] A. Considere, *Ann. ponts Chanssees*, 6, 9 (1885).

[14] R. H. Wagoner and J.-L. Chenot, *Fundamentals of Metal Forming* (Wiley, New York, 1997), Chap. 1.

[15] K. Chung and R. H. Wagoner, "Effects of Work-Hardening and Rate Sensitivity on the Sheet Tensile Test," *Metal. Trans. A.* **19A**, 293 (1988).

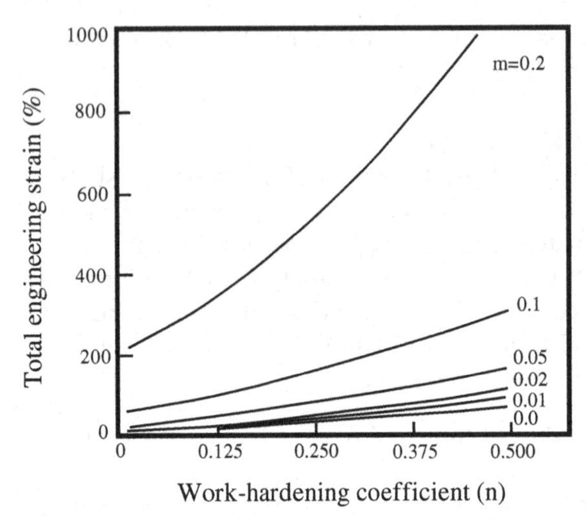


Figure 8.7. The total elongation increases with $n$ under the effect of various $m$ values.

Compared with the simplified one-dimensional analysis performed elsewhere,[16] the FEM two-dimensional analysis predicts additional deformation and a higher failure strain (e.g., ~10% when $m = 0$), reflecting the stabilizing effect of biaxial tension induced by the inhomogenous axial deformation. The lateral stresses at each element in this case are positive and grow steadily during the test. This increasing *biaxiality* in the sheet specimen corresponds to the triaxiality that develops in a round bar tensile specimen. This growth of biaxiality delays the maximum load point and produces a higher rate of increase of axial stress than expected in pure uniaxial loading. The development of biaxiality is caused by the slight geometric neck formation around the center, although the effect begins almost immediately because of the inhomogeneity of the shoulder regions. The results suggest that longer tensile specimens should be used for failure elongation testing to avoid such end effects.

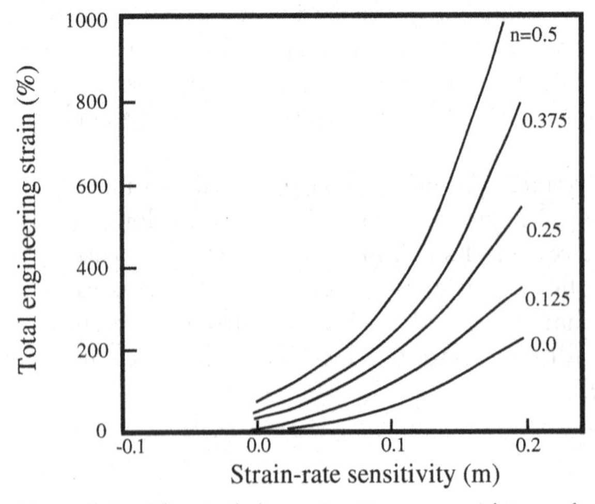


Figure 8.8. The total elongation increases with $m$ under the effect of various $n$ values.

[16] R. H. Wagoner and J.-L. Chenot, *Fundamentals of Metal Forming* (Wiley, New York, 1997), Chap. 1.

### Combined Influence of Geometric Defects and Thermal Gradients on Tensile Ductility[17]

A series of tensile specimen meshes was designed with varying sizes and shapes of imposed taper to model a geometric defect. Two-dimensional finite-element modeling, assuming local adiabatic or isothermal conditions, was carried out to analyze the simultaneous influence of geometric notches and thermal gradients on tensile ductility. Neither of these extreme conditions requires the solution of heat flow problems.

Specimen shapes with varying magnitudes and shapes of geometric defect were produced by using a cosine function:

$$Y(x) = [Y_0 - (\Delta Y/2)] - (\Delta Y/2)\cos\frac{\pi x}{\lambda}, \tag{8.15a}$$

where

$Y(x)$ is the width of the specimen at position $x$ along the axial direction,
$Y_0$ is the width of reference specimen (defect free),
$\Delta Y$ is the difference in width at position $x = \lambda$ and $x = 0$, and
$\lambda$ is the half-wavelength of the cosine function.

The defect size or relative specimen taper is defined to be

$$\text{taper} = \frac{\Delta Y}{Y_0} \times 100. \tag{8.15b}$$

Typical specimens with this initial defect shape are shown in Fig. 8.9. A half-wave-length of 26 mm was chosen.

A two-dimensional FEM formulation was used with the following material model:

- a constitutive law simultaneously including strain, strain rate and temperature effects, as shown in Eq. (8.8), for IF steel, and
- a yield function of Hill's new theory[18] with $M = 2$ and $r = 1.5$.

In the sheet tensile test, gripping at either end induces biaxial stresses during elongation. Simplified one-dimensional models may produce different simulation results since they ignore the biaxial constraint. Two simple one-dimensional analyses, based on an axial force equilibrium in each element carried out the axial loading, are used

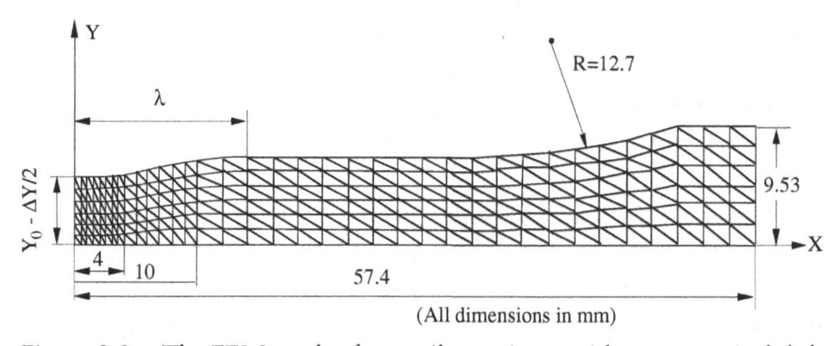


Figure 8.9. The FEM mesh of a tensile specimen with a geometrical defect in the center.

[17] K. S. Raghavan and R. H. Wagoner, "Combined Influence of Geometric Defects and Thermal Gradients on Tensile Ductility," *Metal. Trans. A.* **18A**, 3143 (1987).

[18] R. H. Wagoner and J.-L. Chenot, *Fundamentals of Metal Forming* (Wiley, New York, 1997), Chap. 7.

for comparison with the two-dimensional simulation:

1. Multi-element analysis: this numerical analysis[19] is similar to a chain under axial loading; each element has its own deformation while bearing the same loading along the chain. The initial cross sections of the elements correspond to the two-dimensional FEM mesh, shown in Fig. 8.9.
2. Two-element analysis: a simplified version of the multi-element analysis with only two elements as shown in Fig. 8.1. One element has a reference width and the other element has a width corresponding to the minimum width, at $X = 0$.

Figure 8.10(a) illustrates the rule of geometric defects on tensile ductility. As the defect size increases, pronounced strain gradients occur from the start of deformation. For tapers larger than 5%, the concentration of strain in the center from the beginning of deformation is manifested by an increase in the ultimate tensile strength. The axial stress is increased because of the influence of lateral stress causing additional hardening and higher strain rate in the center. Isothermal ductility loss was predicted for the specimen with various defects, from a final engineering strain of 0.555 for a 0% defect to 0.116 for a 50% defect, as shown in Fig. 8.10(a).

In another series of simulations with the same shape and defect size, the tensile ductility was lowered by the presence of thermal gradients. The corresponding predicted ductility loss under adiabatic conditions was from 0.460 for 0% defect to 0.099 for a 50% defect size. The influence of geometric defects under adiabatic thermal gradients can be seen in Fig. 8.10(b). The thermal gradients were found to decrease ductility by a nearly constant percentage of 15.5%, independent of initial defect size.

A comparison of FEM results to the two-element and multi-element analysis, shown in Fig. 8.11, emphasizes the effect of biaxial stresses for various initial defect sizes. Development of a biaxial stress state in the neck region, ignored in the one-dimensional analysis, accounts for a notable increase in the predicted ductility. The two-element simulation fails to capture the effect.

#### Inertial Effect on Tensile Ductility[20]

One-dimensional numerical simulations of dynamic tensile tests have been carried out over a wide range of test velocities (from $2.54 \times 10^{-6}$ to 127.0 m/s) for materials having a Hollomon-type constitutive law with power-law strain rate sensitivity, as in Eq. (8.1). A variety of values of strain-hardening exponent and strain-rate sensitivity index have been used to analyze the effect of inertia on tensile ductility. A range of material parameters, six cases of $n$ values ($n = 0.0$, 0.0625, 0.125, 0.25, 0.375, and 0.5), each having seven cases of $m$ values ($m = 0.0$, 0.01, 0.02, 0.05, 0.1, 0.2, and 0.3), have been simulated.

Figure 8.12 shows schematically the spatial discretization of the tapered specimen with one-dimensional equilength elements along the $x$ axis. Each element is assumed to have a constant true axial stress, true axial strain, true axial strain rate, and cross-section area. The equilibrium between a resultant internal force, $F_i$, and an inertial

[19] R. H. Wagoner and J.-L. Chenot, *Fundamentals of Metal Forming* (Wiley, New York, 1997), Chap. 1.
[20] X. Hu, R. H. Wagoner, G. S. Daehn, and S. Ghosh, "The Effect of Inertia on Tensile Ductility," *Metal. Trans. A.* **25A**, 2723–2735 (1994).

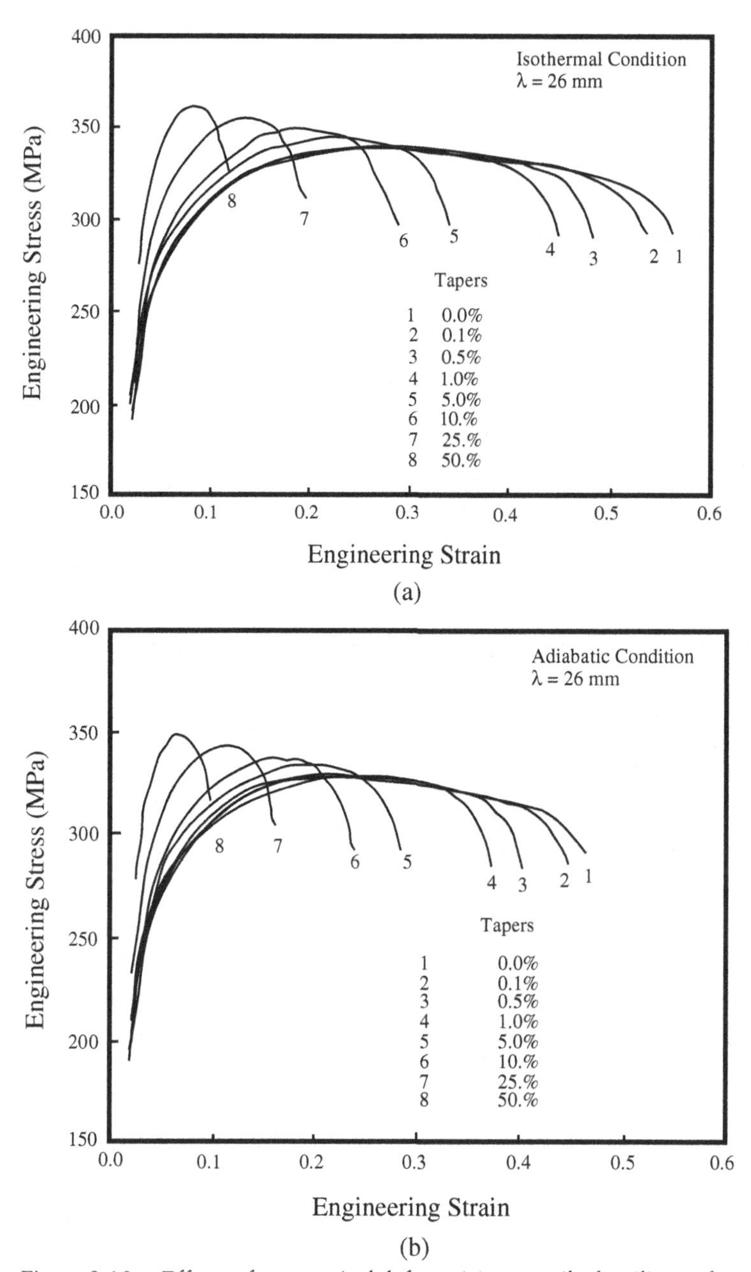


Figure 8.10. Effects of geometrical defects: (a) on tensile ductility under an isothermal condition, and (b) combined with thermal gradients on tensile ductility.

force, $M_i\ddot{u}_i$, in the $x$ direction on each node is

$$M_i\ddot{u}_i + F_i = 0, \tag{8.16}$$

where $\ddot{u}_i$ and $M_i$ are the nodal acceleration and mass, respectively.

The nodal force, $F_i$, is obtained by summing the internal forces of two adjacent elements associated with this node, $\alpha$ and $\alpha + 1$. The internal force of each element is simply equal to the product of its true stress, $\sigma_\alpha$, and its current cross-section area, $a_\alpha$.

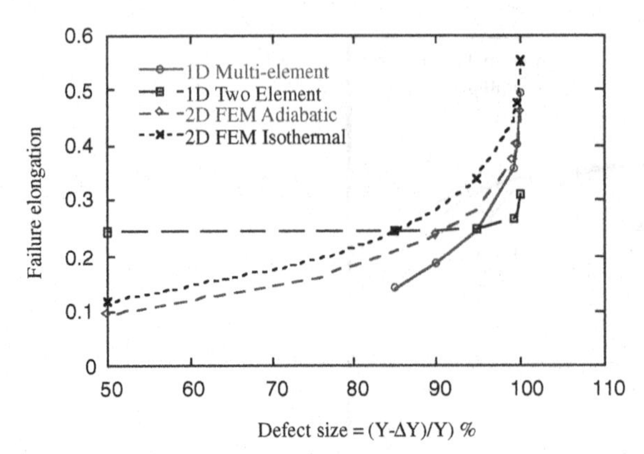


Figure 8.11. Comparison of FEM predicted ductility in terms of failure elongation to the two-element and the multi-element one-dimensional analysis.

Thereby, $F_i$ is expressed as

$$F_i = \sigma_\alpha a_\alpha - \sigma_{\alpha+1} a_{\alpha+1}. \tag{8.17}$$

Using the $n$ and $m$ values, we can write Eq. (8.17) in an alternative form:

$$F_i = k\left[\varepsilon_\alpha^n \dot{\varepsilon}_\alpha^m A_\alpha \exp(-\varepsilon_\alpha) - \varepsilon_{\alpha+1}^n \dot{\varepsilon}_{\alpha+1}^m A_{\alpha+1} \exp(-\varepsilon_{\alpha+1})\right]. \tag{8.18}$$

Equation (8.18), although it is derived for dynamic response, can be also applied in the quasi-static case; $F_i$ is set to zero.

The nodal mass, $M_i$, is obtained by lumping the mass of all connecting elements at associated nodes. For the element in Fig. 8.12, $M_i$ is expressed as

$$M_i = \frac{\rho L}{2}(A_\alpha + A_{\alpha+1}), \tag{8.19}$$

where $\rho$ is the material density. By using the relation of true strain $\varepsilon_\alpha$ and the nodal displacement $u_i$, and assuming $\dot{\varepsilon} = \Delta\varepsilon/\Delta t$ during each time step $\Delta t$, we find a substitution of Eqs. (8.18) and (8.19) into Eq. (8.16) yields

$$\rho_n \ddot{u}_i - \frac{1}{\Delta t^m} \sum_{\kappa=\alpha}^{\kappa=\alpha+1} \xi_\kappa \left[\ln\left(1 + \frac{u_\kappa}{L}\right)\right]^n \left[\Delta \ln\left(1 + \frac{u_\kappa}{L}\right)\right]^m \frac{\partial \ln(1 + u_\kappa/L)}{\partial u_i} = 0. \tag{8.20}$$

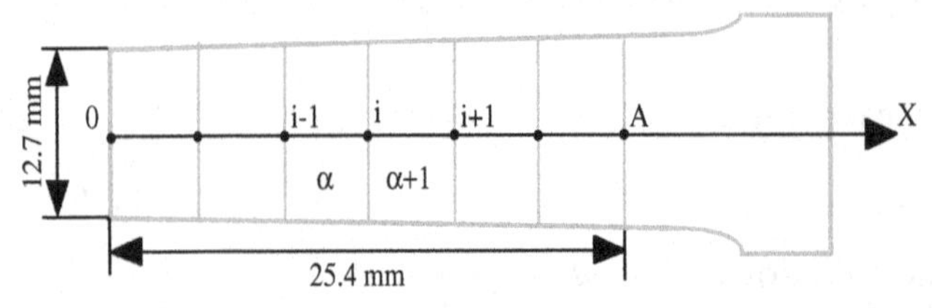


Figure 8.12. Schematic illustration of the spatial discretization of a sheet tensile specimen with 1% taper.

Equation (8.20) is a system of nonlinear equations with respect to $u_i$ at time $t$ with the following definitions:

$$\rho_n = \frac{\rho}{k}, \tag{8.21}$$

$$u_\alpha = u_i - u_{i-1}, \tag{8.22}$$

$$u_{\alpha+1} = u_{i+1} - u_i, \tag{8.23}$$

$$\xi_\alpha = \frac{2A_\alpha}{L(A_\alpha + A_{\alpha+1})}, \tag{8.24}$$

$$\xi_{\alpha+1} = \frac{2A_{\alpha+1}}{L(A_\alpha + A_{\alpha+1})}, \tag{8.25}$$

where $\rho_n$ is called the normalized material density. The solution of Eq. (8.20) can be solved with the progression of time by numerical time integration.

The simulated engineering stress–strain curves over a range of the test velocities are plotted in Fig. 8.13 for various values of $n$ and $m$. The engineering stress, normalized by $k$, is obtained from the tensile force at the mobile end divided by the original cross-section area, while the engineering strain, $e$, is calculated by the total end displacement divided by the gage length $l_0$.

From these curves, it can be seen that the total elongation of the specimen, $e_T$, defined at the failure, remains the same when the specimen deforms at low test velocities from $2.54 \times 10^{-6}$ to 2.54 m/s. When this range of velocities is exceeded, however, increasing test velocity increases $e_T$. As was noted earlier, quasi-static simulations yield results independent of test velocities, corresponding to the low test velocities computed here. Therefore, the results indicate that the increase of the total elongation at high test velocities must be attributed to the inertia term in Eq. (8.16), $M_i\ddot{u}_i$. It can be concluded that the presence of inertia can enhance tensile ductility, and this inertial effect becomes significant beyond a critical test velocity.

The suppression of the strain gradient by inertia would be expected to take place primarily in postuniform deformation, because significant acceleration is developed

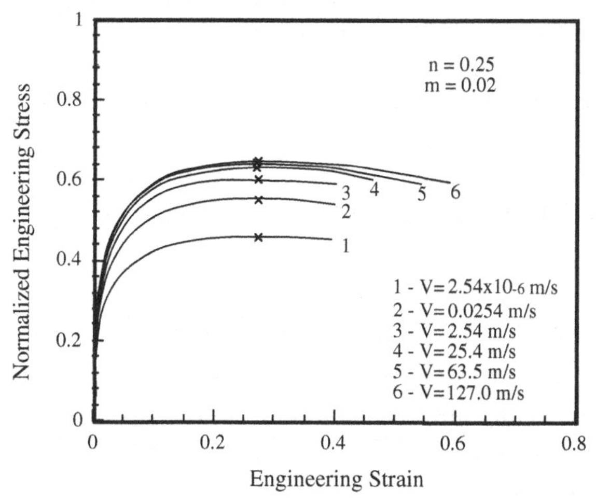


Figure 8.13. Engineering stress–strain curves at different test velocities for $n = 0.25$ and $m = 0.02$.

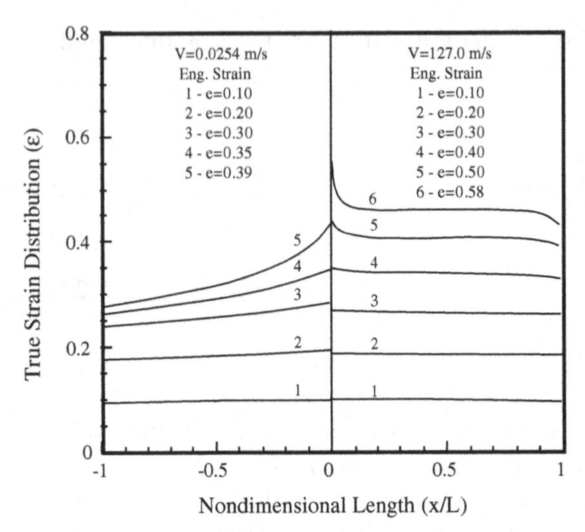


Figure 8.14. Comparison of the true strain distributions for specimens deformed at low and high test velocities with $n = 0.25$ and $m = 0.02$.

only after necking commences. In fact, when the test velocity is changed, there is no significant change of the maximum load elongation ($e_u$, shown by $X$ in Fig. 8.13), while postuniform elongation varies considerably.

Figure 8.14 shows the true strain distributions in the specimen at various engineering strain levels for low and high test velocities. The left-hand side corresponds to the strain distributions obtained at $V = 0.0254$ m/s, and the right-hand side at $V = 127.0$ m/s. When $e \leq 0.2$, the strain distributions at the two test velocities are similar. However, when the elongation exceeds this value, the strain distribution starts to concentrate in the center for the low-velocity specimen whereas, for the high velocity, the concentration of the central strain is retarded, and a relatively uniform strain distribution throughout the specimen is retained at a higher engineering strain.

These results show that the total elongation of the specimen is enhanced by inertia at high test velocities. This inertial effect varies with the strain-hardening exponent and strain-rate sensitivity index, and it can be scaled with the normalized material density and the test velocity as follows:

$$\rho_n V^{2-m} = C, \tag{8.26}$$

where the $\rho_n$ is the normalized material density, and $C$ is a constant. This relation states that the inertial effect produced by a change in the normalized material density is equivalent to that by a change in the test velocity.

## 8.2 The Plane-Strain Tension Test[21]

For sheet tests other than pure uniaxial tension, where the strains are nearly uniform over some range of strain, it is necessary to measure strains from local regions of the

[21] R. H. Wagoner, "Measurement and Analysis of Plane-Strain Work Hardening," *Metal. Trans. A.* **11A**, 165 (1980).
R. H. Wagoner and N-M. Wang, "An Experimental and Analytical Investigation of In-Plane Deformation of 2036-T4 Aluminum Sheet," *Int. J. Mech. Sci.* **21**, 255–264 (1979).

surface of the sheet. The plane-strain test is introduced, as an example. The analysis is based on measurement of surface strains using grids printed by a photographic process on the surface.

### Circle Grid Surface Strain Measurements

*Circle grid measurement* is a common laboratory and industrial method for use in formability analysis and die tryout. This method has improved the understanding of failure in sheet-metal forming and provides a basis for solving such problems. In this technique, grids are marked on the original sheet surface, either by photographic printing or electrochemical etching. The surface strain can be measured by comparing the grid before and after the forming operation. Experimental strains measured by circle grids will be used for comparison with the predicted strains from the FEM.

A homogeneous deformation will change a circle to an ellipse, and the major and minor surface strains can be calculated from the measured lengths of the major and minor axes, $d_1$ and $d_2$, as illustrated in Fig. 8.15. The principal natural strains under incompressibility conditions are defined from the circle as follows:

$$\begin{aligned} \varepsilon_1 &= \ln(d_1/d_0), \\ \varepsilon_2 &= \ln(d_2/d_0), \\ \varepsilon_3 &= -\varepsilon_1 - \varepsilon_2. \end{aligned} \tag{8.27}$$

The measurements can be performed manually by using a measuring microscope, or by means of dividers and a ruler, or by graduated transparent tapes.

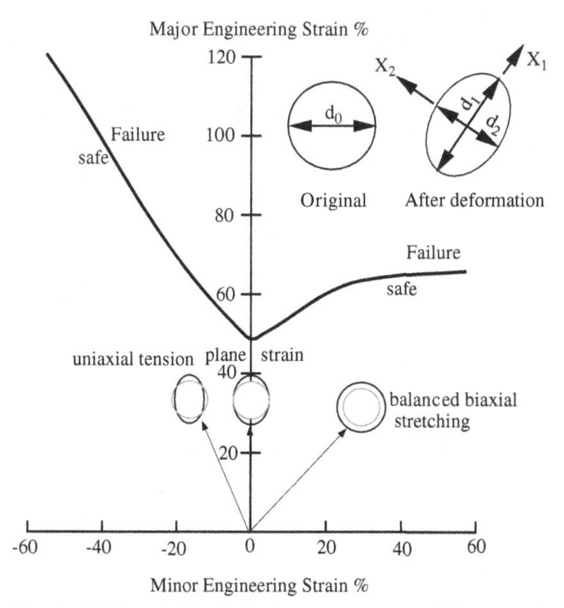


Figure 8.15. Grid circle measurement and forming limit diagram.[22]

[22] S. S. Hecker, "Simple Technique for Determining Forming Limit Curves," *Sheet Metal Ind.* 52, 671–676 (1975).

### Plane-Strain Tensile Testing

In the conventional uniaxial tensile test, the minor or width strain is negative. The test does not provide information on the response of sheet metal in other states, for example, the **plane-strain** state, in which the minor strain is zero.[23] The uniaxial tensile test can be modified by using a very wide, short sample. Increasing the ratio of width to gage length changes the strain state from the one with large negative minor strain (uniaxial tension) toward the one with small or no minor strain (plane strain). However, there is always some error because the edge regions are inherently in uniaxial tension. Designing an effective specimen shape requires balancing the fraction of the width in plane strain with localization failure, which occurs near the edges.

In addition to the more standard uniaxial tensile test and balanced biaxial tension test, the plane-strain test can provide a third independent path to probe the yield surface. It can be used to establish directly "Hill's $M$" value. The approach used in the plane-strain test is to devise an experiment that relies on computer analysis for data interpretation but that eliminates most of the experimental uncertainties inherent in punch or hydraulic bulge testing: friction, boundary constraints, through thickness strain gradients, bending, and multistage deformation paths. The continuous evaluation of stress and strain throughout the test allows monitoring of incremental work hardening.

In order to optimize the specimen geometry that yields the highest center strain at failure with a large region of plane strain, the shapes of eight specimens, shown in Fig 8.16 and Table 8.1, were machined and simulated to determine the effect of edge profile and width-to-length ratio.

Specimens A–G were photogridded with circles and measured after deformation by using a microscope. Specimen B was found to work well for steels ($r \approx 1.5$), but none was suitable for aluminum ($r \approx 0.7$). The H specimen was optimized by iteration for aluminum by using the FEM.

The FEM model of specimen B is shown in Fig. 8.17.[24] Because of the symmetry of the specimen, only one quarter has to be analyzed. The boundary conditions are such that symmetry conditions are enforced along the left and lower edges while a

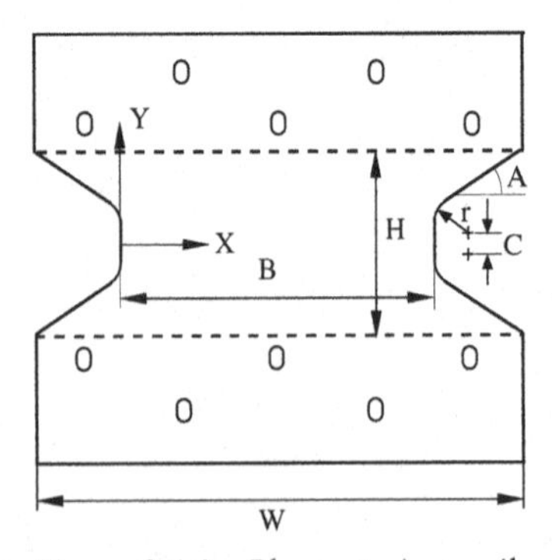


Figure 8.16. Plane-strain tensile specimen.

[23] The plane-strain state is particularly important in sheet forming. For example, estimates that up to 80% of failures in automotive stampings occur near plane strain have been presented: R. A. Ayres, W. G. Brazier, and V. F. Sajewski, *J. Appl. Metalwork*. 3, 272–280 (1984).

[24] R. H. Wagoner and N-M. Wang, "An Experimental and Analytical Investigation of In-Plane Deformation of 2036-T4 Aluminum Sheet," *Int. J. Mech. Sci.* **21**, 255–264 (1979).

**Table 8.1. General Plane-Strain Specimen Geometry with Dimensions for Each Specimen**

| Specimen | *A* (deg) | *r* (mm) | *C* (mm) |
|---|---|---|---|
| A | 0 | 19.1 | 12.7 |
| B | 45 | 19.1 | 12.7 |
| C | 30 | 12.7 | 12.7 |
| D | 45 | 12.7 | 12.7 |
| E | 30 | 19.1 | 12.7 |
| F | 45 | 25.4 | 0.0 |
| G | 0 | 12.7 | 12.7 |
| H | 0 | 68.4 | 0.0 |

rigid-body motion is prescribed on the upper edge, which physically corresponds to the clamped end.

Experimental and theoretical strain distributions at two displacements are presented in Fig. 8.18 for a B specimen of 2036-74 aluminum. A similar agreement for all specimen designs was obtained. The only differences occur at the outer edges, where in some cases the experimental strains tend to be low, probably as a result of minor slippage in the gripped region near the edge of the specimen. Based on a maximum strain criterion, similar to the criterion used in the two-dimensional uniaxial tensile test simulations, the FEM results predict approximately the same displacement to failure for all the specimens.

The plane-strain region in this experiment, which is arbitrarily defined as the region where $|\varepsilon_2/\varepsilon_1|$ is less than 0.2, occupies about 80% of the specimen width. The outer part of the specimen deforms in a similar manner to a standard tensile test specimen; that is, the cross-head displacement rate was chosen so that the average axial strain rate in the plane-strain region is similar to that of the uniaxial tensile test.

A computer program was written to evaluate the effective strain and stress in the plane-strain region at each incremental step. The width of the specimen along the transverse center line is divided into three regions: the center section, which has a strain state close to plane strain, and two edge sections near uniaxial tension. The load supported by the edge regions is calculated from Hill's old theory, using the strain distribution data and the work hardening curve obtained in tension. The load carried by the plane-strain portion was found by subtracting the estimated load carried by the edge regions from the total measured load. The axial stress was obtained simply by dividing the plane-strain load by the corresponding area in the plane-strain portion. When a yield function is assumed, an average effective strain can be found

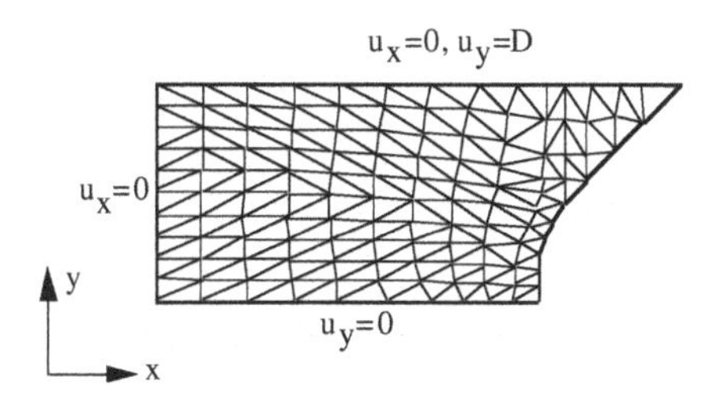


Figure 8.17. Mesh and boundary conditions for one quarter of the B-type plane-strain sample.

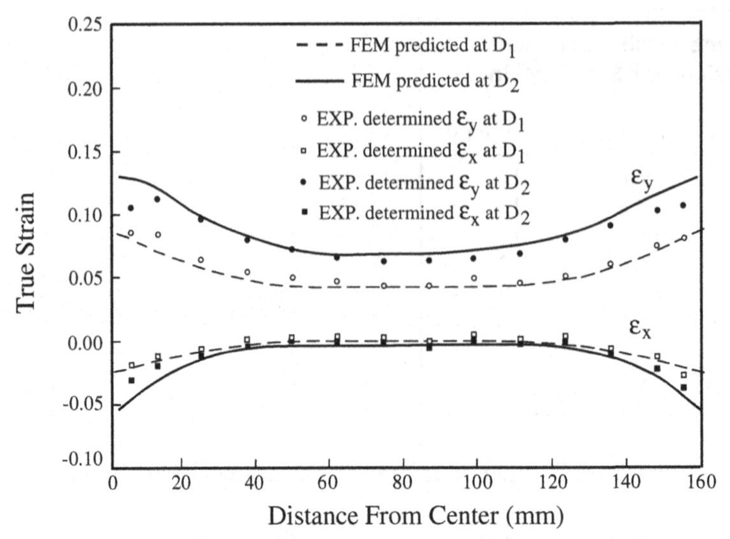


Figure 8.18. Comparison of experimental strain distribution measured at the transverse center line of the sheet specimen with FEM predicted strain at two displacements; $D_1 = 2$ mm, $D_2 = 4$ mm.

from the measured grid strain components in the plane-strain region, and an average effective stress was calculated from the axial stress. The effective stress–strain curve obtained in this way, based on Hill's quadratic yield function, Eq. (8.28), was compared with the effective stress–strain obtained in the uniaxial tensile test, as shown in Fig. 8.19.

$$\bar{\sigma} = \left( \sigma_1^2 + \sigma_2^2 - \frac{2r}{1+r}\sigma_1\sigma_2 \right)^{1/2}. \tag{8.28}$$

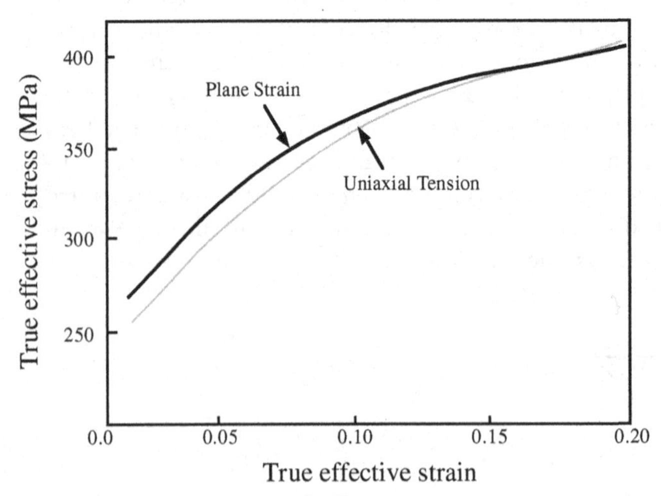


Figure 8.19. Comparison of effective stress–strain curves obtained in uniaxial tension and in plane strain by using Hill's "old" yield theory.

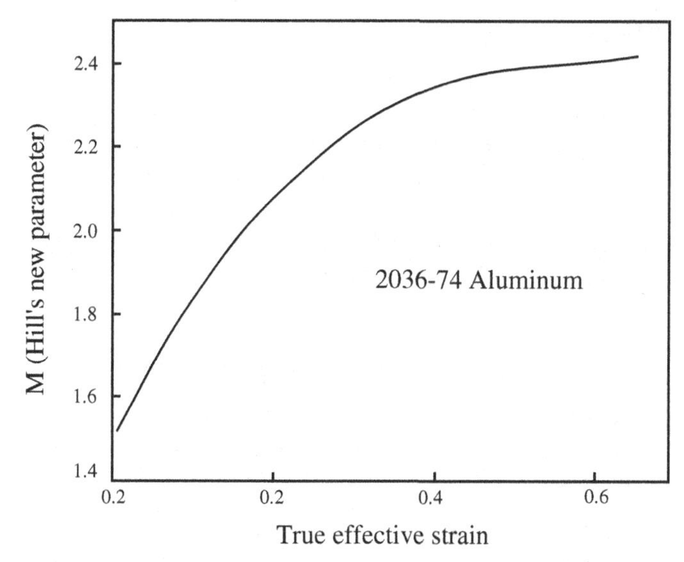


Figure 8.20. Variation of Hill's new parameter (M) with strain.

It was observed that at low strain the plane-strain stress was higher than that of uniaxial tension. At an effective strain near 0.18 the plane strain and tensile curves cross, and only at that point can the yield surface shape be consistent with Hill's "old" theory. If the yield theory was precisely obeyed, these curves would be coincident along their lengths. There are two divergences from the old yield theory:

1. Isotropic hardening does not hold because the work-hardening rate varies along different strain paths.
2. Different effective stresses are found at the same effective strain in different strain paths.

These results imply a distortion of the real yield surface with strain, thereby violating the idea of isotropic hardening. A new plasticity theory was devised to incorporate this result by allowing the $M$ in Hill's nonquadratic theory to vary with effective strain:

$$\bar{\sigma} = \left\{ \frac{1}{2(1+r)} \left[ (1+2r)(\sigma_1 - \sigma_2)^M + (\sigma_1 + \sigma_2)^M \right] \right\}^{1/M}, \tag{8.29}$$

where $M = M(\bar{\varepsilon})$.

In order to make the two effective stress–strain curves (Fig. 8.19) coincide, Hill's $M$ parameter should vary as shown in Fig. 8.20. The variation of $M$ is calculated from these two curves, by numerically searching for $M(\varepsilon)$, which maps the effective stress for each effective strain point. However, the variation curve found in this way may be not true for other strain paths, for example, the balanced biaxial tension in the bulging experiment.

Later work[25] showed that 70/30 brass behaves similarly to aluminum, with lower hardening rates in plane strain. However, steels show nearly identical hardening rates, which implies good accuracy for plasticity laws based on isotropic hardening.

[25] R. H. Wagoner, "Plastic Behavior of 70/30 Brass Sheet," *Metal. Trans. A.* **13A**, 1491 (1982).

## 8.3 In-Plane Forming Limits

The forming limit diagram (FLD), also known as the Keeler–Goodwin diagram,[26] was originally devised as an experimental, semiquantitative tool to aid die designers and sheet-forming specialists in the optimization of die shapes, boundary conditions, and material properties. It followed in part the instability analysis of Hill.[27] Further developments have been based on one-dimensional simulations using a defect of infinite length introduced by Marciniak and Kuczinski.[28] More recently, finite-element simulations have been used to compute FLDs using finite defects.

### Forming Limit Diagram

Sheet metal can be deformed by in-plane tension to a certain level before *instability* or fracture occurs. This final level depends principally on the material itself and also on the stress and strain states imposed. There are two modes of instability that can be distinguished in a uniaxial tensile test:

1. In *diffuse necking*, the instability occurs when the load reaches a maximum.
2. In *localized necking*, the instability occurs when a groove appears in the middle of the diffuse neck.

For sheet tensile specimens, unlike a round bar, a through-thickness thinning or localized necking usually occurs after diffuse necking. This is of particular interest in sheet-metal forming operations, because the limit strain is mostly determined by this phenomenon for wide specimens.

By using circle grid measurement, a sheet-metal FLD can be determined experimentally. The FLD systematically represents the limit strain, based on localized necking and fractures for a wide region of stress paths, as illustrated in Fig. 8.15. The FLD is expressed as a curve in the space of principal engineering strains, $e_1$ and $e_2$, delineating the maximum safe strain, $e_1$, and the strain perpendicular to it, $e_2$, at the onset of necking. Table 8.2 lists three typical stress and strain states, assuming monotonic and proportional loading and isotropic von Mises yielding behavior.[29] The

**Table 8.2. Typical Stress and Strain States in Sheet-Metal Forming**

| Tests | Stress State | Strain State |
|---|---|---|
| Uniaxial tension | $\sigma_2 = 0$ | $\varepsilon_2 = -0.5\varepsilon_1$ |
| Plane strain | $\sigma_2 = 0.5\sigma_1$ | $\varepsilon_2 = 0$ |
| Balanced biaxial stretching | $\sigma_2 = \sigma_1$ | $\varepsilon_2 = \varepsilon_1$ |

[26] S. P. Keeler and W. A. Backofen, "Plastic Instability and Fracture in Sheets Stretched over Rigid Punches," *Trans. ASM* 56, 25–48 (1963).
G. M. Goodwin, "Application of Strain Analysis to Sheet Metal Forming Problems in the Press Shop," *SAE Paper 680093*, 1968.

[27] R. Hill, "On Discontinuous Plastic States, with Special Reference to Localized Necking in Thin Sheets," *J. Mech. Phys. Solids* 1, 19–30 (1952).

[28] Z. Marciniak and K. Kuczynski, "Limit Strain in the Processes of Stretch-Forming Sheet Metal," *Int. J. Mech. Sci.* 9, 609 (1967).

[29] Presented ratios in Table 8.2 assume plastic isotropy.

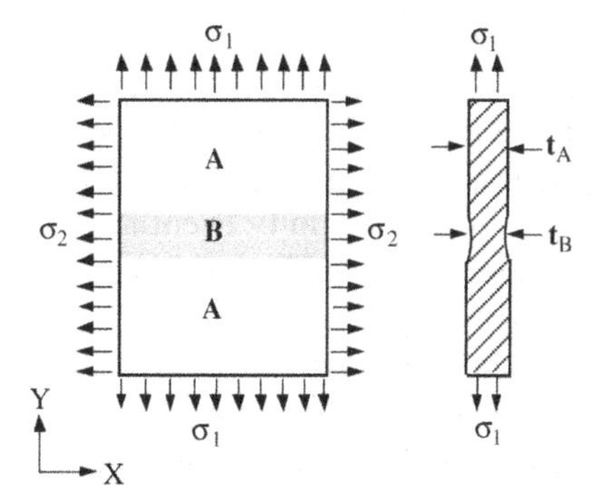


Figure 8.21. The defect used in the M–K model.

FLD can be used to determine how close a certain stamping operation is to failure by splitting.

The FLD is considered a material limit that cannot be exceeded by proportional straining. The "formability" of a sheet is a reflection of its FLD. The lowest level on a typical FLD occurs at or near plane strain, that is, when the minor strain is zero. For most low-carbon steels, the FLDs are similar to this shape. The vertical positions depend on the individual sheet's thickness and its $n$ value and $m$ value. Other important effects may be the history of prestrain, temperature and tool velocity, tool–sheet interfacial condition, and tool curvature. The plane-strain point on an FLD is called $FLD_0$.

### Simulation of Forming Limit Diagrams

In-plane forming limit analysis of sheet metal is a theoretical study to explain strain localization and splitting failure in stretching. Hill found theoretically that localized necking occurs along a "zero extension" direction when the minor strain is less than or equal to zero. His theory adequately predicts observed behavior for failures in this negative $\varepsilon_2$ region. However, for the other half of principal plastic strain space, where the minor strain $\varepsilon_2$ is positive, there is no in-plane direction of zero extension, so the strain localization phenomena cannot be explained by Hill's theory.

Marciniak and Kuczynski (M–K)[30] provided the first analytical model to predict the occurrence of localized necking under biaxial tension or the positive half of the principal plastic strain space. They introduced an intrinsic inhomogeneity in load-bearing capacity throughout a deforming sheet and demonstrated mathematically how this inhomogeneity can lead to an unstable growth of strain in the weaker regions and subsequently lead to localized necking and failure.

M–K considered a sheet material uniform in mechanical properties but with a notch in thickness (region B) representing an initial defect. This notch extends across the sheet, along the minor principal stress direction $x_2$, as shown in Fig. 8.21. The width of this notch is not important and, therefore, not related to this analysis. The geometrical notch only serves as a mechanical analog for the hypothesized initial local weakness in the sheet, and it provides for conceptual and analytical clarity. The source of an intrinsic inhomogeneity may have no relation to thickness but to other material property or geometrical variations. Extrinsic inhomogeneity introduced by external

[30] Z. Marciniak and K. Kuczynski, "Limit Strain in the Processes of Stretch-Forming Sheet Metal," *Int. J. Mech. Sci.*, Vol. 9, 1967, pp. 609–620.

factors such as friction, localized heating, and strain localization over tools may also be present and serve to initiate the localization.

In order to calculate FLDs by the M–K method, an isostrain condition is imposed on the sheet in Fig. 8.21 along $x$ and an isoload condition along $y$. These constraints produce an accelerating $\dot{\varepsilon}_1$ in the notch, which will tend toward infinity eventually, while $\dot{\varepsilon}_2$ both in the notch and in the bulk (region A) remains constant. Thus, the strain state in the notch tends toward plane strain and the $\varepsilon_1$ in the notch becomes very large relative to the constant $\varepsilon_2$. M–K showed that $\varepsilon_1$ outside of the notch approaches a limiting value (together with $\varepsilon_2$ defining a point on a FLD curve). The limiting strain found in this way is characteristic of material properties, the bulk strain state, and the "strength" of the inhomogeneity defined by $f(f = t_B/t_A)$.

In the M–K theory, instability is considered a process in which the strain state in a region of local weakness evolves to the plane-strain state while the global deformation proceeds uniformly in the bulk. While this characteristic agrees with that observed for many sheet failures, the M–K analysis is based on an overly simple model. The M–K defect is unrealistically assumed to be infinite in length. Furthermore, the M–K analysis imposes nonphysical boundary conditions through constant bulk stress paths. While these two simplifications allow the closed-form solutions that M–K sought, they overstate the impact of a real material defect. The FEM can be used to relax these two restricting assumptions and treat the FLD more realistically.

The FEM is used to analyze a strain localization and failure process by introducing a finite-length defect in a sheet.[31] The deformation was imposed by displacement boundary conditions far removed from the defect. The defect is characterized by its aspect ratio, as well as by its thickness ratio $f$. A work-hardening and strain-rate-sensitive material model expressed in Eq. (8.13) was used. The effective stress is based on Hill's quadratic yield function for normal anisotropy.

Figure 8.22 displays the full, undeformed finite-element mesh used in this study. The shaded region represents 22 elements of reduced thickness, which form an artificial notch with an aspect ratio of 11:1. Other defect aspect ratios, including 5:1, were considered. The mesh consists of 1066 constant strain triangular elements and 580 nodes. The shaded region in the center of the mesh represents the location of a thickness defect with an aspect ratio of 11:1. The mesh represents a sheet 50 mm square and 1 mm thick.

The parameters were chosen such that a direct comparison with early M–K results could be made. Calculations are based on the following parameter values, where

$f = 0.98$ (consider the thickness defect only),
$n = 0.2$,
$m = 0.0$ (no strain rate sensitivity),
$k = 1.0$,
$r = 1$ (isotropy),
$M = 2.0$ (Hill's old yield function), and
$\bar{\varepsilon}_0 = 0.0014$.

[31] K. Narasimhan and R. H. Wagoner, "Finite Element Modeling Simulation of In-plane Forming Limit Diagrams of Sheets Containing Finite Defects," *Metal. Trans. A.* **22A**, 2655 (1991).
R. H. Wagoner, K. S. Chan, and S. P. Keeler, eds., *Forming Limit Diagrams: Concepts, Methods, and Applications* (The Metallurgical Society, Warrendale, PA, 1989).

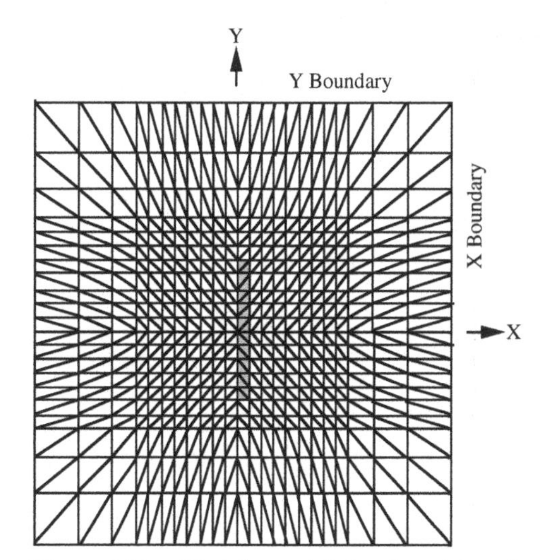


Figure 8.22. Undeformed finite-element mesh.

Various biaxial strain states can be analyzed by imposing constant displacement rates at $x$ and $y$ boundaries. If the mesh is defect free, the FEM will predict a uniform deformation for all strains. Once an inhomogeneity is introduced in the mesh, the deformation of the sheet in and adjacent to the defect region is perturbed locally, while the element distant from the inhomogeneity deforms relatively undisturbed by the presence of the defect. During the FEM stretching of a sheet containing a defect, the strain grows most rapidly in the weakest area of the sheet at an accelerating rate with respect to deformation in the bulk. The final major and minor strains in the uniform area, when the notch strain rates exceeded 10 times the distance strain rates, were drawn as the limit strain in a FLD.

Figure 8.23 depicts the forming limit curves for both the 5:1 and 11:1 aspect ratio notches along with the M–K curve for $f = 0.98$. From a comparison of the relative positions of the various curves, it is apparent that the aspect ratio strongly influences the level of FLD. The rate of strain localization is reduced for a decreased aspect ratio, and this influence is found to be of a comparable magnitude to the influence of $f$. By predicting lower forming limits for a given $f$ value, the M–K curve represents a more severe forming condition than either of the finite-length notch FEM results.

### Statistical Simulation of FLDs

In fact, strain localization processes are statistical in nature, presumably because of statistical distributions of defects in the form of grain orientation and strength. Therefore, when several stretching experiments are repeated under similar conditions (in terms of loading conditions and material properties), the measured critical limit strains from these experiments are found to fall within a narrow range. On this basis, it is useful then to present the FLD as a band rather than a single curve. Then one can expect that a randomly measured limit strain will fall within this band with a high probability.

A *Monte Carlo method* was used to study the statistical deformation behavior of sheet metal under biaxial tension. An initial random thickness distribution of the sheet was used as the origin for the statistical nature of the localization process. This same distribution of thickness was assigned randomly to the FEM mesh shown in Fig. 8.24.

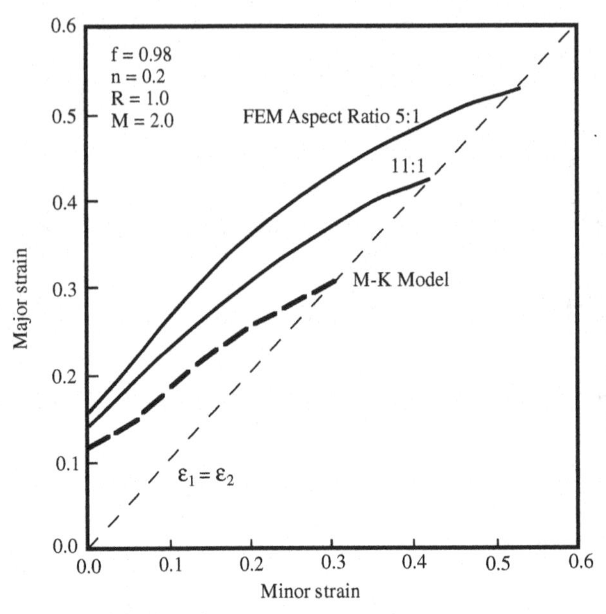


Figure 8.23. FEM predicted FLD, using two aspect ratios. The corresponding M–K curve is also shown as a dashed curve.

In order to establish the statistics of the strain localization process, we need to repeat experiments several times. For our FEM modeling, an overall normal distribution of element thicknesses was constructed by randomly selecting thickness values between 0.98 and 1.00. Then, elements having these thicknesses were randomly assigned positions in the FEM mesh; 34 such initial mesh–thickness patterns were used in this analysis. For each strain path, the simulation of in-plane stretching was then repeated 34 times to generate the statistics of the limit strains. Figure 8.24[32] shows the mesh that represents one quarter of the deforming sheet. The mesh consists of 131 nodes and 220 elements, generated in a "free mesh" style. Boundary conditions were imposed as in the previous section.

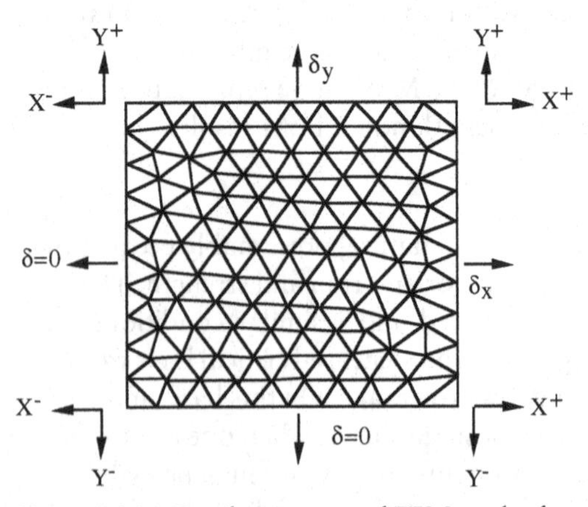


Figure 8.24. Random generated FEM mesh of a quarter of the sheet with a boundary condition.

[32] K. Narasimhan, D. Zhou, and R. H. Wagoner, "Application of the Monte Carlo and Finite Element Method to Predict the Scatter Band in Forming Limit Strain," *Scripta Metal.* 26, 41–46 (1992).

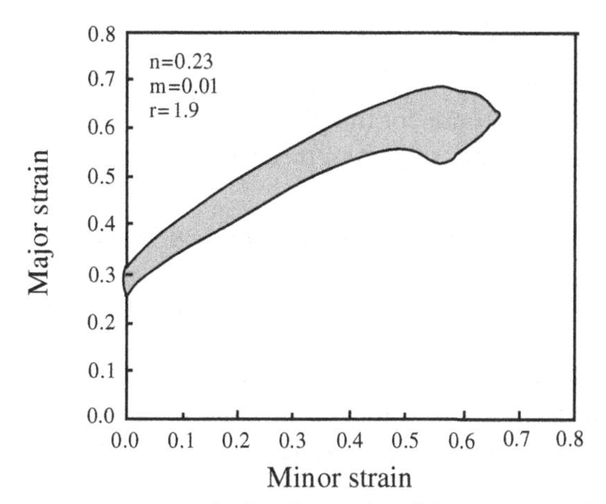


Figure 8.25. Calculated FLD band from 34 initial thickness distributions by the Monte Carlo method from a random mesh.

Figure 8.25 presents the results for strain paths in the familiar FLD form. This figure shows the scatter band produced in the limit strain by the repeated simulation. The FLDs predicted in this way have similar shapes to the FLDs predicted by M–K and finite-element analysis. The main advantages of the Monte Carlo method are as follows.

1. There is realistic representation of the initial sheet.
2. No assumption was made on the defect shape, its aspect ratio, and its orientation.
3. There is prediction of a statistical scatter band in the limit strain rather than a single curve. Such a band agrees with experimental experience.

## PROBLEMS

### A. Proficiency Problems

1. How will the invariance principle be affected by Coulomb friction boundary conditions, where the friction force is a fixed fraction of the contact force? (Contact itself is a displacement boundary condition.) What about for Tresca friction, where the friction force is a fraction of the material flow stress?
2. Considering Exercise 8.1, compute the temperature rise and change in flow stress for a stainless steel that obeys the same hardening law except that $k$ is twice as large.
3. Standard tensile tests are often conducted at a strain rate of $10^{-3}$/s. In view of Fig. 8.6, why does testing at this rate for total elongation involve not only mechanical behavior but also thermal characteristics? Discuss how strength and conductivity would affect measured elongations at these rates.
4. Does deformation heating have a larger effect on the uniform elongation of post-uniform elongation? Why? Under what conditions would the invariance principle apply to nonisothermal deformation?
5. Why does the presence of growing biaxiality stabilize the tensile test? Describe how you might tailor a material's behavior to maximize this effect.

6. Why does inertia stabilize tensile deformation (Fig. 8.13)? For experimental measures, what other effects must be taken into account as the strain rate is increased?
7. Based on the FLD shown in Fig. 8.15 and the strains for the plane-strain specimen presented in Fig. 8.18, where do you expect the specimen to fail? Explain your reasons.
8. How would the plane-strain and tensile hardening laws shown in Fig. 8.19 affect the formability in uniaxial tension versus plane strain?

## B. Depth Problems

9. Austenitic stainless steel was found to form much better for producing catalytic converters with "flood" lubrication, where a water-based lubricant is sprayed onto the sheet during forming. Austenitic is approximately twice as strong and has a thermal conductivity one tenth that of plain carbon steels. What do you think is the major effect of such lubrication, and why?
10. It is often said that $n$ affects primarily the uniform elongation and $m$ affects primarily the postuniform elongation. Do you agree? Discuss why or why not.
11. What are the real-world origins of nonuniformities in materials that are represented as a notch in simulation schemes?
12. FLDs are usually measured using a fixed grid size of −0.1-in. (2.54 mm) diameter circles. Why is this choice important? How would the measured FLDs differ for other grid sizes?

## C. Numerical Problem

13. Construct a one-dimensional numerical scheme such as the one described for investigation on inertia, but use it to predict the role of deformation heating on tensile elongation. First solve isothermal and adiabatic cases, and compare isothermal results with low-speed results in Fig. 8.13. Then investigate the roles of conduction and convection on intermediate cross-head rates.

CHAPTER NINE

# Steady-State Forming Problems

Theoretically, steady-state flow is fully established in a given domain $\Omega$ when, for any spacial point in $\Omega$, all the mechanical and physical variables that are necessary to describe the problem are independent of time $t$. Of course, the position of a material point is not constant, but follows a fixed trajectory as shown in Fig. 9.1.

From the practical point of view, the steady-state idealization is always an approximation, as a true stationary flow would require an infinite time to be established. However, in industrial processes such as rolling, extrusion, drawing, and so on, the steady-state assumption is a rather accurate approximation as soon as the length of the deforming zone is less than half of the rigid parts lying before or after this zone. At this stage the end effects, that is, the discrepancies from the stationary shape that occur at the beginning and at the end of the process, have a vanishingly small influence on the flow. In the sheet-rolling process, the length of the deformed zone is of the order of a few centimeters, and the length of the rolled sheet is several hundred meters. Conversely, during the extrusion process, the length of the billet is only of the order of ten times the diameter so that the steady-state part of the process is more approximate, and represents a smaller percentage of the whole process.

When it is possible, the stationary hypothesis can be tested on representative examples by a transient analysis, which allows the estimation of the time that is necessary to reach the state when all the parameters remain approximately constant. Another possibility is to check the influence of the initial conditions of the process on the steady-state solution, if any. As an example, the cold-rolling process can be initiated (at least theoretically) by two different starting conditions, as shown in Fig. 9.2. In that case the incremental computation shows that the steady-state solution is the same.[1]

At this point the choice between an incremental approach with an updated Lagrangian formulation (see Section 5.4), or an Eulerian formulation can be decided. The first one seems much more straightforward at first sight. However, it is generally much more expensive, with respect to computer time, if an accurate solution is required. The increased cost comes from a large number of time steps, together with a larger domain to mesh, in order to reach the stationary state. Moreover, one must be aware that a large number of time steps introduce accumulated error, which can affect the precision of the computed free surface and other parameters.

[1] For more details, the interested reader is referred to P. Gratacos, "A Problem of Coupled Deformation: Elastoplastic Finite Element Modelling of Cold Rolling of Thin Strips," Ph.D. dissertation (Ecole des Mines de Paris, Sophia-Antipolis, 1991; in French).

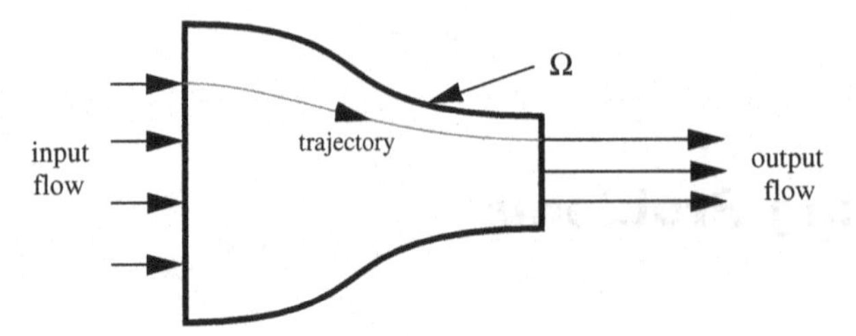


Figure 9.1. Definition of a steady state.

In many instances these considerations orient the choice toward the steady-state formulation, which is generally available for purely viscoplastic materials. The main issues are then:

- calculation of the free surface when the material is not in contact with the tools, which will allow the determination of the actual shape of the workpiece, and
- transport of convective terms such as temperature, accumulated strain, or any other physical parameter, taking into account that a coupling with the material flow can be necessary.

However, when elastoplastic or elastoviscoplastic constitutive equations are considered, despite some attempts to build an analog of the flow theory,[2] the Euler formulation is not easy to use. In this latter case most calculations are performed with the incremental approach, by modeling the starting stage of the process.

## 9.1 Slab Analysis Versus the Finite-Element Solution

The slab method (SM) can be applied only to flat and thin workpieces, but with various degrees of sophistication, including elastoplastic or elastoviscoplastic behavior and strain hardening. In cold sheet rolling the coupling was also introduced with a simple model[3] or with a finite-element model for the elastic deformation of the rolls. The finite-element method can cope with these problems as well, but the difference between the two approaches depends mainly on the geometry of the deformed zone and of the magnitude of the deformation. The discrepancy between the methods will be illustrated in the cases of cold and hot rolling of flat products.

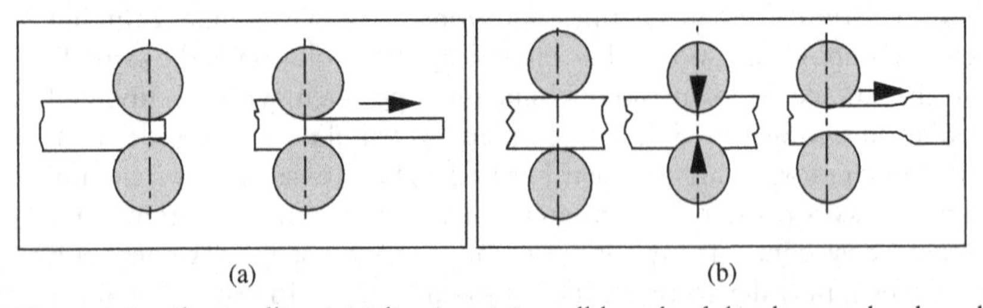


Figure 9.2. Sheet-rolling initialization: (a) small length of the sheet under the rolls and rolling; (b) indentation of the sheet by the rolls followed by rolling.

[2] S. Yu and E. Thompson, "A Direct Eulerian Finite Element Method for Steady State Elastic Plastic Flow," in *Numerical Methods in Industrial Forming Processes, NUMIFORM 89*, E. G. Thompson, R. D. Wood, O. C. Zienkiewicz, and A. Samuelsson, eds. (A. A. Balkema, Rotterdam, 1989), pp. 95–103.

[3] Hitchcock based on Hertzian theory, influence function, and "foundation" models.

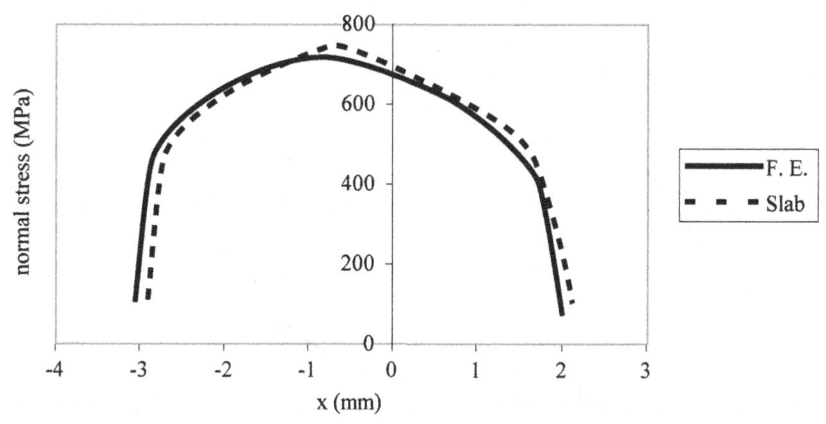


Figure 9.3. Comparison of the SM and the FEM for sheet rolling ($x$ corresponds to the rolling direction).

### Comparison of Slab and Finite-Element Methods for Cold Rolling

In most cases the thickness of the sheet is small enough with respect to the length of the deformed zone, so that the slab method is a very good approximation, as is shown in Fig. 9.3. For this comparison, the initial thickness of the sheet was 0.4 mm, the radius of the rolls was 285 mm, and the reduction ratio was 5%. The material parameters were chosen according to the following:

elastic constants: $E = 210 \times 10^9$ Pa and $n = 0.3$,
yield stress: $\sigma_0 = 400 \times 10^6$ Pa,
Tresca friction coefficient: $\bar{m} = 0.3$.

The computations were made with deformable rolls, the profile of which is shown in Fig. 9.4. We observe that the results are almost identical by the two methods.

By the slab analysis we get a stationary solution at a relatively low computational cost,[4] making this method appear more attractive than the incremental finite-element approach. However, when very small reductions of thickness are involved (skin pass),

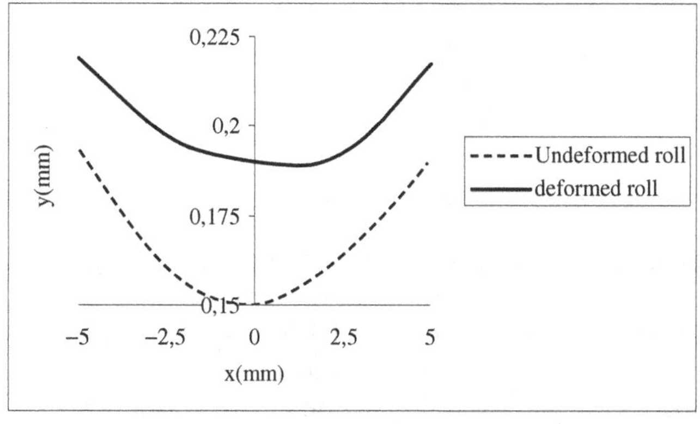


Figure 9.4. The deformed profile of the roll during cold rolling; $x$ corresponds to the rolling direction and $y$ to the vertical axis, with $y = 0$ being the coordinate of the symmetry plane of the sheet.

[4] Although a long iterative process may be necessary to couple roll deformation.

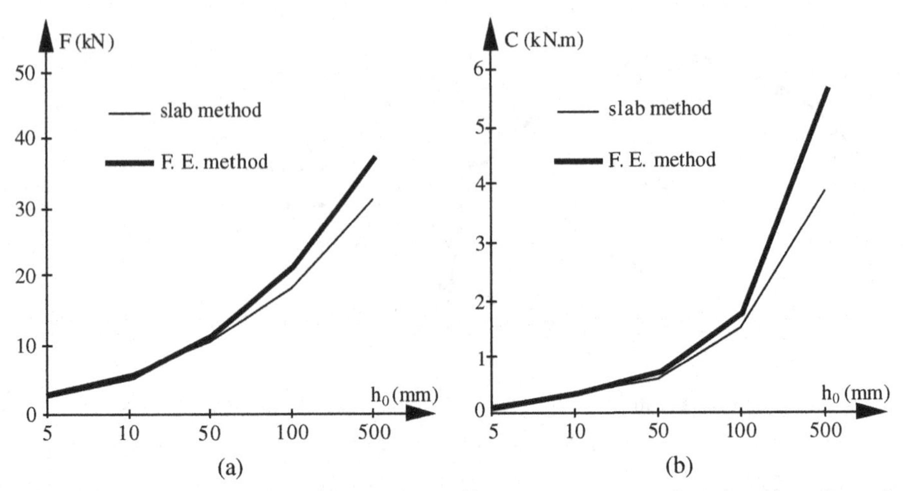


Figure 9.5. The SM vs. the FEM for hot rolling: comparisons of (a) the force $F$ on the rolls, (b) the torque C.

the basic hypotheses of the slab method are no longer valid: namely, the deformation cannot be considered constant in the thickness. In this latter case the finite-element method allows us to take more precisely into account the deformation pattern.

#### Comparison of Slab and Finite-Element Methods for Hot Rolling

For thin sheets the slab method can be used with a generalization taking into account a viscoplastic constitutive equation. But when relatively thick slabs are rolled, the strain is heterogeneous in the thickness and the slab method produces inaccuracies when the slab method is utilized. Figure 9.5 shows an example of the difference between the slab method and a reference finite-element solution (using a large number of elements), as a function of the initial thickness $h$ of the slab, for a constant reduction of 30%. The slab method computation was made for a material with a yield stress of $\sigma_0 = 150$ MPa and a friction coefficient of $\bar{m} = 0.5$. The finite-element results were obtained with a quasi-rigid-plastic material, the parameters of which were

$$K = 86 \text{ MPa s}^{-m}, \quad m = 0.05, \quad \alpha = 0.5, \quad p = 0.05.$$

For both cases the slab width is $L = 1$ m, the roll radius is $R = 0.5$ m, and the velocity is 1 m/s.

## 9.2 Rolling

The rolling process can be defined as a continuous process of plastic deformation for long parts of constant cross section, in which a reduction of the cross-sectional area is achieved by compression between two rotating rolls (or more). We have already made the distinction between *cold rolling*, in which the material is deformed at room temperature (but it can be slightly higher with heat dissipation due to plastic work) and *hot rolling*, in which the temperature is high (more than half of the absolute melting temperature). Another important distinction is made according to geometric considerations.

*Flat rolling* is performed with cylinders: it is the case for sheet rolling or strip rolling (in which the thickness is very small: of the order of a millimeter or less), or

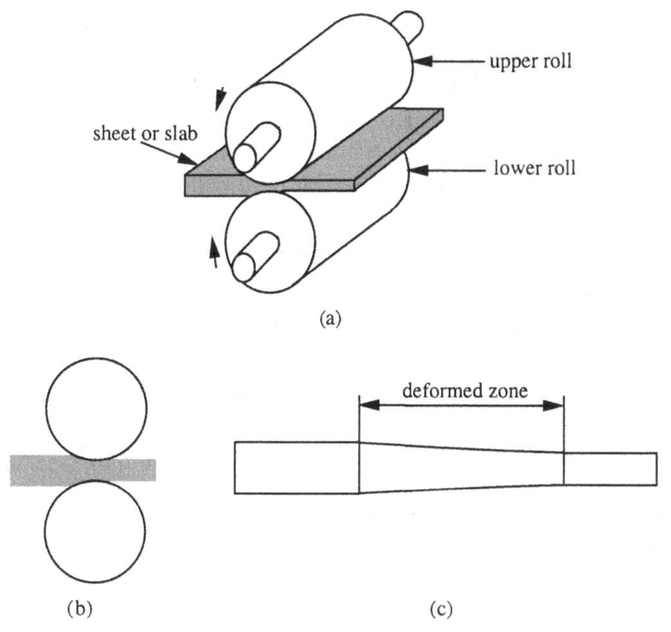


Figure 9.6. Flat rolling: (a) general, (b) cross section for slab rolling, and (c) cross section for sheet rolling.

slab rolling (in which the slab thickness is of the order of 0.1 m) and any intermediate situation.

*Shape rolling* allows the production of more complex workpieces by using appropriate roll geometries: the cross section of the part can be a round, an oval, various beams, a rail, and so on. For tube rolling a mandrel is introduced in the tube to prevent a collapse caused by the rolling pressure.

Figure 9.6 represents the general principle for flat rolling and two examples of geometries in a vertical cross section by a plane parallel to the rolling direction.

For cold or hot rolling of any geometry, the desired reduction of cross-sectional area is too important to be feasible in one pass. The final deformation is progressively applied by using several stands so that the same part is successively deformed by several pairs of cylinders as is shown in Fig 9.7. There are thus interacting forces between two successive stands, which induce tensions either in the direction of rolling or in the opposite direction.

In Fig. 9.7 the smaller rolls are in contact with the sheet and produce successive reductions of thickness. Their small diameters limit the width of the deforming region and thus the roll-separating force. The bigger rolls are designed to prevent excessive bending of the work roll.

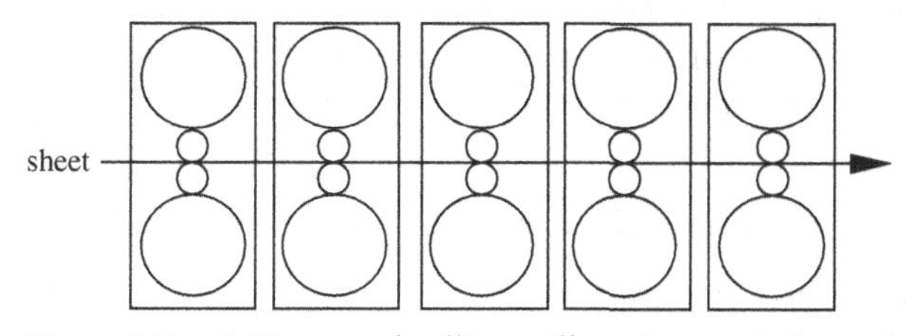


Figure 9.7. A Five-stand rolling mill equipment (schematic).

### The Flow Formulation Applied to Rolling

The basic formulation for a general viscoplastic material, the constitutive equation of which is expressed in term of a potential $\varphi$, has been presented elsewhere.[5] In most cases the viscoplastic power law (also called the Norton–Hoff law) is used according to Eq. (1.92). It is a satisfactory approximation for hot rolling, and the rate sensitivity index $m$ ranges between 0.1 and 0.2 for usual hot metals. For cold rolling, the same formulation can be used with $m = 0.01$–$0.02$, but, for numerical reasons linked to the convergence of the Newton–Raphson iterative method, a mathematical regularization must be introduced (see Exercise 9.1).[6]

---

**Exercise 9.1: Show that the second derivative of the viscoplastic potential of the one-dimensional power law ($m < 1$) is infinite for vanishing strain rate, and that it remains finite for the regularized law given by**

$$\varphi^r = \frac{\sigma_0}{m+1}\left(\dot{\varepsilon}^2 + \dot{\varepsilon}_0^2\right)^{(m+1)/2}, \tag{9.1–1}$$

**where $\dot{\varepsilon}_0$ is a constant that is chosen much smaller than the average strain rate in the part, in order to avoid a significant perturbation.**

For the unperturbed viscoplastic power law we have obviously

$$\varphi = \frac{\sigma_0}{m+1}\dot{\varepsilon}^{m+1}. \tag{9.1–2}$$

The first derivative gives the stress according to

$$\sigma = \frac{d\varphi}{d\dot{\varepsilon}} = \sigma_0 \dot{\varepsilon}^m, \tag{9.1–3}$$

and the second derivative of the viscoplastic potential is

$$\frac{d^2\varphi}{d\dot{\varepsilon}^2} = m\sigma_0\dot{\varepsilon}^{m-1}. \tag{9.1–4}$$

Then it is clear that if the rate sensitivity index $m$ is less than unity, the second derivative tends to infinity, when the strain rate tends to zero.

Alternatively, when Eq. (9.1–1) is differentiated, we get

$$\frac{d\varphi^r}{d\dot{\varepsilon}} = \sigma_0\left(\dot{\varepsilon}^2 + \dot{\varepsilon}_0^2\right)^{(m-1)/2}\dot{\varepsilon}, \tag{9.1–5}$$

and the second derivative is

$$\begin{aligned}\frac{d^2\varphi^r}{d\dot{\varepsilon}^2} &= (m-1)\sigma_0\left(\dot{\varepsilon}^2 + \dot{\varepsilon}_0^2\right)^{(m-3)/2}\dot{\varepsilon}^2 + \sigma_0\left(\dot{\varepsilon}^2 + \dot{\varepsilon}_0^2\right)^{(m-1)/2},\\ &= \sigma_0\left(\dot{\varepsilon}^2 + \dot{\varepsilon}_0^2\right)^{(m-3)/2}\left(m\dot{\varepsilon}^2 + \dot{\varepsilon}_0^2\right).\end{aligned} \tag{9.1–6}$$

The second derivative remains finite for a zero strain rate and is equal to

$$\frac{d^2\varphi^r}{d\dot{\varepsilon}^2}(0) = \sigma_0\dot{\varepsilon}_0^{m-1}. \tag{9.1–7}$$

---

For the definition of the domain in which the computation is made, we shall at first consider that there is no spread, that is, no horizontal flow perpendicular to the rolling direction and the vertical direction. With this hypothesis the deformed domain can be determined only with geometric considerations.

[5] R. H. Wagoner and J.-L. Chenot, *Fundamentals of Metal Forming* (Wiley, New York, 1997), Chap. 5, Sect. 7.

[6] The same regularization is also used for higher values of $m$, as it improves convergence.

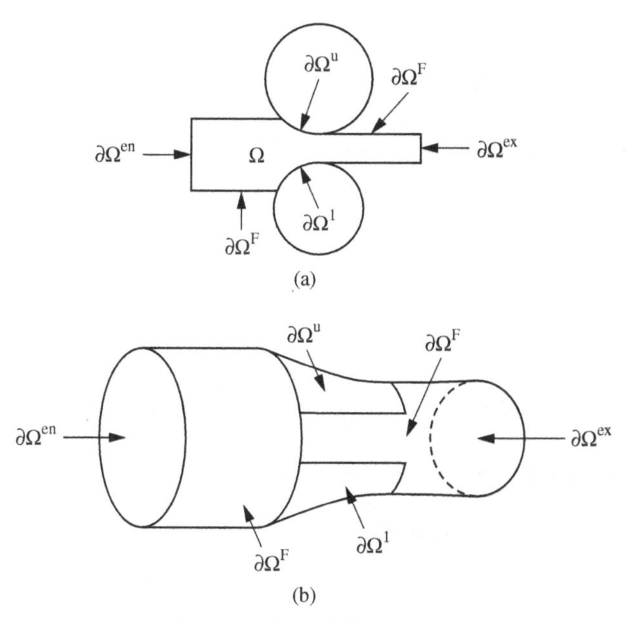


Figure 9.8. Definition of the deformed part in rolling: (a) two-dimensional approximation for flat rolling and (b) three-dimensional case for shape rolling (here a round oval pass).

We introduce the following notation (see Fig. 9.8).

1. The $ox_1$ axis corresponds to the rolling direction.
2. $v^{\mathrm{en}}$ and $v^{\mathrm{ex}}$ are respectively the entry velocity of the workpiece and the exit velocity at first.
3. $\partial\Omega^{\mathrm{en}}$, $\partial\Omega^{\mathrm{ex}}$ are respectively the entry and exit cross sections (orthogonal to $ox_1$), which are chosen at an appropriate distance from the zone under the rolls.
4. $F^{\mathrm{ex}}$ and $F^{\mathrm{en}}$ are the tensions in the direction of rolling, and in the opposite direction, respectively (by definition both are positive when there is a tension). They act on $\partial\Omega^{\mathrm{en}}$ and $\partial\Omega^{\mathrm{ex}}$ uniformly; that is, the corresponding stress components $f^{\mathrm{en}}$ and $f^{\mathrm{ex}}$ are assumed to be constant so that we have

$$F^{\mathrm{en}} = f^{\mathrm{en}} S^{\mathrm{en}}, \qquad F^{\mathrm{ex}} = f^{\mathrm{ex}} S^{\mathrm{ex}}. \tag{9.1}$$

5. $S^{\mathrm{en}}$ and $S^{\mathrm{ex}}$ are the areas of the entry and exit cross sections.
6. $\partial\Omega^{u}$, $\partial\Omega^{l}$ are the contact surfaces with the upper roll and the lower roll, respectively; the total surface of contact is denoted by $\partial\Omega^{f}$.
7. $\partial\Omega^{F}$ is the free surface, where the stress vector is assumed to be nil.

This notation applies as well to the two-dimensional approximation in which surfaces are replaced by curves.

With these hypotheses, according to the penalty method outlined in Section 6.3, the unknown velocity field $\mathbf{v}$ minimizes the functional:

$$\Phi(v) = \int_{\Omega} \frac{K}{m+1} (\sqrt{3}\dot{\bar{\varepsilon}})^{m+1}\, \mathrm{d}V + \frac{1}{2}\int_{\Omega} \rho_p K[\mathrm{div}(\mathbf{v})]^2 \mathrm{d}V$$

$$+ \int_{\partial\Omega^f} \frac{\alpha K}{p+1} |\Delta \mathbf{v}|^{p+1}\, \mathrm{d}S + \int_{\partial\Omega^{\mathrm{en}}} f^{\mathrm{en}} v^{\mathrm{en}}\, \mathrm{d}S - \int_{\partial\Omega^{\mathrm{ex}}} f^{\mathrm{ex}} v^{\mathrm{ex}}\, \mathrm{d}S, \tag{9.2a}$$

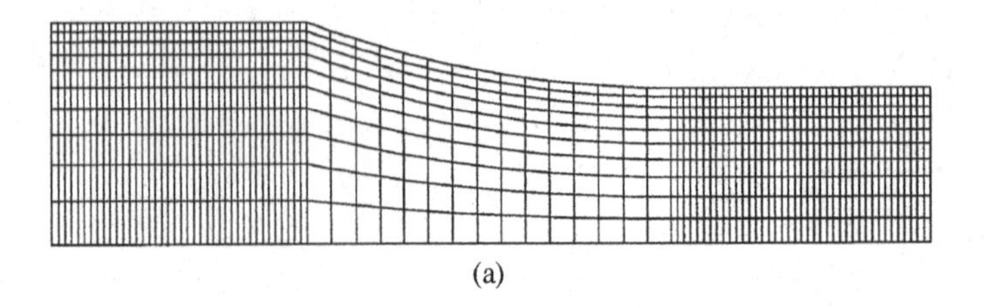

(a)

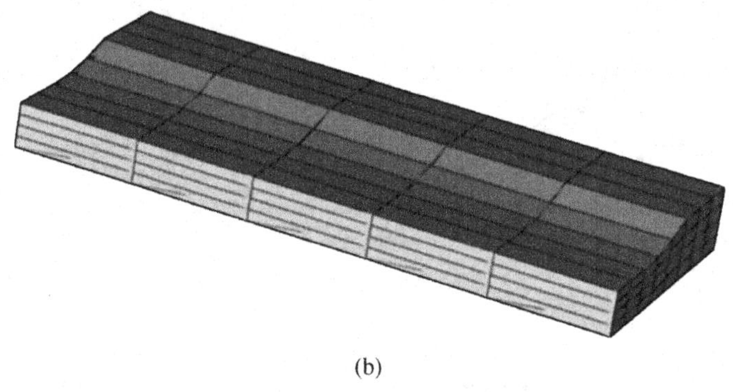

(b)

Figure 9.9. Finite-element meshing of the domain: (a) two-dimensional approximation and (b) three-dimensional discretization.

subject to the condition of contact on the upper and lower rolls:

$$\mathbf{v} \cdot \mathbf{n} = 0 \quad \text{on } \partial\Omega^f = \partial\Omega^u \cup \partial\Omega^l. \tag{9.2b}$$

In the functional $\Phi$ of Eq. (9.2a), the first term of the right-hand side corresponds to the mechanical behavior, the second term to the penalty contribution, which was added in order to enforce approximately the incompressibility condition, the third term comes from the viscoplastic friction law, and the last two terms take into account the back tension and the front tension, respectively.

In order to build the finite-element solution, the domain $\Omega$ is meshed with, for example, four-node quadrilaterals in two dimensions, or eight-node brick elements in three dimensions. Two examples are presented in Fig. 9.9.

At this stage, the unknown velocity field $\mathbf{v}$ can be discretized in the usual way with isoparametric finite elements:

$$\begin{aligned} \mathbf{v} &= \sum_n \mathbf{V}_n N_n = \mathbf{N} \cdot \mathbf{V}, \\ \mathbf{x} &= \sum_n \mathbf{X}_n N_n = \mathbf{N} \cdot \mathbf{X}. \end{aligned} \tag{9.3}$$

The gradient of the functional is set equal to zero. The resulting set of nonlinear equations is denoted by $\mathbf{R}(\mathbf{V}) = 0$, and it is solved by the Newton–Raphson iterative procedure (see Section 3.5). In Fig. 9.10 an example of a solution is given for a two-dimensional rolling geometry of a thick slab. The longitudinal component of the velocity field is pictured in Fig. 9.10(a): it shows that the perturbation is only important in the roll gap. The isoequivalent strain rate map indicates that the most

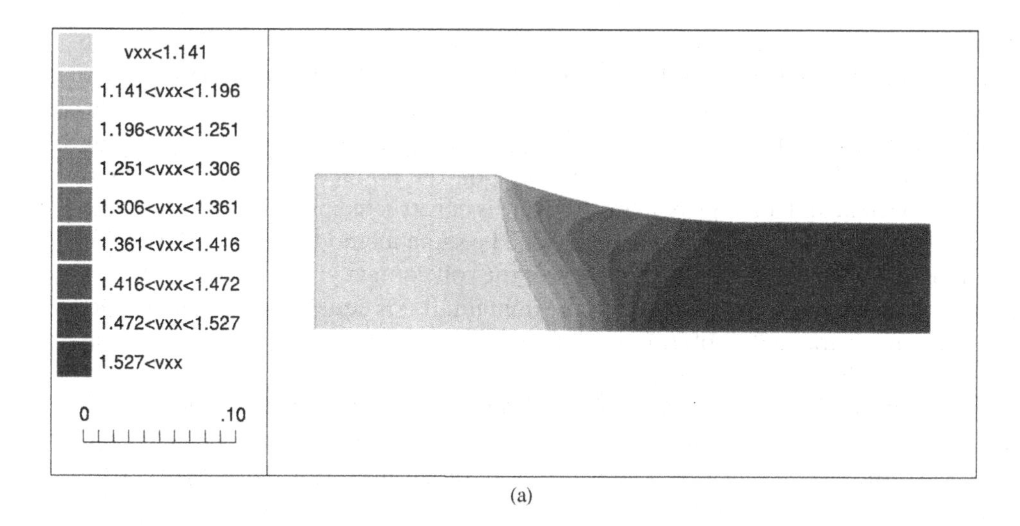


(a)

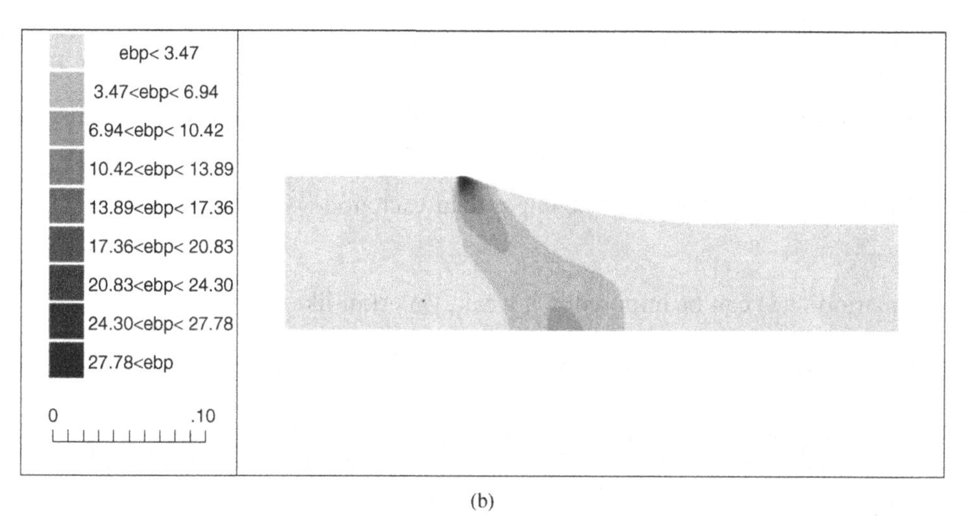


(b)

Figure 9.10. Finite-element solution of the metal flow in rolling: (a) longitudinal velocity distribution in the workpiece, and (b) isostrain rate map.

intense deformation zone is located at the entry, when the surface of the slab meets the roll.

### Determination of the Free Surface

The problem is extremely important in three-dimensional simulations of steady-state processes (for metals as well as polymers or other materials). For rolling, the workpiece is not only deformed as desired in the vertical and rolling directions, but also in the orthogonal transverse direction, or the width, as it generally exhibits a parasitic spread while deformation takes place under the rolls. The actual shape of the part can be accurately predicted only if the spread is taken into account. The general problem of the free surface in stationary processes was briefly outlined in Section 4.4.

More precisely, an iterative algorithm is proposed here, which allows computation of the free surface in a relatively easy way. At first a velocity field is computed on a trial domain $\Omega^{(0)}$, which is, for example, the geometric domain with no spread. We suppose that at the end of iteration $(k-1)$ the computed domain $\Omega^{(k-1)}$ is known, so that a new velocity field $\mathbf{v}^{(k)}$ can be calculated, with zero stress vector on the free surface as the boundary condition on $\partial\Omega^{F(k-1)}$. A new free surface is then obtained, using the flow line method and Eq. (5.54). For that, any node on the boundary of the entry cross section is selected, and an approximate flow line is constructed progressively, until the surface of a roll, or the exit plane, is reached. The same method can also be used for streamlines, starting when the material leaves the roll contact. This procedure allows building of a new approximation $\Omega^{(k)}$ of the domain. If it is equal to the previous one, up to a given accuracy, the solution is obtained. If not, the mesh of the new domain is updated and the iterations continue.

Other methods were proposed to solve free-surface problems numerically: they are generally based on a discretization of the free surface $\partial\Omega^F$ into surface elements and nodes, the positions of which are defined by a set of parameters $\mathbf{a}$. The nonlinear set of equations obtained by writing the minimum condition on the functional $\Phi$ includes the $\mathbf{a}$ parameters and therefore must be written as

$$\mathbf{R}(\mathbf{V}, \mathbf{a}) = 0. \tag{9.4}$$

In order for this to be solved, the boundary conditions must be added; this can be achieved in different ways, for nodes on the free surface:

a. The boundary condition can be imposed at each node $n$ by simply putting

$$\mathbf{V}_n \cdot \mathbf{n} = 0. \tag{9.5}$$

b. Equation (9.5) can be imposed in a weak, Galerkin-like form:

$$\int_{\partial\Omega^F} \mathbf{v} \cdot \mathbf{n} N_\nu^S \, dS = 0, \tag{9.6}$$

where $N_\nu^S$ is a surface shape function corresponding to the node number $\nu$ on the free surface. In cases a and b, an augmented system is thus obtained that can be solved again by the Newton–Raphson method, but the linear system obtained at each iteration is no longer b.

c. A new functional $J$ of the parameters $\mathbf{a}$ can be built, the minimum of which should be zero and corresponds to the free surface. We define $J$ by

$$J(\mathbf{a}) = \int_{\partial\Omega^F} (\mathbf{v} \cdot \mathbf{n})^2 \, dS. \tag{9.7}$$

However the numerical minimization of Eq. (9.7) is rather complicated if the Newton–Raphson method is used. Some simplifications must be made for the problem to be tractable: the modified Newton–Raphson method is utilized, so that the derivatives are calculated approximately.

For illustration of the free-surface computation, three examples will be presented. In the first one, a slab of thickness $h = 0.2$ m and width $L = 1$ m is rolled with a 30% reduction of thickness. The material parameters are $K = 100$ MPa s$^{-m}$, $m = 0.1$, $\alpha = 0.5$, and $p = 0.1$. The rolled slab possesses two symmetry planes, which allow to mesh only one quarter of the workpiece. In Fig. 9.11 the lateral spread is easily seen on the deformed mesh after the free-surface computation; the map of iso normal stress on the rolls is also pictured.

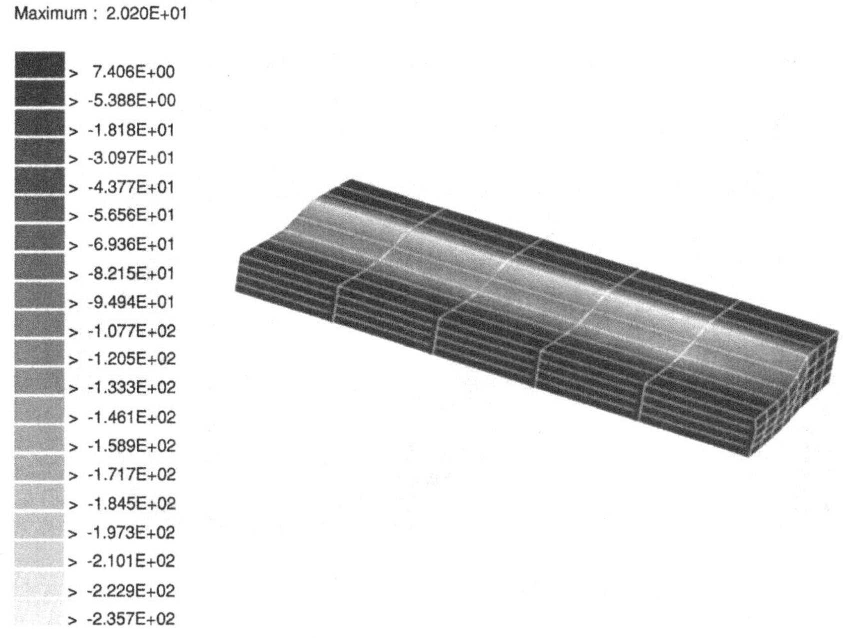


Figure 9.11. Computation of spread in slabbing.

The next example is a square–diamond sequence, pictured in Fig. 9.12, where the entry section is a square with side $a = 0.08$ m. The material parameters are the same as those of the previous example, except for the friction coefficient, which is taken as $\alpha = 0.3$.

The last example describes tube rolling: the basic principle is represented for one stand in Fig. 9.13(a). The thickness of the tube is progressively decreased by two rolls with circular shapes, while an internal mandrel (not represented here) is designed

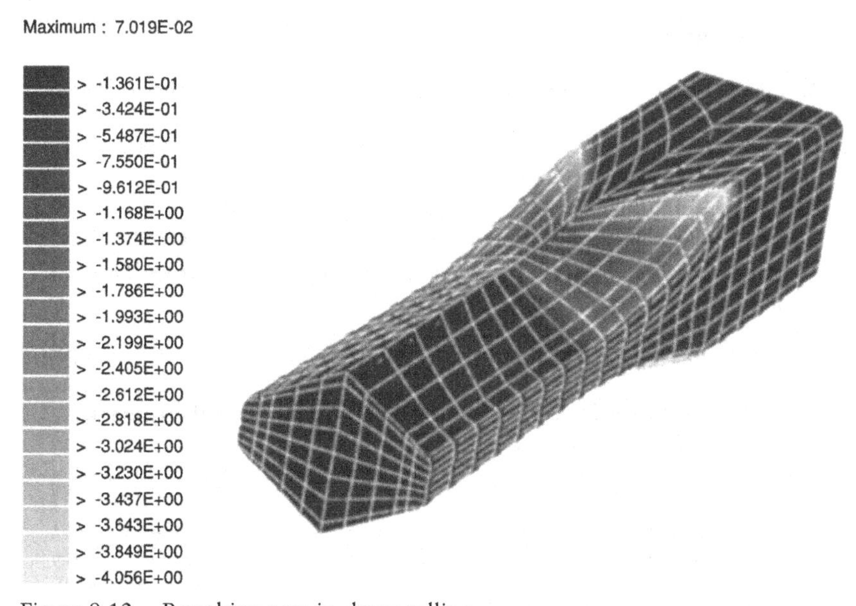


Figure 9.12. Roughing pass in shape rolling.

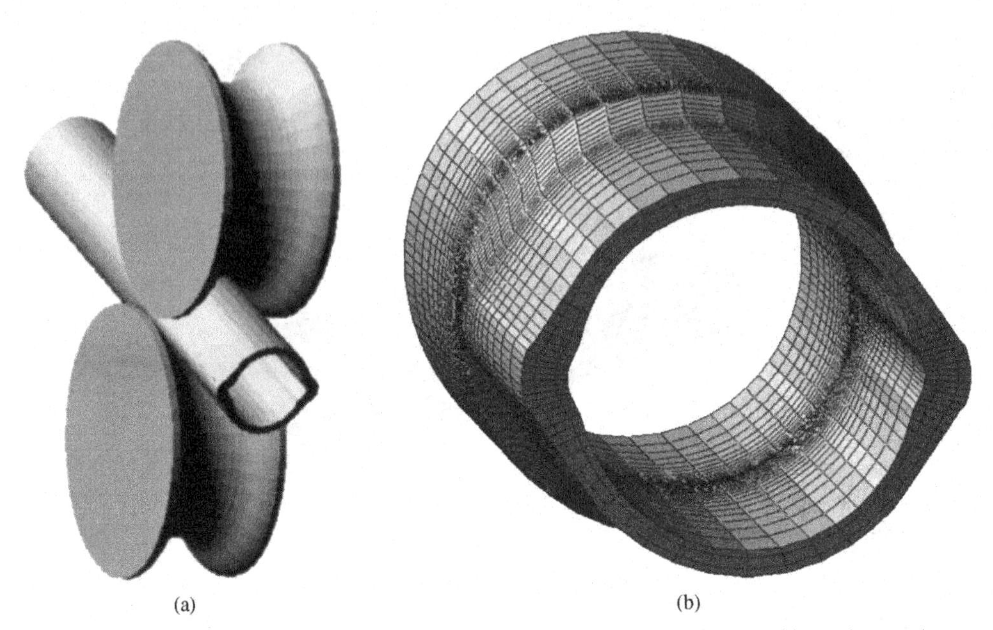

(a) (b)

Figure 9.13. Tube rolling: (a) principle of tube rolling, and (b) prediction of lateral spread.

to prevent collapse of the tube and to fix the internal diameter of the roll. For the level of stress on the roll to be limited, the material is allowed to spread on the sides, producing the shape that is computed and plotted in Fig. 9.13(b).

**Thermal Coupling**

The heat equation for deformable bodies was presented in Section 7.3 with the Lagrangian form. Here the Euler description is used, for which the steady-state equation becomes

$$\rho c\, \mathbf{grad}(T) = \mathrm{div}[k\, \mathbf{grad}(T)] + \dot{w}, \tag{9.8}$$

where $\dot{w}$ is the viscoplastic work dissipated into heat, with the usual expression[7]:

$$\dot{w} = K(\sqrt{3}\dot{\bar{\varepsilon}})^{m+1}.$$

The following boundary conditions are rather complicated, but it is essential to note that the final results will depend heavily on the accuracy of the approximations we shall introduce at this stage. For simplicity, a first approximation will be considered here. The simpler condition is on the entry section:

$$T = T^{\mathrm{en}} \quad \text{on } \partial\Omega^{\mathrm{en}}, \tag{9.9}$$

which means that the temperature in known at the entry section, and in fact may result from the calculation of the previous stand.

[7] For simplicity, we shall assume here that the entire viscoplastic work is transformed into heat. It is often considered that only a fraction $r$ of the total viscoplastic work is transformed into heat; the remaining energy represents about 5–10% and is stored in the material to produce defects such as dislocations, microvoids, etc.

On the exit section we shall have

$$\frac{\partial T}{\partial n} = 0 \quad \text{on } \partial\Omega^{\text{ex}}, \tag{9.10}$$

which means that the temperature will be considered as constant after the exit section. On the free surface, a heat flux condition is assumed (also see Section 7.2). It takes the form

$$-k\frac{\partial T}{\partial n} = h^F(T - T^a) \quad \text{on } \partial\Omega^F, \tag{9.11}$$

and it represents a convection plus a radiation term. The value of the global convection term $h^F$ can actually be decomposed into the air convection contribution, and the radiation one according to

$$h^F = h^a + \varepsilon^r\sigma^r(T + T^a)(T^2 + T^{a2}), \tag{9.12}$$

where $\varepsilon^r$ and $\sigma^r$ are respectively the emissivity factors corresponding to the surface of the workpiece, and the Stefan constant: $\sigma^r = 5.67 \times 10^{-8}$ kg s$^{-3}$ $T^{-4}$.

On the rolls the heat flux is

$$-k\frac{\partial T}{\partial n} = h^f(T - T^R) + \phi^f \quad \text{on } \partial\Omega^f. \tag{9.13}$$

In Eq. (9.13), the first contribution is the conduction term between the part and the roll at temperature $T^R$. The friction heat flux $\phi^f$ is a fraction of the surface heat dissipation that is due to contact with friction, so the approximation may be used:

$$\phi^f = \frac{b}{b + b^{\text{to}}}\alpha_f K|\Delta\mathbf{v}|^{p+1}, \tag{9.14}$$

where $b$ and $b^{\text{to}}$ are the *effusivity coefficients*,[8] corresponding respectively to the workpiece and to the tool. They control the sharing of the total friction heat flux between the part and the roll. In some instances this analysis can be simplified and the conditions given by Eqs. (9.11), (9.13), and (9.14) are replaced by a prescribed temperature on the free surface and on the rolls.

Now the Galerkin formulation can be used: it consists of multiplying Eq. (9.8) by the weighting functions $N_n$, and integrating by parts the diffusion term. We thus obtain

$$\int_\Omega \rho c\, \mathbf{grad}(T) N_n \,\mathrm{d}V + \int_\Omega k\, \mathbf{grad}(T)\cdot\mathbf{grad}(N_n)\,\mathrm{d}V - \int_\Omega \dot{w} N_n \,\mathrm{d}V$$
$$+ \int_{\partial\Omega^F} h^F(T - T^a)N_n \,\mathrm{d}S + \int_{\partial\Omega^f} [h^f(T - T^R) + \phi^f]N_n \,\mathrm{d}S = 0, \tag{9.15}$$

with the prescribed temperature boundary condition given by Eq. (9.10).

From the computational point of view, a more satisfactory formulation is often preferred. It is called the streamline upwind Petrov Galerkin method (SUPG). It consists mainly in replacing the usual weighting functions $N_n$ by modified functions $N'_n$,

[8] The effusivity coefficient is defined by $b = \sqrt{\rho c k}$.

in which the direction of the flow is taken into account:

$$N'_n = N_n + \frac{1}{2}\sum_i v_i^c \frac{\partial N_n}{\partial x_i}, \tag{9.16}$$

where $\mathbf{v}^c$ is the velocity at the center of the element.

The interaction between the temperature evolution and the flow comes from two coupling effects. The first is the density evolution caused by heating (or cooling), which is expressed by

$$\mathrm{div}(\mathbf{v}) = -\frac{\dot{\rho}}{\rho} = 3\alpha_d \dot{T}, \tag{9.17}$$

where $\alpha_d$ is the one-dimensional dilatation coefficient. Taking into account the expression of the material derivative for steady-state flows, we find it also to be

$$\mathrm{div}(\mathbf{v}) - 3\alpha_d\, \mathbf{grad}(T) \cdot \mathbf{v} = 0. \tag{9.18}$$

It can be introduced into the penalty expression of Eq. (9.3) with the expression

$$\frac{1}{2}\int_\Omega \rho_p K[\mathrm{div}(\mathbf{v}) - 3\alpha_d\, \mathbf{grad}(T) \cdot \mathbf{v}]^2 \,\mathrm{d}V. \tag{9.19}$$

The second is the effect of temperature change on the physical parameters entering the constitutive law; the consistency function is generally

$$K = K_0 \exp(\beta/T), \tag{9.20}$$

and the rate sensitivity index $m$ is often assumed as constant or is a linear function of the temperature:

$$m = m_0 + m_1 T. \tag{9.21}$$

The solution of the thermomechanical problem would require the simultaneous resolution of the nonlinear system merging the mechanical equations and the thermal equations.

As the coupling is not extremely strong, an alternate algorithm can be easily imagined:

1. Initialize the temperature distribution with a reasonable approximation.
2. Compute the velocity field, keeping the temperature constant.
3. Introduce the heat dissipated by plastic work into the heat equation and solve it with the last approximation of the velocity field.
4. Check if the velocity and the temperature fields have varied since the previous iteration: if yes, go back to 2; if no, the procedure has converged.

For relatively fast rolling processes, the diffusion term in Eq. (9.8) can be neglected by putting $k = 0$; we thus obtain the adiabatic approximation:

$$\rho c\, \mathbf{grad}(T) \cdot \mathbf{v} = \dot{w}. \tag{9.22}$$

Moreover, the temperature can be prescribed approximately on the whole boundary as previously mentioned. With these hypotheses, and if the nodes follow the flow lines (at least approximately), the temperature calculation can be greatly simplified. The procedure is schematically illustrated in Fig. 9.14.

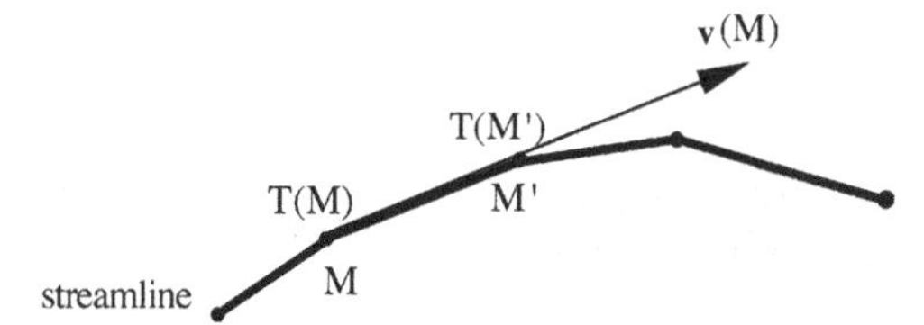


Figure 9.14. Temperature calculation along the discretized streamlines in the adiabatic case.

We suppose that the node $M'$ is the next to $M$ on the same streamline; then we have (see Exercise 9.2)

$$\rho c\, \mathbf{grad}[T(M') - T(M)] = \dot{w}(M)\Delta t, \tag{9.23}$$

with the time increment given by

$$\Delta t = \frac{MM'}{|\mathbf{v}(M)|}. \tag{9.24}$$

---

**Exercise 9.2: Establish the approximate relations given by Eqs. (9.23) and (9.24).**

If points $M$ and $M'$ are associated with the same streamline, we have obviously the approximate relation:

$$\overrightarrow{\mathbf{MM}'} = \mathbf{v}(M)\Delta t. \tag{9.2–1}$$

$\overrightarrow{\mathbf{MM}'}$ is the vector with origin $M$ and terminus $M'$. From Eq. (9.2–1) we deduce immediately Eq. (9.24). Now the first-order Taylor expansion of the temperature field allows us to write

$$T(M') - T(M) = \mathbf{grad}(T) \cdot \overrightarrow{\mathbf{MM}'}\,. \tag{9.2–2}$$

If we multiply Eq. (9.24) by $\Delta t$ and use Eqs. (9.2–1) and (9.2–2), we obtain Eq. (9.23).

---

A more precise scheme can also be derived in a similar manner when Eq. (9.23) is replaced by

$$\rho c[T(M') - T(M)] = \frac{1}{2}[\dot{w}(M) + \dot{w}(M')]\Delta t, \tag{9.25}$$

and Eq. (9.24) by

$$\Delta t = \frac{MM'}{\frac{1}{2}|\mathbf{v}(M) + \mathbf{v}(M')|}, \tag{9.26}$$

We have thus a semi-implicit one-dimensional finite-difference scheme along streamlines.

### Strain Hardening

The duration of the deformation under the rolls is short enough to allow the neglect of annealing[9] that is due to the high temperature of processing, so that the usual laws for work hardening can be used. For example, strain hardening can be

[9] This annealing effect is due to recrystallization. Most of the time, no dynamic recrystallization can occur during the short deformation time; recrystallization would then be limited to interpasses or cooling stages.

introduced by means of the consistency with the law:

$$K(\bar{\varepsilon}, T) = K_1(1 + a\bar{\varepsilon})^n \exp(\beta/T), \tag{9.27}$$

where $K_1$ and $a$ are experimental parameters, and $\bar{\varepsilon}$ is the equivalent strain. For stationary processes it is clear that $\bar{\varepsilon}$ satisfies the following equation:

$$\frac{\mathrm{d}\bar{\varepsilon}}{\mathrm{d}t} = \mathbf{grad}(\bar{\varepsilon}) \cdot \mathbf{v} = \dot{\bar{\varepsilon}}. \tag{9.28}$$

One can remark immediately that Eq. (9.28) is similar to the heat equation with the adiabatic expression (i.e., $k = 0$). The boundary conditions are much simpler; a prescribed equivalent strain on the entry section is

$$\bar{\varepsilon} = \bar{\varepsilon}^{\mathrm{en}} \quad \text{on } \partial\Omega^{\mathrm{en}}. \tag{9.29}$$

There is no generation or loss of strain hardening on the other parts of the boundary:

$$\frac{\mathrm{d}\bar{\varepsilon}}{\mathrm{d}n} = 0. \tag{9.30}$$

The complete problem can again be solved by introducing the accumulated strain as a new unknown at each node, or an alternate scheme can be used, similar to the adiabatic thermomechanical problem.

Finally, both the temperature effect and the work hardening can be taken into account in the model. An example of this coupled analysis is presented in Fig. 9.15 for the prediction of equivalent strain and temperature during a sequence of tube rolling.

### Cold Rolling

As we already pointed out in the introduction, the elastoplastic constitutive equation is more easily computed by an incremental approach, which will be fully described in the next chapter. Here we shall mainly focus on the different levels of approximation and give some examples. For **sheet rolling,** the approximation is most often two dimensional with plane strain,[10] corresponding to a vertical section of the process by a plane parallel to the rolling direction. The sections of the rolls are no longer circles in the vicinity of the contact with the sheet, so that the deformation of the rolls under the rolling pressure must be taken into account if one desires to obtain realistic results. The roll mechanical behavior is then assumed to be linear elastic according to Hooke's law, and the rolls are meshed as well as the workpiece. The notation is outlined in Fig. 9.16, in which, as a result of the symmetry of the problem, only the upper half of the process is represented. The two domains are $\Omega^1$, the part, and $\Omega^2$, the upper roll, while $\partial\Omega^f$ is the roll–part interface.

We denote the displacement fields by $\Delta\mathbf{u}^i$ and the stress tensors by $\boldsymbol{\sigma}^i$, where $i = 1$ corresponds to the sheet domain, and $i = 2$ is for the roll domain. The coupling conditions between the two domains can be written as follows.

- continuity of the normal displacement:

  $$\Delta\mathbf{u}^1 \cdot \mathbf{n} = \Delta\mathbf{u}^2 \cdot \mathbf{n}; \tag{9.31}$$

  if $\mathbf{n}$ is the normal to the interface $\partial\Omega^f$;
- continuity of the stress vector:

  $$\boldsymbol{\sigma}^1 \cdot \mathbf{n} = \boldsymbol{\sigma}^2 \cdot \mathbf{n}. \tag{9.32}$$

[10] However, if the bending of the roll is to be taken into account, and the flatness of the sheet predicted, a true three-dimensional analysis must be performed.

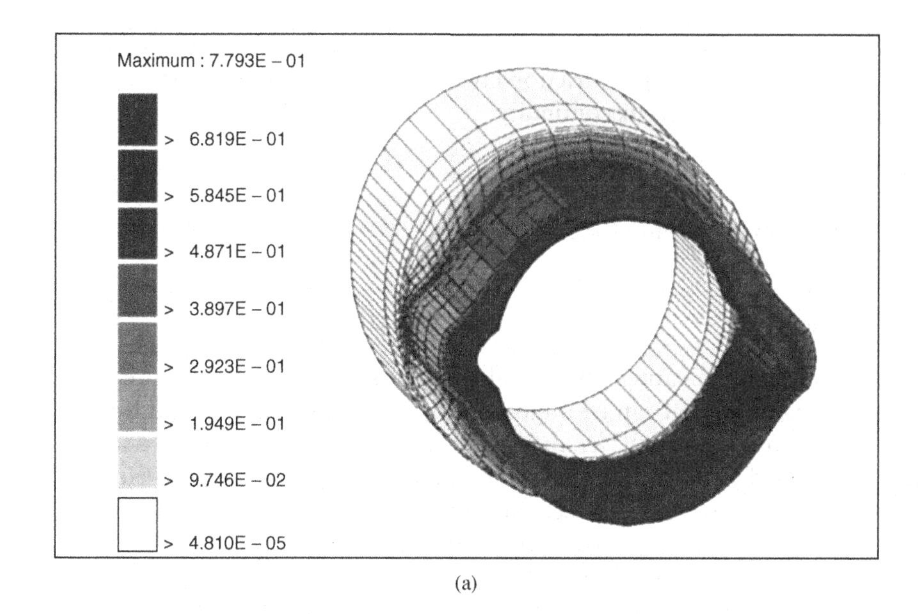


(a)

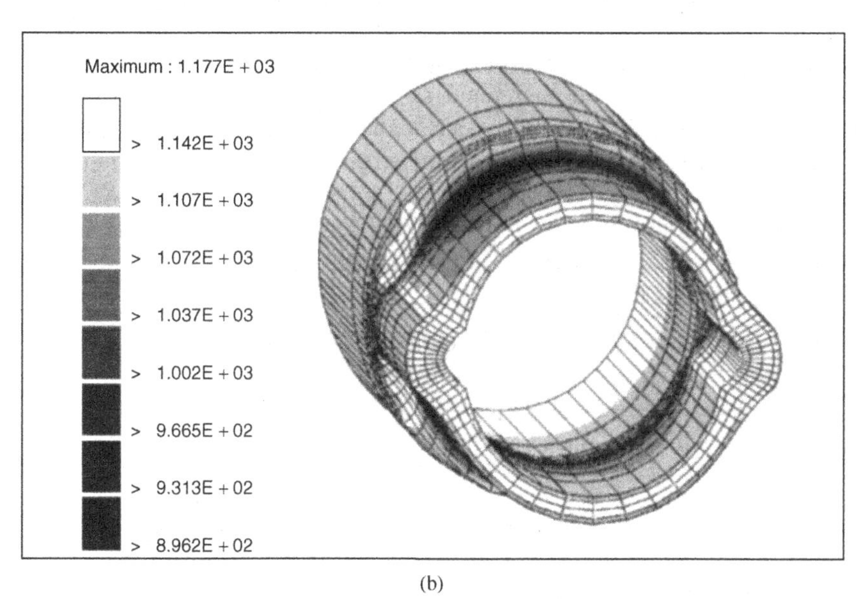


(b)

Figure 9.15. Coupled analysis with temperature and strain hardening effects: maps of (a) isoequivalent strain and (b) isotemperature.

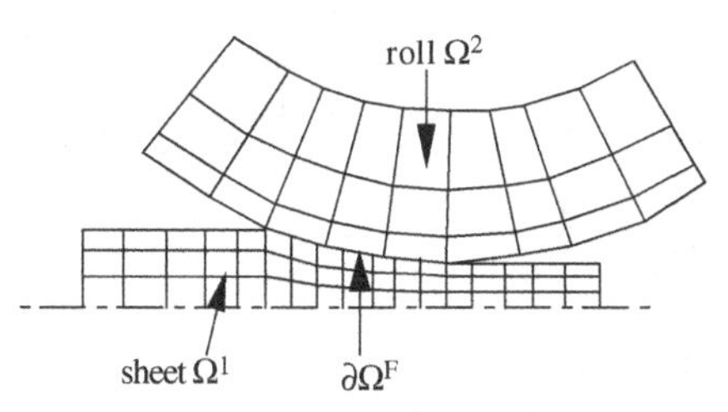


Figure 9.16. Meshes of the roll and the sheet for a coupled analysis (schematic).

The stress vector can be decomposed into a normal component $\sigma_n^i$ and a tangential component $\tau^i$. The latter is given by the friction law as a function of the normal stress and of the tangential displacement discontinuity:

$$\tau = g(\sigma_n, \Delta\mathbf{u}), \tag{9.33}$$

with a special form for the Coulomb law:

$$\tau = -\mu\sigma_n \frac{\Delta\mathbf{u}}{|\Delta\mathbf{u}|}. \tag{9.34}$$

As it will be described in Chapter 10, the calculation is incremental, but at each increment the coupling condition can be imposed differently. We can distinguish for each increment the *strong or exact*[11] *coupling*, for which we introduce the entire domain $\Omega$ that is obtained by assembling both subdomains. The resolution procedure allows us to compute directly the displacement and stress fields on the domain $\Omega$, taking into account the coupling conditions and the friction law.

We can also distinguish the *weak or alternate coupling* during one increment: the deformation of the sheet is calculated with a given shape of the roll; then the contact stress vector is introduced into the elastic finite-element procedure for the computation of the roll shape, which is then updated at the end of the increment. No iteration is performed during one increment so that, for strong physical coupling, this scheme may be insufficient and produce artificial oscillations. If iterations are performed until convergence, then the previous strong coupling scheme is obtained.

An example of global resolution is given in Fig. 9.17 for a thin sheet undergoing a skin pass, that is, a rolling pass with a very small reduction of thickness. For this example the input data were as follows:

entry thickness of the sheet 0.66 mm, exit thickness 0.644 mm;
rolls: $E = 210 \times 10^9$ Pa, $\nu = 0.3$, diameter = 600 mm;
sheet: $E = 210 \times 10^9$ Pa, $\nu = 0.3$, $\sigma_0 = 400 + 20\,\varepsilon$ (MPa);
front tension 20 MPa, back tension 37 MPa.

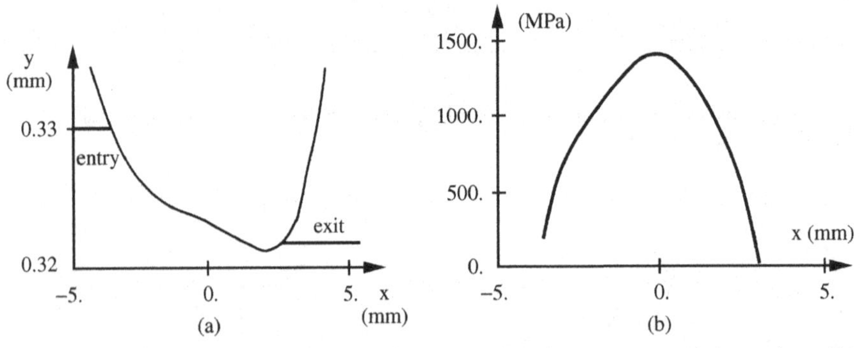


Figure 9.17. Skin pass of a thin sheet with high roll deformation: (a) deformed profile of the roll and (b) distribution of the normal stress at the roll–sheet interface.

[11] Up to the accuracy of the finite-element discretization.

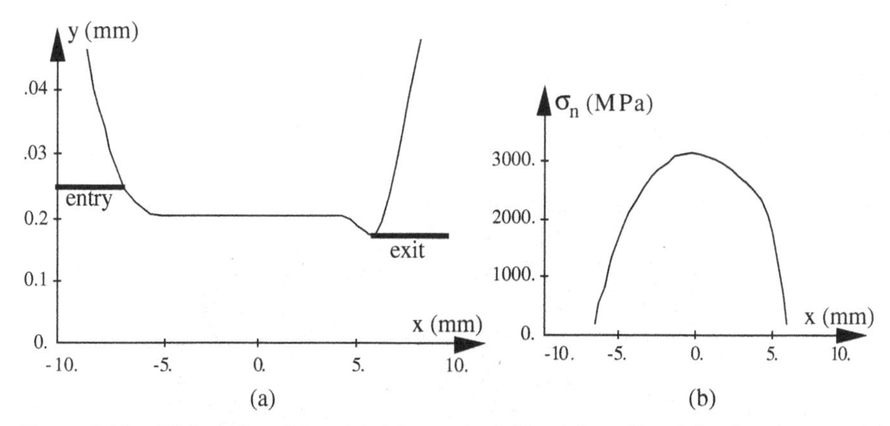


Figure 9.18. Thin strip rolling: (a) deformed profile of the roll and (b) distribution of the normal stress at the roll–sheet interface.

In this example we can see that the deformed profile of the roll is rather different from the initial one before loading.

However, even more complicated shapes can be obtained by computation,[12] as shown in Fig. 9.17. The process parameters for the computation were as follows:

entry thickness of the sheet 0.05 mm, reduction ratio 32%;
rolls: $E = 210 \times 10^9$ Pa, $\nu = 0.3$, diameter $= 570$ mm;
sheet: $E = 210 \times 10^9$ Pa, $\nu = 0.3$, $\sigma_0 = 410 + 325\varepsilon$ (MPa);
front tension 247 MPa, back tension 265 MPa.

On the last example it is possible to find an elastic zone under the rolls.

More complex three-dimensional *cold shape rolling* is also utilized in industry to produce high strength parts with various cross sections. This process must also be modeled by an incremental elastoplastic approach, but the workpiece is thicker so that the rolling pressure is lower than for thin strip rolling, and the rolls' deformation can be neglected as a first approximation. In the example presented in Fig. 9.19, the computation allows us to predict the exit profile, the possible bending or twisting of the workpiece, and the residual stress distribution.

## 9.3 Extrusion

The principle of the process is schematically outlined in Fig. 9.20. A round billet is heated and introduced into a container, and a ram pushes the material through a die in order to produce a workpiece with a constant cross section.

The process is used for various metals or alloys: aluminum, copper, steel, and so on. The most complicated shapes are obtained with aluminum or aluminum alloys, which exhibit a high ductility and thus allow extremely large deformation without defects. The major problems in metal extrusion of complicated shapes are related to the material flow and the many factors that can influence it. A small perturbation of the shape of the die, or of the friction coefficient, or the temperature distribution, may

[12] At the present time, it seems difficult to observe the experimental profiles with enough accuracy. However, the other parameters computed by the codes can be more easily compared to experiments and are used to validate the approach.

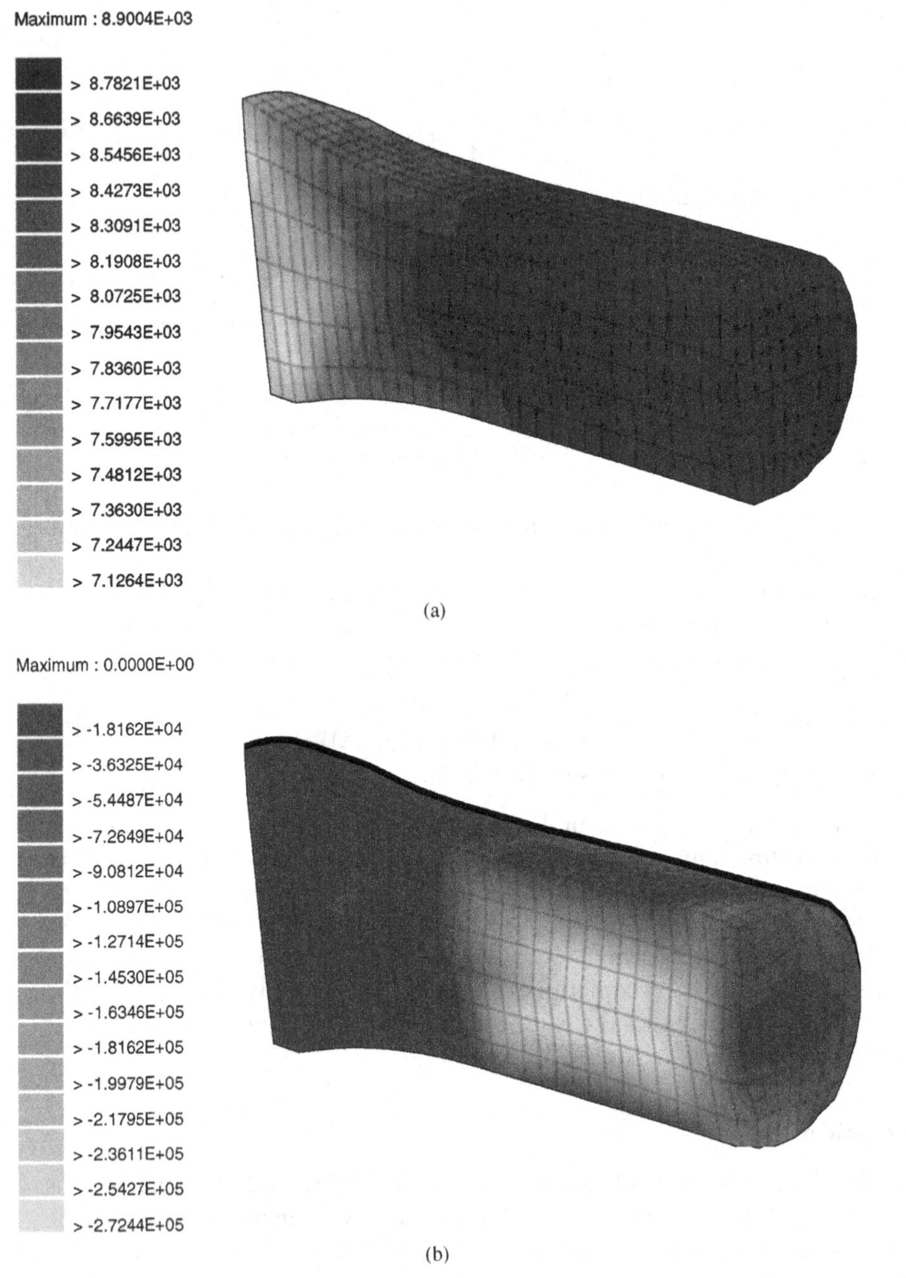


Figure 9.19. Cold shape rolling. (a) iso longitudinal velocities in mm/s (b) iso vertical stress component in kPa.

result in a large variation of the workpiece profile, or induce a curved or warped part. Moreover the deformation is complicated by quasi-shear surfaces that are present close to the die exit. Finally, the temperature coupling is very high, especially for aluminum alloys.

For these reasons very few computer codes are available for three-dimensional simulation, even at the research level; moreover, they can cope only with a simplified approach of the problem, and they use a limited number of elements instead of

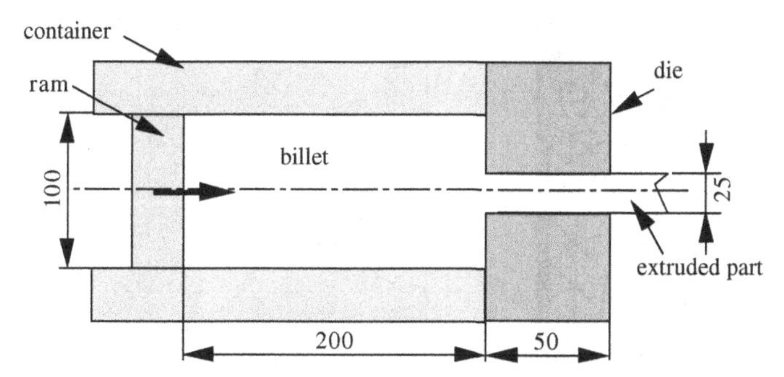


Figure 9.20. Principle of extrusion.

extremely refined ones that should be necessary to represent the complex flow with quasi-discontinuities in the vicinity of the zones of very intense shear.

More studies were devoted to *axisymmetric extrusion*, which involves more tractable computations, at least from the geometric point of view. Hot extrusion of aluminum alloy is approximated by a steady-state approach, which presents some similarities with the corresponding approach for rolling. However, the heat coupling can be considered stronger, and other physical parameters related to the evolution of the material microstructure can also play an important role in the flow computation and then necessitate a coupled analysis. A simple example of isothermal steady-state computation is carried out for a material of rate sensitivity index $m = 0.15$ and with a sticking contact between the billet and the die or the container, as a first approximation of high friction shear stress. The geometry is the same as that of Fig. 9.20, for which the mesh must be refined in the corners and at the exit of the workpiece (see Fig. 9.21).

The velocity field is given in Fig. 9.22(a), showing a dead zone in the corner between the die and the container. Accordingly, the map of isoequivalent strain rate of Fig. 9.22(b) exhibits a narrow zone of very intense shear close to the exit and on the bearing of the die.

In the future, three-dimensional configurations for complex profiles, which are the more important from the industrial point of view, will be treated accurately. A relatively simple example of extrusion of a *T* section can be treated approximately by using the mesh represented in Fig. 9.23, taking advantage of the symmetry plane to mesh only half of the part.

The computation is again based on a stationary, purely mechanical approach of the flow, with the same parameters as those used for the axisymmetric case. The map of isoequivalent strain rate (Fig. 9.24) indicates that most of the deformation occurs in the vicinity of the die exit and on the upper part of the T.

## 9.4 Drawing

The drawing process consists of pulling a preform with a constant section through a die to give a precise profile to the workpiece. In fact, it is identical to extrusion except with a front tension instead of a back pressure. This process also improves the strength of the material and the straightness of the bars, even if a further operation of straightening is often necessary. This forming process can be done at high temperatures and more likely at room temperature where strain hardening is higher. It is used for sheets, wires, bars, or more complex sections in one or several operations. The general

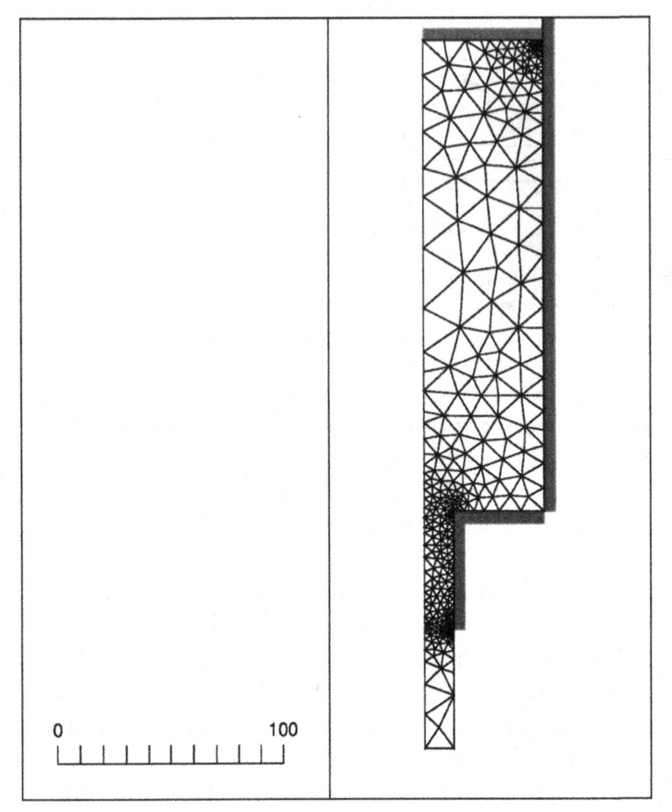


Figure 9.21. Axisymmetric extrusion – mesh of the part.

limiting factors are the drawing force, the onset of internal or external defects, and the final level of residual stresses, which can have a negative effect for the part use.

An example of wire drawing is presented, the importance of which is clear when we consider the application of high strength wires to tire reinforcement. The wire cross-sectional area is reduced by pulling the wire through successive conical dies of decreasing diameter. The mesh of the wire is represented in Fig. 9.25, where a sufficient length was considered for drawing in order to reach the stationary state.

The reduction ratio in cross-sectional area is 30% with a die half-angle of 4°. An accurate description of the material flow during wire drawing can be performed by an incremental work hardening elastoplastic approach. The material parameters are defined by

$$E = 210{,}000 \text{ MPa}, \quad \nu = 0.3, \quad \sigma_0 = 1000 \text{ MPa}.$$

We used a Tresca friction law with $\bar{m} = 0.033$. The prediction of residual stresses is of major importance, as they can determine the lifetime of the wire in service. In the previous example, the result of the calculation gives the distribution of residual stresses as pictured in Fig. 9.26.

For a different condition of deformation (geometry and friction coefficient), the internal state of stress can be highly tensile and even cause a cup and cone fracture.

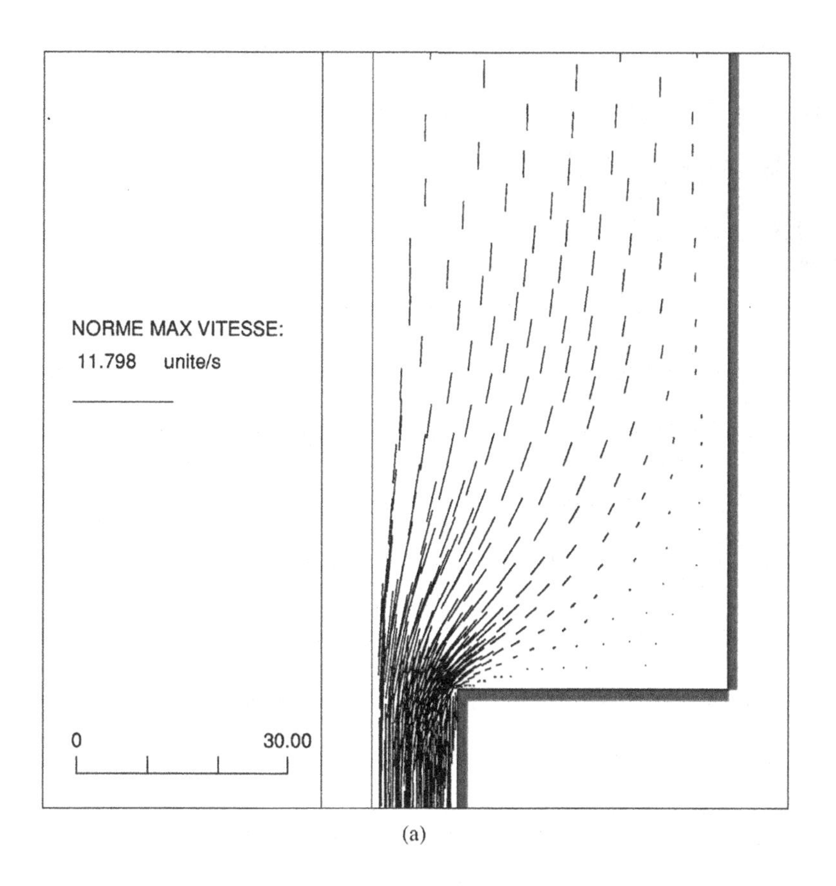


(a)

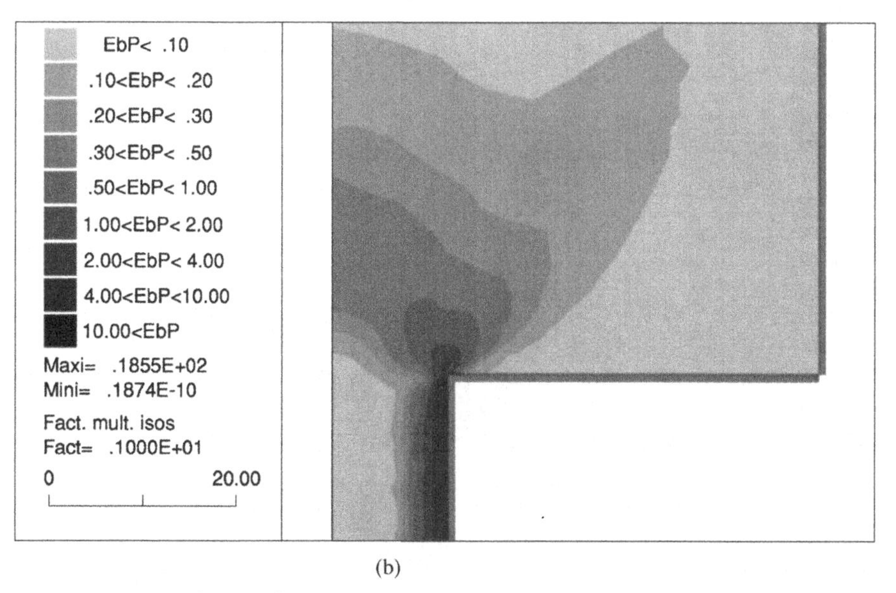


(b)

Figure 9.22. Simulation of axisymmetric extrusion: (a) velocity field and (b) map of isoequivalent strain rate.

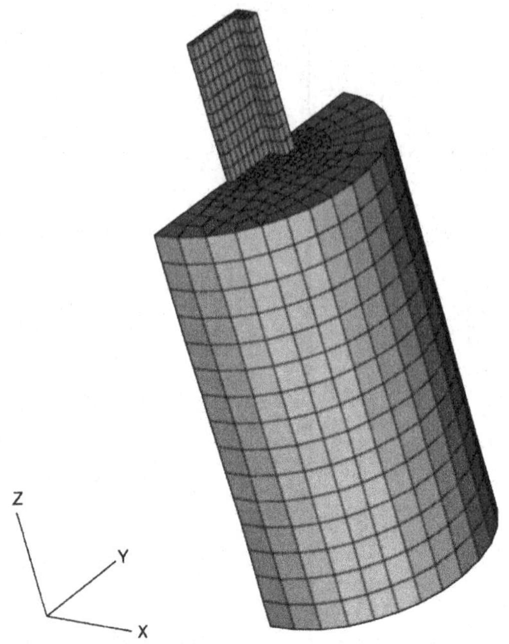


Figure 9.23. Three-dimensional finite-element extrusion of a T profile – mesh of the part.

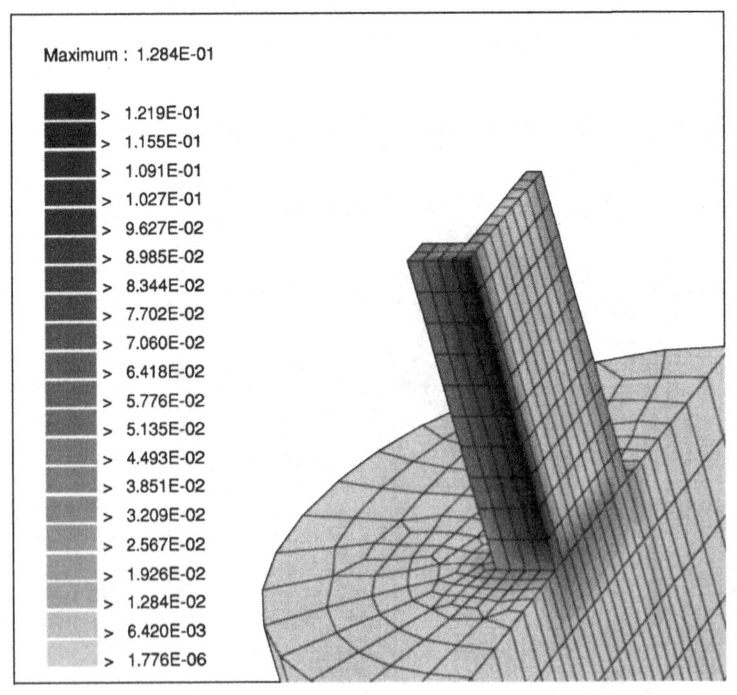


Figure 9.24. Three-dimensional finite-element extrusion of a T profile – map of isoequivalent strain rate.

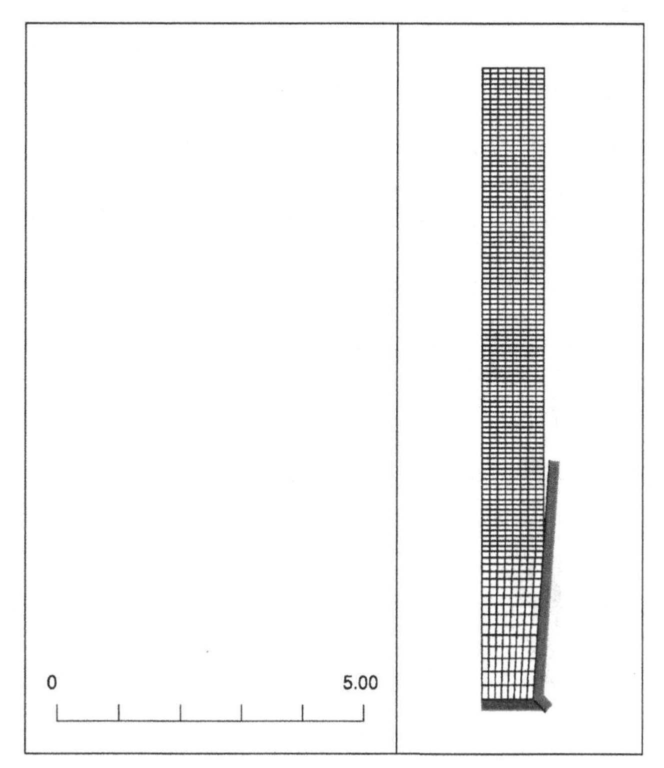


Figure 9.25. Mesh for the simulation of wire drawing.

Figure 9.27 shows an example of such a situation, which should be avoided in industrial practice. In this latter case the material parameters are the same, except the friction coefficient, which is $\bar{m} = 0.05$, and the die imposes a reduction of 10% with an half-angle of 10°. We can observe in Fig. 9.27(a) the map of iso $z$ component of the stress tensor. Close to the die, $\sigma_{zz}$ is compressive and its minimum value is about

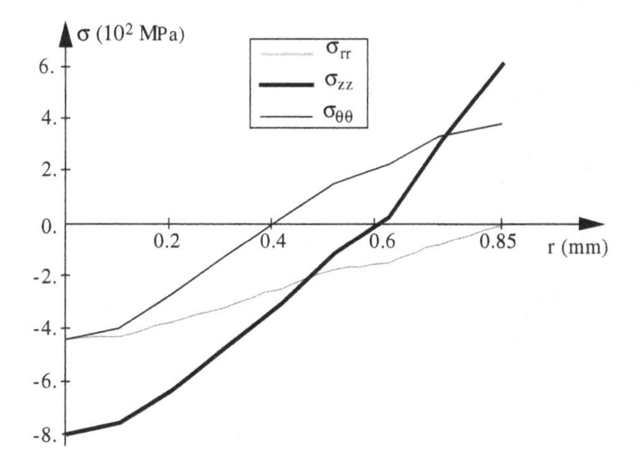


Figure 9.26. Residual stress after wire drawing; the reduction ratio is 30%, and the die angle is 4°.

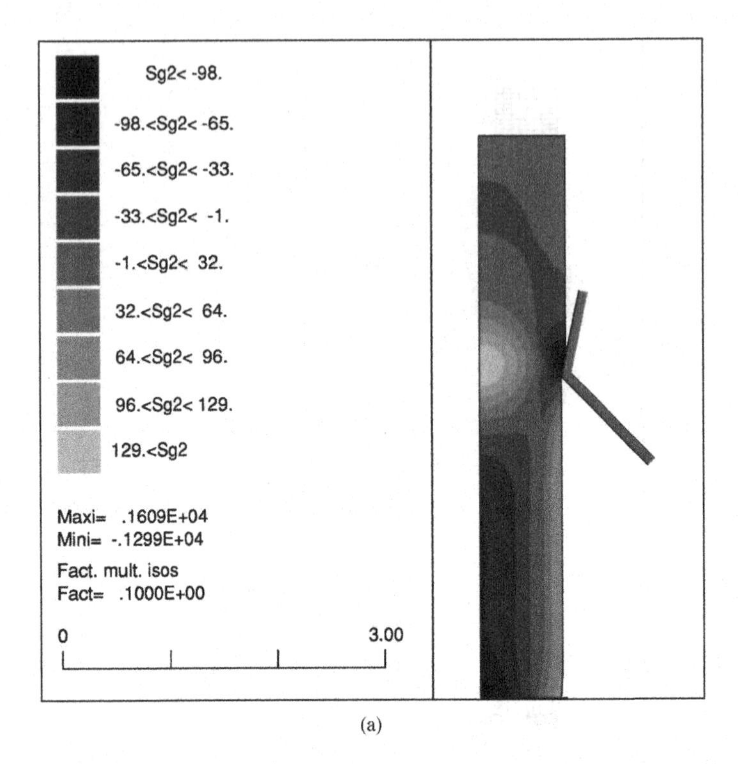


(a)

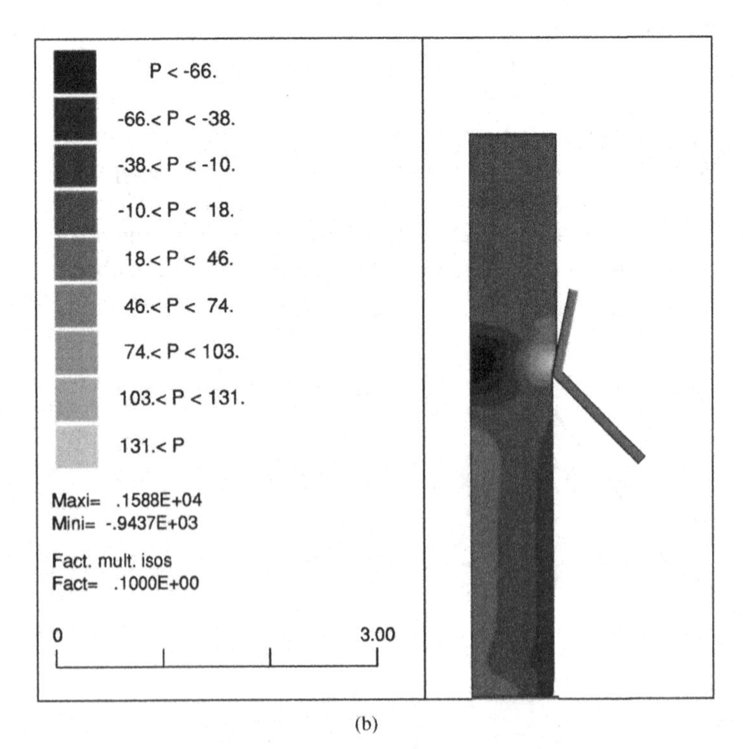


(b)

Figure 9.27. Tensile stress during wire drawing: distributions of (a) $\sigma_{zz}$ and (b) the hydrostatic pressure $p$.

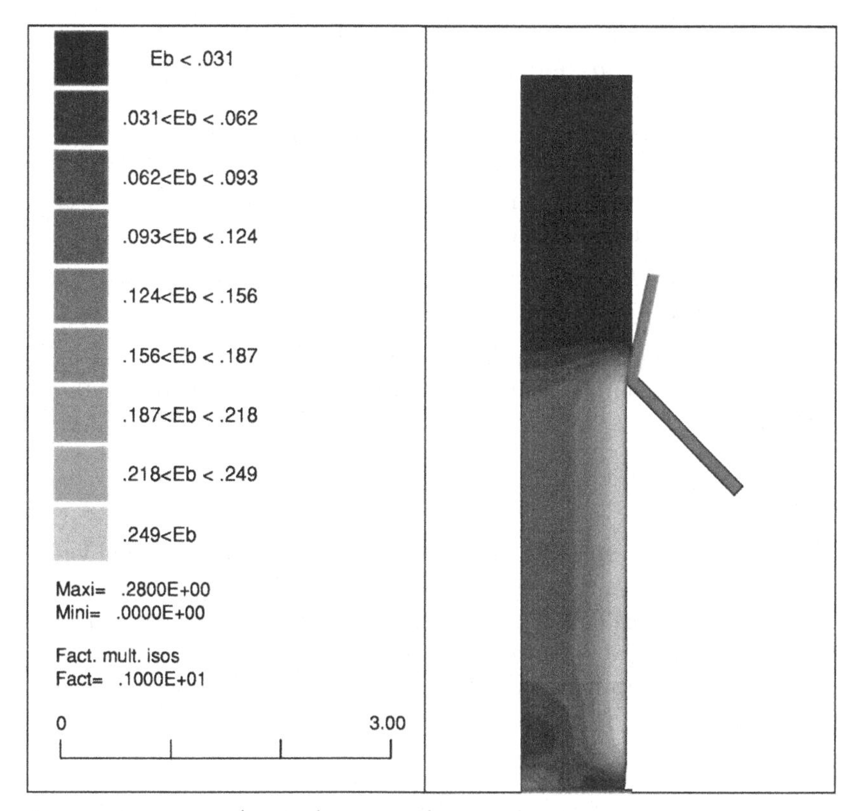


Figure 9.28. Map of equivalent strain for wire drawing.

−1300 MPa. However, we see in the zone of deformation that the stress $\sigma_{zz}$ is highly tensile, its maximum value is 1600 MPa, and it is reached on the symmetry axis. Even the hydrostatic pressure can become negative so that the material can exhibit internal decohesion. In Fig. 9.27(b) the pressure map indicates a zone with large negative values close to the symmetry axis , with a minimum equal to −940 MPa.

Finally, for this badly designed example, Fig. 9.28 shows that there is an important heterogeneity of deformation along the radius of the wire. The maximum value of $\bar{\varepsilon} = 0.28$ is observed on the outer surface, while the central part, close to the symmetry axis, is much less deformed with $\bar{\varepsilon} \cong 0.14$.

## PROBLEMS

### A. Proficiency Problems

1. Compute the local adiabatic heating for a plastic material that undergoes deformation defined by the equivalent strain $\bar{\varepsilon}$. We assume here that the mechanical energy is completely transformed into heat and the yield stress $\sigma_0$, the density $\rho$, and the specific heat $c$ are first assumed independent of the temperature and there is no strain hardening.

   Compute the adiabatic heating for 10 steps of wire drawing of steel. The reduction of area at each step is 10%; the material constants are

$$\rho = 7.8 \times 10^3 \text{ kg/m}^3, \quad c = 4.6\ 10^2 \text{ J/kg/°C}, \quad \sigma_0 = 700 \text{ MPa}.$$

   Examine separately the case of strain hardening with a power law.

2. Compute the adiabatic heating of a viscoplastic material for which the consistency $K$ and the rate sensitivity index $m$ are independent of the temperature. Show that the problem can be easily solved only if an assumption is made about the evolution of the equivalent strain rate during the process. Examine the case in which only the consistency $K$ is an exponential function of temperature according to Eq. (7.94).

   Compute the adiabatic heating for a rolling sequence with 30% reduction of thickness in 0.01 s of a steel plate at about 900 °C with

$$\rho = 7.8 \times 10^3 \text{ kg/m}^3, \quad c = 6.3 \times 10^2 \text{ J/kg/°C}, \quad K_1 = 1.2 \text{ MPa s}^{-m},$$

$$m = 0.1, \quad \beta = 0.4000\,°\text{K}^{-1}.$$

3. The viscoplastic functional for hot rolling is given in Eq. (9.2a). Justify the terms corresponding to back tension and to front tension, utilizing the general form of the viscoplastic functional.

## B. Depth Problem

4. Suppose that the free surface problem is expressed in terms of the discretized velocity vector $\mathbf{V}$ and of the free-surface parameters vector $\mathbf{a}$ by Eq. (9.4) and the functional of Eq. (9.7). Write the equations the solution vector $\mathbf{a}$ must satisfy. The Newton–Raphson method is used to solve this problem: express the associated matrix and show that its exact computation would be very costly. Propose a simplified evaluation of the matrix.

CHAPTER TEN

# Forging Analysis

Within the name of forging, a lot of different processes are referred to, so that a classification must be endeavored. As for other metal-forming processes, the deformation temperature can be the first criterion, and we shall therefore distinguish between the usual three cases.

In *hot forging* the part is heated above half the fusion absolute temperature to reduce the forging force, take advantage of the higher material ductility, and allow large deformation. However, the tools are submitted to high thermal stresses as the temperature difference between the part and the tool is high (even when the tools are preheated). For example, steel is forged at about 1100–1200 °C, while the tools have a temperature ranging from room temperature to 250 °C. For simplicity we shall introduce here free forging, in which the shape of the tools is rather simple, and which is generally used for a small series of large or very large parts; open-die forging, in which the preform is squeezed between two dies to give a prescribed shape; and closed-die forging (see Fig. 10.1).

*Cold forging* is performed at room temperature, but the dissipated plastic work may raise the temperature of the part up to about 250 °C. Cold forging is used to achieve better tolerance, higher mechanical properties, and a better surface aspect that can avoid further machining. But as cold metals are less ductile in general than hot ones, cold forging allows moderate deformation, unless a heat treatment is introduced to restore a deformation capability.

At intermediate temperature, *warm forging* can be used to avoid or reduce heat treatments and obtain a reasonable geometrical accuracy with moderate forging forces.

*Forging limitations*: the designer of the forging process has to find a proper sequence of deformation, select the optimum forging press, find an adequate lubricant, choose the material for the tools, and so on. The main practical constraints he must keep in mind are as follows.

- limitation of the maximum forging force on the press;
- control of the stress distribution on the tools caused by forging pressure, and possibly by thermal transfer, in order to increase the tool life;
- avoidance of geometrical defects of the part such as poor filling of the die, folding, excessive waste of material in the flash;
- avoidance of material defects of the workpiece such as rupture, crack opening, porosity, surface imperfections;

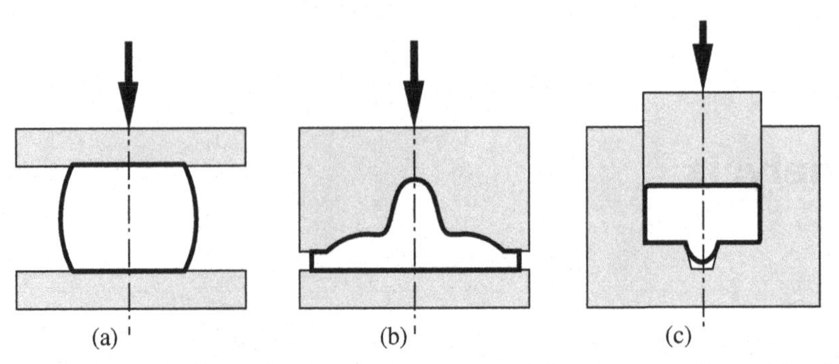


Figure 10.1. Different forging configurations: (a) free, (b) open die, and (c) closed die.

- optimization of the final microstructure of the part in order to obtain the best possible structural properties; and
- decrease of the overall cost of the forging operation.

The conventional approach of this problem is mainly based on one hand on previous practical experience on real presses and toolings and on the other hand on trial and error. However, in many occasions the industrial requirements can be extremely demanding with respect to design delay, workpiece quality, and cost. New tools were soon asked for such as physical simulation on model materials: Plasticine or other wax blends and lead alloys. Today the development of relatively inexpensive powerful computers and work stations allows us to use finite-element computer codes in industry. These codes are often considered as new design tools that are faster than real trials, and more cost effective. This evolution seems very important in an increasing number of plants, where it is associated with other modern uses of the computer: drawing, computer-aided machining, management, and so on.

## 10.1 Non-Finite-Element Results

Much work has been devoted to the development of relatively simple numerical models – as compared with finite-element codes – to simulate cold forging or hot forging. Using a rigid plastic approximation for the constitutive equation of the workpiece material, most of these models use either the slip line field (SLF) theory, the upper bound method (UBM), or the slab method (SM). Despite the obvious advantages of the finite-element method, which lie mainly in its flexibility for the treatment of complex parts or complex constitutive laws, the older methods still have some interest, as they provide a basis for comparison on specific cases for which they are known as reliable.

### The Slip Line Field Method

A simple introduction to the SLF theory was given in Chapter 10 of a previous book.[1] Three examples of increasing complexity will be quoted here.

In the *plane-strain indentation test*, as pictured in Fig. 10.2(a), several solutions are known, which allow us to compute the indentation force as

$$F = \frac{4\sigma_0}{\sqrt{3}} a \left(1 + \frac{\pi}{2}\right),$$

[1] R. H. Wagoner and J.-L. Chenot, *Fundamentals of Metal Forming* (Wiley, New York, 1997).

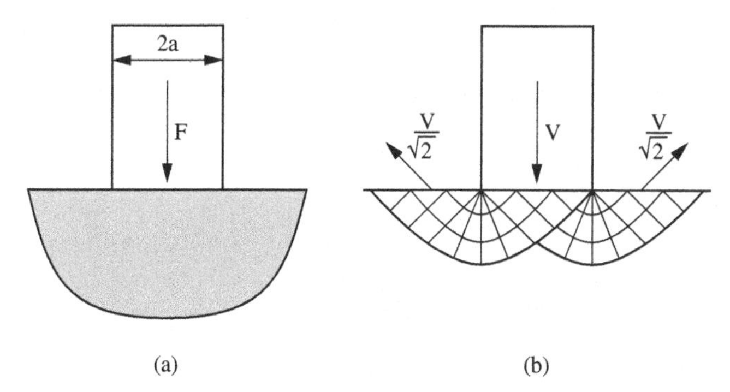


Figure 10.2. Plane-strain indentation: (a) geometrical configuration and (b) SLF solution.

and the velocity field and strain-rate distribution. In Fig. 10.2(b) the Prandtl solution is schematically represented: for the derivation see Hill.[2]

The plane-strain compression test was also treated by several authors; the solution depends on the aspect ratio $a/h$ and on the friction coefficient, as shown in Fig. 10.3, where only the deformed region is outlined.

Forward extrusion through a conical die is a deformation process that is often used in cold or hot forging for the manufacture of a shaft with different diameters. When the friction coefficient on the conical die is negligible, a rather simple SLF solution can be built; the procedure is much more complicated when a Coulomb friction law is assumed. However, the calculation can be performed approximately[3] on a computer and gives the results shown in Fig. 10.4.

In this case the computation allows us to evaluate the hydrostatic pressure $p$ on the axis, which can be negative and can be responsible for the classical central bursting defect when the associated deformation $\bar{\varepsilon} = 2\ln(R_1/R_2)$ exceeds the ductile limit $\bar{\varepsilon}^R$ of the material. With the help of this computation it is possible to draw the safe region in the plane $\alpha$ – half-angle of the die, $r$ – reduction in area, for a given material when the experimental result $\bar{\varepsilon}^R(P)$ is known (see Fig. 10.5).

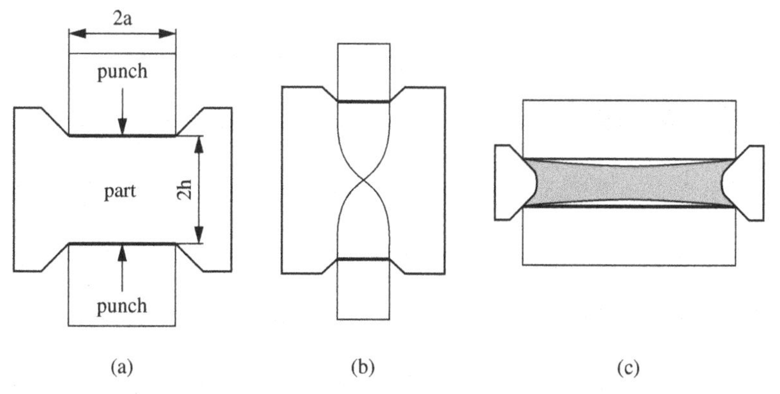


Figure 10.3. Plane-strain compression test: (a) geometry of the process, (b) SLF when $a/h \ll 1$, and (c) SLF for $a/h \gg 1$.

[2] R. Hill, *The Mathematical Theory of Plasticity* (Clarendon Press, Oxford, 1989; 1st ed. 1950), pp. 254–255.

[3] J.-L. Chenot, L. Felgeres, B. Lavarenne, and J. Salençon, "A Numerical Application of the Slip Line Field Method to Extrusion Through Conical Dies," *Int. J. Eng. Sci.* **16**, 263–273 (1978).

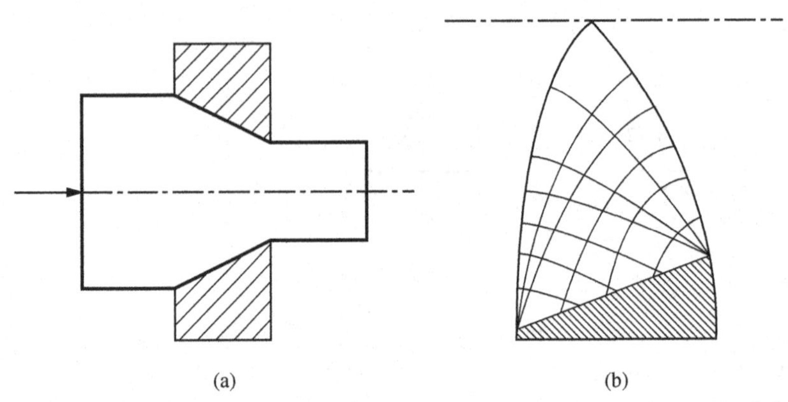


Figure 10.4. SLF solution of axisymmetrical extrusion through a conical die: (a) geometry of the process and (b) SLF solution (schematic).

## Upper Bound Method Results

### *Upsetting of a Cylinder with Flat Dies*

The geometry is pictured in Fig. 10.6(a); the lower die is still and the upper one is moving down with a constant vertical velocity $V^0$. The simplest velocity field in an axisymmetrical coordinate system has the following form:

$$\begin{aligned} v_r &= \frac{1}{2} V^0 \frac{r}{h}, \\ v_z &= -V^0 \frac{z}{h}, \end{aligned} \tag{10.1}$$

so that we can deduce immediately the strain-rate tensor:

$$\dot{\varepsilon} = \begin{bmatrix} \frac{1}{2}\frac{V^0}{h} & 0 & 0 \\ 0 & \frac{1}{2}\frac{V^0}{h} & 0 \\ 0 & 0 & -\frac{V^0}{h} \end{bmatrix}. \tag{10.2}$$

The equivalent strain rate is immediately obtained:

$$\dot{\bar{\varepsilon}} = \frac{V^0}{h}. \tag{10.3}$$

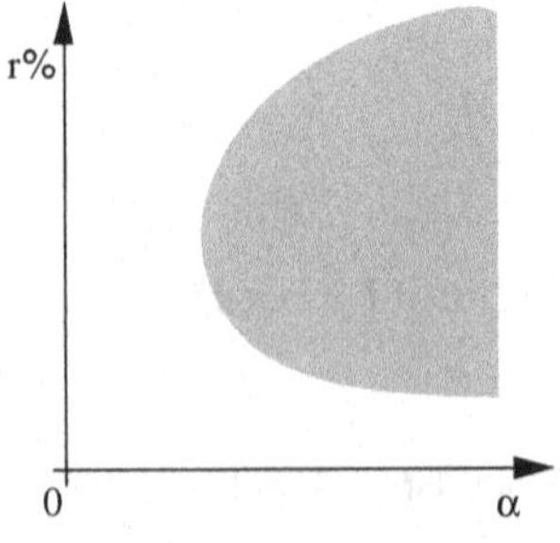


Figure 10.5. Schematic representation of the safe region for cold forging with a conical die (the unsafe region is hatched).

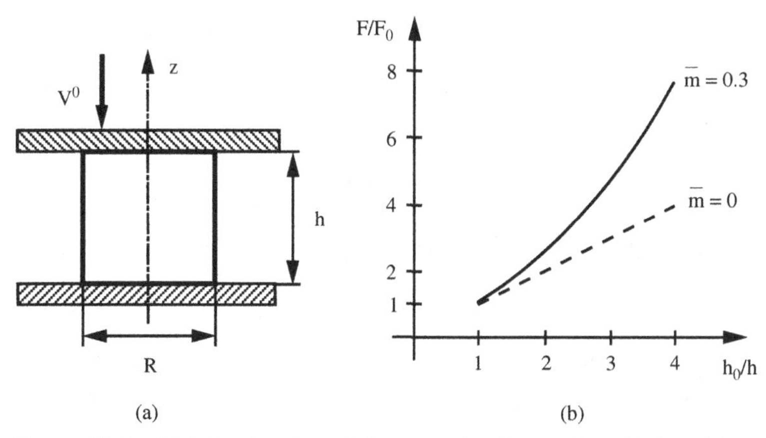


Figure 10.6. UBM estimation of the upsetting force of a cylinder: (a) geometry of the process and (b) forging force as a function of the height.

The plastic work rate is then

$$\dot{W}_p = \int_{\Omega} \sigma_0 \dot{\bar{\varepsilon}} \, d\mathcal{V} = \pi R^2 V^0 \sigma_0. \tag{10.4}$$

If the friction is given by a Tresca law with coefficient $\bar{m}$, then the friction work rate (for the two dies) is

$$\begin{aligned} \dot{W}_f &= \int_{\partial\Omega f} \bar{m} \frac{\sigma_0}{\sqrt{3}} |\Delta \mathbf{v}| dS = 2\bar{m} \frac{\sigma_0}{\sqrt{3}} 2\pi \int_0^R r \frac{1}{2} R^0 \frac{r}{h} \, dr, \\ &= \frac{2}{3} \bar{m} \frac{\sigma_0}{\sqrt{3}} \pi \frac{R^3}{h}, \end{aligned} \tag{10.5}$$

so that the estimated forging force will be

$$F^{\text{UBM}} = \pi R^2 \sigma_0 \left( 1 + \frac{2}{3\sqrt{3}} \bar{m} \frac{R}{h} \right). \tag{10.6}$$

As a result of the incompressibility condition, if $R_0$ and $h_0$ are respectively the initial radius and height of the cylinder, we have obviously

$$\pi R_0^2 h_0 = \pi R^2 h. \tag{10.7}$$

Equations (10.6) and (10.7) allow us to express the force $F^{\text{UBM}}$, estimated by the upper bound method, as a function of the height:

$$F^{\text{UBM}} = \pi \frac{R_0^2 h_0}{h} \sigma_0 \left( 1 + \frac{2}{3\sqrt{3}} \bar{m} \frac{R_0 \sqrt{h_0}}{h \sqrt{h}} \right). \tag{10.8}$$

The resulting curve is plotted in Fig. 10.6(b) for $R_0/h_0 = 1$ with $\bar{m} = 0$ and $\bar{m} = 0.3$ ($F_0$ is the initial force).

### *Forward Extrusion*

Many velocity fields were proposed to study this problem. The most popular one is probably the Avitzur one, which is expressed in spherical coordinates defined in Fig. 10.7. The entry and exit velocities are denoted by $V^0$ and $V^1$ and the

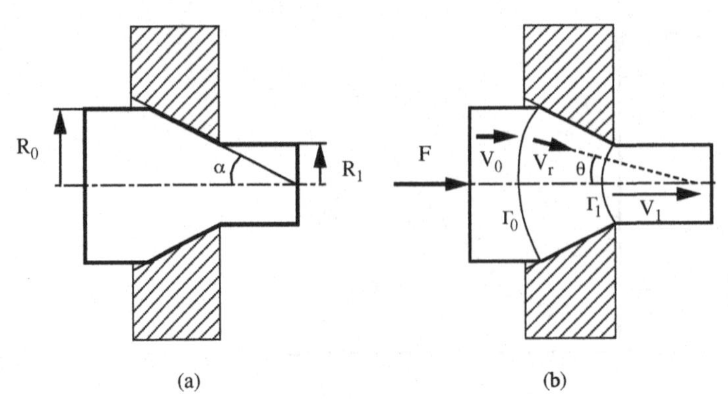


Figure 10.7. Extrusion through a conical die: (a) geometry of the process and (b) velocity field.

corresponding radii of the part $R_0$ and $R_1$. The volume conservation compels us to put

$$R_0^2 V_0 = R_0^2 V_1. \tag{10.9}$$

The deformed region is limited by the boundary of the conical die and by two portions of spheres, $\Gamma_0$ at the entry and $\Gamma_1$ at the exit.

The proposed velocity field is assumed to have only a radial component:

$$v_r = -V_1 R_1^2 \frac{\cos\theta}{r}, \qquad v_\theta = v_\varphi = 0. \tag{10.10}$$

It is easy to verify that this field ensures

- incompressibility in the deformed region,
- the velocity is tangential to the dies,
- the normal velocity discontinuity is null on $\Gamma_0$ and $\Gamma_1$,

so that the hypotheses of the UBM are satisfied.

The strain-rate tensor expressed in spherical coordinates is

$$\dot{\varepsilon} = -V_1 R_1^2 \begin{bmatrix} -\dfrac{2\cos\theta}{r^3} & -\dfrac{\sin\theta}{2r^3} & 0 \\ -\dfrac{\sin\theta}{2r^3} & \dfrac{\cos\theta}{r^3} & 0 \\ 0 & 0 & \dfrac{\cos\theta}{r^3} \end{bmatrix}, \tag{10.11}$$

and the equivalent strain rate is easily computed:

$$\dot{\bar{\varepsilon}} = \frac{2}{\sqrt{3}} V_1 R_1^2 \sqrt{3\cos^2\theta + \frac{1}{4}\sin^2\theta}. \tag{10.12}$$

The forging force is computed by taking into account the rate of plastic work in the deformed region, the rate of work at the velocity discontinuities $\Gamma_0$ anf $\Gamma_1$, and the friction contribution with a Tresca law on the dies. The final expression of the

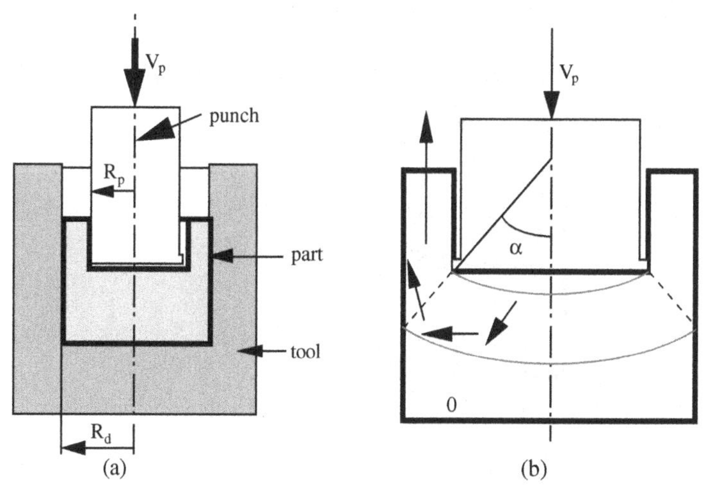


Figure 10.8. Backward extrusion: (a) geometry of the process and (b) velocity field.

UBM estimation of the extrusion force is[4]

$$F^{\mathrm{UBM}} = 2\pi R_0^2 \sigma_0 f(\alpha) \ln\left(\frac{R_0}{R_1}\right) + \frac{2\pi R_0^2 \sigma_0}{\sqrt{3}} \left[\frac{\alpha}{\sin^2\alpha} - \cot\alpha + \bar{m}\cot\alpha \ln\left(\frac{R_0}{R_1}\right) + \bar{m}\frac{L}{R_1}\right]. \tag{10.13}$$

The $f$ function is given by

$$f(\alpha) = \frac{1}{\sin^2\alpha}\left(1 - \cos\alpha\sqrt{1 - \frac{11}{12}\sin^2\alpha}\right) + \frac{1}{\sqrt{132}\sin^2\alpha}\ln\left(\frac{1+\sqrt{\frac{11}{12}}}{\sqrt{\frac{11}{12}}\cos\alpha + \sqrt{1 - \frac{11}{12}\sin^2\alpha}}\right). \tag{10.14}$$

### *Backward Extrusion*

The geometry of the process is illustrated in Fig. 10.8, where it must be emphasized that friction occurs only on a small vertical part of the punch. At the beginning of the process, a velocity field can be built from the previous. We remark that if we take

$$V_1 = -V_p,$$

and if an axial velocity equal to

$$v_z = V_0 = \frac{R_d^2}{R_p^2} V_p$$

is added to the velocity field of Eq. (10.10), then the new velocity field is still incompressible. Moreover, it is easy to verify that the new discontinuity surfaces are upper and lower circles, and two straight lines making an $\alpha$ angle with the axis. The $\alpha$ angle

[4] B. Avitzur, *Metal Forming: Processes and Analysis* (McGraw-Hill, New York, 1968), pp. 153–217.

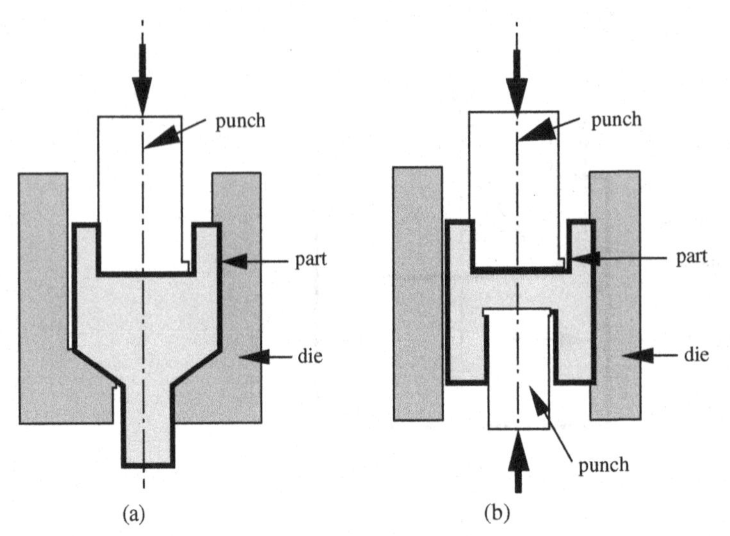


Figure 10.9. Combined extrusion: (a) forward-backward extrusion and (b) backward-backward extrusion.

is unknown and can be determined so that the total power consumption is minimized. At the end of the process a new velocity field must be chosen, which can be composed of a zone 1, with uniform deformation zone under the punch and a velocity field similar to that used for cylinder upsetting, and a rigid zone 2. The boundary between zones 1 and 2 is determined so that the velocity has no normal discontinuity on it.

### *Combined Extrusion*

The combined forward-backward[5] and backward-backward extrusions (see Fig. 10.9) were studied with velocity fields of various levels of complexity.[6] Different velocity fields can be introduced in the same model to cope with the evolution of geometry as deformation is progressing. The more physical velocity field is selected at each step by computing the total energy rate and choosing the velocity field that corresponds to the lower energy rate.

For the combined forward-backward extrusion, a velocity field is defined with two plastic zones and six rigid-body zones at the beginning of the process. For illustration it is represented schematically in Fig. 10.10 (the plastic zones are in gray). This approach can be more systematically designed to fit more complex geometries: this is the upper bound elemental technique method.

### *The Upper Bound Elemental Technique Method*

UBET is short for upper bound elemental technique.[7] The part (mainly two-dimensional or axisymmetrical) is automatically divided into rectangular or triangular

[5] B. Avitzur, *Metal Forming: The Application of Limit Analysis* (Marcel Dekker, New York, 1980), pp. 127–134.

[6] H.-J. Braudel, "Contribution to the Practical Study and to the Numerical Modelling of Combined Extrusion in the Cold Forging Process," Ph.D. dissertation (Sophia-Antipolis, 1981, in French).

[7] A. N. Bramley and F. H. Osman, "The Upper Bound Method," in *Numerical Modelling of Material Deformation Processes. Research, Development and Applications*, P. Hartley, I. Pillinger, and C. Sturgess, eds. (Springer-Verlag, London, 1992), pp. 124–130.

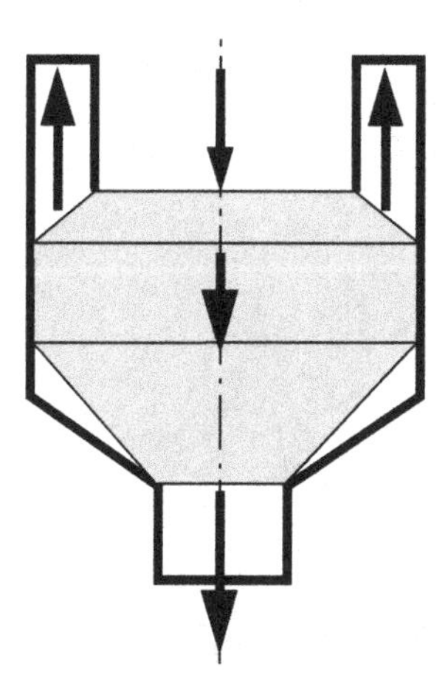

Figure 10.10. Velocity field for combined forward-backward extrusion.

zones, where the velocity field has a prescribed form. At the interelement boundaries the normal velocity is continuous, and the elements are designed to ensure appropriate boundary conditions on the tools, which are discretized with straight lines. When the number of elements is small enough, the computing time is rather low for the computation of the following:

- forging force,
- material flow,
- and even the local normal stress with an approximate method.

In its present form the UBET method seems to be close to the FE method, with the advantage that the velocity field is truly incompressible. However, it appears that for very complex parts, with complicated flow patterns, the UBET method could hardly incorporate as much flexibility as the FEM. Moreover, the generalization of the UBET method to three-dimensional problems seems to be a very difficult issue.

### Slab Method Results

We shall present here the methodology[8] for the determination of the vertical force and the approximate metal flow on a turbine blade of an aircraft engine. This example is typical of flat products for which the slab method gives satisfactory results. The geometry is presented in Fig. 10.11: the blade itself is divided into several sections

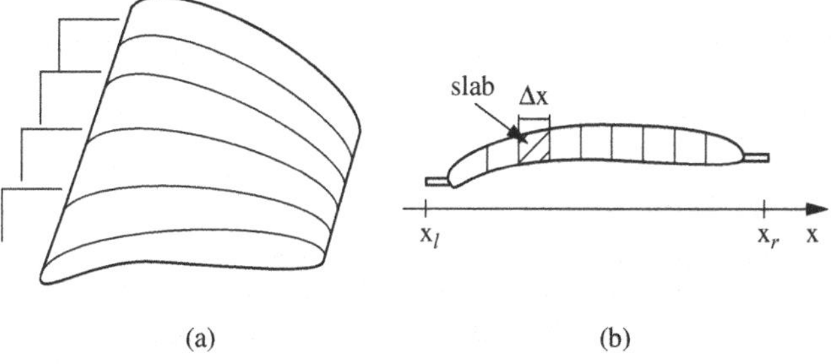


Figure 10.11. Discretization of a turbine blade: decomposition into (a) sections and (b) slabs.

[8] T. Altan, F. W. Boulger, and J. R. Becker, "Forging Equipment, Materials and Practices," sponsored by the U.S. Air Force Materials Laboratory, Ohio: Metals and Ceramic Information Center, 1973.

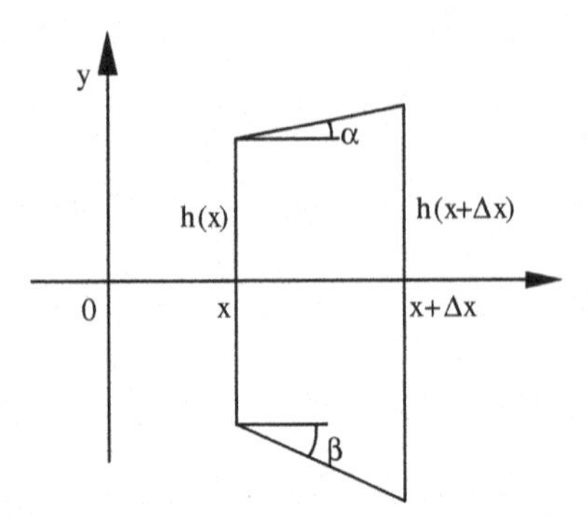


Figure 10.12. Definition of a slab.

by vertical planes normal to the direction of its length, shown in Fig 10.11(a), and each section is again divided into small vertical slabs of thickness $\Delta x$, as shown in Fig. 10.11(b).

Each slab is approximated by upper and lower straight lines, as indicated in Fig. 10.12.

The $x$ and $y$ stress components can be approximated by functions of $x$ only, while the shear component ($xy$) is neglected. The material is assumed rigid plastic, obeying the von Mises criterion, and a Tresca friction law is introduced so that the friction shear stress is written as

$$\tau = \varepsilon_s \bar{m} \frac{\sigma_0}{\sqrt{3}}. \qquad (10.15)$$

In Eq. (10.15), $\varepsilon_s = 1$ if the material flow is oriented toward $x < 0$, and $\varepsilon_s = -1$ when it is oriented toward $x > 0$. With the use of these hypotheses, it is easy to show with the slab method approximation that the stress tensor verifies

$$\sigma_{yy}(x + \Delta x) - \sigma_{yy}(x) = A \ln \left[ \frac{h(x + \Delta x)}{h(x)} \right], \qquad (10.16)$$

$$\sigma_{xx}(x) = \sigma_{yy}(x) - \frac{2\sigma_0}{\sqrt{3}}, \qquad (10.17)$$

with the $A$ parameter defined by

$$A = \frac{\sigma_0}{\sqrt{3}} \left( -2 + \varepsilon_s \bar{m} \frac{\tan^2 \alpha + \tan^2 \beta + 2}{\tan \alpha + \tan \beta} \right). \qquad (10.18)$$

Now the boundary conditions are

$$\sigma_{xx}(x_r) = \sigma_{xx}(x_l) = 0, \quad \sigma_{yy}(x_r) = \sigma_{yy}(x_l) = -\frac{2\sigma_0}{\sqrt{3}}.$$

Here $x_r$ and $x_l$ are respectively the right-hand-side and left-hand-side $x$ coordinates of the ends of the part. We see that Eqs. (10.16)–(10.18) can be written with $\varepsilon_s = 1$ for each slab successively, starting from the left-hand side, with the corresponding boundary condition. The same can be done with $\varepsilon_s = -1$ when we start from the right-hand side. Two curves can be drawn, and their intersection determines

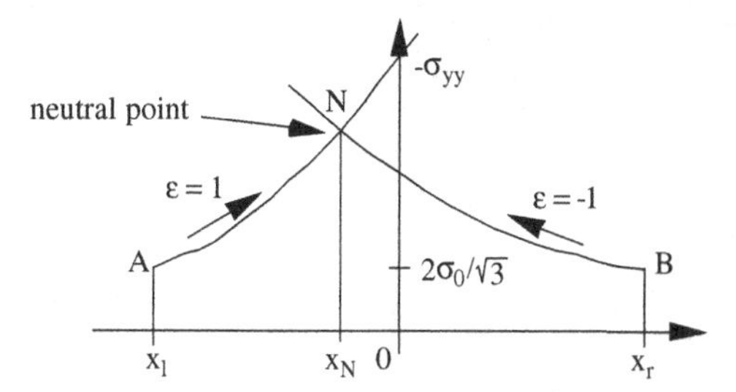


Figure 10.13. Determination of the neutral point.

the neutral point $N$ where the material flow parts into a left and a right flow (see Fig. 10.13).

The final solution for the $y$ component of the stress is given by the curve $ANB$, the integral of which is the forging force contribution of the section. The total forging force is readily obtained by adding each force associated to the sections. The neutral point concept, and the material incompressibility, can also be used to propose an approximate scheme for the calculation of the flow for each incremental displacement $\Delta d$ of the upper die on the vertical axis $oy$, as shown in Fig. 10.14.

## 10.2 Two-Dimensional Finite-Element Results

All the steps of a forging code are summarized on the flow chart in Fig. 10.15. The first approach for the finite-element modeling of forging is the purely mechanical simulation of viscoplastic or quasi-rigid-plastic deformation of a part by rigid tools. With these hypotheses the flow formulation is used in order to compute the velocity field and the strain and stress distribution at each time step. The time-integration scheme for the nodal coordinates vectors can be performed by one of the schemes previously described in Section 5.3. Most often the explicit scheme is chosen, as it is the simplest one to implement; see Eq. (5.39). Many authors selected four-node elements with reduced integration for the penalty term. But the meshing and remeshing requirements are not easily tackled by using these elements, so there is a tendency to prefer the six-node triangle. This latter element can be used with one reduced integration point. The $P^{2+}/P^1$ six-node triangular element was shown to be more satisfactory from the theoretical point of view, and togive better

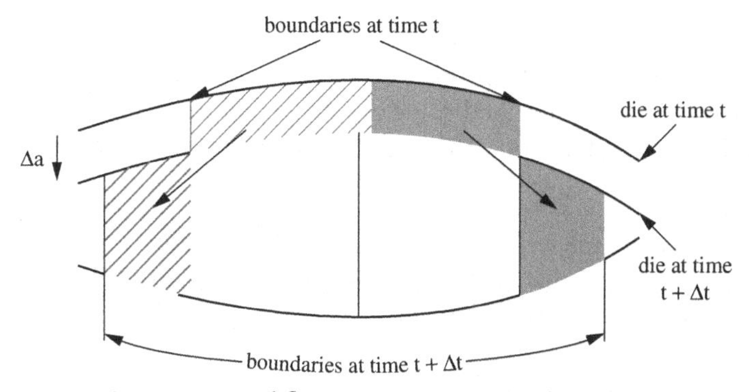


Figure 10.14. Material flow approximation by the slab method.

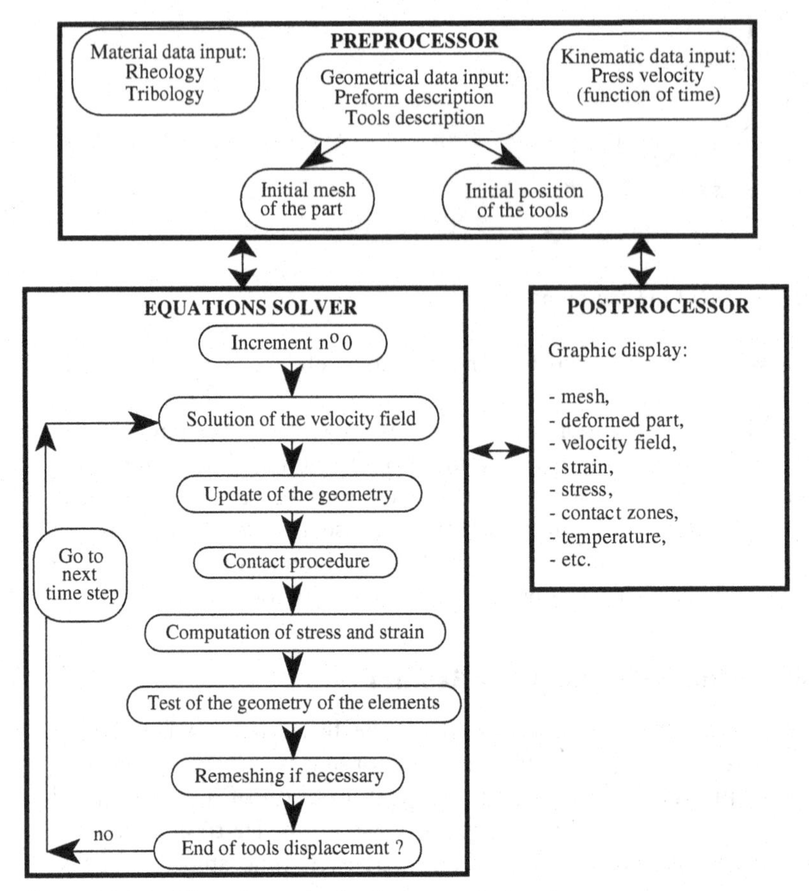


Figure 10.15. Flow chart of a forging numerical code.

results.[9] However, we shall not give its description here for simplicity. At each increment the resulting nonlinear problem is solved by the Newton–Raphson method, and subincrementation, or even optimization in the direction of displacement, may be necessary to achieve convergence.

The contact conditions are analyzed at the end of each increment with a nodal approach. In order to simplify the procedure, the following assumptions were made.

1. If a node penetrates inside a tool, it is projected on the tool surface only (that corresponds to only one iteration of the process described in Section 6.1).
2. If at the end of the increment the computed value of the normal stress $\sigma_n$ is no longer negative (i.e., becomes a traction), the node is allowed to lose contact at the next time step.

Finally, it is necessary to check at each time step if the geometry of the elements is good enough for accurate F.E. computation. If this is not the case, a remeshing procedure is triggered in order to regenerate non distorted elements.

[9] M. Fortin and A. Fortin, "Newer and Newer Elements for Incompressible Flow," *Fin. Element Fluid* 6, 171–187 (1985).

### Synopsis of the Procedure

The general procedure for generating the data for a finite-element computation is essentially the same. It can be divided into the following steps:

- description of the different tools by their contour,
- law of motion of the tools,
- definition of the initial shape of the part by its boundary and meshing of the part (some information on the size of element is required),
- introduction of possible symmetry axes,
- initial relative position of the tools and of the part,
- constitutive law parameters of the material, and
- friction coefficients.

A visualization of the initial positions of the tools, of the possible symmetry axes, and of the meshed part is necessary to check that no error was introduced before the finite-element computation is launched.

In most of the following examples, the upper tool will have a velocity of 1 m/s while the lower tool is assumed to remain fixed.

### Upsetting of a Square Block Between Flat Dies

The length of contact between the flat dies and the workpiece is continuously increasing for two reasons.

1. There is first a relative slip of material points already in contact with the tools.
2. Material points previously on the free surface come into contact with the dies, as indicated schematically in Fig. 10.16.

This phenomenon may cause particularly large distortion of the local mesh, so that remeshing is necessary; other elements can also be distorted by shearing in the vicinity of the tool–part interface.

The initial mesh of the part is pictured in Fig. 10.17(a), where the size of the specimen is given with the velocities of the upper die. The material parameters are $K = 1$, $m = 0.1$, $\alpha = 0.3$, and $p = 0.1$. The shapes of the deformed part for 30% and 60% reduction of height with the corresponding meshes are given respectively in Figs. 10.17(b) and 10.17(c).

The computation of the equivalent strain shows at the beginning the classical shape of a cross, which is apparent in Fig 10.18(a), while at further stages the deformation tends to be more and more homogeneous, as can be seen in Fig 10.18(b).

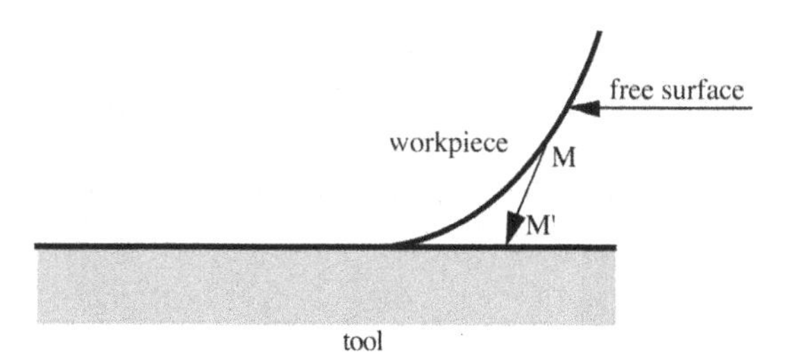


Figure 10.16. A material point $M$ comes into contact with the tool in $M'$.

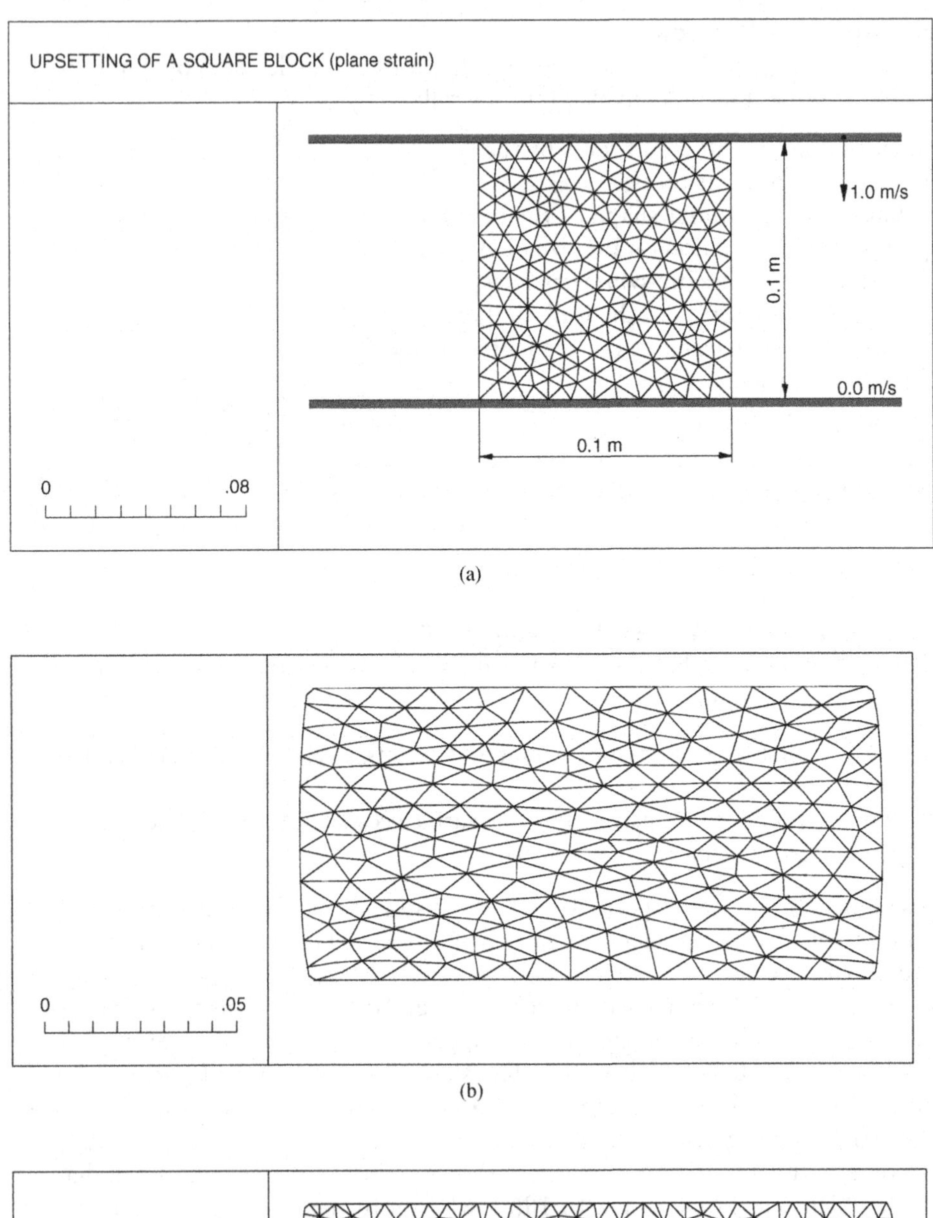


(a)

(b)

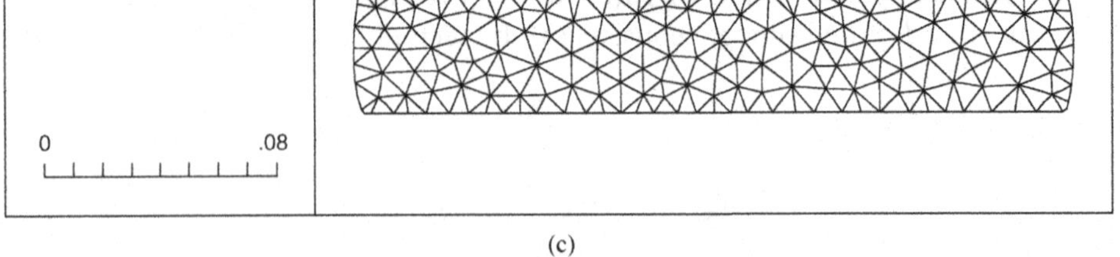


(c)

Figure 10.17. Plane-strain upsetting of a square block: (a) initial mesh, (b) after 30% reduction, and (c) after 60% reduction.

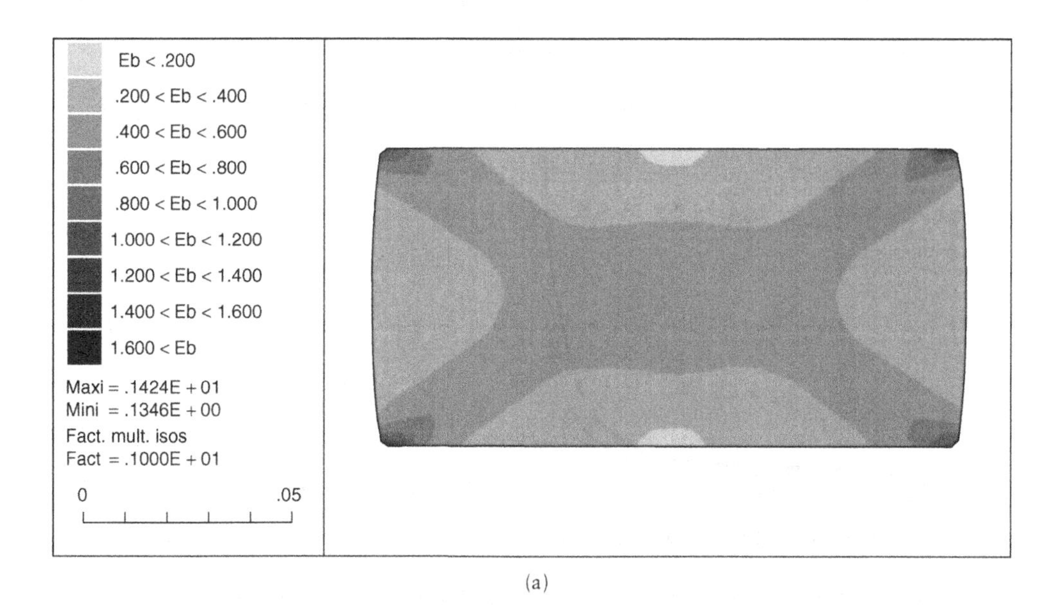


(a)

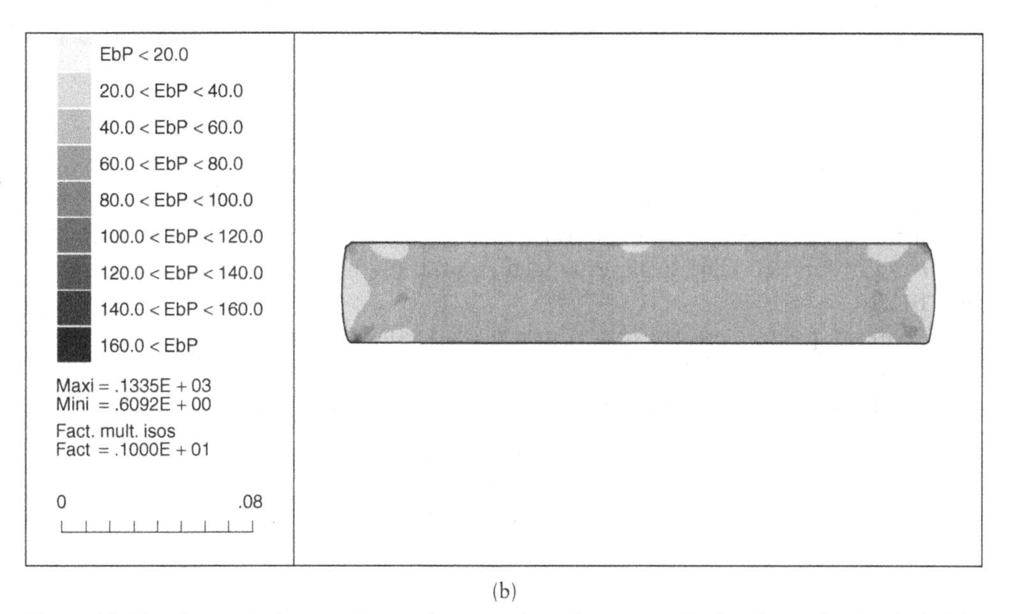


(b)

Figure 10.18. Isoequivalent strain rate for upsetting of a square block: after reduction in height of (a) 30% and (b) 60%.

The forging force is also computed during all of the forging process until a 90% reduction of height was achieved. In Fig. 10.19, the influence of the rate sensitivity parameter $m$ is analyzed (the friction rate sensitivity parameter $p$ was given the same value as $m$). Its effect is shown for the quasi-plastic material with a cold metal for which $m = 0.02$, then for a classical hot material with $m = 0.1$, and finally for a material with $m = 0.3$, the behavior of which is relatively close to superplasticity. The other parameters are $K = 1$ and $\alpha = 0.3$. We observe that the vertical force is an increasing function of $m$ (and $p$, which is taken equal to $m$).

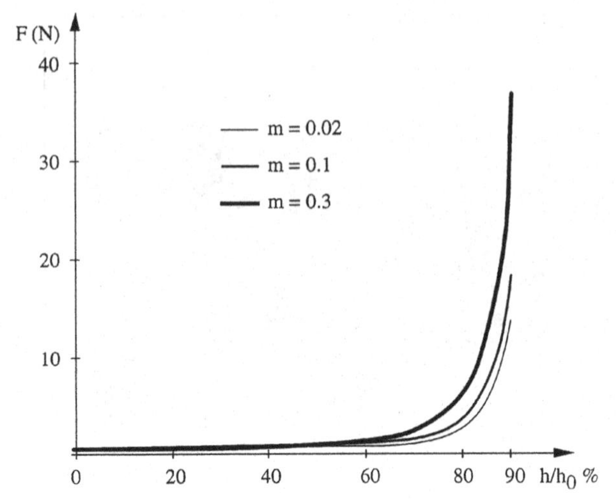


Figure 10.19. Influence of the rate sensitivity index $m$ on the forging force $F$.

The friction factor $\alpha$ is also very important for the determination of the forging force, especially when the surface in contact with the dies becomes important, as it appears clearly in Fig. 10.20, where the other parameters correspond to $m = p = 0.1$ and $K = 1$.

### Upsetting of a Rectangular Block

This example is close to the previous one, but we shall focus our attention on the bulging phenomenon that takes place as soon as the deformation proceeds. The width of the workpiece will be kept constant and equal to 0.1 m, while the initial height $H$ will be varied. The situation and the initial mesh for one case are pictured in Fig. 10.21.

The material parameters are $K = 1$, $m = p = 0.1$, and the friction parameter was given a medium value of $\alpha = 0.5$. The height $H$ was given successively the values 0.15, 0.2, 0.3, 0.4, and 0.5, and the results can be observed after 30% reduction of $H$ in Fig. 10.22.

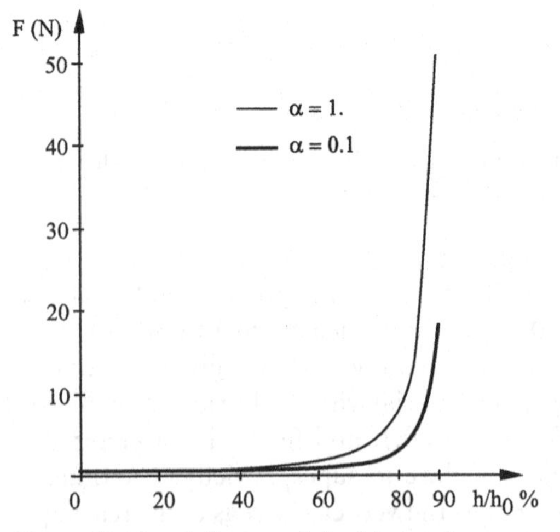


Figure 10.20. Influence of the friction factor $\alpha$ on the forging force $F$.

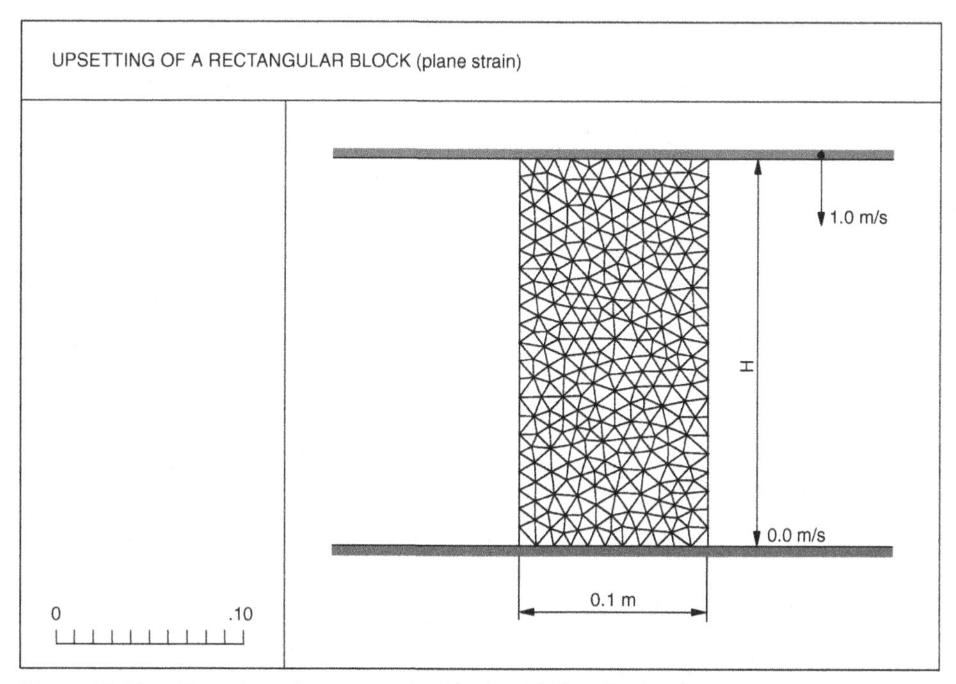


Figure 10.21. Upsetting of a rectangular block with flat dies in plane strain.

We remark that when the initial height $H$ is 1.5 or 2, Figs. 10.22(a) and 10.22(b), a classical bulging occurs. For higher values of $H$ a double bulging appears, which can be attributed to a more local deformation pattern in the vicinity of the dies. When the initial height $H$ is too large; the deformation process becomes unstable, as it appears in Fig. 10.22(e).[10]

**Indentation with Loss of Contact**

In Section 6.1 the full contact problem was analyzed and an algorithm for evolving contact was presented. An essential aspect of contact analysis is the possibility of losing contact during forging. This loss of contact occurs when the contact normal stress is no longer compressive, as no traction can be exerted by the tools on the part. This situation is well illustrated by the plane-strain indentation of a square block by a rectangular punch (see Fig 10.23).

The material parameters are selected according to $K = 1$, $m = 0.1$, $\alpha = 0.5$, and $p = 0.1$. When the deformation proceeds, we can distinguish several stages.

1. At the very beginning, there is a loss of contact on the vertical edges of the punch, as shown in Fig. 10.24(a).
2. The upper part of the workpiece comes again into contact with the vertical edges of the punch on a very narrow zone, as shown in Fig. 10.24(c).
3. The lower side of the workpiece loses contact with the lower die, and curves itself on an increasing zone, as shown in Figs. 10.24(d)–10.24(f).

[10] In this case the instability deformation pattern is enhanced by the mesh, which does not respect exactly the symmetries of the part. In a real situation the unstable path is more likely to be chosen, as small imperfections always exist in real materials.

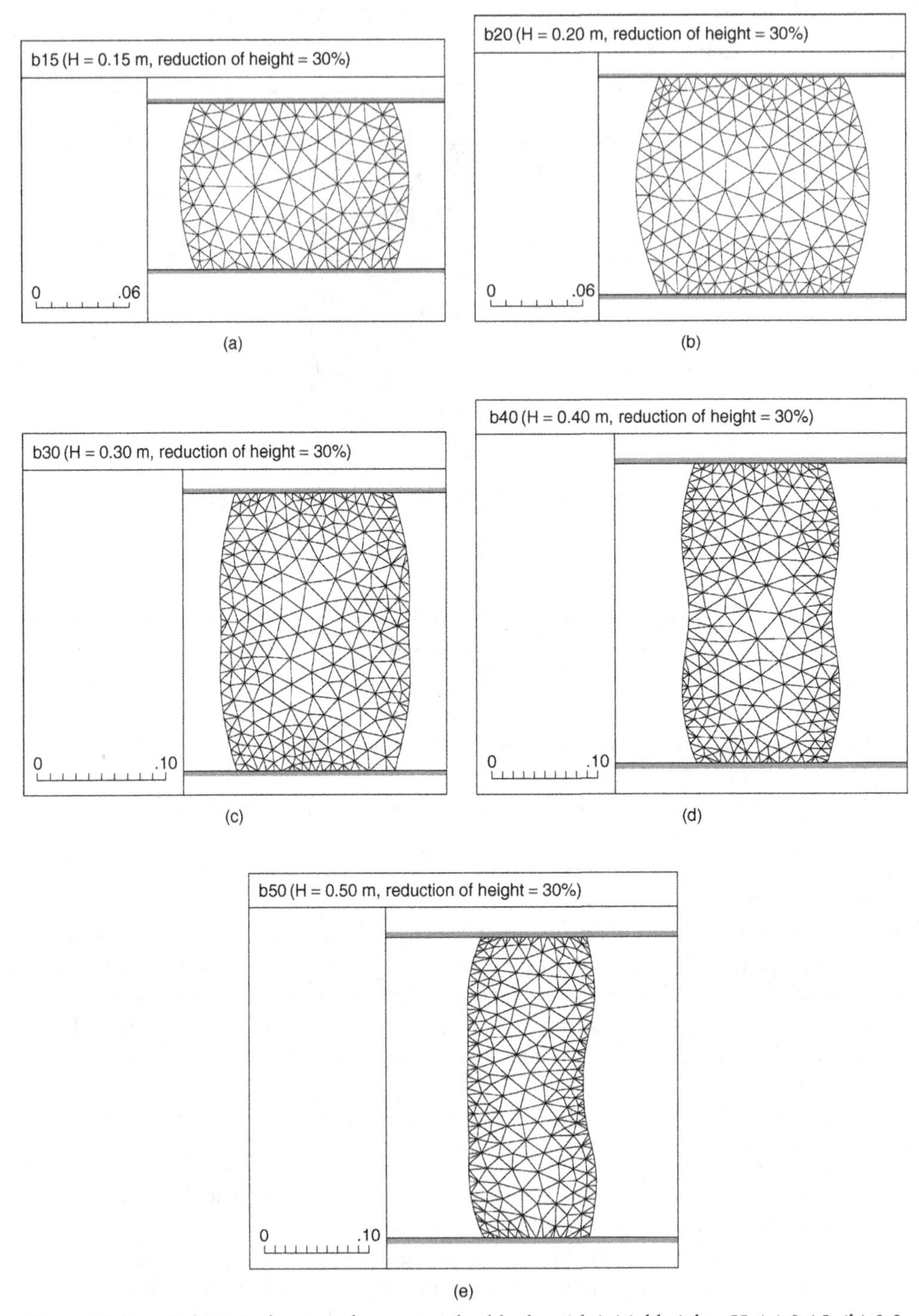


Figure 10.22. Bulging in forging of a rectangular block, with initial heights $H$: (a) 0.15, (b) 0.2, (c) 0.3, (d) 0.4, and (e) 0.5.

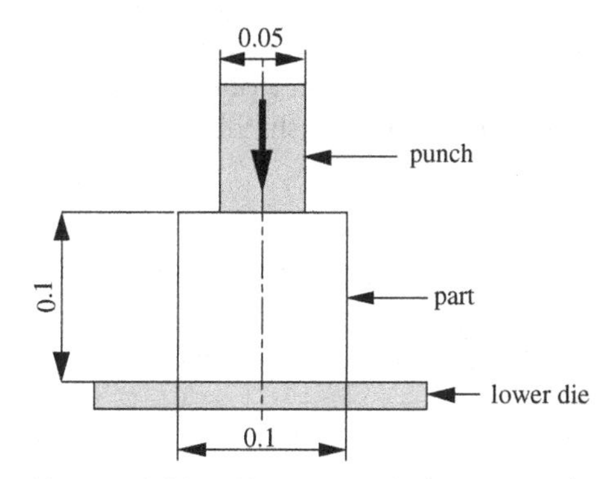


Figure 10.23. Plane-strain indentation of a square block (initial geometry).

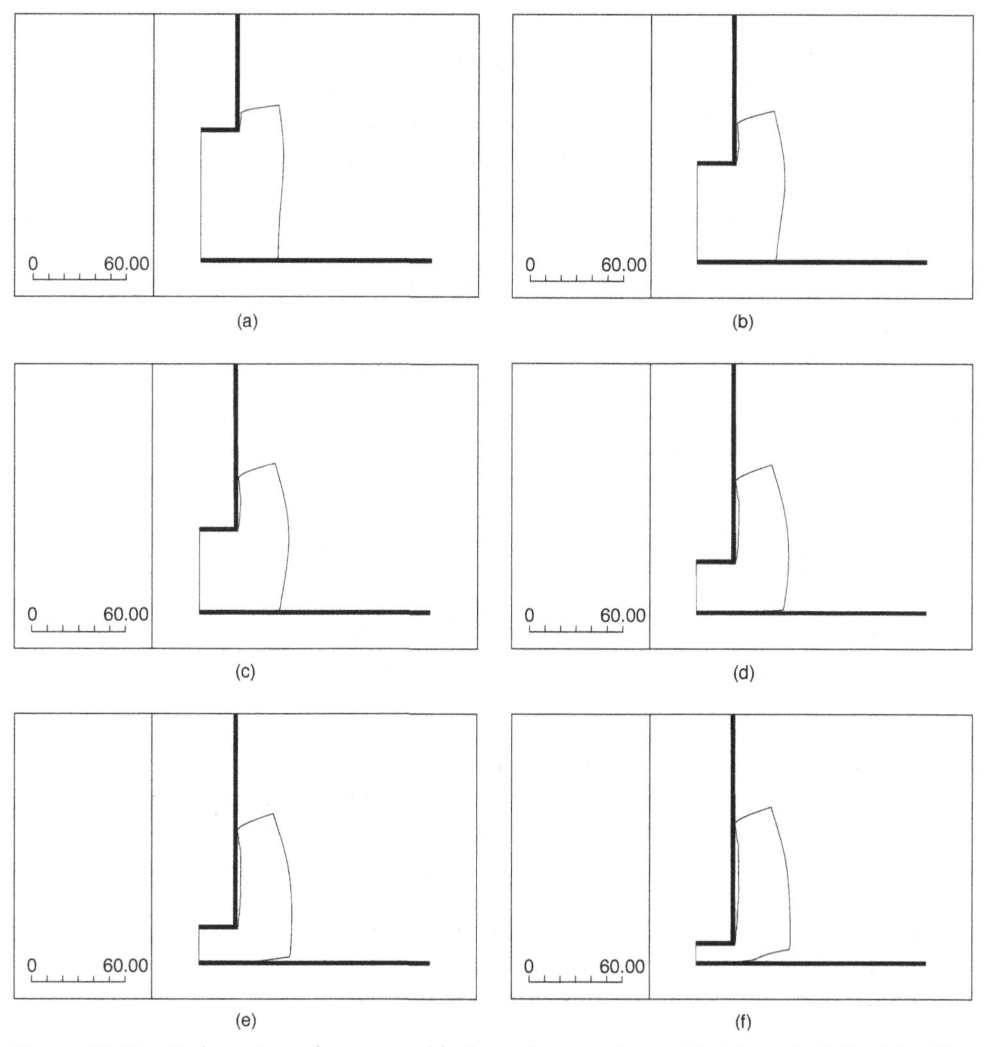


Figure 10.24. Indentation of a square block, with reductions of height: (a) 20%, (b) 40%, (c) 50%, (d) 70%, (e) 80%, and (f) 90%.

The map of isostrain rate is given at the end of the process for a reduction of 90% in Fig. 10.25. It shows two different zones of high strain rate: one is under the punch, and the other is associated with loss of contact and with the resulting bending.

### Forging of a Complex Part

The geometry of the dies and the initial mesh of the part are given in Fig. 10.26.

This example was chosen to illustrate the competition between two possible material flows:

- filling of the upper rib,
- upsetting of the workpiece by the horizontal part of the upper die.

The material parameters are selected according to $K = 1$, $m = 0.15$, $\alpha = 0.5$, and $p = 0.15$, and the initial height of the square block is $h = 70$ mm. The results are given in Fig. 10.27, where we observe the following.

1. There is combined filling of the rib and upsetting with a beginning of a local bulging, which could lead to the defect of material folding, as shown in Fig. 10.27(a).
2. There is partial contact of the workpiece with the lower lateral part of the die and progressive filling of the rib, as shown in Fig. 10.27(b).
3. There is formation of a flash, which allows the rib to fill completely at the end of the process, as shown in Fig. 10.27(c).

The map of isoequivalent strain rate at height of 9 mm shows a complicated pattern. In Fig. 10.28 we observe that the maximums are located in three narrow contact zones where the curvature of the boundary is high, and that they are connected by region of medium strain rate.

### Influence of Strain Hardening

When the material exhibits strain hardening, the zones that would undergo the higher strain rate become harder and harder, so that deformation takes place partly in other regions. The plane-strain tension test of a notched sample is appropriate to show the influence of strain hardening on the shape, the strain, and the force. The geometry of the specimen and the initial mesh are depicted in Fig. 10.29 (because of symmetries, only one quarter of the sample is represented).

The initial height of the sample is $h = 180$ mm; the final length is 196 mm. Three strain hardening viscoplastic laws are introduced with the same rate sensitivity index $m = 0.1$:

material a, $K = K_0$ (no strain hardening)
material b, $K = K_0(1 + 0.3\bar{\varepsilon})$ (medium strain hardening)
material c, $K = K_0(1 + \bar{\varepsilon})$ (high strain hardening)

The final shapes of the samples are drawn in Fig. 10.30. It appears clearly that increasing strain hardening sensitivity decreases notably the thinning of the specimens

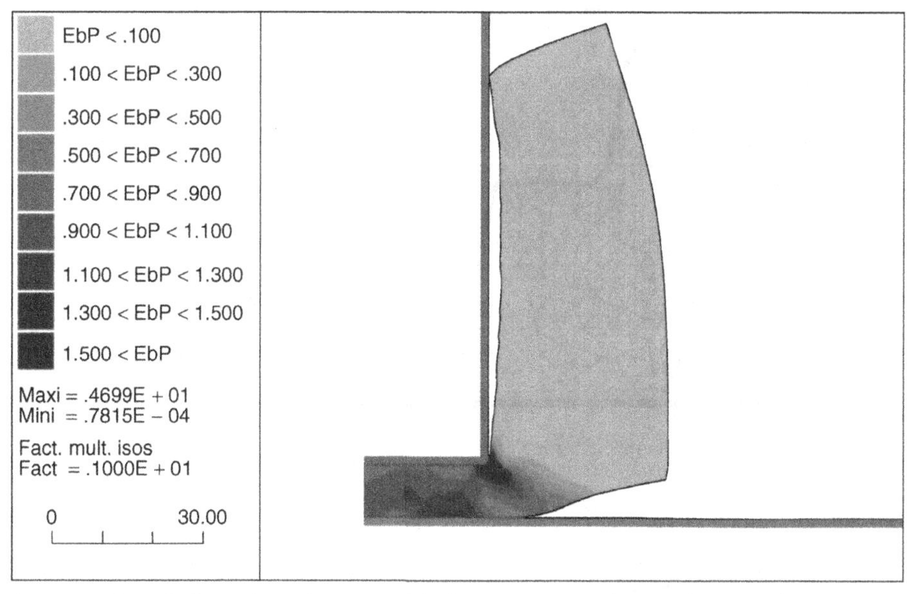


Figure 10.25. Indentation of a square block isostrain rate for a reduction of 90%.

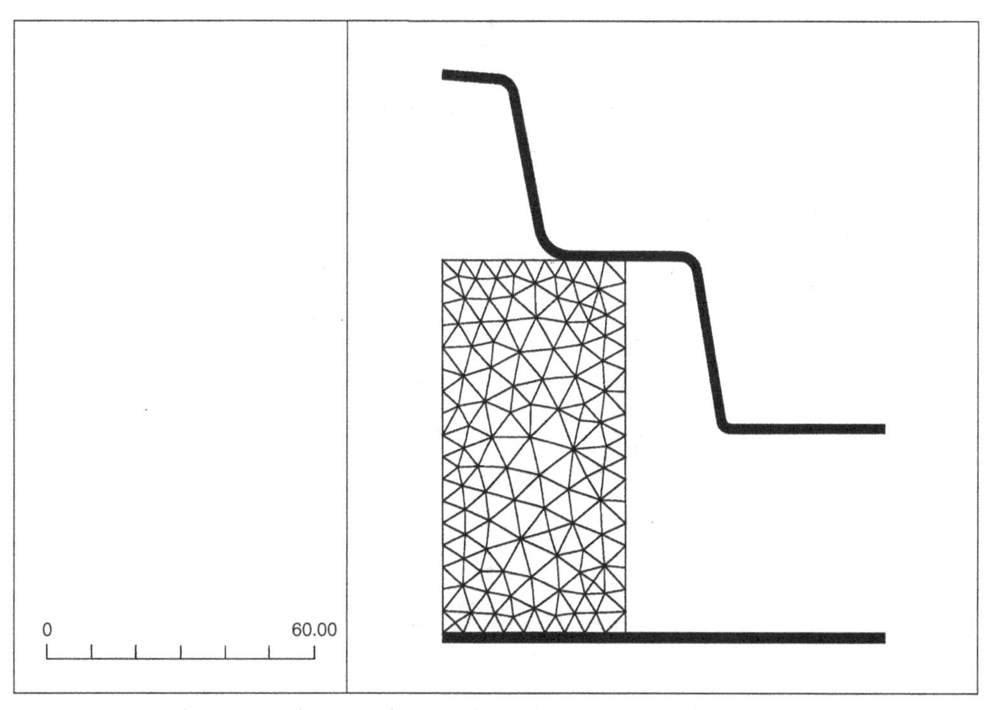


Figure 10.26. Plane-strain forging of a complex part.

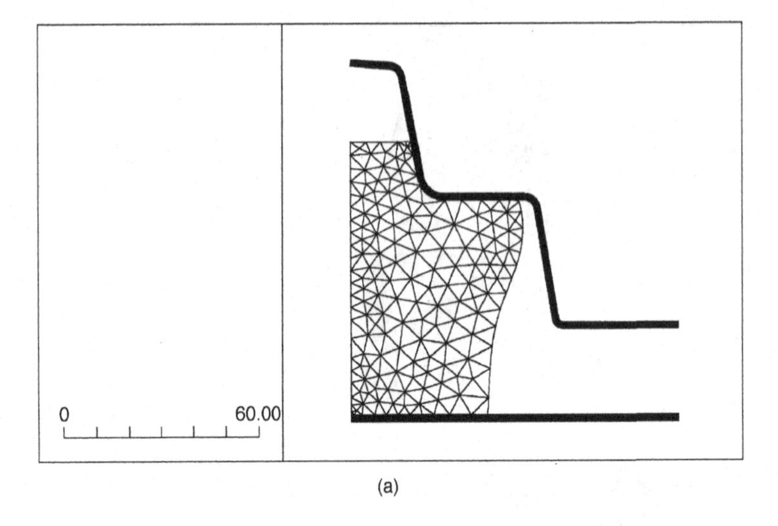


(a)

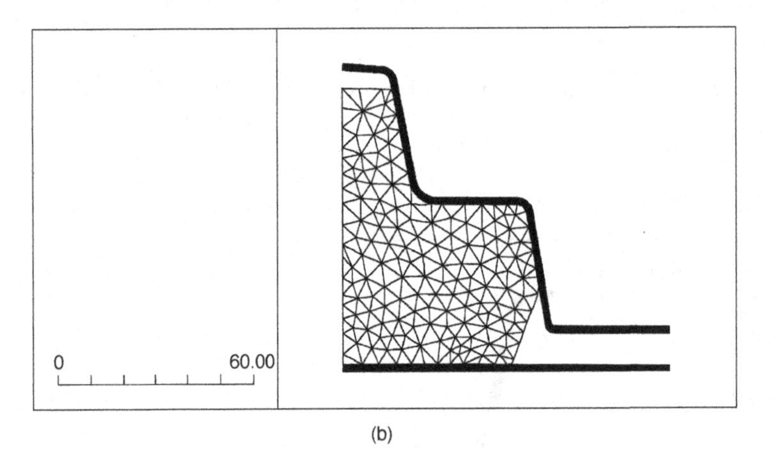


(b)

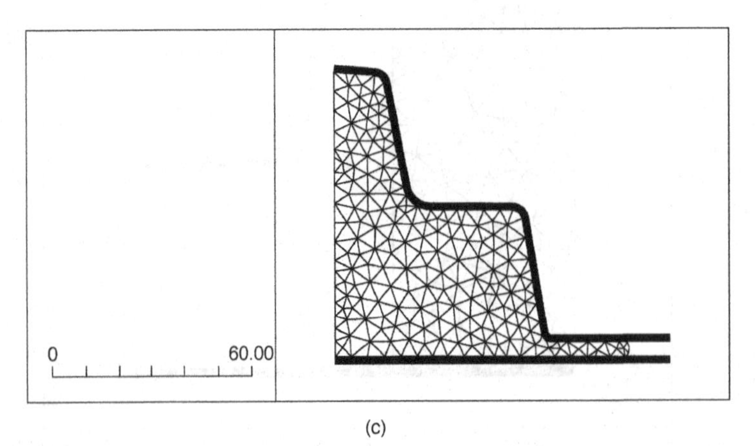


(c)

Figure 10.27. Plane-strain forging of a complex part, with tool heights of (a) 25, (b) 9, and (c) 4 mm.

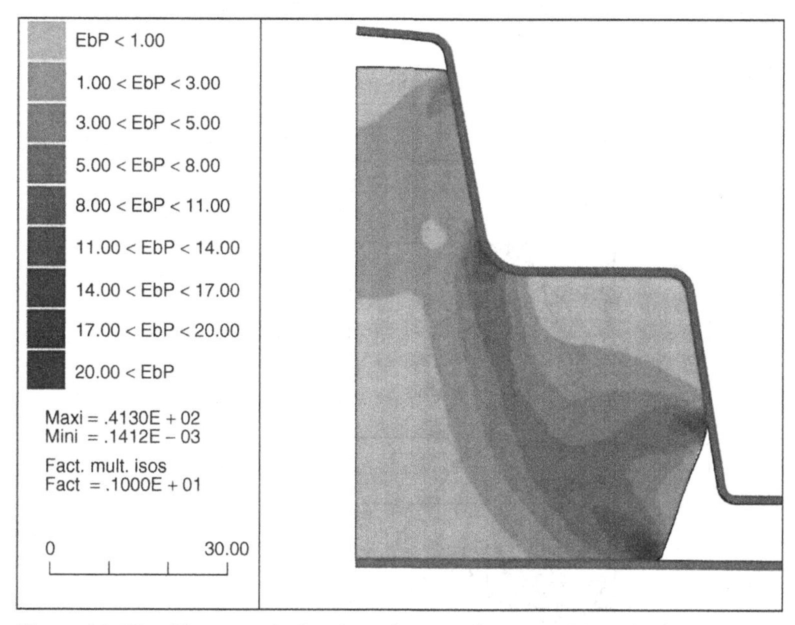


Figure 10.28. Plane-strain forging of a complex part, isoequivalent strain rate.

in the notched region, where the deformation is mostly concentrated in the nonstrain hardening material.

To have a better insight in the deformation process, a zoom of the notched region is performed, with a plot of the map of isoequivalent strain. In Fig. 10.31 we see that when no strain hardening is present, the isocurves are approximately horizontal. The material is deformed nearly as if it was made of horizontal slabs, and the maximum value of the equivalent strain is $\bar{\varepsilon} = 2.9$. For a medium strain hardening sensitivity, the isocurves are inclined and the maximum value is decreased to $\bar{\varepsilon} = 2.3$. Finally, for a high strain hardening sensitivity, the deformation is much less heterogeneous and the maximum equivalent strain is then $\bar{\varepsilon} = 1.6$.

Finally, the effect of strain hardening on the tension force can be visualized in Fig. 10.32. The plots of forces versus displacements for the three materials are approximately linear and have the same value at the origin. However, we observe that the final tension forces are extremely sensitive to the strain hardening: the higher strain hardening material imposes a force that is about six times that for the nonstrain hardening material.

## 10.3 Finite-Element Axisymmetrical Results

### The Tension Test

The tension test is probably the most widely used laboratory test because of its simplicity, despite the limitation in strain that is imposed by necking and rupture. The geometry of the sample can vary according to the characteristics of the tension machine: an example is pictured in Fig. 10.33. In our example the useful region is a cylinder with a diameter of 6 mm, where the deformation is assumed to occur

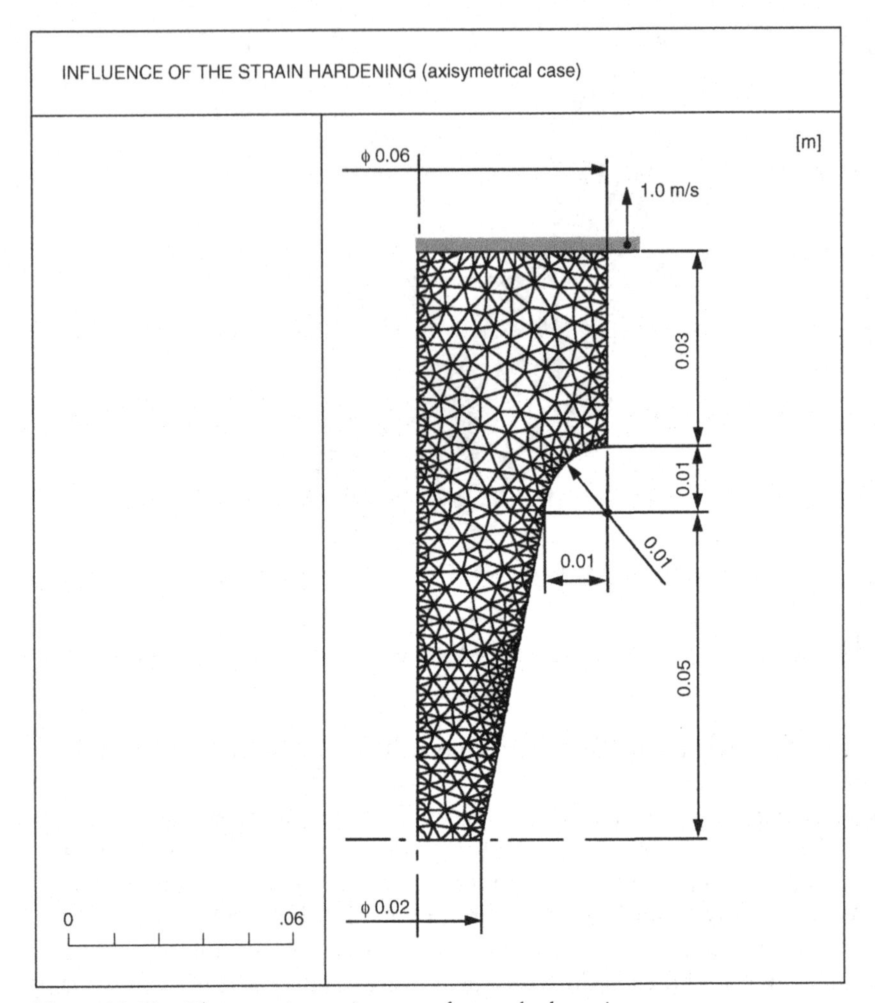


Figure 10.29. Plane-strain tension test of a notched specimen.

mainly with a uniform distribution at the beginning. This geometry is slightly simplified, and only the quarter of a longitudinal cross section is meshed according to Fig. 10.34.

At first the behavior of a purely viscoplastic material is simulated with $m = 0.08$. The evolution of the shape of the sample with the progression of the equivalent strain can be seen in Fig. 10.35. We observe that the necking phenomenon occurs relatively soon and is allowed to develop here until the elongation of the specimen is 20 mm. For real metallic materials, the total elongation is much less, as it is limited by failure caused by progressive damage in the region of necking.

The vertical force is plotted as a function of the total elongation of the specimen. In Fig. 10.36, for a velocity of 1 mm/s, we have compared

- the purely viscoplastic material for which the force is continuously decreasing,
- the strain hardening viscoplastic material where the force curve exhibits a maximum that is due to the competition between hardening of the material and geometric weakening of the sample, caused by the decrease of the cross-sectional area.

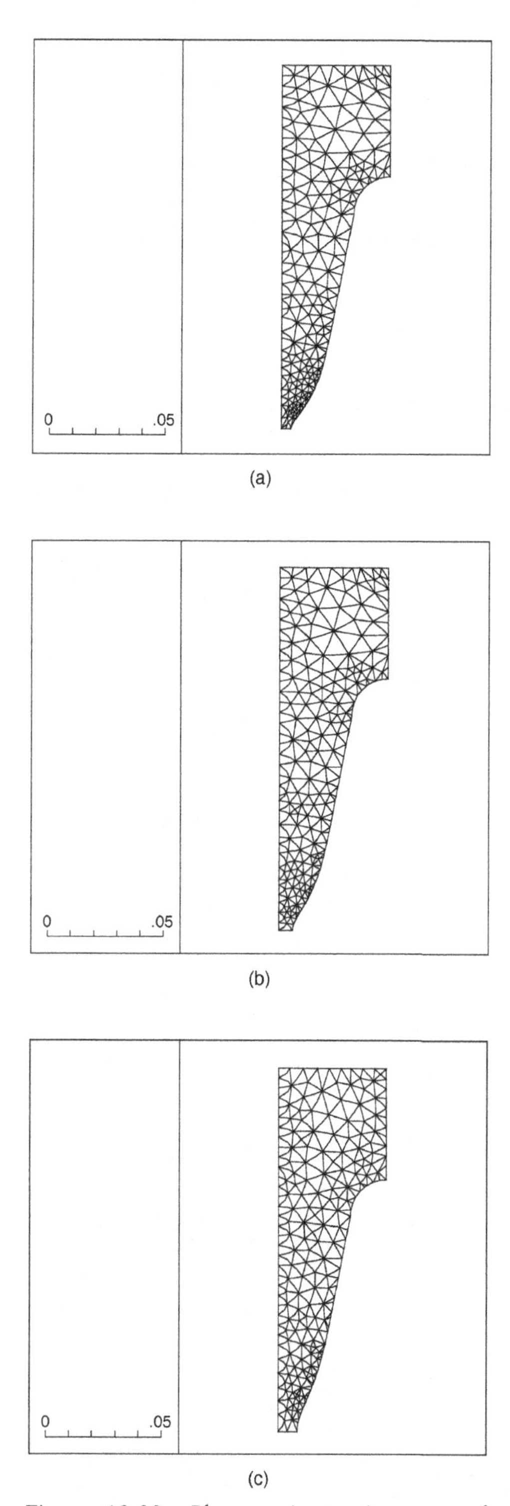


Figure 10.30. Plane-strain tension test of notched specimens: (a) nonstrain hardening, (b) medium strain hardening, and (c) high strain hardening.

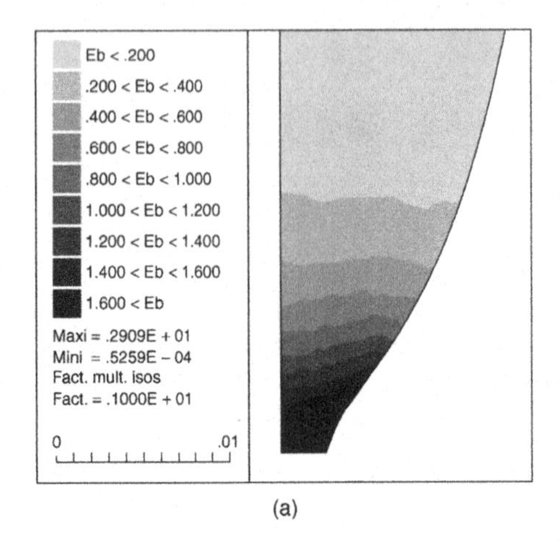


(a)

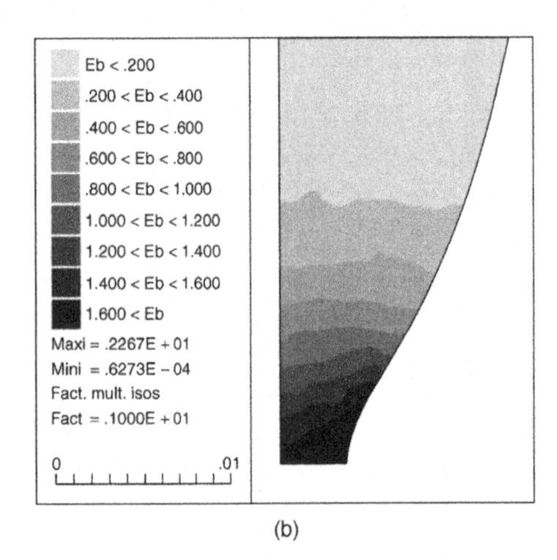


(b)

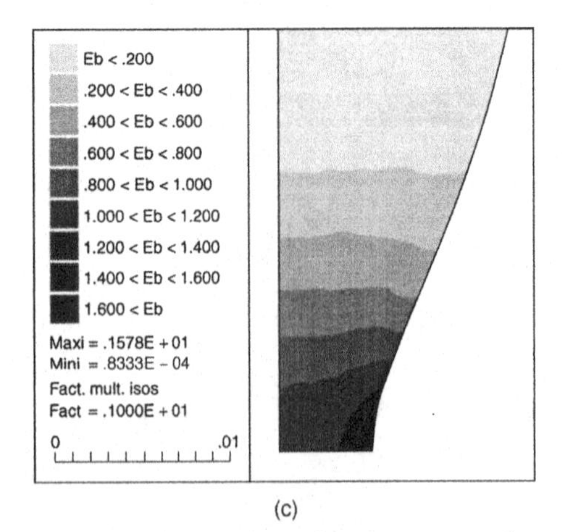


(c)

Figure 10.31. Isoequivalent strain in a plane-strain tension test of notched specimens: (a) non-strain hardening, (b) medium strain hardening, and (c) high strain hardening.

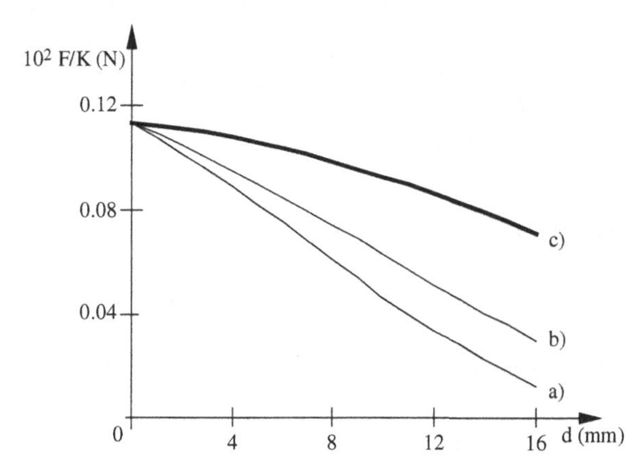


Figure 10.32. Force $F$ vs. displacement $d$, in plane-strain tension test of notched specimens: (a) nonstrain hardening, (b) medium strain hardening, and (c) high strain hardening.

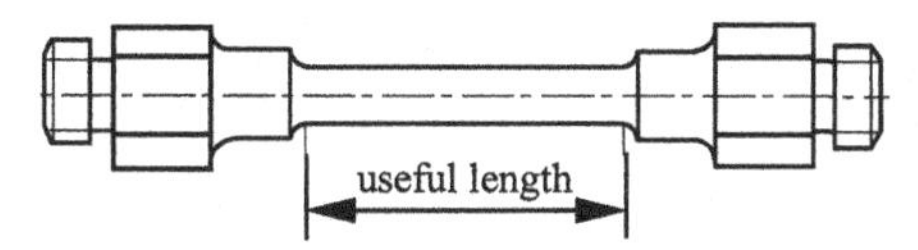


Figure 10.33. Geometry of the tension test specimen.

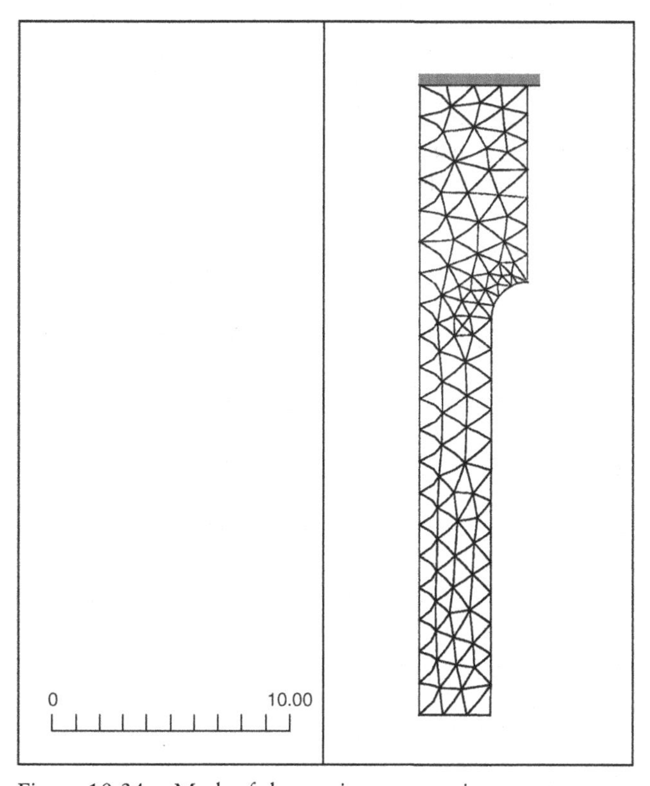


Figure 10.34. Mesh of the tension test specimen.

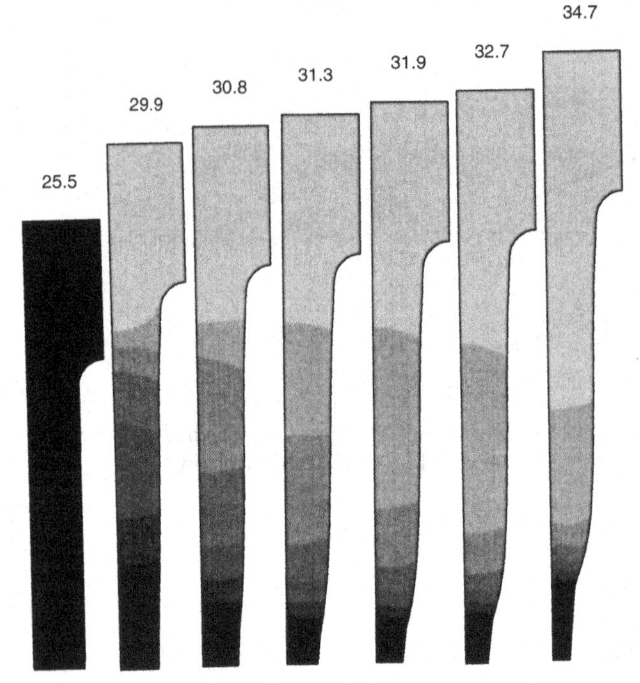


Figure 10.35. FE simulation of the tension test for a viscoplastic metal, isoequivalent strain distribution.

### Forward Extrusion: The Piping Defect

The geometry given in Fig. 10.7(a) corresponds mainly to cold forging, in which deformation generally does not take place in the region of the part with the higher diameter. For hot forging the process also consists of pushing a cylindrical preform with a ram through a conical die, but the latter also possesses a cylindrical zone in order to contain the part; see Fig. 10.37.

This process differs from the extrusion described in Chapter 9 mainly by the reduction ratio, which is much smaller here. A first geometry is simulated numerically

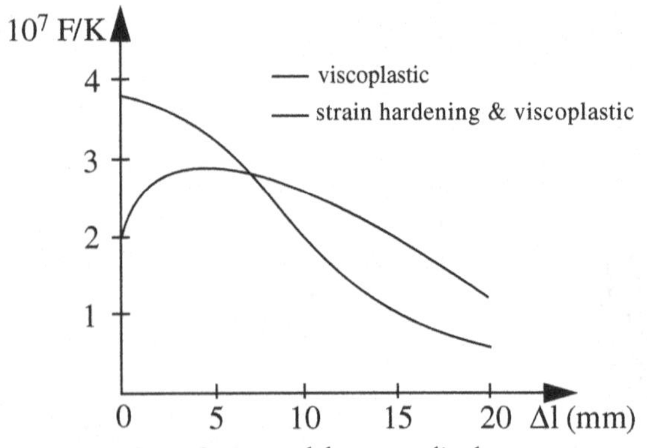


Figure 10.36. Computed force vs. displacement curve in the tension test.

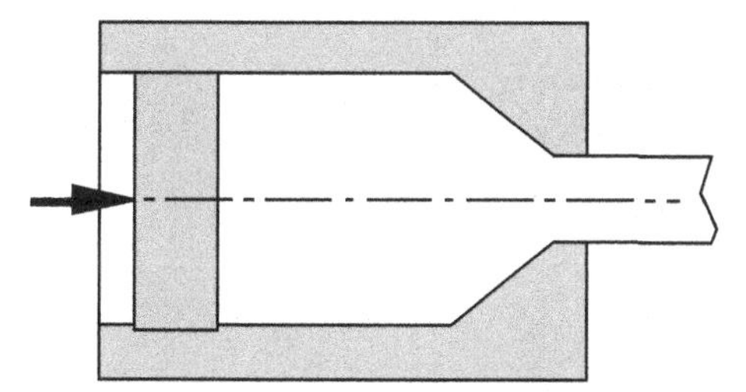

Figure 10.37. Forward extrusion.

with a die angle of 120° and a material with behavior and friction defined by $K = 1$, $m = 0.1$, $p = 0.1$, and $\alpha = 0.5$. The results at the end of the process are given in Fig. 10.38. We observe in this example that the extrusion process can be performed very far, regarding the ram displacement, without observing any material flow defect. With a flat die, that is, a die angle of 180°, the pattern is very different, as can be observed in Fig. 10.39.

We see that a major flow defect appears in the vicinity of the symmetry axis: it is called the piping defect. The material is driven by the main flow and loses contact with the die. In Fig. 10.40 a zoom is made, showing the velocity field with straight lines, the length of which is proportional to the norm of the local velocity vector.

### Backward Extrusion

The process was already presented in Section 10.2 and the geometry was represented in Fig. 10.8. A finite-element computation is made here with a punch of radius $R_p = 135$ mm, and a die of radius $R_d = 215$ mm. The preform is first upset until its maximum radius is also 215 mm and its height is $h = 350$ mm. This initial geometry and the mesh are pictured in Fig. 10.41. The material parameters are $K = 1$, $m = 0.15$, $p = 0.35$, and $\alpha = 0.35$.

The maps of isoequivalent strain rate are given in Fig. 10.42 for an intermediate step and at the end of the process. On the intermediate step of forging, Fig. 10.42(a), we can remark that the zone that is mostly deformed corresponds qualitatively to the assumed deformed zone of the velocity field, which is used in Section 10.2 for the upper bound approach.

### Combined Forward-Backward Extrusion

The geometry of the process with the initial mesh of the cylindrical preform is represented in Fig. 10.43. The computation is made with a strain hardening viscoplastic material with rate sensitivity index $m = 0.16$, a consistency given by $K = K_0(\varepsilon_0 + \bar{\varepsilon})^n$, and with the parameters defined by $K_0 = 0.194$ MPa s$^{-m}$, $\varepsilon_0 = 1.10^{-4}$, and $n = 0.1$. In addition, the friction behavior is defined by $\alpha = 0.05$ and $p = 1$, which corresponds to a Newtonian lubricant; the ram velocity is 1 m/s. The evolution of the shape of the part can be seen in Fig. 10.44, with the progression of the deformed region. We observe the following.

1. At the beginning of forging, two deformation zones are well separated, as shown in Fig. 10.44(a).

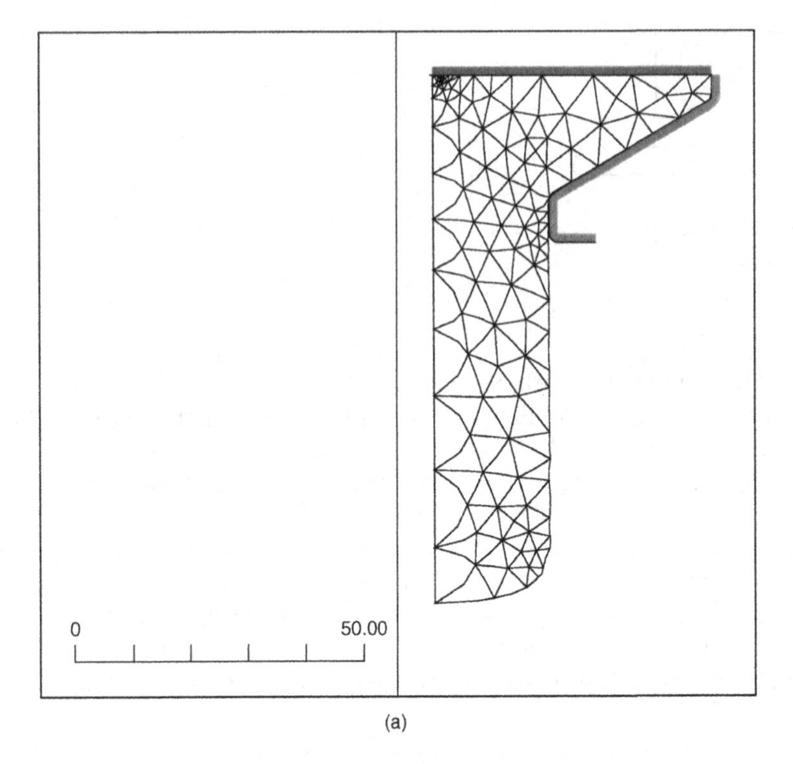


(a)

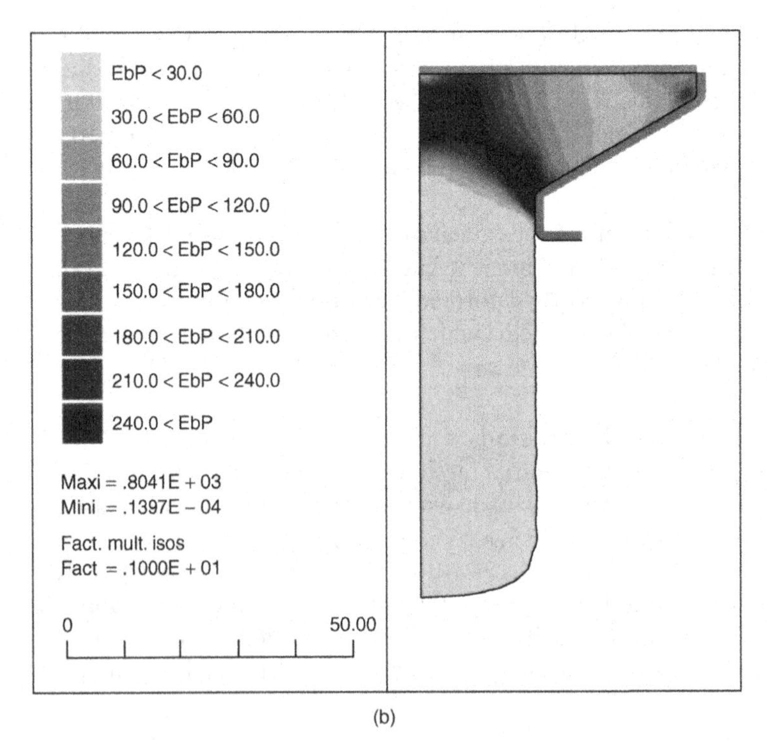


(b)

Figure 10.38. Forward extrusion with a die angle of 120°: (a) mesh and (b) map of isoequivalent strain rate.

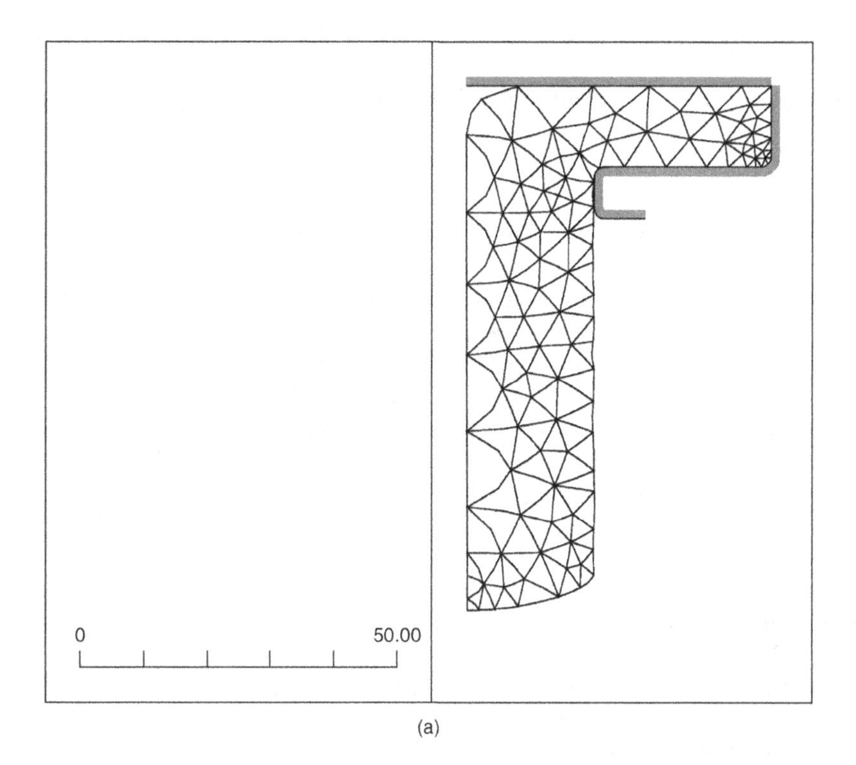


(a)

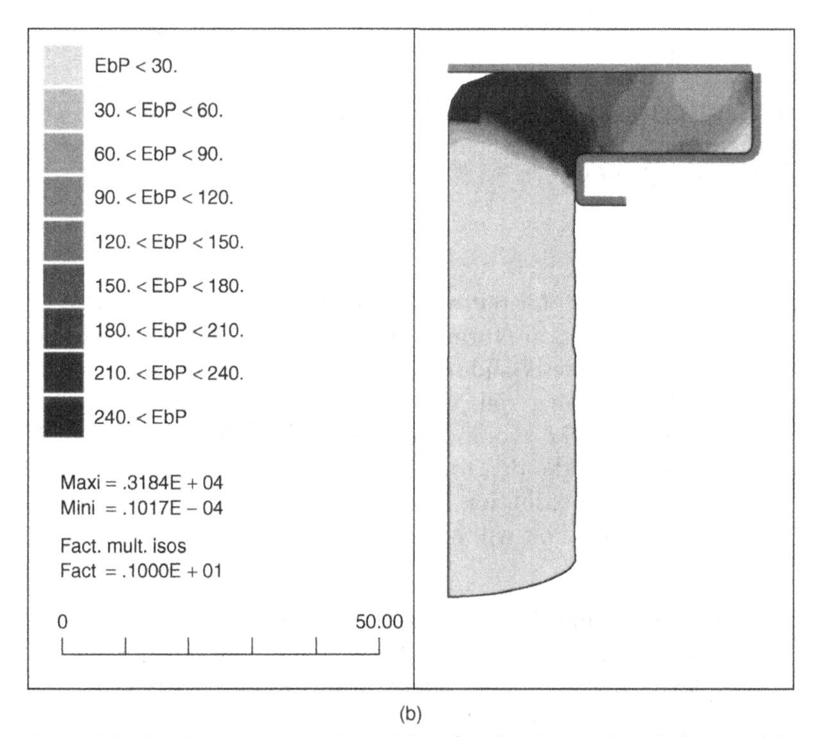


(b)

Figure 10.39. Forward extrusion with a flat die: (a) mesh and (b) map of isoequivalent strain rate.

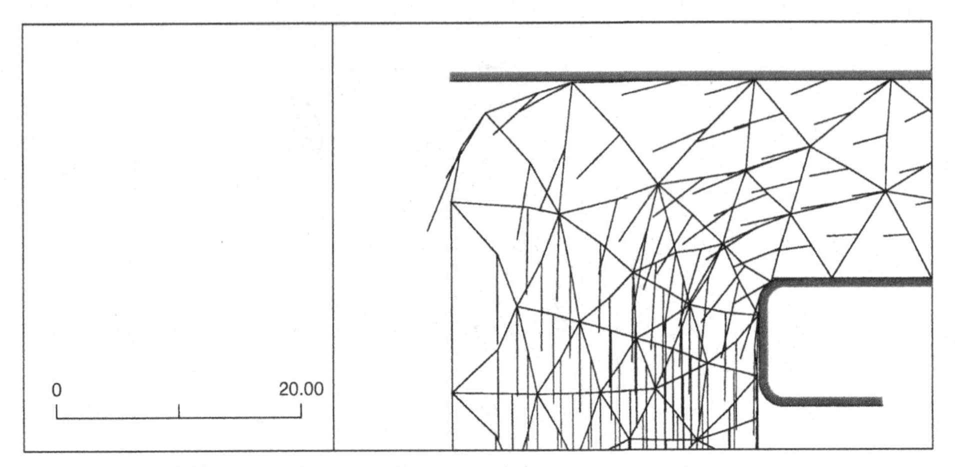


Figure 10.40. Velocity field around the piping defect.

2. At an intermediate stage, the two main zones of maximum equivalent strain rate tend to join, as shown in Fig. 10.44(b).
3. At the end of the process, there is only one zone of intense strain rate, which extends from the punch to the die, as shown in Fig. 10.44(c).

In order to evaluate the influence of the die angle, three computations were made with angles of 180°, 127°, and 90°, keeping constant all the other parameters. The comparison is made in Fig. 10.45 with the final shapes and the maps of isoequivalent strain. The main differences are

- the length of backward extrusion, which is the greatest for the die angle of 180°,
- the appearance of the maximum equivalent strain distribution, which is more widespread for the die angle of 90°.

#### Bolting

Bolting is a process that is often used for putting together two parts, for example, two sheets, as shown in Fig. 10.46. Bolts are formed by two spherical dies that upset a cylinder. A quarter of the initial geometry and mesh are pictured in Fig. 10.47.

For simplicity, we chose the same material parameters as those of the previous example. It is interesting to visualize the evolution of the mesh, which is automatically designed to fit the geometry of the die, together with the shape of the part; see Fig. 10.48. The equivalent strain distribution, which is represented in Fig. 10.49, shows the zones where the head of the bolt will be hardened by plastic work.

#### Forging of a Complex Part: Several Sequences of Forging

For complex parts that are actually produced by forging in industry, it is generally impossible to obtain complexity with only one sequence of forging. Starting from a simple preform, which is generally a cylinder for axisymmetrical parts, a set of several sequences of forging with different dies is necessary to deform the workpiece progressively. It is then necessary to simulate these sequences numerically, keeping in

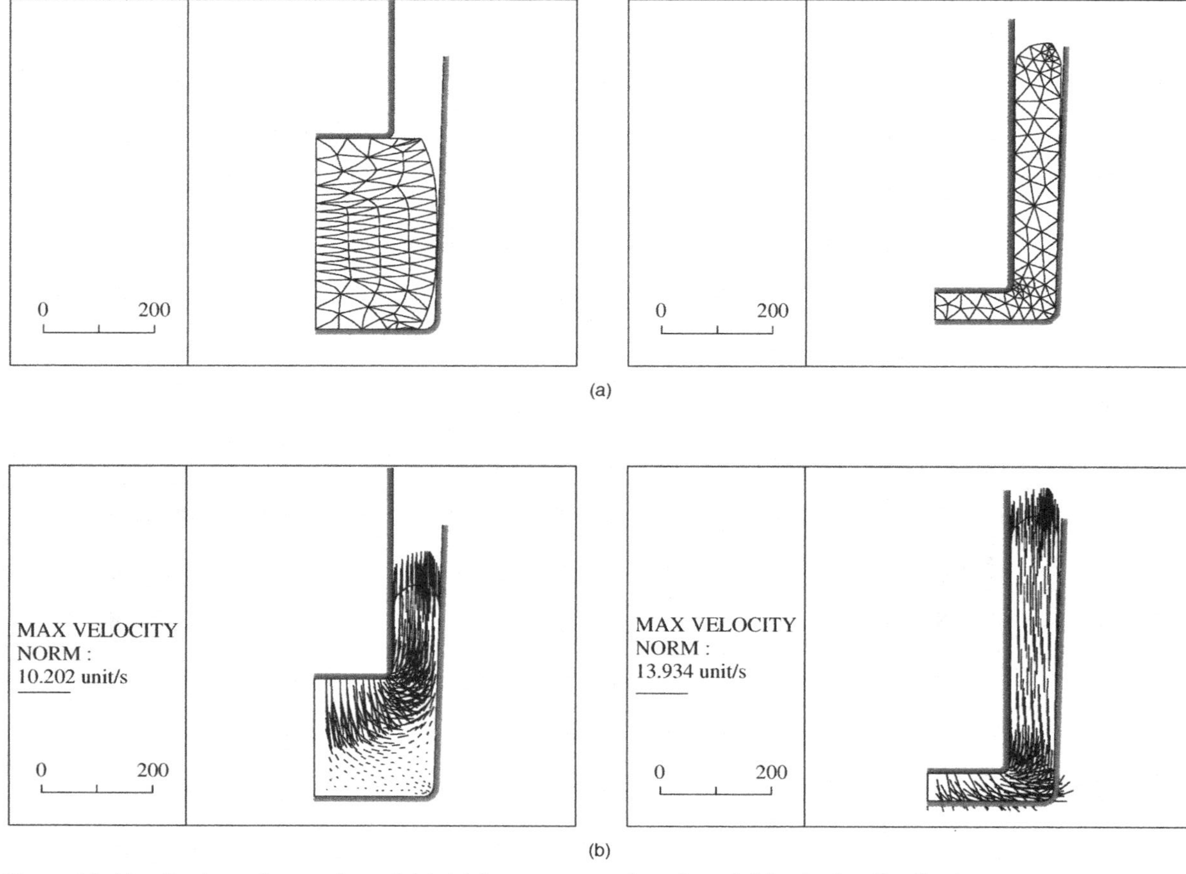


Figure 10.41. Backward extrusion: (a) initial geometry and mesh and (b) velocity distribution.

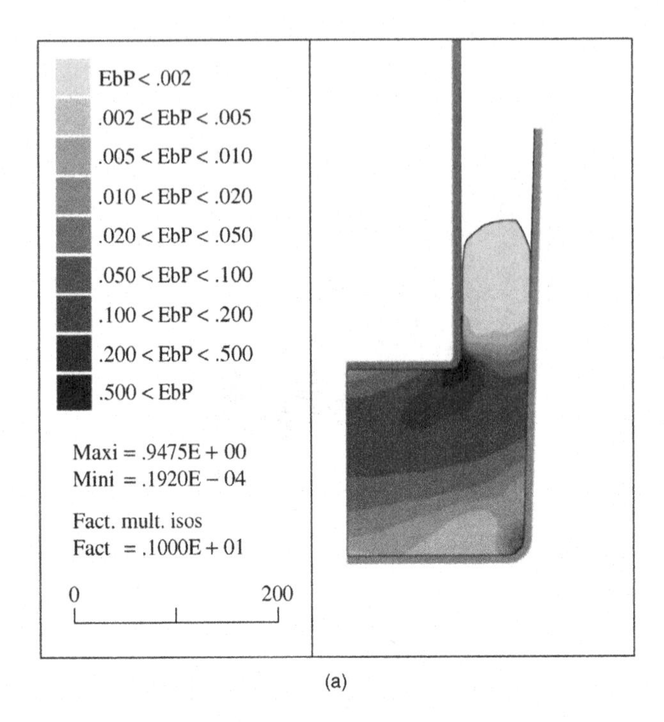


(a)

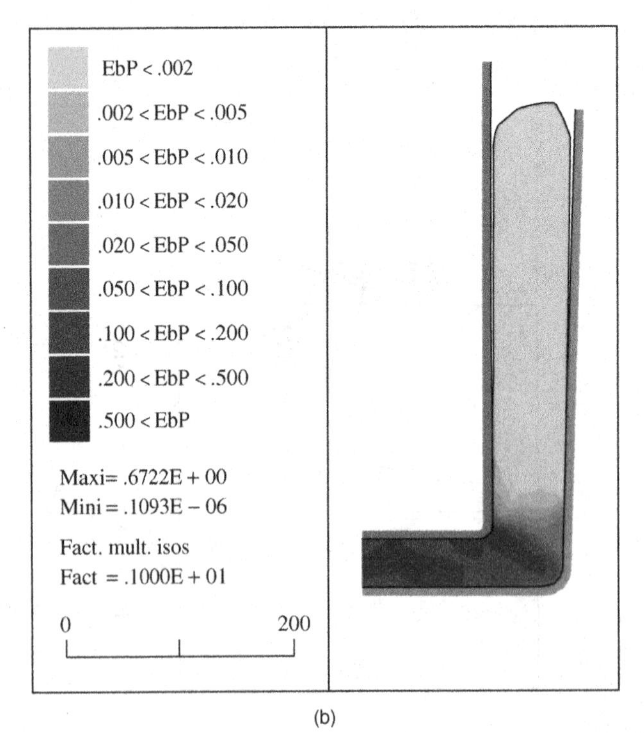


(b)

Figure 10.42. Isoequivalent strain rate in backward extrusion: (a) intermediate step ($h =$ 218 mm) and (b) final step ($h = 50$ mm).

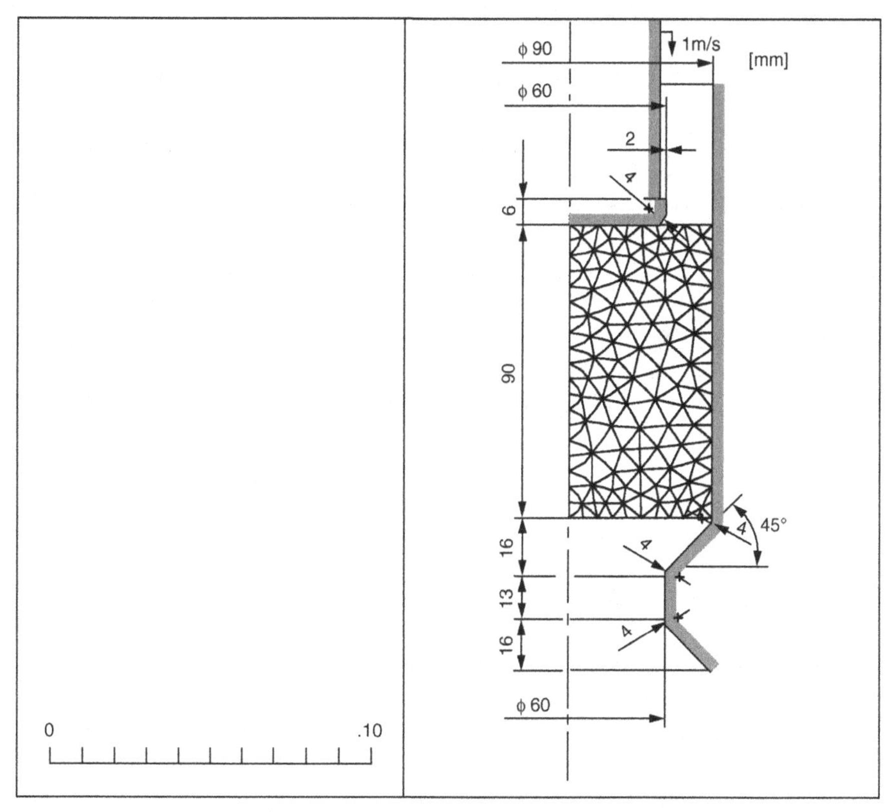


Figure 10.43. Combined forward-backward extrusion (die angle of 90°).

mind that

- the shape of the part at the end of a sequence is the input shape of the next sequence so that it must be transferred, and
- the history must be saved from a sequence to another one, mainly the equivalent strain distribution (and possibly the temperature map, or the physical evolution).

An example involving a set of three sequences of forging is given in the following. The material is characterized by the hardening law $K = K_0(\varepsilon_0 + \bar{\varepsilon})^n$, and the parameters $K_0 = 1.34$ MPa $\mathrm{s}^{-m}$, $\varepsilon_0 = 10^{-4}$, $n = 0.022$, $m = 0.113$, $\alpha = 0.35$, and $p = 0.113$. In Fig. 10.50 the successive shapes of the part, and the corresponding meshes, are represented at the beginning of each sequence and at the end of the process. As usual, the first sequence is a simple upsetting of the initial cylinder, as shown in Fig. 10.50(a). The second sequence, the beginning of which is shown is Fig. 10.50(b), produces a more accurate preform. The final sequence, starting in Fig. 10.50(c), involves globally a smaller deformation, but it is aiming a precise final shape, as shown in Fig. 10.50(d). This is possible thanks to the formation of a thin flash on the right of the part, which induces a pressure field high enough to fill the part.

At the end of the whole process it is important to know the location of the zones of high or low equivalent strain in the final part. This is shown on the map of isoequivalent strain in Fig. 10.51.

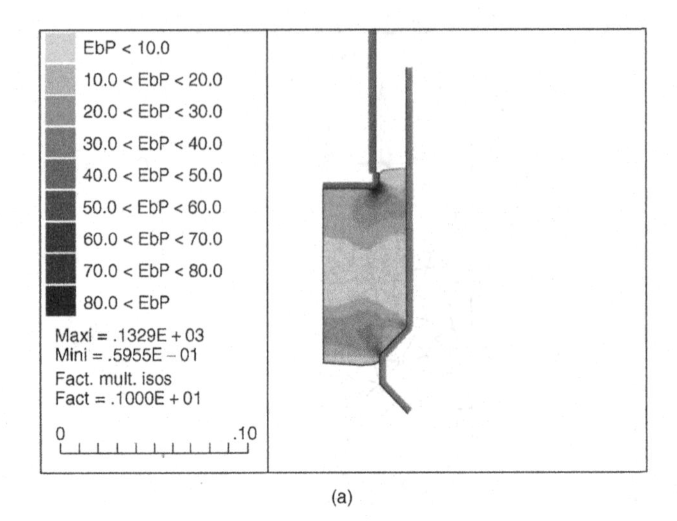


(a)

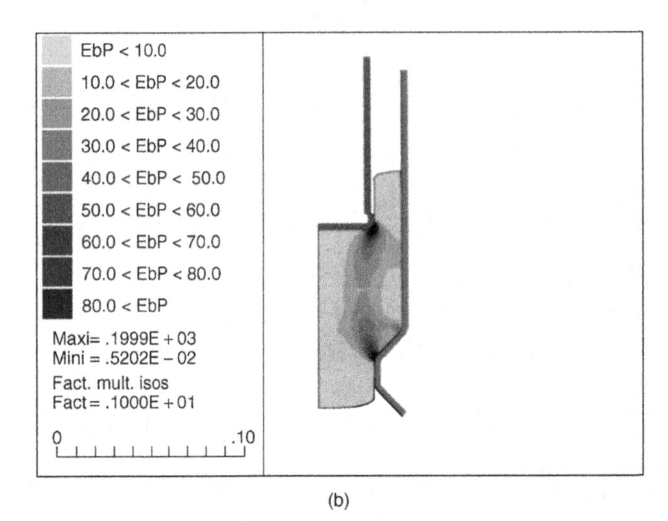


(b)

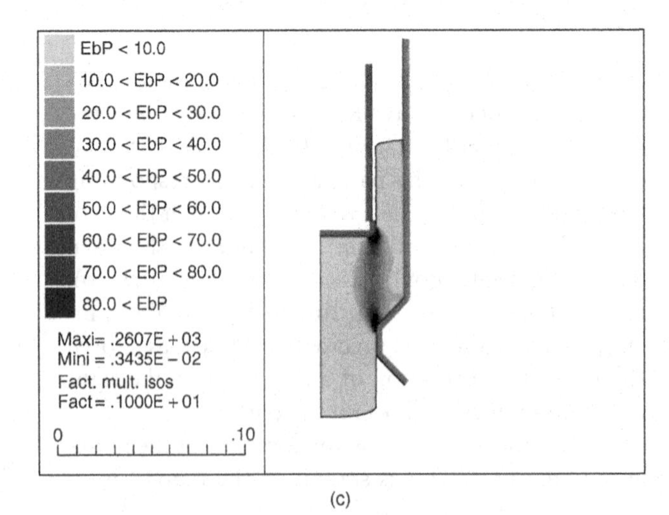


(c)

Figure 10.44. Combined forward-backward extrusion, isoequivalent strain rate for a displacement of the punch of (a) 20, (b) 40, and (c) 60 mm.

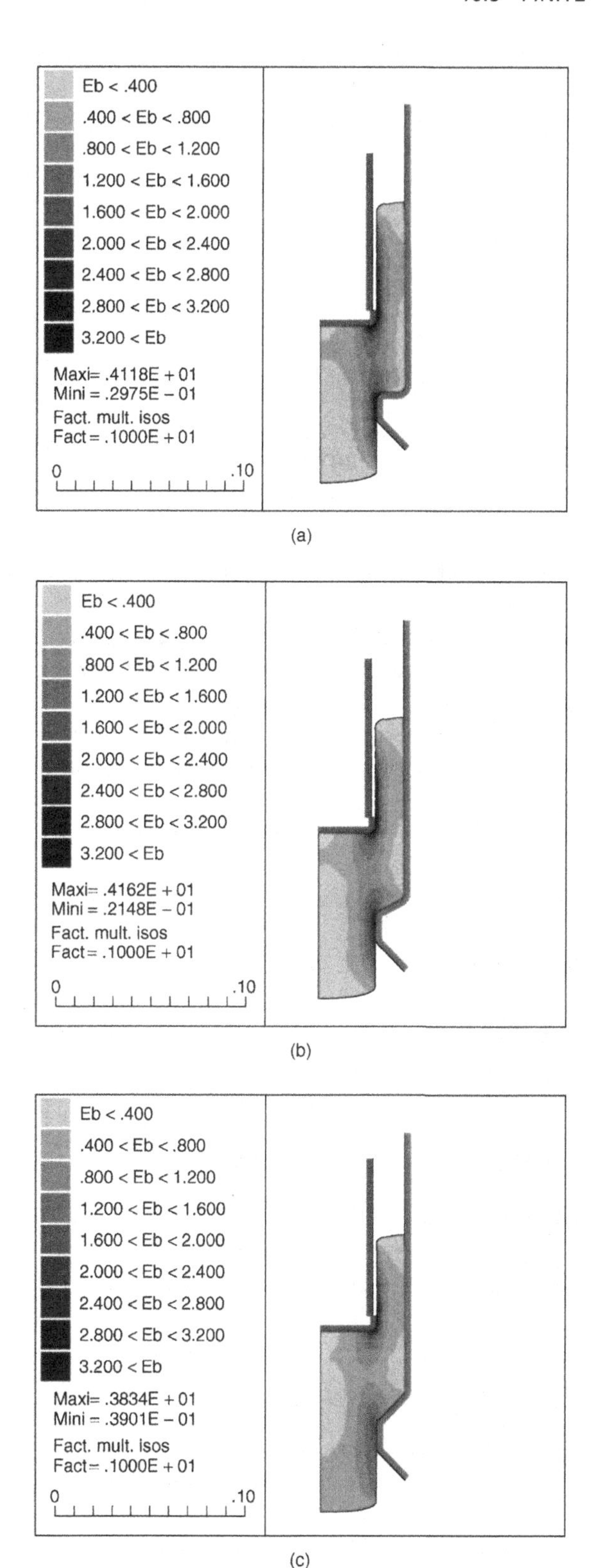


Figure 10.45. Combined forward-backward extrusion with different die angles. Isoequivalent strain for a die angle of (a) 180°, (b) 127°, and (c) 45°.

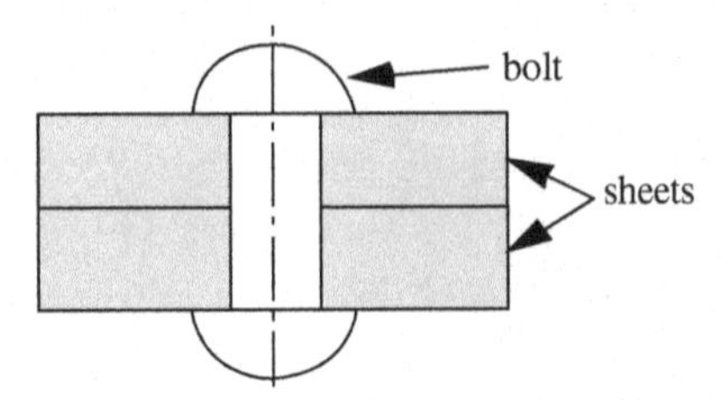


Figure 10.46. Assembly of two sheets with a bolt.

## 10.4 Nonisothermal Effects

### Thermomechanical Coupling

The general problem of thermal and mechanical coupling was presented in Section 7.6. We shall recall here the main steps. The governing equations are as follows.

1. There are the mechanical equations, for which the constitutive equation depends on temperature, Eqs. (7.94) and (7.95) for a viscoplastic material; the friction law is a function of the temperature, Eq. (7.96) for viscoplastic friction; and the material density evolution is a function of the temperature rate, Eq. (7.89) for a viscoplastic material.
2. The heat equation itself depends on the temperature (for nonconstant thermal parameters) and includes a volume dissipation term, caused by plastic or

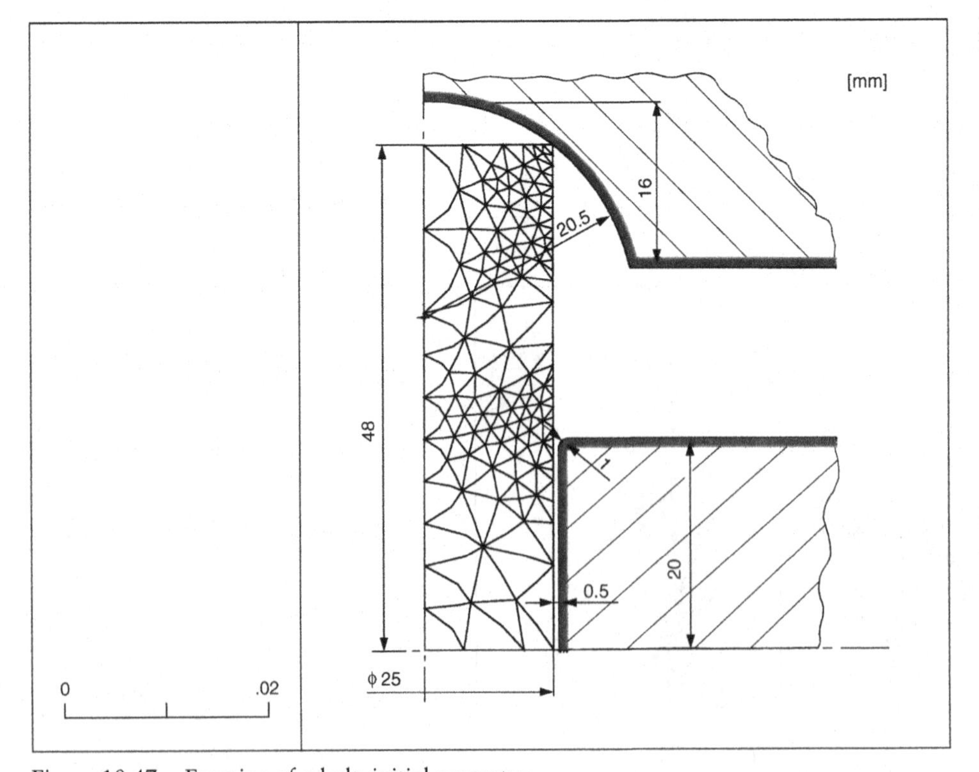


Figure 10.47. Forming of a bolt: initial geometry.

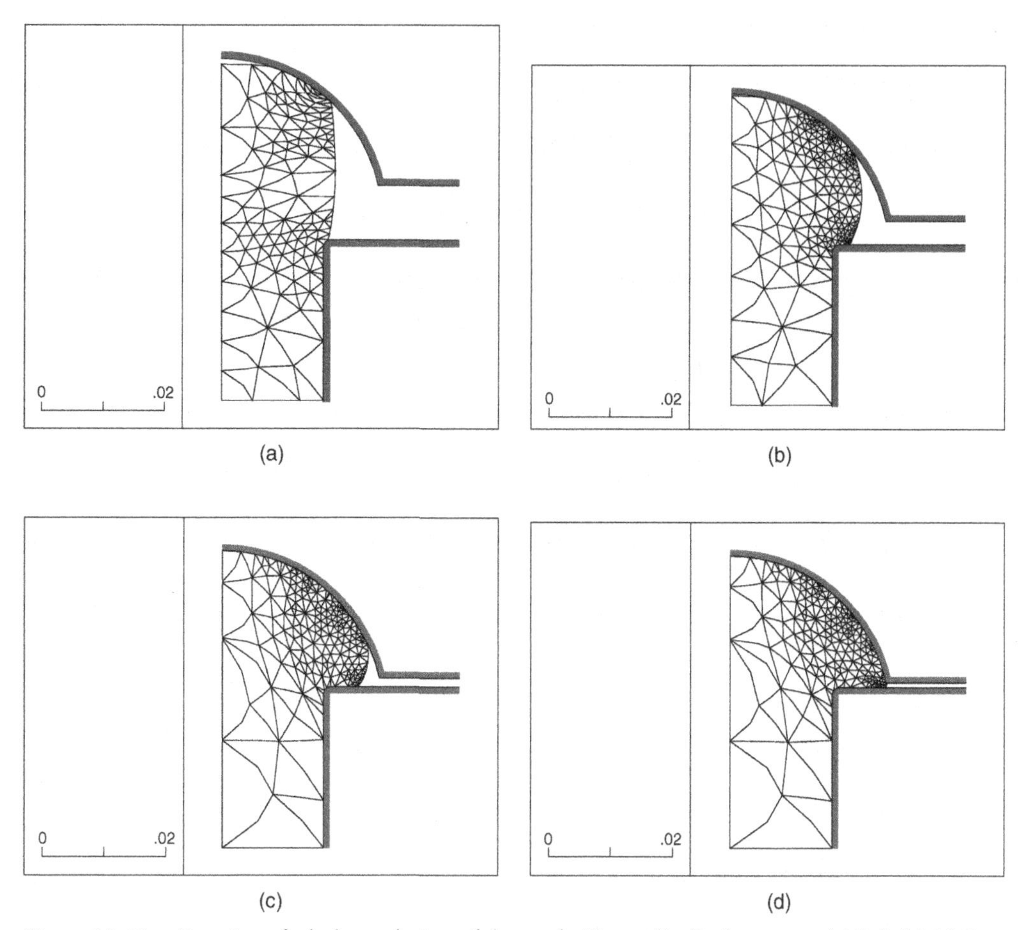


Figure 10.48. Forming of a bolt: evolution of the mesh. Upper die displacement: (a) 9.5, (b) 13.3, (c) 15.2, and (d) 15.8 mm.

viscoplastic work, Eq. (7.93); a surface dissipation contribution, corresponding to friction at the tool–part interface, Eq. (7.97); and several boundary conditions, describing convection, Eq. (7.18), or radiation, Eq. (7.19).

The finite-element discretization of the mechanical problem is performed in the usual way, except that the penalty term is replaced by

$$\int_{\Omega} \rho_p K[\mathrm{div}(\mathbf{v}) - 3\alpha \dot{T}]^2 \, d\mathcal{V} \tag{10.19}$$

in order to satisfy Eq. (7.89) in an average sense over each element.

The heat equation for deformable bodies is discretized in space and time according to Section 4.6. We finally obtain a set of nonlinear equations involving both the nodal velocity vector and the nodal temperature vector at the end of the increment as unknowns, as shown in Section 7.6. We recall that, theoretically, these equations can be solved directly by the Newton–Raphson method, performed on the whole set of equations and unknowns. Another method is to compute the velocity and the temperature alternately until convergence of the process. But most often it is sufficient to compute

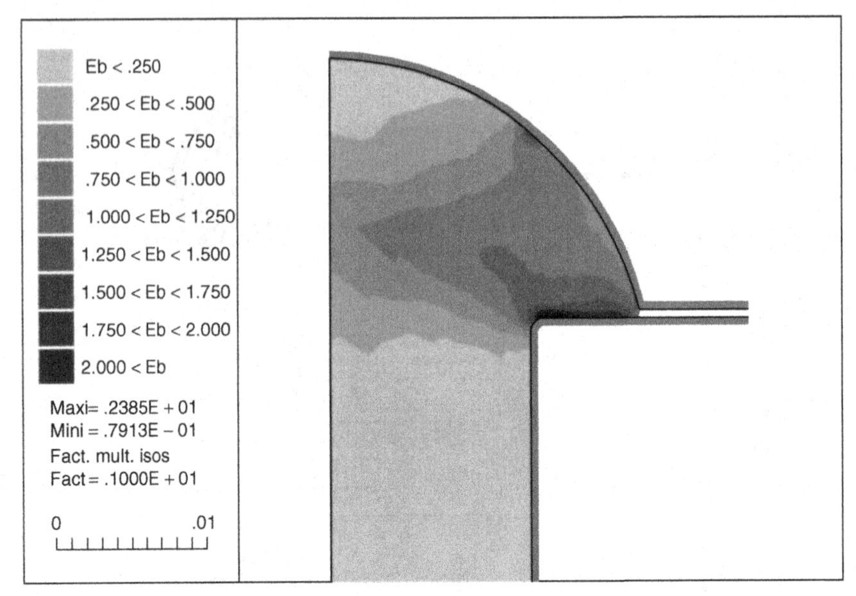


Figure 10.49. Forming of a bolt: isoequivalent strain.

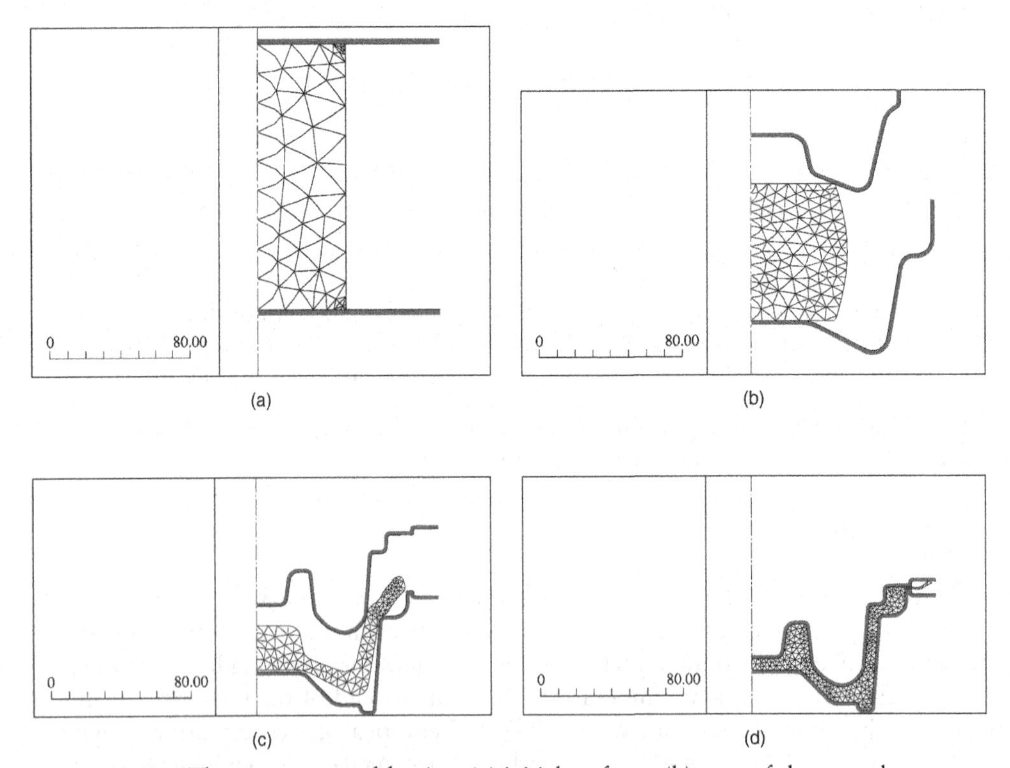


Figure 10.50. Three sequences of forging: (a) initial preform, (b) start of the second sequence, (c) start of the third sequence, and (d) end of forging.

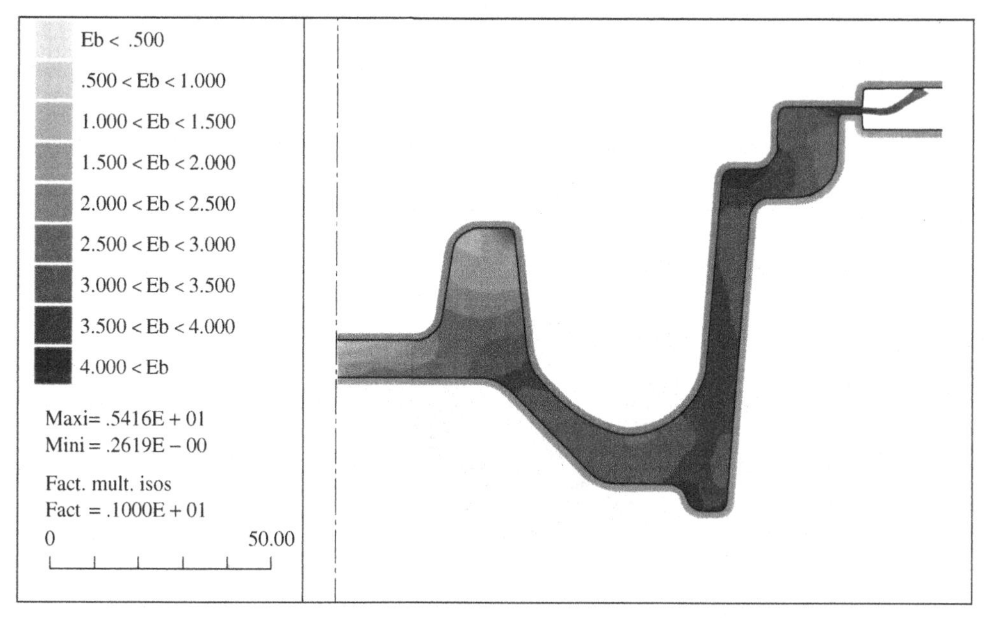


Figure 10.51. Map of isoequivalent strain after three sequences of forging.

first the velocity field, and the temperature field just after, without strong coupling; that is, without iteration during a time increment. The result is satisfactory when the time step is small enough, and we shall make use of this simplified scheme in the following.

### Example of Thermal Localization

The first example of forging with thermal coupling corresponds to the same part as that pictured in Fig. 10.26. The constitutive law is viscoplastic with a constant rate sensitivity index $m = 0.15$, while the consistency $K$ depends on temperature according to $K = 5.86 \times 10^3 \exp(2.814 \times 10^3/T)$ MPa $s^{0.15}$. The thermal parameters are $\rho = 7.8 \times 10^3$ kg m$^{-3}$, $c = 7 \times 10^2$ J/ kg/°C, and $k = 23$ W/m/°C.

The initial temperature of the part is assumed homogeneous and equal to 1200 °C, while the dies are supposed to remain at a uniform temperature of 200 °C. In Fig. 10.52 the map of isotemperature distribution is plotted: the maximum values lie in the zone of intense shear where the heat generation that is due to viscoplastic work and friction is greater than the loss of thermal energy that is due to contact with cool dies.

### Numerical Approach of Thermal Coupling with the Tools

The constant temperature approximation in the tools is rather crude, especially for relatively low speed forging when the time of contact between the hot part and the cold tools is long enough for significant heat exchange to take place. A more realistic approach consists of computing the temperature field in the part and in the tools. The heat exchange between the part and the tool that is due to the temperature difference must be introduced. This is done with the definition of a contact resistance $R$, so that

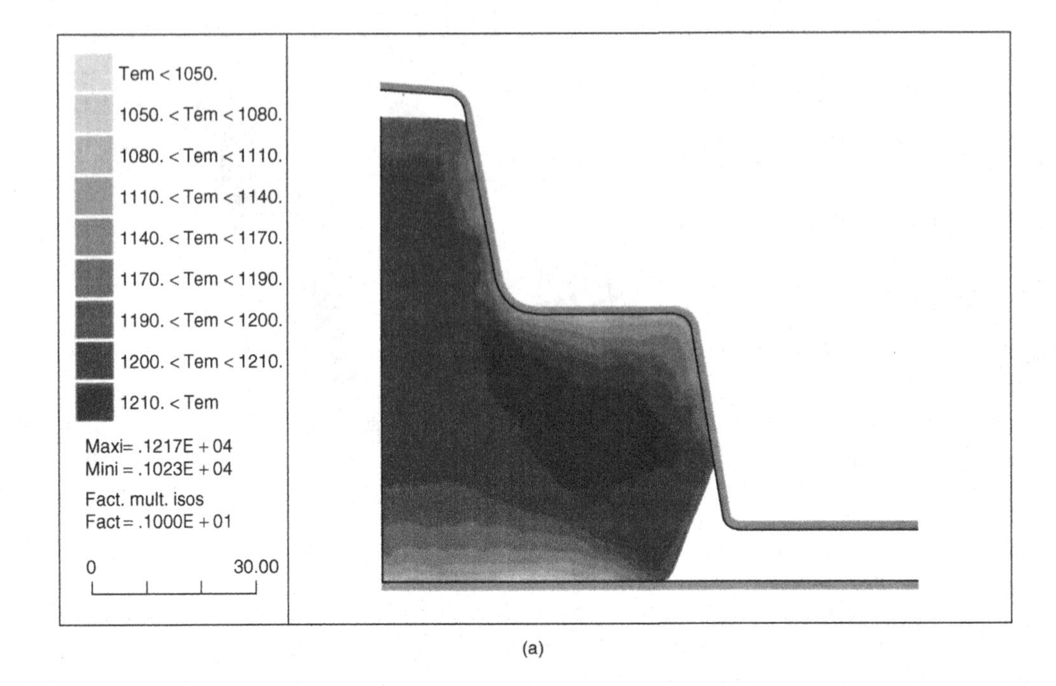


(a)

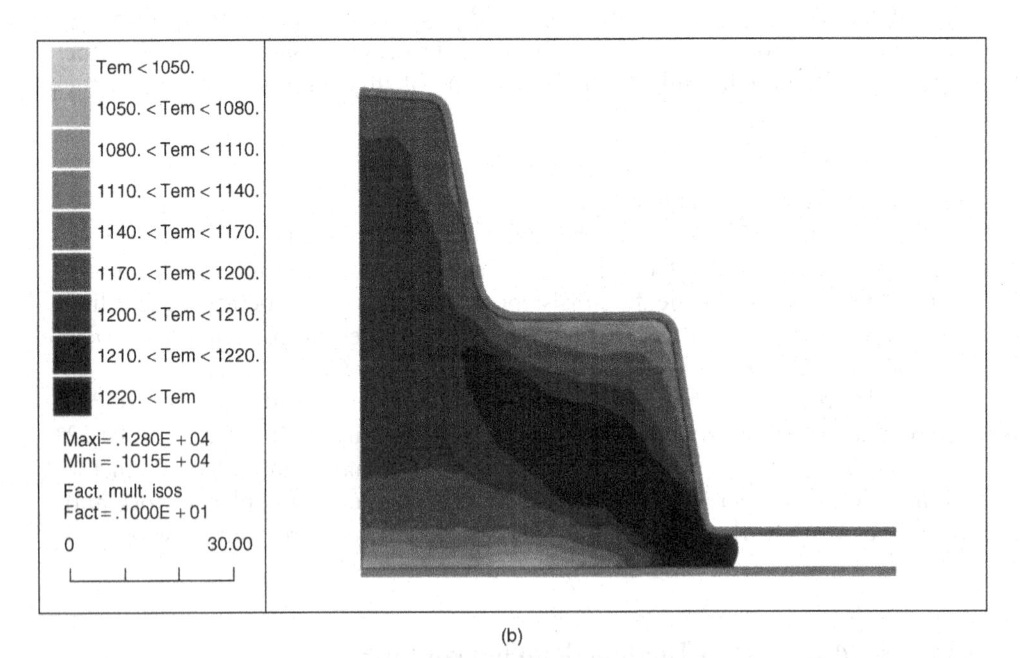


(b)

Figure 10.52. Temperature contours during forging of a two-dimensional complex part. Height of the tool: (a) 9 and (b) 4 mm.

the heat that flows throughout the interface from the tool to the part is given by

$$\phi_{t-p} = -k\frac{\partial T}{\partial n} = \frac{1}{R}(T - T_{\text{to}}), \tag{10.20}$$

where $R$ is the contact resistance, $n$ is the normal at the tool–part interface, and $T$ and $T_{\text{to}}$ are respectively the temperature of the part and of the tool at the interface. Obviously the opposite heat exchange contribution flows from the part to the tool.

The general methodology is to write the heat equation in each domain, that is, in the part and in the tools separately. A finite-element discretization and the Galerkin method are utilized to obtain a discrete set of equations. These equations are coupled as a result of the introduction of heat exchange between the domains, according to Eq. (10.20). It is then possible to solve them as a whole. Another method, which proved more efficient as far as computer time is concerned, is to consider each domain successively. The temperature field is computed in each domain, assuming that the temperature in the others remains constant for the calculation of the heat exchange contribution. When the temperature of the part and the temperature of the tools are very different, it is necessary to iterate the process until convergence.

**Example of Coupled Thermal Calculation in the Part and in the Tools**

The preform of the part, and the upper and the lower die are meshed according to Fig. 10.53. It is worthy to note the present formulation does not impose coincident nodes at the tool–part interface: the heat flux condition of Eq. (10.20) is interpolated at the integration points.

The material parameters for the part are chosen to represent the behavior of a super alloy: $K = 0.9526$ MPa $s^{-0.2738}$, $m = 0.2738$, $\beta = 5430$ °K, $\alpha = 0.32$, $p = 0.2738$, $\rho = 4.65 \times 10^3$ kg/m$^3$, $c = 8.95 \times 10^2$ J/kg/°C, $k = 17$ W/m/°C, and $b = 8.4 \times 10^3$ J/m$^2$/s$^{1/2}$/°C. The initial temperature of the part is $T = 920$ °C.

Both tools have the same thermal properties: $\rho_{\text{to}} = 7.5 \times 10^4$ kg/m$^3$, $c_{\text{to}} = 6.69 \times 10^8$ J/kg/°C, $k_{\text{to}} = 275.8$ W/m/°C, and $b_{\text{to}} = 1.18 \times 10^3$ J/m$^2$/s$^{1/2}$/°C. Their initial temperature is $T_{\text{to}} = 785$ °C.

On each free surface the same thermal parameters are used: $\alpha_c = 2 \times 10^4$ W/m$^2$/°C, $T_{\text{ext}} = 25$ °C, $\varepsilon_r = 0.7$. The contact resistance is $R = 1/\alpha_c$.

The velocity of the upper die is $v^{\text{to}} = 5 \times 10^{-3}$ m/s: this is a rather low forging velocity for which the influence of thermal coupling with the tool is more important. The temperature field at two steps of forging is pictured in Fig. 10.54.

We observe that the maximum temperature in the part is 940 °C, which is higher than the initial temperature, while the minimum temperature in the tools is 626 °C. For comparison, a similar computation was made with the hypothesis of a uniform and constant temperature of the tools equal to 785 °C. The results are plotted in Fig. 10.55: they show a very different temperature distribution in the vicinity of the symmetry axis. The difference in this zone is mainly due to the important heating of the tools in the coupled calculation. However, for this example, the differences on the metal flow, induced by the different temperature distributions, are very small and therefore can be neglected.

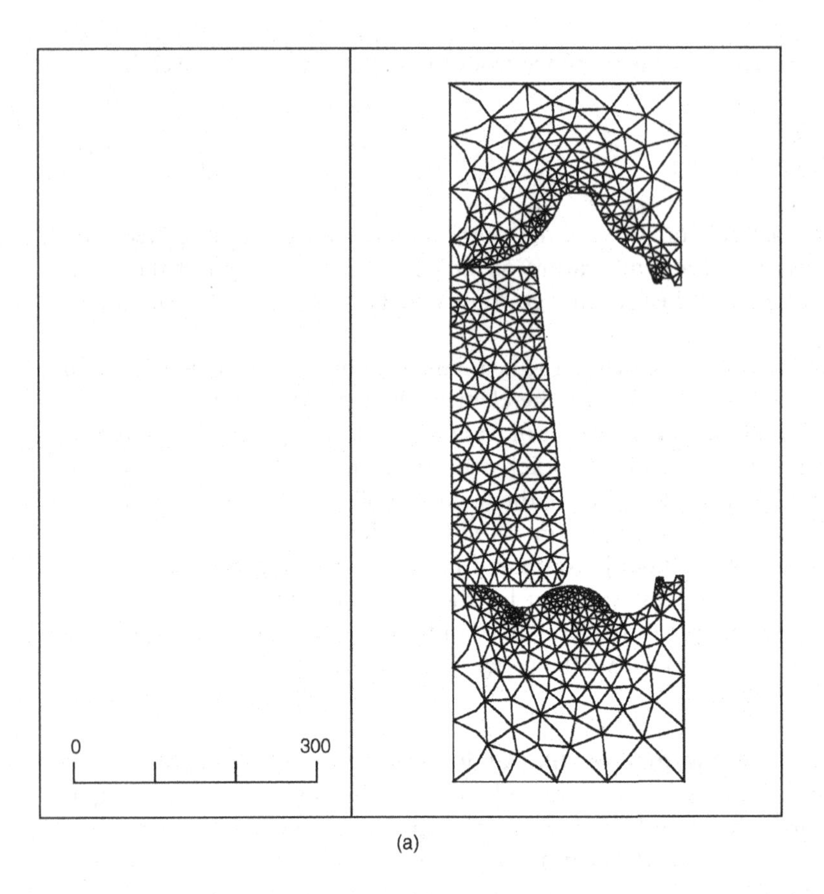


(a)

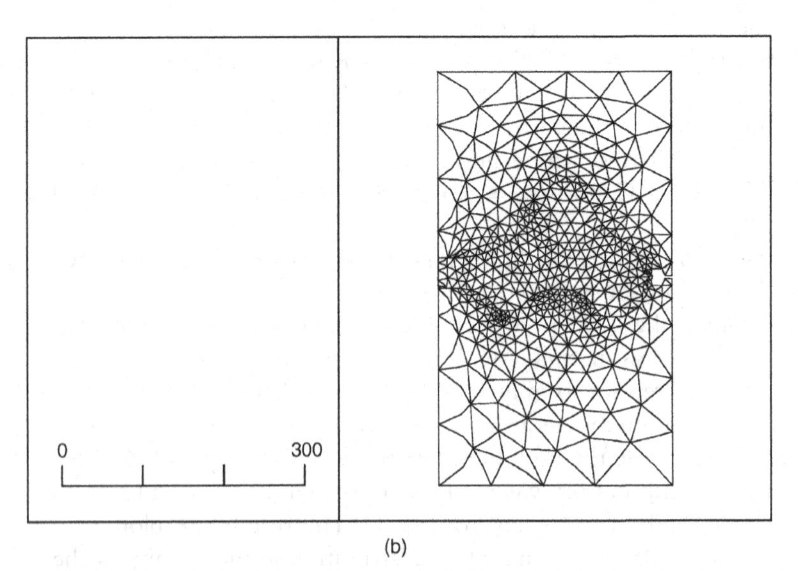


(b)

Figure 10.53. Mesh of the preform and of the tools: (a) before forging and (b) at the end of the process.

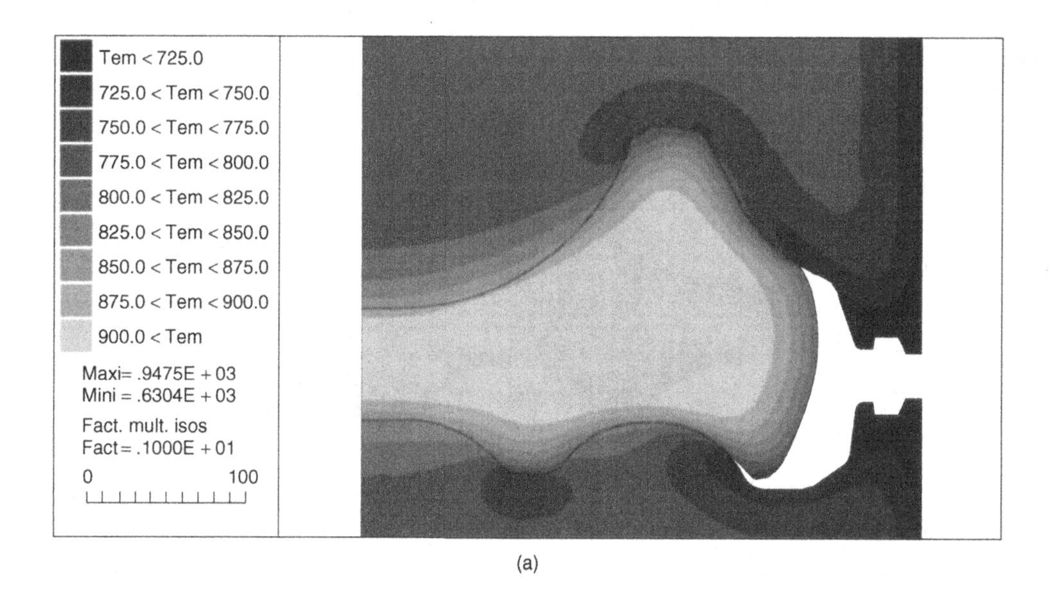


(a)

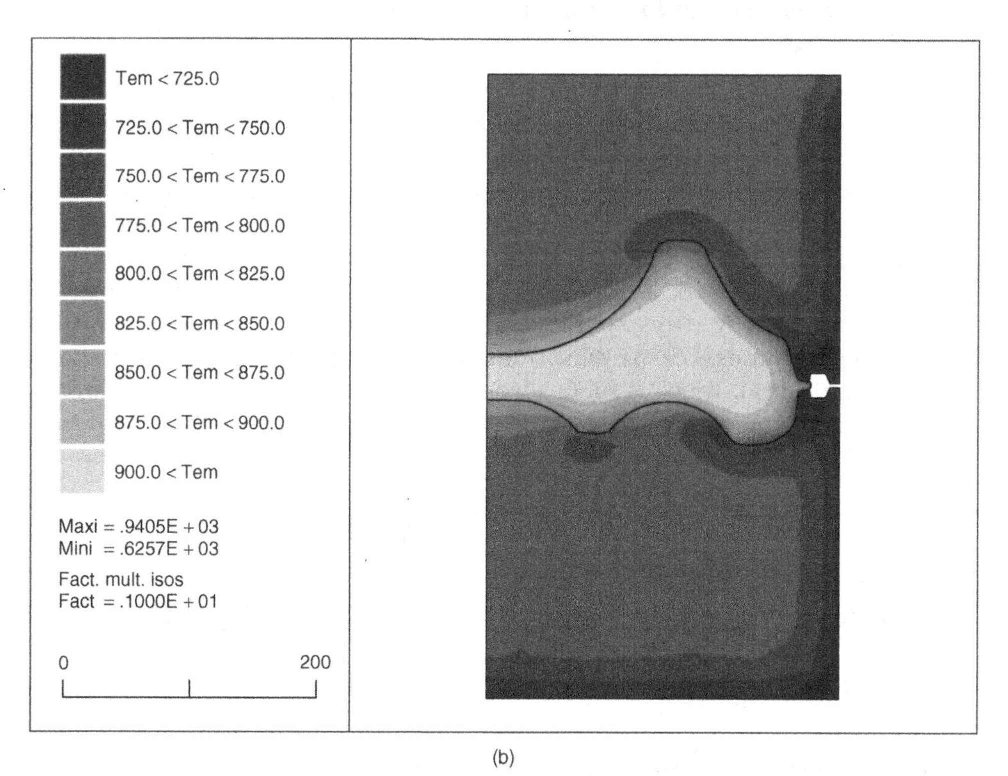


(b)

Figure 10.54. Coupled thermal computation in part and tools: (a) intermediate step (25 mm before the end) and (b) final configuration (T °C).

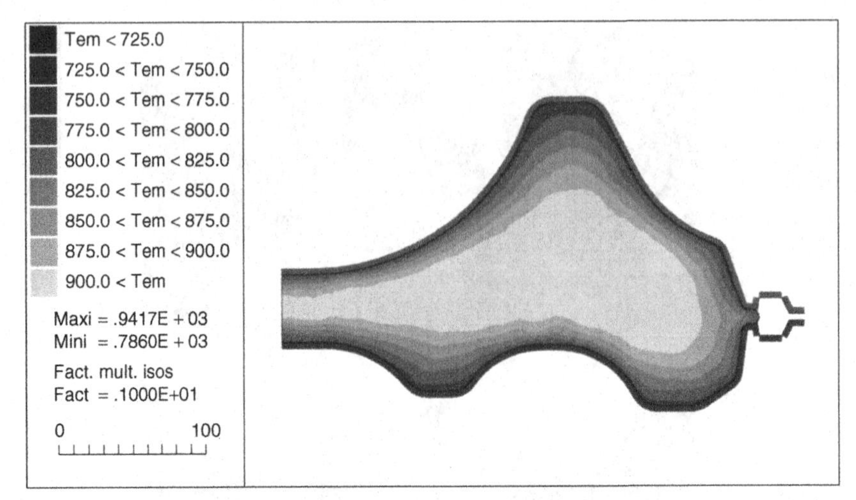


Figure 10.55. Noncoupled thermal computation (T °C).

## 10.5 Three-Dimensional Finite-Element Computation of Complex Parts

### Three-dimensional Elements for Incompressible Flows

The most widely used three-dimensional elements are the eight-node hexahedral elements with one reduced integration point for the incompressibility. However, despite several attempts, the remeshing problem is extremely difficult to solve in a proper way, and if one wants to use completely automatic remeshing, it appears almost impossible. For this reason tetrahedral elements seem to be more convenient. The quadratic ten-node tetrahedron can be used with one reduced integration point for the penalty term. Another way is to use a mixed formulation with special linear tetrahedrons. For the velocity discretization a node is added at the centroid of the tetrahedron, which is subdivided into four subtetrahedral, as indicated in Fig. 10.56. Doing that the velocity at the centroid of the tetrahedron is added as an unknown. The pressure field is discretized in the classical way in the initial linear tetrahedron.

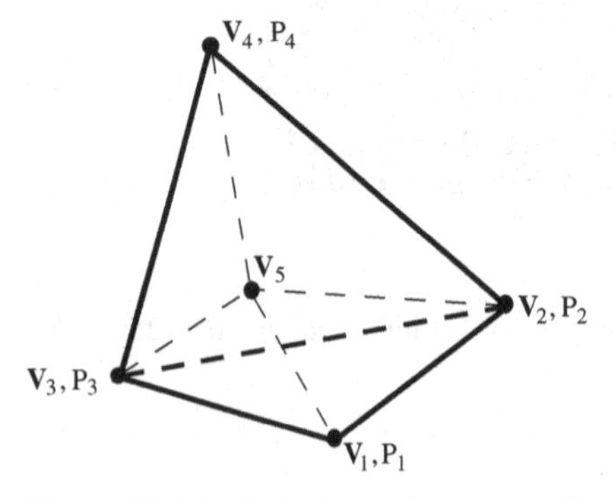


Figure 10.56. Velocity and pressure discretization in a tetrahedron. The local numbering of velocity and pressure is used.

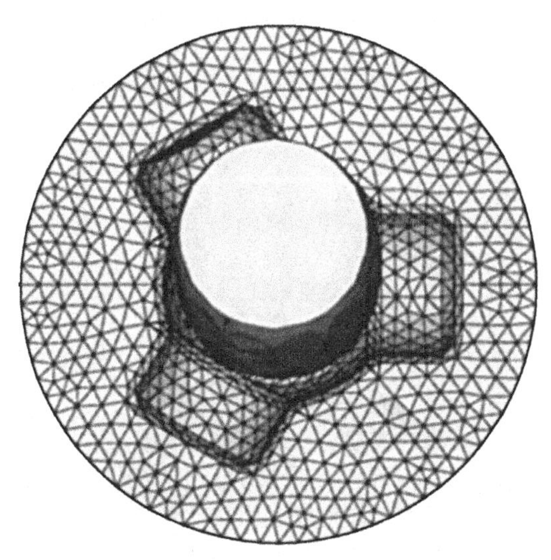

Figure 10.57. Forging of a part with three axles: initial configuration.

All of the following examples are computed with the mixed formulation, using the special tetrahedron described above. The calculations are purely mechanical, with material parameters corresponding to hot steel.

### Forging of a Part with Three Axles

The preform is a cylinder, which is represented with the surface mesh of the lower die in Fig. 10.57. The upper and lower dies are symmetrical with respect to a horizontal plane. During forging, a hemispherical hole is first formed in the center of the part, as can be seen in Fig. 10.58(a). At the end of forging the three axles are formed and a thin flash is produced, which corresponds to the excess material necessary in order to provide accurate filling of the die.

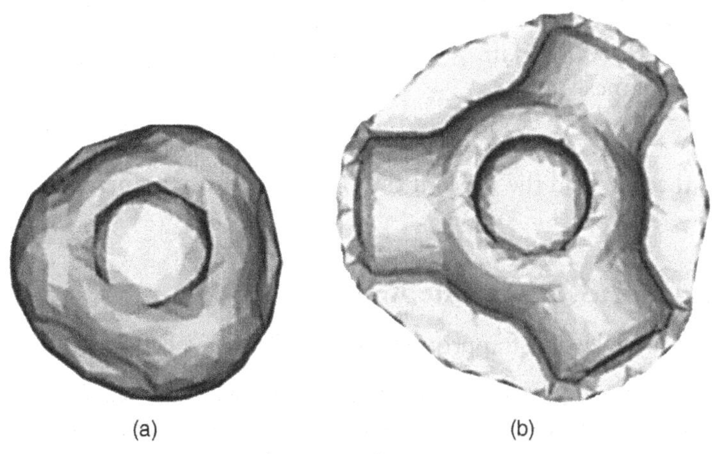


Figure 10.58. Forging of a part with three axles: (a) intermediate step of forging, and (b) final shape.

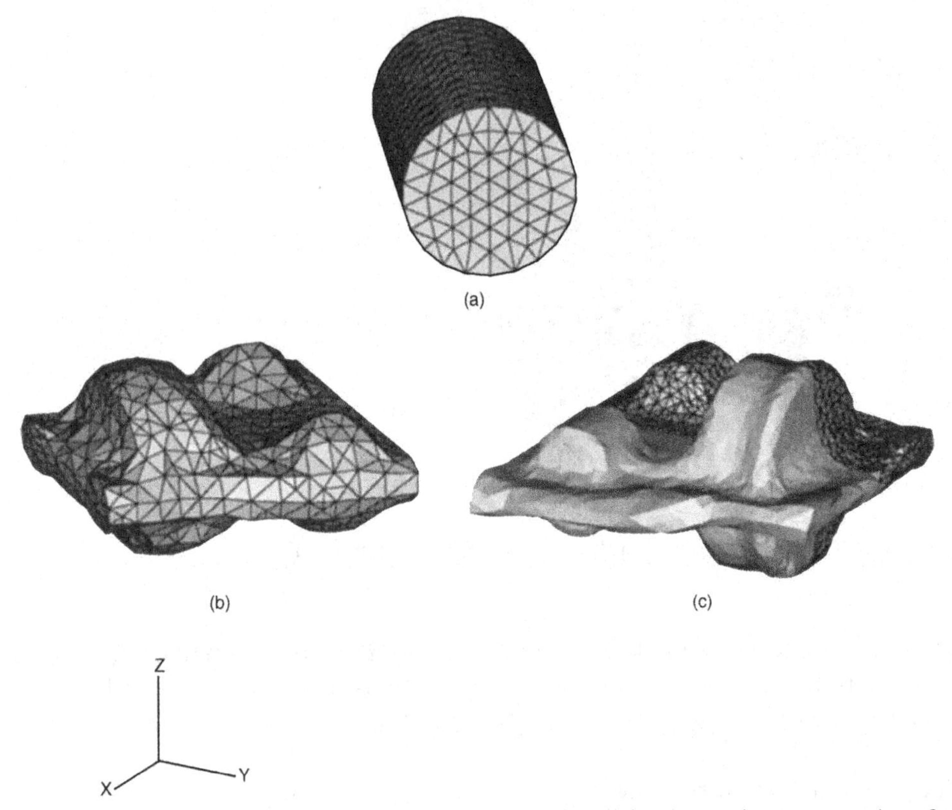


Figure 10.59. Forging of an automotive part: (a) preform, (b) intermediate step, and (c) final step.

**Forging of an Automotive Part**

This is a part used in the gear box that is obtained by forging a horizontal cylinder between two symmetrical dies as pictured in Fig. 10.59.

**Forging of a Cardan Joint**

The preform is again a cylinder that is forged between a punch and a die, the meshes of which are pictured in Fig. 10.60(a). Half of the part is drawn in Fig. 10.60(b), showing the shape and the final mesh.

**Forging of a Gear**

The final shape of half the gear and the mesh at the end of the forging operation is given in Fig. 10.61.

## 10.6 Elastoplastic and Elastoviscoplastic Analysis

For cold forging simulation, three levels of approximation can be utilized.

1. This is a purely viscoplastic approximation; that is, it neglects the elasticity effects, with a small enough rate sensitivity index $m = 0.01$–$0.05$ for usual metals.
2. An elastoplastic or elastoviscoplastic approach can be used during the whole process.

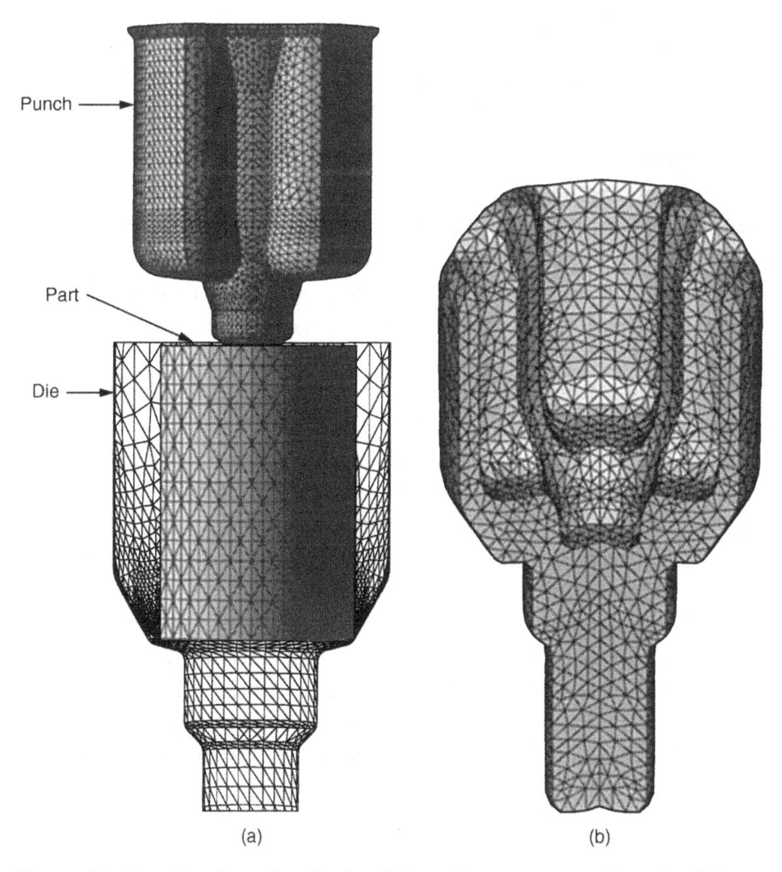


Figure 10.60. Forging of a Cardan joint: (a) geometry and mesh of the part and the tools, and (b) final part.

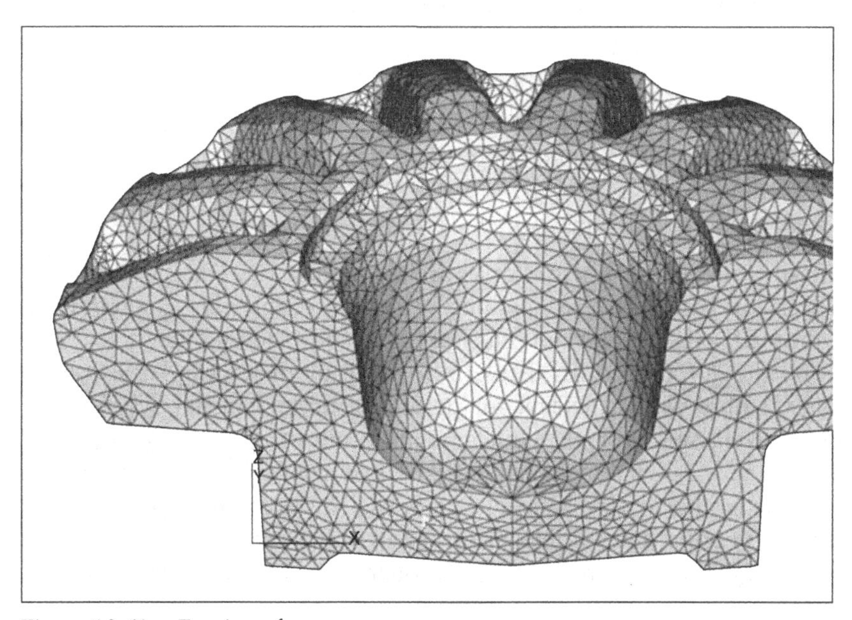


Figure 10.61. Forging of a gear.

3. An intermediate approach can be used, in which most of the forging process is computed with a purely viscoplastic behavior; the end of forging and tool removal are calculated with an elastoplastic or elastoviscoplastic law.

It was shown on several practical examples that methods 1 and 2 both give satisfactory results as far as the forging force and metal flow are concerned. However, the distribution of residual stress, which is of major importance for the prediction of the part life after forging, can be computed only by methods 2 or 3, which generally give similar results. When selecting an approximation, we must keep in mind that, for medium size cases, the introduction of elasticity often increases the computation time at least by a factor of 3 or 4.

### The Discrete Formulation

The classical incremental approach for the analysis of elastoplastic material was discussed in Section 5.3. For a more general scheme to be built, the general principles for the derivation of an incremental formulation for an elastoplastic or an elastoviscoplastic material were described in Section 5.6. Here we shall give only some details regarding the FE discretization of the elastoplastic problem.

At the beginning of each time increment, we suppose that the stress field $\sigma^t$ is known. Because of the integral formulation we utilize for the FE discretization, this stress field has to be defined and stored only at the integration points.[11] In the incremental formulation, it is usual to choose the incremental displacement $\Delta\mathbf{u}$ during the time interval between $t$ and $t + \Delta t$ as the principal unknown. With the usual small increment approximation, the incremental strain tensor is immediately deduced:

$$\Delta\varepsilon_{ij} = \frac{1}{2}\left(\frac{\partial \Delta u_i}{\partial x_j} + \frac{\partial \Delta u_j}{\partial x_i}\right). \tag{10.21}$$

The incremental elastoplastic constitutive equation is given by Eqs. (5.19)–(5.24) so that we can compute the stress at the end of the increment:

$$\sigma^{t+\Delta t} = \sigma^t + \Delta\sigma. \tag{10.22}$$

Three cases are considered according to Section 5.3:

- purely elastic loading a), or a′), unloading,
- elastoplastic loading b),
- transition from elastic to elastoplastic behavior c).

For the sake of illustration, all the previous cases are pictured in Fig. 10.62 in the deviatoric stress space. This latter choice is possible for incompressible plastic materials for which the hydrostatic pressure has no influence on plastic yielding.

If we summarize the different forms that can be taken by the elastoplastic constitutive equation, we observe that, except for purely elastic increments, when the strain increment is given, we have no explicit relation allowing us to determine the stress increment. The stress increment is then computed from the strain increment by solving the implicit nonlinear elastoplastic equations. This computation must be done at each integration point and can be performed by the Newton–Raphson procedure; the unknowns are the stress increment components and the plastic multiplier.

[11] Some kind of smoothing is only necessary for visualization of the isostress components.

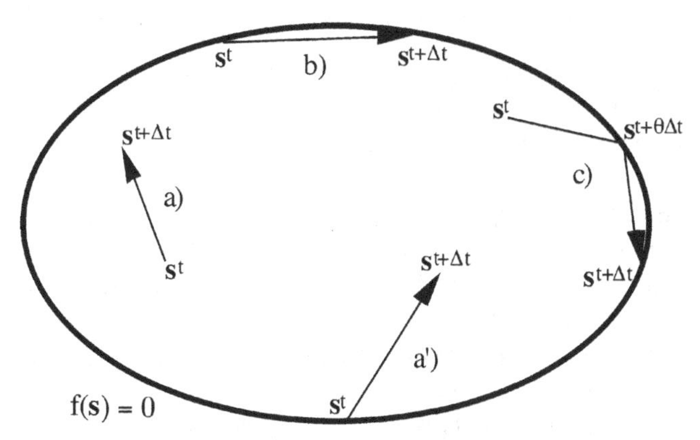


Figure 10.62. Incremental evolution of the deviatoric stress for an elastoplastic material.

### Finite-Element Discretization

The incremental displacement $\Delta\mathbf{u}$ is discretized in the usual way in terms of the nodal displacement vectors $\Delta\mathbf{U}$:

$$\Delta\mathbf{u} = \sum_n \Delta\mathbf{U}_n N_n. \tag{10.23}$$

Using the notations included in the introduction to Chapter 5, we can write the integral form of the equilibrium equation, Eq. (5.46), here:

$$\int_{\Omega_{t+\Delta t}} (\sigma^t + \Delta\sigma[\Delta\varepsilon(\Delta\mathbf{U}), \bar{\varepsilon}^t + \Delta\bar{\varepsilon}]: \mathbf{B}\,\mathrm{d}V - \int_{\partial\Omega^s_{t+\Delta t}} \mathbf{T}^{(t+\Delta t)d} \cdot \mathbf{N}\,\mathrm{d}S = 0. \tag{10.24}$$

This set of nonlinear equations is often solved by the Newton–Raphson method, which forces us to compute the matrix derivative of Eq. (10.23). The different contributions to this derivative are

- the derivative of the constitutive law with respect to the displacement:

$$\frac{\partial\Delta\sigma_{ij}}{\partial\mathrm{U}_{kn}} = \sum_{\lambda\mu} \frac{\partial\Delta\sigma_{ij}}{\partial\Delta\varepsilon_{\lambda\mu}} B_{\lambda\mu kn} + \frac{\partial\Delta\sigma_{ij}}{\partial\Delta\bar{\varepsilon}} \frac{\partial\Delta\bar{\varepsilon}}{\partial\Delta U_{kn}}, \tag{10.25}$$

- the terms corresponding to the domain and the boundary at time $t + \Delta t$,
- the derivative of $\mathbf{B}^{t+\Delta t}$.

Most often only the first contribution is introduced, as it is the major contribution, and even the second term on the right-hand side of Eq. (10.25) is often neglected. With the implicit formulation of the incremental constitutive law we have given in Eq. (5.23), it can be shown that the matrix derivative is symmetrical.

### The Remeshing Problem

A new problem arises for the elastoplastic computation when remeshing is necessary, which happens almost always when modeling forging operations. The difficulty comes from the necessity to save the previous state of stress. When the stress field is transported from an old mesh to a new one, we must perform numerical interpolations, and the stress state defined on the new mesh is slightly different from the initial one. Because of this supplementary level of approximation, the new stress state does not verify the integral form of the equilibrium equation on the new mesh. Moreover, it does not satisfy exactly the yield criterion in the regions of elastoplastic deformation on the old mesh. This problem is solved by an approximate remedy consisting of performing a fictitious elastoplastic increment, in order to obtain a new stress field, which will satisfy again all the mechanical equations.

**Three-Dimensional Example of Cold Forging**

A cold forging simulation of a complex part was carried out. The preform is a cylinder with a diameter of 55.7 mm and a height of 84 mm. The initial geometry of half the part and the surface of upper and lower tools can be visualized in Fig. 10.63.

The material parameters were selected to represent a classical carbon steel according to $E = 21 \times 10^4$ MPa, $\nu = 0.28$, and $\sigma_0 = 146 \times 10^6(1 + 500\bar{\varepsilon})^{0.25}$.

The friction law is a Tresca one with $\bar{m} = 0.1$, which corresponds to a normal lubrication in cold forging. The computation was made with ten-node tetrahedrons and was completed after six remeshing procedures. The final geometry of the part is displayed in Fig. 10.64(a), showing more refined elements in the zones of important geometrical evolution of the tools. The residual stress distribution is plotted in Fig. 10.64(b) in terms of the relative equivalent stress defined by

$$\bar{\sigma}^2 \doteq \frac{1}{\frac{2}{3}\sigma_0^2(\bar{\varepsilon})} \sum_{i,j} s_{ij}^2 .$$

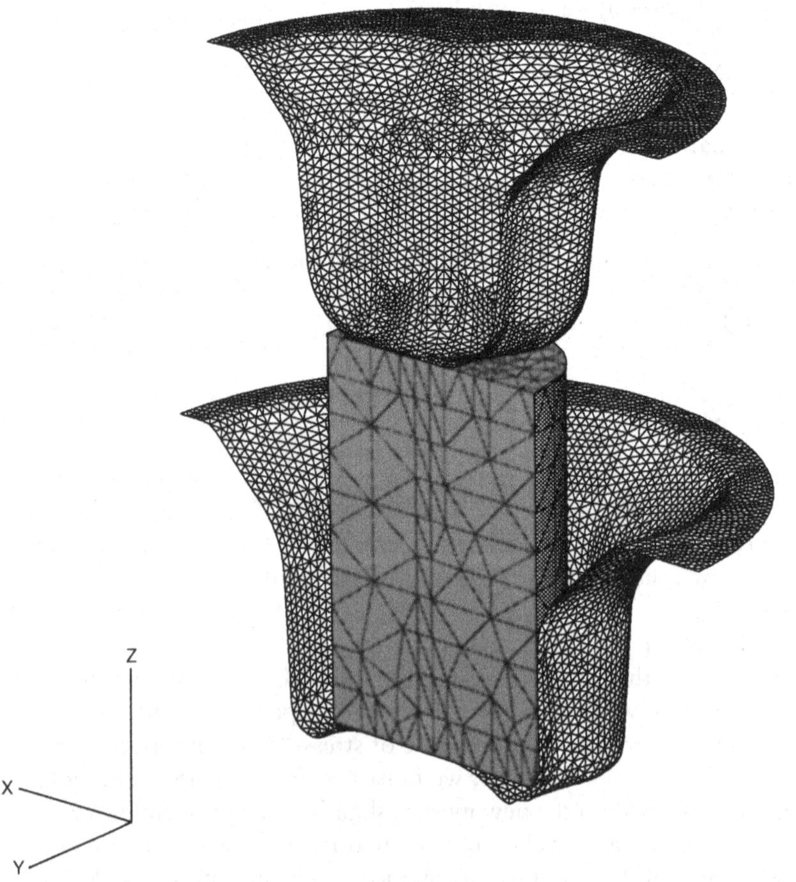


Figure 10.63. Mesh of half of the part and of the tools for three-dimensional cold forging.

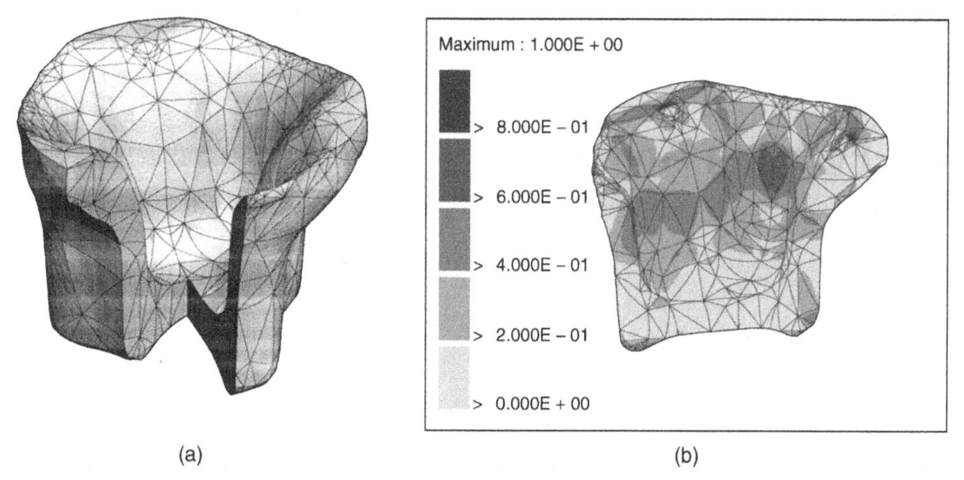


Figure 10.64. Simulation of three-dimensional elastoplastic cold forging: (a) predicted final geometry and (b) distribution of the relative equivalent residual stress.

## PROBLEMS

### A. Proficiency Problems

1. Adapt the flow chart of a FE forging code in Fig. 10.12 to take into account explicitly the following: strain hardening and thermal coupling.
2. Using the curves in Fig. 10.19 or 10.20, compute approximately the maximum forging force for the upsetting of a square block with initial height $h = 50$ mm and a length $L = 300$ mm for a reduction of height of 85% with a constant die velocity of 0.1 m/s. The material is viscoplastic with coefficients of

   $$K = 40 \times \text{MPa s}^{-m}, \quad m = 0.2, \quad \alpha = 0.3, \quad p = 0.2.$$

### B. Depth Problems

3. Consider the plane-strain upsetting of an initially square block as defined in the first example of Section 10.3. We consider first a frictionless contact for which the strain rate will be homogeneous, so that the material parameters are

   $$K = 1, \quad m = 0.1, \quad \alpha = 0.$$

   Find a linear velocity field that satisfies the symmetry of the problem and the conditions on the flat dies. Compute the viscoplastic dissipated energy and deduce the forging force as a function of the reduction in height. Compare the result with the corresponding curve in Fig. 10.6. If we suppose that the same velocity field is assumed when the friction coefficients are

   $$\alpha = 0.3, \quad p = 0.1,$$

   compute again the forging force and compare it with the appropriate curve in Fig. 10.6. Explain the difference.
4. We consider the example of full thermal coupling of Section 10.5, including the heat exchange between the part and the tool. Explain why the minimum temperature in the part is higher than the temperature that is computed when the temperature of the tools is assumed uniform and constant.

CHAPTER ELEVEN

# Sheet-Forming Analysis

Sheet-metal forming[1] is one of the most important manufacturing processes for mass production in industry. It is inexpensive in large quantity production, but a great deal of time and expense are devoted to the design and production of reliable stamping dies. Since the mid-1970s, important advances have been made in the development of computer simulation codes for deformation modeling during sheet-metal forming.[2]

Typical sheet-forming operations carried out in standard presses are characterized by quasi-static loading,[3] principal loading by in-plane tension (with some bending influence), irregularly shaped and continuously changing contact surfaces, and large relative movements of material on the contacting surfaces. Nearly all FEM nodes in the sheet mesh will come into contact with the tooling at some stage of the simulation. (This contrasts with forming or bulk forming, in which the majority of mesh nodes are internal.) Therefore, sheet forming is in general much more dependent on friction and contact than most other forming operations. It is essential to handle contact and friction accurately and consistently in sheet-forming analysis; otherwise, large errors can accumulate. Furthermore, contact algorithms must be sufficiently stable that small perturbations do not induce numerical divergence and instability into the techniques.

Current numerical research focuses on whole-part analysis by FEM. The principal obstacles remain stability, time, knowledge of physical boundary conditions and friction, and limited experimental verification. While contact is more complex for sheet forming, the process offers some simulation advantages in terms of the possibility of neglecting through-thickness stresses and, in some cases, bending stresses. Experimental measurements of strains can be performed throughout the simulation domain. These characteristics produce a more severe test of simulation techniques than the usual bulk-forming comparisons of shapes and boundary strains only.

[1] Chapter 11 was written in draft form by Dr. Dajun Zhou, Department of Materials Science and Engineering, The Ohio State University. It is based on a series of papers by R. H. Wagoner and co-workers as cited throughout.

[2] R. H. Wagoner and M. J. Saran, "Finite Element Modeling of Sheet Stamping Operations," in *Material Innovations and Their Application in the Transportation Industry, ATA MAT '91* (Assoc. Ind. Metall. Meccd. Affini, Torino, Italy, 1991), pp. 12–22.

[3] That is, inertial forces may be neglected without loss of accuracy.

## 11.1 Overview

The development of a particular formulation for sheet-forming analysis is presented in this chapter, based on two FEM programs developed at the Ohio State University: SHEET-S and SHEET-3. Examples of simulations carried out with various evolutions of these programs will be presented, but it is useful to first put these programs in the context of other developments in other laboratories.

Sheet-forming FEM codes generally are based on an updated Lagrangian scheme in which nodal displacements, velocities, or accelerations are the primary variables. Programs are often grouped into "implicit" and "explicit" categories,[4] depending on the solution algorithm used at each time step. A more precise terminology makes use of the qualifiers "static" vs. "dynamic" in conjunction with "implicit" and "explicit." (See Chapter 5 for a more thorough presentation.) Even among these simple divisions, there are many alternate approaches; however, it is useful to focus on the mainstream decisions that are followed by several commercial and research programs. Other classification schemes are by element type and by constitutive equation.

### Static Implicit

This is the traditional approach to nonlinear, quasi-static problems. Formulation of the equilibrium problem at a given time involves finding a set of nonlinear equations corresponding to each degree of freedom at each node. This set of nonlinear equations is linearized for solution typically by a Newton–Raphson technique, which involves defining derivatives at a current trial state and updating repeatedly until either the force residual or the magnitude of the update variable (or both) falls below some tolerance. Each successive linear set is solved by either a direct solver (such as Gaussian reduction), or an alternative. Once convergence is obtained, the time is incremented, the new boundary conditions are applied, and a new solution is sought.

#### *Characteristics*

The static implicit approach may be made as accurate as required (in principle) in a given step, limited only by the digital accuracy of the computer (which is usually not an issue). Within the basic spatial and temporal discretization inherent to the FEM, there are no further systematic sources of error. There are three major drawbacks to this approach: (1) it may be impossible to find a converged solution; (2) for large problems, the storage of the large matrix may be troublesome; and (3) the time for solving the linear equations increases rapidly for large problems (roughly proportional to $n^3$, where $n$ is the number of degrees of freedom). The convergence often depends on finding a trial solution at each time step sufficiently close to the real solution. Usually the solution at the previous time step is used as the trial solution.

#### *Variations*

Other techniques may be used to solve the sets of linear equations to avoid the time and storage problems of large problems. Typically, computation times favor direct solvers for problems with fewer than a few thousand degrees of freedom,

[4] N. Rebelo, J. C. Nagtegaal, L. M. Taylor, and R. Passmann, "Comparison of Implicit and Explicit Finite Element Method in the Simulation of Metal Forming Processes," in *NUMIFORM92*, J-L. Chenot, R. D. Wood, and O. C. Zienkiewicz, eds. (A. A. Balkema, Rotterdam, 1992), pp. 99–108.

while iterative solvers are more efficient for larger problems. In some velocity-based implementations, convergence may be based on a "rate-of-equilibrium" (i.e., only $dF/dt = 0$ within a given tolerance, rather than $F = 0$). In these formulations, force equilibrium cannot be guaranteed over many time steps because the solution may drift away. This situation can be improved by checking true equilibrium at each step or after several steps and then correcting by adding fictitious forces such that the total force acting at each node is zero.

### Static Explicit

The static explicit approach is very similar in concept to the static implicit one; the principal difference is that iteration is not performed at each time step. The quasi-static equilibrium equation is formulated again by a set of linear equations that are solved by direct or iterative solver. There is no update or iteration based on the current solution at a given time; thus the solution is one given "explicitly" from the condition at the previous time step. Although the formulation is usually quite different, one can envision the static explicit method as the static implicit one in which only a single Newton–Raphson iteration is carried out at each time step, starting from the trial solution from the previous time step. A simple variant could be envisaged in which, instead of a single linear solution, a fixed number of iterations was specified. In either case, the program can always proceed forward in time without concern about convergence.

#### *Characteristics*

The static explicit scheme eliminates the need to worry about convergence. For example, in some simulations there may be a point of real instability or bifurcation that cannot be solved by normal implicit methods because the solution is not unique or doesn't exist. The explicit scheme will proceed through this point. The penalty is that equilibrium is not checked, so that errors may accumulate. In fact, if a small error is introduced, the solution can rapidly deteriorate. The static explicit method has the same numerical drawbacks as the static implicit one in that large computation times and storage requirements accompany problems requiring many degrees of freedom.

#### *Variations*

This method is used infrequently, but it is very robust when used with sufficiently small time steps and careful checking for nonphysical answers. It is possible to introduce limited iteration, of the contact condition for example, into the procedure. Also, it is possible to use a predictor-corrector approach to adjusting the solution at each step or at certain steps to reflect total equilibrium.

### Dynamic Explicit

Originally intended for problems dominated by inertial effects, the dynamic explicit method has recently found favor for simulating sheet-metal forming operations, which are generally quasi-static. In this approach, the force imbalance at each node is computed from adjacent nodal positions by means of elemental, quasi-static, stiffness matrices. The acceleration is computed by using Newton's law and a lumped-mass matrix, which, being diagonal, allows node-by-node calculation. The computed acceleration and previous acceleration, velocity, and position are used to obtain the updated quantities. Because the calculation is carried out element by element and node by node, there is no need to formulate, store, or solve a set of equations.

### *Characteristics*

As with the static explicit technique, on one hand, very small steps should be taken so that nonlinearities of geometry and material do not corrupt the step-by-step answer. On the other hand, the solution time per step is much lower than that for solving the full set of static equations. While static solution times rise rapidly with the number of degrees of freedom, dynamic explicit solution times rise more slowly. Therefore, for problems with a large enough number of variables, the dynamic explicit method should always be more efficient than direct solution of static implicit equations. However, the accuracy is generally not known. In fact, explicit solutions that require solving for equilibrium are generally much slower than implicit solutions.

Unfortunately, there doesn't seem to be any mathematical proof available to demonstrate that the dynamic explicit solution approaches the static solution for small time steps. However, many benchmark results have shown similar real-world accuracy in the two techniques, presumably a result of careful choice of tuning parameters (damping, mass multiplication, speed-up) and because many sheet-forming problems are dominated by surface contact conditions.

### *Variations*

As with other sheet-forming simulations, the contact and friction implementation can dominate the solution characteristics. In the basic implementation of dynamic explicit methods, a penalty approach is used to apply a fictitious force to nodes in contact (or at least those that penetrate the tooling). The details of this procedure are critical to the stability and accuracy of the method. Some recent advances use a locally implicit contact implementation within the dynamic explicit framework to make a more regular and stable constraint because of contact.

## Element Types

Several finite elements were formulated in Chapter 4. For a sheet geometry, with one very small dimension relative to the other two characteristic lengths, profitable approximations can be made. In-plane tension and out-of-plane bending are the two most common deformation patterns of sheet-metal forming processes. There are three basic element types used in sheet-forming numerical modeling: membrane elements, shell elements, and solid elements.

### *Membrane Elements*

These neglect bending effects by heat treating stress and strain as independent of position through the thickness. This gives great efficiency in computation and adequate approximation for most stretching types of deformation with a large ratio of tool radius over sheet thickness. However, for a ratio less than 6 for stretch-type operations (and greater ones for loosely constrained draw), the accuracy of the membrane theory may deteriorate.

### *Shell Elements*

These allow a simple variation in shape and an assumed linear dependence of strain on position through the thickness. Additional degrees of freedom corresponding to the tangents at the nodes are usually introduced, often with a large penalty in computation time (typically 3–20 times greater than a similar membrane calculation), both because of the increased degrees of freedom and also because of the need to use smaller steps for numerical stability. Depending on the formulation, shell elements may have a limitation of radius of curvature, perhaps of the order of $r/t > 3$.

#### *Three-Dimensional Solid Elements*

These have been little tested as yet in large-scale sheet-forming analyses because of the huge time penalties. (Some two-dimensional models, such as for drawbeads, are the exception.) Solid elements allow contact on both sides of the sheet with different tools, with through-thickness compression, and allow complicated interactions between surface shear (friction) stresses with bending. The usual shell assumption that the plane section remain plane is thus relaxed. Many commercial programs have suitable three-dimensional elements that may be tried for small problems, but the need to provide several rows of elements through the sheet thickness, and to have element aspect ratios near 1, effectively limit their usefulness for whole-part calculations of sheets with today's computer speeds. In addition, three-dimensional elements can introduce spurious deformation patterns (e.g., hourglassing) that require special approaches. However, there is no inherent limit to the thickness, radius of curvature, or other geometric quantity beyond the time of the computation.

### Constitutive Equations

According to the constitutive equations used in process modeling, the finite-element method can be grouped into typical categories; these are rigid-plastic analysis and elastic-plastic analysis. Either approach may incorporate viscous effects by means of a flow stress that depends on strain rate, although this modification is more readily made in the rigid-plastic formulation.

#### *Elastic-Plastic Programs*

These can model both the elastic and the plastic response of materials, with the possibility to predict springback, which is mainly an elastic bending effect upon unloading. The penalty is that the elastic-plastic constitutive equation is usually formulated in a rate form and small time steps must be taken in order to make a good approximation of the elastic behavior, particularly at the transition between loading and unloading in each element. Estimates show that elastic-plastic programs require a CPU time that is longer by a factor of 3–20 than that of rigid-plastic programs to obtain a result for the same problem. For problems dominated by plastic deformation (positive loading), there is little difference in simulation results.

#### *Rigid-Plastic Programs*

These make use of a plastic stress–strain curve that begins at zero strain. Throughout the calculation, any elastic effects are ignored. This approach allows for significant improvements in efficiency because no abrupt transitions in the constitutive equation require small time steps and careful monitoring. However, rigid-plastic programs cannot handle springback, and they can become inaccurate when the problem is dominated by very small strains in large areas of the sheet. Although this is not usual in standard sheet forming, one can imagine a slender beam, which has very significant deflections and shape changes but which has very small strains, that is dominated by elastic effects.

It is possible, by numerical modifications, to combine characteristics of the two approaches. For example, elastic-plastic programs can treat the transition point and elastic response in an approximate way in order to speed up the computation and allow larger time steps. Conversely, rigid-plastic programs can introduce an approximate elastic response at each time step to allow convergence for unloading cases.

## 11.2 Elements Used in SHEET-S and SHEET-3

We now begin to outline the procedures and algorithms used in SHEET-S and SHEET-3: static implicit, rigid viscoplastic, finite-element programs for two- and three-dimensional analyses of sheet-forming operations.

### Plane-Strain Line Element

For a section analysis like that performed by SHEET-S, a plane-strain line element[5] is the simplest one, as shown in Fig. 11.1. The longitudinal strain increment, $\Delta\varepsilon_1$, during the incremental time step can be written as

$$\Delta\varepsilon_1 = \ln(l/L), \tag{11.1}$$

where $L$ and $l$ are respectively element lengths in the current and subsequent configurations, expressed in the $x$–$z$ plane in terms of nodal coordinates, as defined in Fig. 11.1:

$$L^2 = (X_2 - X_1)^2 + (Z_2 - Z_1)^2, \tag{11.2}$$

$$l^2 = (x_2 - x_1)^2 + (z_2 - z_1)^2. \tag{11.3}$$

The nodal coordinates are related by the incremental displacement, $\Delta\mathbf{u}$.

$$\mathbf{x} = \mathbf{X} + \Delta\mathbf{u} \tag{11.4}$$

or

$$\begin{pmatrix} x_1 \\ z_1 \\ x_2 \\ z_2 \end{pmatrix} = \begin{pmatrix} X_1 \\ Z_1 \\ X_2 \\ Z_2 \end{pmatrix} + \begin{pmatrix} \Delta u_1 \\ \Delta u_2 \\ \Delta u_3 \\ \Delta u_4 \end{pmatrix}. \tag{11.5}$$

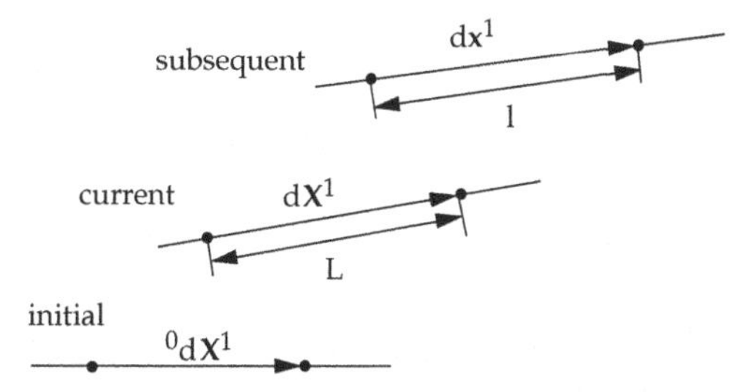


Figure 11.1. Definition of material line vectors at time $t = 0$, $t = t_0$, and $t = t_0 + \Delta t$ for a line element.

[5] S. Kobayashi and J. H. Kim, "Deformation Analysis of Axisymmetric Sheet Metal Forming Processes by the Rigid-Plastic Finite Element Method," in *Mechanics of Sheet Metal Forming*, D. P. Koistinen and N. M. Wang, eds. (Plenum, New York, 1978), p. 341.

The derivative of the strain increment, $\Delta\varepsilon_1$, with respect to the incremental displacement $\Delta\mathbf{u}$, is needed to find the internal force for equilibrium. The required form can be obtained from Eqs. (11.4) and (11.5) as follows:

$$\frac{\partial\Delta\varepsilon_1}{\partial\Delta\mathbf{u}} = \frac{1}{l}\frac{\partial l}{\partial\Delta\mathbf{u}} = \frac{1}{l^2}\begin{bmatrix} x_1 - x_2 \\ z_1 - z_2 \\ x_2 - x_1 \\ z_2 - z_1 \end{bmatrix}. \tag{11.6}$$

In plane-strain analysis, and either rigid plasticity (or ignoring elastic volume change), the strains are simply related: $d\varepsilon_1 = -d\varepsilon_3$, $d\varepsilon_2 = 0$. For any yield function with isotropic hardening, we can write

$$\Delta\bar{\varepsilon} = F_{ps}\Delta\varepsilon_1, \tag{11.7}$$

where we define $F_{ps}$ as the ratio of effective strain increment to the major strain increment as derived from the given yield function. By virtue of Eqs. (11.6) and (11.7), therefore, the derivative of the effective strain increment with respect to nodal incremental displacement is expressed as follows:

$$\frac{\partial\Delta\bar{\varepsilon}}{\partial\Delta\mathbf{u}} = \frac{\text{sign}(\Delta\varepsilon_1)}{l^2}F_{ps}\begin{bmatrix} x_1 - x_2 \\ z_1 - z_2 \\ x_2 - x_1 \\ z_2 - z_1 \end{bmatrix}, \tag{11.8}$$

where the sign has a value of $+1$ for $\Delta\varepsilon_1 \geq 0$ and $-1$ for $\Delta\varepsilon_1 \leq 0$. Similarly, differentiation of Eq. (11.8), needed by the element stiffness matrix, with respect to the incremental displacement, $\Delta\mathbf{u}$, yields

$$\frac{\partial^2\Delta\bar{\varepsilon}}{\partial\Delta\mathbf{u}\,\partial\Delta\mathbf{u}} = \frac{\text{sign}(\Delta\varepsilon_1)}{l^4}F_{ps}\begin{bmatrix} K_{11} & K_{12} & K_{13} & K_{14} \\ K_{12} & K_{22} & K_{23} & K_{24} \\ K_{13} & K_{23} & K_{33} & K_{34} \\ K_{14} & K_{24} & K_{34} & K_{44} \end{bmatrix}, \tag{11.9}$$

where

$$\begin{aligned}
K_{11} &= l^2 - 2(x_2 - x_1)^2,\\
K_{12} &= -2(x_2 - x_1)(z_2 - z_1),\\
K_{13} &= -l^2 + 2(x_2 - x_1)^2,\\
K_{14} &= 2(x_2 - x_1)(z_2 - z_1),\\
K_{22} &= l^2 - 2(z_2 - z_1)^2,\\
K_{23} &= 2(x_2 - x_1)(z_2 - z_1),\\
K_{24} &= -l^2 + 2(z_2 - z_1)^2,\\
K_{33} &= l^2 - 2(x_2 - x_1)^2,\\
K_{34} &= -2(x_2 - x_1)(z_2 - z_1),\\
K_{44} &= l^2 - 2(z_2 - z_1)^2.
\end{aligned}$$

### Constant Strain Triangular Membrane Element

The constant strain triangular membrane element[6] is commonly used for three-dimensional FEM simulations, as discussed in Chapter 4. More details will be

[6] Y. Germain, K. Chung, and R. H. Wagoner, "A Rigid-Viscoplastic Finite Element Program for Sheet Metal Forming Analysis," *Int. J. Mech. Sci.* 31, 1–24 (1989).

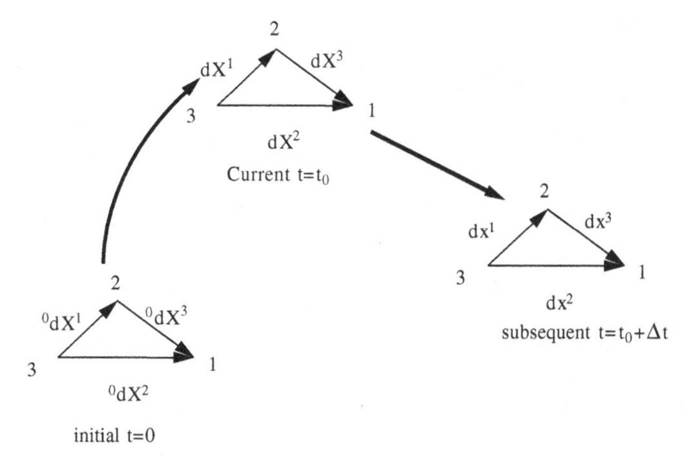


Figure 11.2. Definition of convected material vectors at times 0, $t$, and $t + \Delta t$ for a triangular element.

presented here specifically for a sheet-metal forming application, as implemented in SHEET-3. The initial, current, and subsequent configurations of the sheet correspond to the time $t = 0$, $t = t_0$, and $t = t_0 + \Delta t$ during the deformation, as displayed in Fig. 11.2.

The deformation during this incremental period is small, but not infinitesimal; that is, it is sufficiently large to require some aspects of large deformation kinematics. Considering an updated Lagrangian description of motion of the continuum with respect to a fixed system of rectangular Cartesian coordinates, we find that Eq. (11.5) becomes, in three dimensions,

$$\begin{Bmatrix} x_1 \\ y_1 \\ z_1 \\ x_2 \\ y_2 \\ z_2 \\ x_3 \\ y_3 \\ z_3 \end{Bmatrix} = \begin{Bmatrix} X_1 \\ Y_1 \\ Z_1 \\ X_2 \\ Y_2 \\ Z_2 \\ X_3 \\ Y_3 \\ Z_3 \end{Bmatrix} + \begin{Bmatrix} \Delta u_1 \\ \Delta u_2 \\ \Delta u_3 \\ \Delta u_4 \\ \Delta u_5 \\ \Delta u_6 \\ \Delta u_7 \\ \Delta u_8 \\ \Delta u_9 \end{Bmatrix}, \tag{11.10}$$

where the subscripts 1, 2, and 3 are the node numbers in a triangular element and $\Delta u_i (i = 1, \ldots, 9)$ is the incremental displacement from the current configuration $\mathbf{X}$ to the subsequent configuration $\mathbf{x}$.

The metric tensors $g_{ij}$ and $G_{ij}$ for a deformed system and fixed system can be defined in terms of curvilinear coordinates, $\mathbf{g}_i$ and $\mathbf{G}_i$, respectively, as introduced elsewhere.[7] The element stretch ratios, defined as the principal values of the square root of the Green's deformation tensor, are obtained as follows:

$$\lambda_{1,2} = \sqrt{\frac{1}{2}\left[G^{ij} g_{ij} \pm \sqrt{(G^{ij} g_{ij})^2 - 4g/G}\right]}, \tag{11.11}$$

[7] R. H. Wagoner and J.-L. Chenot, *Fundamentals of Metal Forming* (Wiley, New York, 1997).

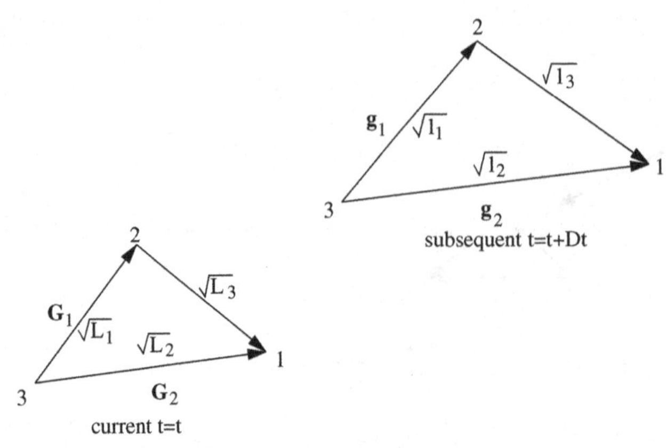


Figure 11.3. Two covariant base vectors on current and subsequent configurations, and the definition of the square of each side length.

where $g$ and $G$ denote the determinants of the metric tensors, $g_{ij}$ and $G_{ij}$, respectively, and $G^{ij}$ is the inverse of $G_{ij}$. Let $\phi_1$ and $\phi_2$ be defined as

$$\phi_1 = G^{ij} g_{ij}, \tag{11.12}$$

$$\phi_2 = g/G. \tag{11.13}$$

Figure 11.3 defines the convected base vectors and the square of side lengths at the current and subsequent triangle element.

The $L_r$ as well as $l_r (r = 1, 2, \text{and } 3)$ in Fig. 11.3 are the squares of the side lengths (not the side lengths themselves) of the current and subsequent triangles whose $r$th side is opposite to the $r$th vertex. Then, the *cosine rule* leads to the equations as follows:

$$g_{ij} = \begin{bmatrix} \mathbf{g}_1 \cdot \mathbf{g}_1 & \mathbf{g}_1 \cdot \mathbf{g}_2 \\ \text{sym} & \mathbf{g}_2 \cdot \mathbf{g}_2 \end{bmatrix} = \begin{bmatrix} l_1 & \frac{1}{2}(l_2 + l_1 - l_3) \\ \text{sym} & l_2 \end{bmatrix}, \tag{11.14}$$

$$G_{ij} = \begin{bmatrix} \mathbf{G}_1 \cdot \mathbf{G}_1 & \mathbf{G}_1 \cdot \mathbf{G}_2 \\ \text{sym} & \mathbf{G}_2 \cdot \mathbf{G}_2 \end{bmatrix} = \begin{bmatrix} L_1 & \frac{1}{2}(L_2 + L_1 - L_3) \\ \text{sym} & L_2 \end{bmatrix}, \tag{11.15}$$

$$G^{ij} = \begin{bmatrix} \mathbf{G}_2 \cdot \mathbf{G}_2 & -\mathbf{G}_1 \cdot \mathbf{G}_2 \\ \text{sym} & \mathbf{G}_1 \cdot \mathbf{G}_1 \end{bmatrix} = \begin{bmatrix} L_2 & -\frac{1}{2}(L_2 + L_1 - L_3) \\ \text{sym} & L_1 \end{bmatrix}, \tag{11.16}$$

where the $l_i$ and $L_i$ can be expressed by referring to Fig. 11.3 and Eq. (11.10) as

$$\begin{Bmatrix} l_1 \\ l_2 \\ l_3 \end{Bmatrix} = \begin{Bmatrix} (x_2 - x_3)^2 + (y_2 - y_3)^2 + (z_2 - z_3)^2 \\ (x_3 - x_1)^2 + (y_3 - y_1)^2 + (z_3 - z_1)^2 \\ (x_1 - x_2)^2 + (y_1 - y_2)^2 + (z_1 - z_2)^2 \end{Bmatrix} \tag{11.17}$$

and

$$\begin{Bmatrix} L_1 \\ L_2 \\ L_3 \end{Bmatrix} = \begin{Bmatrix} (X_2 - X_3)^2 + (Y_2 - Y_3)^2 + (Z_2 - Z_3)^2 \\ (X_3 - X_1)^2 + (Y_3 - Y_1)^2 + (Z_3 - Z_1)^2 \\ (X_1 - X_2)^2 + (Y_1 - Y_2)^2 + (Z_1 - Z_2)^2 \end{Bmatrix}. \tag{11.18}$$

Eventually, the $\phi_i$ in terms of $l_i$ and $L_i$ are obtained by substituting Eqs. (11.14)–(11.16) into Eqs. (11.12) and (11.13) as follows:

$$\phi_1 = \frac{4(l_1L_2 + l_2L_1) - 2(l_1 + l_2 - l_3)(L_1 + L_2 - L_3)}{4L_1L_2 - (L_1 + L_2 - L_3)^2}, \tag{11.19}$$

$$\phi_2 = \frac{4l_1l_2 - (l_1 + l_2 - l_3)^2}{4L_1L_2 - (L_1 + L_2 - L_3)^2}. \tag{11.20}$$

These two invariants, $\phi_1$ and $\phi_2$, are assumed to be constant throughout the element, and their expressions are simply

$$\phi_1 = \frac{1}{2G}\sum_{i=1}^{i=3} B_i \cdot l_i, \tag{11.21}$$

$$\phi_2 = \frac{1}{4G}\sum_{i=1}^{i=3} b_i \cdot l_i, \tag{11.22}$$

where

$$b_i = \left(\sum_{j=1}^{j=3} l_j\right) - 2l_i, \quad B_i = \left(\sum_{j=1}^{j=3} L_j\right) - 2L_i, \quad G = \frac{1}{4}\sum_{i=1}^{i=3} B_i \cdot L_i.$$

Let $\xi$ be defined as follows:

$$\xi_1 = \ln(\lambda_1\lambda_2) = \ln(\lambda_1) + \ln(\lambda_2) = \varepsilon_1 + \varepsilon_2 = \frac{1}{2}\ln(\phi_2), \tag{11.23}$$

$$\xi_2 = \ln\left(\frac{\lambda_1}{\lambda_2}\right) = \ln(\lambda_1) - \ln(\lambda_2) = \varepsilon_1 - \varepsilon_2 = \frac{1}{2}\ln\left[\frac{\phi_1 + (\phi_1^2 - 4\phi_2)^2}{\phi_1 - (\phi_1^2 - 4\phi_2)^2}\right]. \tag{11.24}$$

The principal strain increments during the time step are

$$\varepsilon_1 = \frac{\xi_1 + \xi_2}{2}, \tag{11.25}$$

$$\varepsilon_2 = \frac{\xi_1 - \xi_2}{2}. \tag{11.26}$$

Using the assumption of constant volume,

$$\varepsilon_3 = -\varepsilon_1 - \varepsilon_2, \tag{11.27}$$

the effective strain can be written generally by

$$\bar{\varepsilon} = f(\varepsilon_1, \varepsilon_2). \tag{11.28}$$

The detailed expression depends on the yield function used.[8]

The above formulations demonstrate how to obtain three principal strain values from the change of side lengths during an incremental time, from $t$ to $t + \Delta t$, for a triangle membrane element. No information on the principal directions and six general strain components is available from this approach. Unlike the FEM formulations

[8] D. Zhou and R. H. Wagoner, "A Numerical Method for Introducing an Arbitrary Yield Function into Rigid-Visco-Plastic FEM Programs," *Int. J. Num. Meth. Eng.* 37, 3467–3487 (1994).

presented in the previous chapters, this approach does not involve an explicit shape function and no numerical integration is necessary. The principal strains inside this triangle element are constant; thus an implicit shape is assumed.

### Expressions of $\partial\Delta\bar{\varepsilon}/\partial\Delta\mathbf{u}$ and $\partial^2\Delta\bar{\varepsilon}/\partial\Delta\mathbf{u}\partial\Delta\mathbf{u}$

In order to calculate the internal force terms, the derivative of the incremental effective strain, $\Delta\bar{\varepsilon}$, with respect to the incremental displacement $\Delta\mathbf{u}$, is a $1 \times 9$ vector, and can be written by using the chain rule as follows:

$$\frac{\partial\Delta\bar{\varepsilon}}{\partial\Delta u_i} = \sum_{\alpha=1}^{2}\Delta\varepsilon_{,\alpha}\sum_{\beta=1}^{2}\xi_{\alpha,\beta}\sum_{r=1}^{3}\phi_{\beta,r}l_{r,i}, \tag{11.29}$$

where $\Delta\varepsilon_{,\alpha} \equiv \partial\Delta\bar{\varepsilon}/\partial\xi_\alpha$, $\xi_{\alpha,\beta} \equiv \partial\xi_\alpha/\partial\phi_\beta$, $\phi_{\beta,r} \equiv \partial\phi_\beta/\partial l_r$, and $l_{r,i} \equiv \partial l_r/\partial\Delta u_i$. These are all found by geometry alone, except $\Delta\varepsilon_{,\alpha}$.

The internal stiffness matrix is a differentiation of Eq. (11.29) with respect to the incremental displacement $\Delta\mathbf{u}$. A $9 \times 9$ matrix is found:

$$\begin{aligned}\frac{\partial^2\Delta\bar{\varepsilon}}{\partial\Delta u_i\partial\Delta u_j} &= l_{s,j}\phi_{\upsilon,s}\xi_{\mu,\upsilon}(\Delta\bar{\varepsilon})_{,\alpha,\mu}\,\xi_{\alpha,\beta}\phi_{\beta,r}l_{r,i} + l_{s,j}\phi_{\upsilon,s}(\Delta\bar{\varepsilon}_{,\alpha}\xi_{\alpha,\beta\upsilon})\phi_{\beta,r}l_{r,i}\\ &\quad + l_{s,j}(\Delta\bar{\varepsilon}_{,\alpha}\xi_{\alpha,\beta}\phi_{\beta,rs})l_{r,i} + \Delta\bar{\varepsilon}_{,\alpha}\xi_{\alpha,\beta}\phi_{\beta,r}l_{r,ij},\end{aligned} \tag{11.30}$$

where the indices after the comma stand for partial differentiation and repeated indices denote the usual summation convention on their respective ranges. All Greek indices range from 1 to 2; $r$ and $s$ range from 1 to 3; $i$, $j$ range from 1 to 9. As before, all second differentiations in Eq. (11.30) with the exception of $\Delta\bar{\varepsilon}_{,\alpha\beta}$ are determined by geometry. The detailed expressions are straightforward.

It remains only to specify $\Delta\bar{\varepsilon}_{,\alpha}$ and $\Delta\bar{\varepsilon}_{,\alpha\beta}$, which depend on the particular yield function used. For Hill's new yield theory,

$$(\Delta\bar{\varepsilon}_{,\alpha}) = D_1\left[\left(\xi_1^2\right)^{[M/2(M-1)]} + D_2\left(\xi_2^2\right)^{[M/2(M-1)]}\right]^{1/M}\begin{bmatrix}\left(\xi_1^2\right)^{[(2-M)/2(M-1)]}\xi_1\\ D_2\left(\xi_2^2\right)^{(2-M)/2(M-1)}\xi_2\end{bmatrix}, \tag{11.31}$$

$$(\Delta\bar{\varepsilon}_{,\alpha\beta}) = \kappa\begin{bmatrix}\xi_2^2 & -\xi_1\xi_2\\ -\xi_1\xi_2 & -\xi_1^2\end{bmatrix}, \tag{11.32}$$

$$\kappa = \frac{D_1D_2}{M-1}\left[\left(\xi_1^2\right)^{[M/2(M-1)]} + D_2\left(\xi_2^2\right)^{[M/2(M-1)]}\right]^{M+1/M}(\xi_1\xi_2)^{[2-M/(M-1)]}, \tag{11.33}$$

$$D_1 = \frac{1}{2}[2(1+R)]^{1/M} \quad \text{and } D_1 = \frac{1}{2}[1+2R)]^{-1/(M-1)}. \tag{11.34}$$

For the Hosford yield function, however, because there is no explicit equation of effective strain available, a numerical method has to be used.

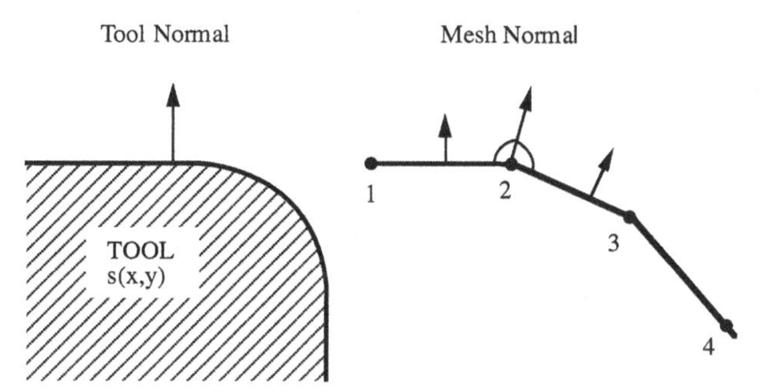


Figure 11.4. The normal vector defined by tool and mesh in two dimensions.

## 11.3 Mesh Normal Formulation

### Two-Dimensional Mesh Normal[9]

The evaluation of external forces (contact and friction) requires the definition of a contact normal direction and contact tangential direction. The stiffness matrix requires the derivatives of these directions with respect to nodal positions.

The almost-universal approach taken by FEM programmers for sheet-analysis programs is shown on the left side of Fig. 11.4; this is the "**tool-normal**" approach. The surface description of the tool surface is used to find the normal, tangent, their derivatives, and perhaps the surface curvatures in terms of spatial coordinates at a point where a contact node is located. Since these quantities depend only on the position of the contacting node, the terms appear only on the diagonal of the stiffness matrix.

After 1988, a different approach was taken in developing SHEET-S and SHEET-3, as shown on the right side of Fig. 11.4. We call this the "**mesh-normal**" approach, where all of the geometric information used in formulating the external forces is derived from the FEM mesh, not from the tooling description. In this approach, the normal is defined by the positions of connecting nodes, so the normal, tangent, and derivatives depend on these nodal positions. The programming is significantly more lengthy, but there are several major advantages.

Figure 11.5 shows the normal vectors and forces calculated from the tool surface ($\mathbf{n}_T$) and the FEM mesh ($\mathbf{n}_M$). The tool-normal vector is computed from the tool description, and the mesh normal is computed from the connected nodes and elements. Since SHEET-S and SHEET-3 now make exclusive use of the mesh-normal formulation, we will assume use of the mesh normal for our discussion. Unless there is a special note, $\mathbf{n}_M$ is simply written by $\mathbf{n}$ hereafter.

The mesh normal is a function of the connected FEM nodal positions and displacements. The tool normal and mesh normal are nearly identical in the regions of constant curvature, unless adjacent element sizes vary widely, but they are different in changing-curvature regions. Considering the force equilibrium at the node in Fig. 11.5, the net internal force, $\mathbf{F}_I(=\mathbf{f}_1 + \mathbf{f}_2)$, obtained from vector summation of those of neighbor elements, is balanced by the contact force, $\mathbf{F}_E$. For the case of zero friction, it is easy to see that the external force $\mathbf{F}_E$ must be antiparallel to $\mathbf{F}_I$.

[9] Y. T. Keum, C.-T. Wang, M. J. Saran, and R. H. Wagoner, "Practical Die Design via Section Analysis," *J. Mat. Processing Tech.* 35, 1–36 (1992).

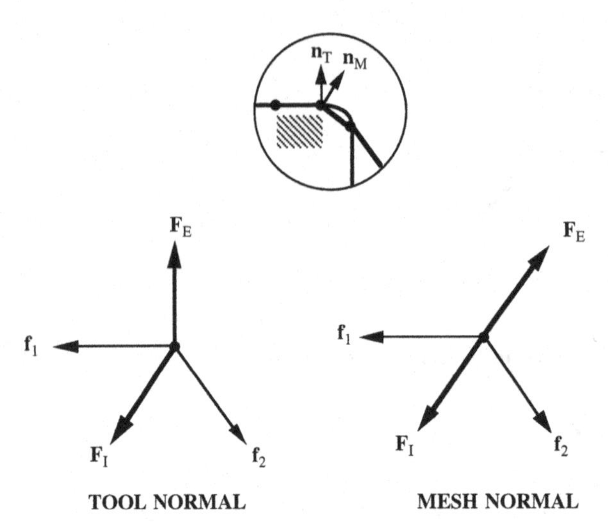


Figure 11.5. Free-body diagram at the contact node for tool and mesh normal.

But this condition implies that the strains and stresses are different on the left-hand and right-hand sides of the contact node if a tool-normal description is followed. In a membrane (no bending) calculation, these forces and deformations must be equal physically for zero friction; thus, a realistic equilibrium state is attainable only by using the mesh-normal description.

The two-dimensional mesh normal and tangential vectors at node $k$ in the FEM mesh can be calculated as follows: Consider, in Fig. 11.6, the adjacent nodes $i$, $k$, $j$ situated in the $x$, $z$ plane that define two line elements between having current lengths denoted by $L_i$ and $L_j$. The unit normal vector at node $k$ depends on the coordinates of the three considered nodes:

$$\mathbf{n}_k = f(\mathbf{x}^i, \mathbf{x}^k, \mathbf{x}^j), \tag{11.35}$$

where $\mathbf{n}_k$ is the mean value (unweighted) of the element normal unit vectors[10] for the connected elements, $\mathbf{N}_i$ and $\mathbf{N}_j$. (In the previous chapters, $\mathbf{N}$ is used for the shape function; in this chapter, the $\mathbf{N}$ is used for element normal.) For elements $i$ and $j$ we can define

$$\mathbf{n}_k = \frac{\mathbf{N}_i + \mathbf{N}_j}{|\mathbf{N}_i + \mathbf{N}_j|}. \tag{11.36}$$

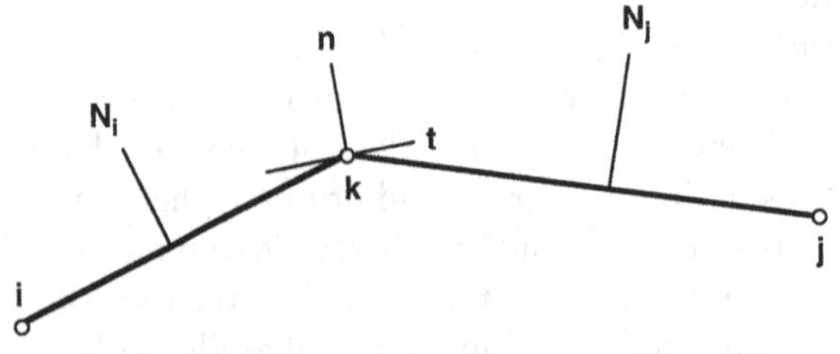


Figure 11.6. Two-dimensional mesh-normal definition. Nodal normal depends on its connected element normals.

[10] The element normals are unambiguously defined because the elements are lines.

When the normal vector, $\mathbf{n}_k$, is rewritten in the following component form,

$$\mathbf{n} = \frac{1}{S_n}\begin{Bmatrix} -S_x \\ 1 \end{Bmatrix}, \tag{11.37}$$

$$\mathbf{t} = \frac{\text{sign}}{S_n}\begin{Bmatrix} 1 \\ S_x \end{Bmatrix}, \tag{11.38}$$

where the sign depends on the relative displacement vector, the sign has a value 1 for the die and $-1$ for the punch, and $S_x$ is a slope defined by

$$S_x = \frac{\partial S(x)}{\partial x}. \tag{11.39}$$

Here the hypotenuse of a triangle having $\mathbf{S_x}$ rise and unit run is

$$S_n = \sqrt{S_x^2 + 1}. \tag{11.40}$$

In fact, although the tool surface is defined by $S(x)$, its derivative, $S_x$ in Eq. (11.39), is found only from nodal positions of the sheet mesh. Thus, there is no need to be concerned about smoothness or continuity of $S(x)$ and its derivatives, which can be a significant advantage in dealing with approximate digital data generated from CAD/CAM systems, piecewise plots, or digitized tool data from coordinate measuring machines.

The required slope at node $k$ can be obtained from Eq. (11.36), in terms of coordinates $(x, z)$, of the nodes $i$, $j$, and $k$, as follows:

$$S_x = \frac{L_j(z_k - z_i) + L_i(z_j - z_k)}{L_j(x_k - x_i) + L_i(x_j - x_k)} = \frac{n_x}{n_z}. \tag{11.41}$$

And, defining another important quantity,

$$S_{xx} = \frac{\partial S_x}{\partial \Delta u_x}, \tag{11.42}$$

we can finally obtain

$$\frac{\partial \mathbf{n}}{\partial \Delta \mathbf{u}} = \frac{-1}{S_n^3}\begin{bmatrix} S_{xx} & S_{xz} \\ S_x S_{xx} & S_x S_{xz} \end{bmatrix}, \tag{11.43}$$

$$\frac{\partial \mathbf{t}}{\partial \Delta \mathbf{u}} = \frac{\text{sign}}{S_n^3}\begin{bmatrix} -S_x S_{xx} & -S_x S_{xz} \\ S_{xx} & S_{xz} \end{bmatrix}, \tag{11.44}$$

where $t$ is the tangent vector.

### Three-Dimensional Mesh Normal[11]

Following the two-dimensional mesh normal presented above, the three-dimensional mesh-based normal vector can be defined by using the normals of the connecting elements, as shown in Fig. 11.7. The nodal normal unit vector $\mathbf{n}$ at the contact node 1, which is connected to elements designated generally by the subscript $k$, is evaluated by an included-angle-weighted average method:

$$\mathbf{n} = \text{sign}\frac{\sum_{k=1}^{n_e} \theta_k \mathbf{N}_k}{\left\| \sum_{k=1}^{n_e} \theta_k \mathbf{N}_k \right\|}, \tag{11.45}$$

[11] Y. T. Keum, E. Nakamachi, R. H. Wagoner, and J. K. Lee, "Compatible Description of Tool Surfaces and FEM Meshes for Analysing Sheet Forming Operations," *Num Meth. Eng.* 30, 1471–1502 (1990).

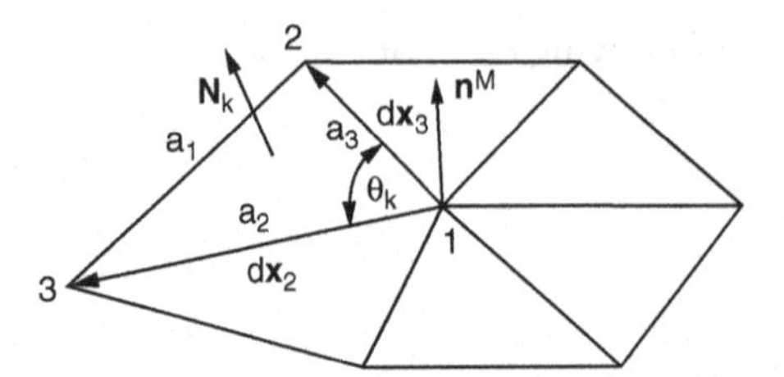


Figure 11.7. Definition of the three-dimensional mesh-normal vector at a node from its adjacent finite elements.

where $n_e$ is the number of elements containing node $j$, $\theta_k$ is the included angle of element $k$ at node 1, and $\mathbf{N}_k$ is the element normal of the element $k$. The sign has the same meaning as in the last section. $\mathbf{N}_k$ is obtained by cross products of two line (or side) vectors lying on element $k$, $\mathbf{dx}_1$ and $\mathbf{dx}_2$:

$$\mathbf{N}_k = \frac{\mathbf{dx}_3 \times \mathbf{dx}_2}{\|\mathbf{dx}_3 \times \mathbf{dx}_2\|}, \tag{11.46}$$

and the included angle of element $k$ at node 1:

$$\theta_k = \cos^{-1}\left(\frac{\mathbf{dx}_2 \cdot \mathbf{dx}_3}{\|\mathbf{dx}_2\|\ \|\mathbf{dx}_3\|}\right), \tag{11.47}$$

where $\mathbf{dx}_1$, the vector from node 1 to node 2, and $\mathbf{dx}_2$, the vector from node 1 to node 3, are expressed by Cartesian basis vectors $\mathbf{e}_x$, $\mathbf{e}_y$, and $\mathbf{e}_z$ as follows:

$$\mathbf{dx}_3 = (x_2 - x_1)\mathbf{e}_x + (y_2 - y_1)\mathbf{e}_y + (z_2 - z_1)\mathbf{e}_z, \tag{11.48}$$

$$\mathbf{dx}_2 = (x_3 - x_1)\mathbf{e}_x + (y_3 - y_1)\mathbf{e}_y + (z_3 - z_1)\mathbf{e}_z. \tag{11.49}$$

When the element normal unit vector, $\mathbf{N}_k$, is rewritten in the following component form,

$$\mathbf{N}_k = (N_{xk}, N_{yk}, N_{zk}), \tag{11.50}$$

the components of $\mathbf{N}_k$ can be computed from Eq. (11.46) as follows:

$$N_{xk} = \frac{C_1}{C}; \quad N_{yk} = \frac{C_2}{C}; \quad N_{zk} = \frac{C_3}{C}, \tag{11.51}$$

where

$$C_1 = (y_2 - y_1)(z_3 - z_1) - (z_2 - z_1)(y_3 - y_1), \tag{11.52}$$

$$C_2 = (z_2 - z_1)(x_3 - x_1) - (x_2 - x_1)(z_3 - z_1), \tag{11.53}$$

$$C_3 = (x_2 - x_1)(y_3 - y_1) - (y_2 - y_1)(x_3 - x_1), \tag{11.54}$$

$$C = \sqrt{C_1^2 + C_2^2 + C_3^2}. \tag{11.55}$$

In Eqs. (11.52)–(11.54), $x_i$, $y_i$, and $z_i$ are the spatial coordinates of the $i$th node in element $k$, and subscript numbers are the nodal sequences.

We assume the nodal points are labeled in counterclockwise order. It is easy to see that the element mesh normal will change to the antiparallel direction if the order is reversed. If $a_1$, $a_2$, and $a_3$ are respectively the edge lengths of element $k$ associated with nodal sequence numbers 1, 2, and 3, then the included angle, $\theta_k$, ($k$ = 1, 2, and 3) can be expressed as follows:

$$\theta_1 = \cos^{-1}\left(\frac{a_2^2 + a_3^2 - a_1^2}{2a_2a_3}\right), \tag{11.56}$$

$$\theta_2 = \cos^{-1}\left(\frac{a_3^2 + a_1^2 - a_2^2}{2a_1a_3}\right), \tag{11.57}$$

$$\theta_3 = \cos^{-1}\left(\frac{a_1^2 + a_2^2 - a_3^2}{2a_1a_2}\right), \tag{11.58}$$

where

$$a_1^2 = (x_3 - x_2)^2 + (y_3 - y_2)^2 + (z_3 - z_2)^2, \tag{11.59}$$
$$a_2^2 = (x_3 - x_1)^2 + (y_3 - y_1)^2 + (z_3 - z_1)^2, \tag{11.60}$$
$$a_3^2 = (x_2 - x_1)^2 + (y_2 - y_1)^2 + (z_2 - z_1)^2. \tag{11.61}$$

After calculating the connected element normals, $\mathbf{N}_k$, by Eqs. (11.50) and (11.51), and the included angle, $\theta_k$, by Eqs. (11.56)–(11.58), we find for every element containing the $j$th node, the normal unit vector at the concerned node, $\mathbf{n}$, is calculated from Eq. (11.45) by the usual assembly procedure.

After obtaining the nodal normal unit vector, $\mathbf{n}$, from Eq. (11.45) in the assembly procedure, the nodal tangential unit vector, $\mathbf{t}$, can be found. The vector $\mathbf{t}$ is normal to the mesh-normal vector and lying in the same plane with the normal unit vector, $\mathbf{n}$, and the nodal displacement, $\Delta\mathbf{u}$. The vector $\mathbf{t}$ is expressed as follows:

$$\mathbf{t} = \frac{\mathbf{n} \times \Delta\mathbf{u} \times \mathbf{n}}{\|\mathbf{n} \times \Delta\mathbf{u} \times \mathbf{n}\|}. \tag{11.62}$$

Thus, the three vectors $\mathbf{n}$, $\mathbf{t}$, and $\Delta\mathbf{u}$ lie in the same plane containing the incremental displacement vector and mesh-normal vector. Employing the normal and tangential vectors in Eq. (11.45) and Eq. (11.62), we find these derivatives with respect to the nodal incremental displacement vector $\Delta\mathbf{u}$, $\partial\mathbf{n}/\partial\Delta\mathbf{u}$ and $\partial\mathbf{t}/\partial\Delta\mathbf{u}$, are obtained from the connectivity. The normal and tangential vectors and their spatial derivatives are related to the neighboring nodes associated with that node. In other words, the surface normal, tangent, and their derivatives at each node are evaluated directly from the FEM mesh and assembled in the same way as that of the global stiffness equation.

### Three-Dimensional Mesh-Based Derivatives

The mesh-normal vector is a function of positions of the contact node and adjacent nodes on the current or deformed mesh. Therefore, the derivatives of a mesh normal vector, $\mathbf{n}$, with respect to the nodal incremental displacement, $\Delta\mathbf{u}$, $\partial\mathbf{n}/\partial\Delta\mathbf{u}$, can only be determined from the connectivity. First, the derivative of the included angle of element $k$, $\theta_k$, and the derivative of the element normal vector, $\mathbf{N}_k$, with respect to the incremental displacements of element nodes, $\Delta\mathbf{u}$, are calculated for all connected elements. Then, the derivatives of the global nodal normal vector components are derived by the assembling procedure.

Similar to the derivative in the two-dimensional case, we can obtain all the detailed equations for the three-dimensional mesh normal simply by following the chain rule.

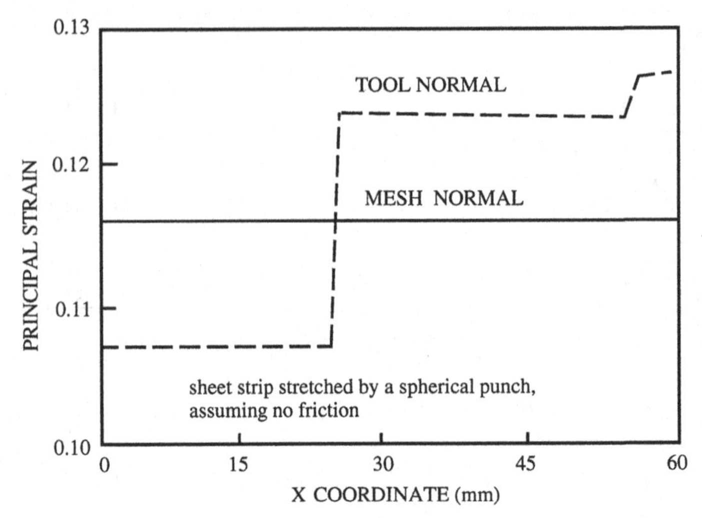


Figure 11.8. Comparison of simulated strain distributions by mesh and tool normals, for a frictionless case. A uniform strain is physically correct.

### Comparison of Mesh Normal with Tool Normal

The mesh normal is more complicated in programming than that of the tool normal, because the tool normal involves only one node with three components of $\Delta\mathbf{u}$. The derivatives of tool-based $\mathbf{n}$ and $\mathbf{t}$ with respect to $\Delta\mathbf{u}$ is simply a $3 \times 3$ matrix. In the mesh normal, however, the $\mathbf{n}$ involves many nodes, each one of which has three components. For instance, if a node has six neighboring nodes, the derivatives of $\mathbf{n}$ and $\mathbf{t}$ for this node with respect to $\Delta\mathbf{u}$ will be a $3 \times 21$ matrix. After assembling, the global stiffness matrix has a wider bandwidth and may require more CPU time for solution unless there are compensating efficiencies.

Figure 11.37, presented later in this chapter, shows the tooling and specimen geometry for simulations used to compare tool-based and mesh-based normals. Figure 11.8 presents one two-dimensional result, for a cylindrical punch-stretch case under frictionless conditions. Under these conditions, it is easy to show that the strain distribution should be uniform as long as bending is unimportant. The tool-based normal approach introduces significant spurious results, which are eliminated by the mesh normal. Surprisingly, the computation times were similar, within 10%, presumably because the additional stability introduced by the mesh normal offsets the additional bandwidth.

## 11.4 Equilibrium Equation

Mechanical equilibrium is achieved by satisfying the equilibrium condition[12] of internal force and external force at each node. Namely, the internal force must be balanced with the external force. The force equilibrium condition also can be derived from an approximation of the classical virtual work theorem. More details were presented in Chapter 3. Since the two forces all depend on the incremental displacement, $\Delta\mathbf{u}$,

[12] Y. Germain, K. Chung, and R. H. Wagoner, "A Rigid-Viscoplastic Finite Element Program for Sheet Metal Forming Analysis," *Int. J. Mech. Sci.* 31, 1–24 (1989).

during a time step, the equilibrium equation can be expressed as

$$\mathbf{F}_I(\Delta\mathbf{u}) = \mathbf{F}_E(\Delta\mathbf{u}), \tag{11.63}$$

where the subscripts $I$ and $E$ stand for internal and external, respectively. For each node, Eq. (11.63) can be written as three equations in component form in the $x$, $y$, and $z$ directions.

For a given yield function, the internal equivalent work is

$$\Delta W_p = \int_V \int_{\bar{\varepsilon}}^{\bar{\varepsilon}+\Delta\bar{\varepsilon}} \bar{\sigma}(\bar{\varepsilon}, \dot{\bar{\varepsilon}})\, \mathrm{d}\bar{\varepsilon}\mathrm{d}V, \tag{11.64}$$

where $\bar{\sigma}$, $\bar{\varepsilon}$, and $\dot{\bar{\varepsilon}}$ denote the effective stress, strain, and the strain rate, respectively. The effective stress is generally a function of current effective strain and effective strain rate, depending on the hardening law chosen. The incremental effective strain $\Delta\bar{\varepsilon}$, whose definition depends on choice of yield function, is defined as the change of the effective strain over the time step $\Delta t$:

$$\Delta\bar{\varepsilon} = \int_t^{t+\Delta t} \dot{\bar{\varepsilon}}\, \mathrm{d}t. \tag{11.65}$$

Adopting an assumption of a proportional path within an incremental time step, we find the increase of the effective strain is a fixed function of the principal values of the incremental strain tensor (as long as an explicit effective strain equation is available for a given yield function):

$$\Delta\bar{\varepsilon} = \Delta\bar{\varepsilon}(\Delta\varepsilon_i), \tag{11.66}$$

where $\Delta\varepsilon_i$ are the incremental principal logarithmic strains ($i = 1, 2, 3$), which can be easily obtained from the element configuration change from time t to time $t + \Delta t$. The specific form of Eq. (11.66) can be obtained explicitly for many commonly used yield functions, or it can be solved numerically for an arbitrary yield function without explicit equation.

After integration over an incremental time, the work increment can be written as a perturbation of the incremental work, Eq. (11.64), in the form of

$$\delta\Delta W_p = \int_V \bar{\sigma} \frac{\partial \Delta\bar{\varepsilon}}{\partial(\Delta\mathbf{u})} \delta(\Delta\mathbf{u})\, \mathrm{d}V, \tag{11.67}$$

resulting finally in the following expression for the internal force vector:

$$\mathbf{F}_I = \int_V \bar{\sigma} \frac{\partial \Delta\bar{\varepsilon}}{\partial(\Delta\mathbf{u})}\, \mathrm{d}V. \tag{11.68}$$

The physical meaning of internal force on a node can be understood as the sum of all forces from its adjacent elements because of their deformation. If the node does not make contact with a tool surface, all its internal forces on this specific node have to be balanced with each other; or, in another sense, the summation of all the internal forces must be zero to satisfy the equilibrium condition. In a two-dimensional line element case, for example, the internal force on a node is simply the vector summation of element forces (in-line principal stress multiplied by the section area) at either side.

## 11.5 Contact and Friction: General Considerations

### Common Methods for Enforcing Contact and Friction Conditions

The contact problem in sheet-metal forming processes is complex. Contact between the tool and sheet is highly nonlinear because of its asymmetry, where at a position in space a node is either free or is rigidly constrained depending on an infinitesimal change of position normal to the tool surface. The boundary conditions dramatically change as a result of change in the contact region, which evolves continually and unpredictably. Three basic constraints for contact arise.

1. Impenetrability: A sheet node may not lie inside a tool.
2. Unilateral contact: A sheet node in contact may not have an external force directed *toward* the tool surface (no adhesion), whereas a noncontacting node has no external contact force.
3. Friction conditions: A node in contact may have an external force whose direction depends on the incremental displacement and sheet normal, and whose magnitude may depend on the normal force and on friction law. If sticking is allowed, the external force and direction may depend entirely on the internal force imbalance.

In sheet forming, these conditions lead to more severe nonlinearity in a given time step than either the material law or the geometry of the problem. Furthermore, because the conditions change abruptly in space, the usual derivatives are not available for use in a Newton–Raphson type of procedure. For these reasons, the contact and friction algorithms are the most critical part of FEM programs for sheet-forming analyses.

A sliding node is physically constrained to move along the tool surface, and may not penetrate. Because this condition cannot be imposed directly for arbitrarily shaped interfaces, an iterative technique is needed to achieve this constraint. In many FEM programs, impenetrability is imposed by what we call the direct node projection (DNP) scheme. The DNP algorithm consists of three major steps.

1. Equilibrium iteration: Contact nodes are constrained to move in a tangent plane, as outlined below. Typically, the equilibrium iteration takes place with a fixed set of contact nodes defined before the Newton–Raphson iteration begins, and it is carried out until the force residual is sufficiently small.
2. Contact check: Free nodes that lie inside tools at the end of the equilibrium iteration are updated to become contact nodes. Contact nodes which have an external force toward the tooling are updated to become free nodes.
3. Nodal projection: Nodes lying inside the tools are projected onto the tool surface along a fixed direction. Typical direction choices include the $z$ direction (punch travel direction), tool-normal direction, or closest-approach direction. A local iteration may be necessary to find the projected position of the node.

Once the projection step is accomplished, the normals and tangents are updated and the equilibrium iteration is begun. Complete convergence is declared when, after the equilibrium iteration, the contact set is consistent with the second and third steps; that is, no changes are made during the contact check.

There are many variations of the DNP scheme. For example, the contact set of nodes may be updated during the equilibrium iteration. In our experience with SHEET-S and SHEET-3, before we replaced the DNP scheme with a more stable one, the sequence listed above is nearly optimal. The best projection direction is usually either the closest approach or the mesh normal, but the instabilities introduced by the projection scheme itself are too severe, particularly for draw-in cases in which we were seldom able to achieve convergence without special numerical tricks.

As mentioned in Chapter 6, the major alternatives to imposing contact conditions are the boundary (gap) element, penalty function method, and Lagrangian multiplier. In SHEET-3 and SHEET-S, we devised a variation of the Lagrangian multiplier method, the consistent full set (CFS) method, which provides much better stability than the DNP method. Our experience with other programs indicates that the gap elements and penalty function methods can either be tuned to be stable or to be accurate, but that achieving both is not certain. Because these methods introduce unknown parameters that affect the solution accuracy, we prefer not to use them for sheet forming, where the contact is critical for physical accuracy.

### Relative Nodal Slip on Tool Surface

No matter how exactly the nodal constraint condition is applied numerically, it is necessary to know the constrained directions and constraint plane. It is necessary to find $\mathbf{F}_E$ for a contact/slipping node, and a constraint equation for $\Delta\mathbf{u}$ that forces the node's movement within the local tool surface tangent plane to be satisfied. Under such a constraint, only two ($x$ and $y$ are chosen) of three components of $\Delta\mathbf{u}$ are independent in a three-dimensional case, and only one of two in a two-dimensional case. In SHEET-3 and SHEET-S we use the mesh normal, as defined above, to determine the tangent vector.

Two-Dimensional Case: The incremental displacement $\Delta\mathbf{u}$ of a contact node is constrained only within the tangential direction, which is determined by the first derivative of the FEM mesh. The relative slip distance for a contact node on the tool surface in two dimensions can be obtained as

$$|\Delta\mathbf{u}_t| = |\Delta\mathbf{u}\cdot\mathbf{t}| = \sqrt{\Delta u_x^2 + \Delta u_z^2}, \tag{11.69}$$

where, by the same reason the $z$-direction slip displacement is

$$\Delta u_z = S_x \Delta u_x. \tag{11.70}$$

By defining the reciprocal of the incremental displacement distance,

$$v = \left(\Delta u_x^2 + \Delta u_z^2\right)^{-1/2}, \tag{11.71}$$

and by substituting Eqs. (11.70) and (11.71), we obtain the derivative of $\partial|\Delta\mathbf{u}_t|/\partial\Delta\mathbf{u}$ as

$$\frac{\partial|\Delta\mathbf{u}_t|}{\partial\Delta\mathbf{u}} = v\begin{bmatrix}\Delta u_x + \Delta u_z(\partial\Delta u_z/\partial x)\\ 0\end{bmatrix}. \tag{11.72}$$

Three-Dimensional Case: The incremental displacement $\Delta\mathbf{u}$ of a three-dimensional contact node is constrained only within the tangential plane passing through the contact point. The tangential plane is determined from the FEM mesh, as described above. The direction of $\Delta\mathbf{u}$ projecting on the tangential plane is needed in order to find the slip direction, which is antiparallel to the friction force. The relative slip distance for the node on the tool surface is

$$|\Delta\mathbf{u}_t| = |\Delta\mathbf{u}\cdot\mathbf{t}| = \sqrt{\Delta u_x^2 + \Delta u_y^2 + \Delta u_z^2}. \tag{11.73}$$

In order to ensure the slip direction is indeed on the tangential plane, $\Delta u_z$ will no longer be an independent variable, but can be expressed as

$$\Delta u_z = S_x \Delta u_x + S_y \Delta u_y. \tag{11.74}$$

Letting the reciprocal of the magnitude of this vector be

$$v = \left(\Delta u_x^2 + \Delta u_y^2 + \Delta u_z^2\right)^{-1/2}, \tag{11.75}$$

and by substituting Eqs. (11.68) and (11.69), we obtain the derivative of $\partial|\Delta\mathbf{u}_t|/\partial\Delta\mathbf{u}$ as

$$\frac{\partial|\Delta\mathbf{u}_t|}{\partial\Delta\mathbf{u}} = v \begin{bmatrix} \Delta u_x + \Delta u_z(\partial\Delta u_z/\partial x) \\ \Delta u_y + \Delta u_z(\partial\Delta u_z/\partial y) \\ 0 \end{bmatrix} \tag{11.76}$$

### Regularization of Friction Force

In order to satisfy the force equilibrium condition with the internal force at a contact node, the external contact nodal force, $\mathbf{F}_E$, must be defined. The external contact force can be resolved into two parts, the normal part and the tangential part:

$$\mathbf{F}_E = \mathbf{F}_n + \mathbf{F}_t. \tag{11.77}$$

The normal part of the external force, along the mesh normal direction, is the reaction force pushing against the node to prevent its penetration into the tool. The tangential part, along a tangential direction antiparallel to the node motion, is solely dependent on the friction effect between the tool surface and the node. For frictionless cases, $\mathbf{F}_t$ is zero. Using Coulomb's friction law and denoting by $P$ the magnitude of the normal stress (contact pressure), we can rewrite Eq. (11.77) as

$$\mathbf{F}_E = (\mathbf{n} - \mu\mathbf{t})P, \tag{11.78}$$

where $\mathbf{n}$ and $\mathbf{t}$ are the mesh-normal and tangential incremental displacement unit vectors at the node, respectively, and $\mu$ is the coefficient of friction. Although there are an infinite number of "tangential unit vectors" perpendicular to the normal vector, only one of them is the "tangential incremental displacement unit vector." We will frequently shorten the terminology to "tangent vector," meaning the more precise definition.

The original Coulomb friction law is an inequality for slip and stick. In order to overcome the discontinuity associated with the stick–slip transition, an ever-slipping friction condition governed by a piecewise linear function was used. Though this function overcomes the discontinuity inherent in the original Coulomb's friction law, it is not a smooth function with a continuous first derivative when it changes from stick to slip. Recently, a modified version of Coulomb's friction law with a smoothing function and a continuous first derivative was used, Fig. 11.9. The smoothing function used is

$$\phi(\Delta\mathbf{u}_t) = \tanh\left(\frac{3|\Delta\mathbf{u}_t|}{\delta}\right), \tag{11.79}$$

where the $\delta$ is a very small number, commonly about 0.0001.

The tangential vector can be constructed by the normal unit vector and the incremental displacement vector,

$$\Delta\mathbf{u}_t = [\Delta\mathbf{u} \times \mathbf{n}] \times \mathbf{n}. \tag{11.80}$$

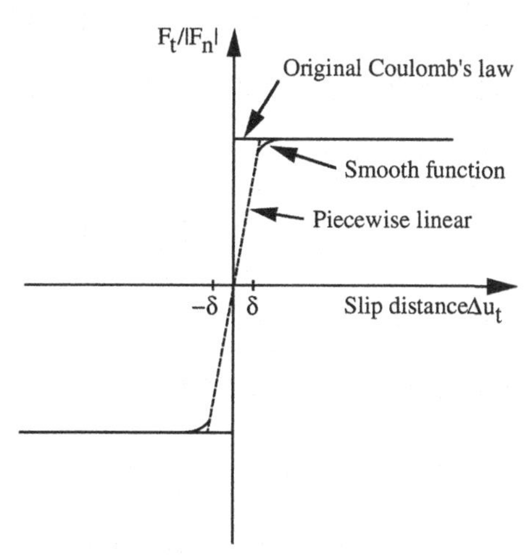


Figure 11.9. The original Coulomb law and its two modifications.

The sticking condition with a discontinuous first derivative is replaced by an ever-slipping condition with a smooth transition region, improving numerical stability of the solution procedure. Thus, the external force at node $k$, indicated by superscript $k$, is written as

$$\mathbf{F}_E^k = [\mathbf{n} - \mu\phi(\Delta\mathbf{u}_t)\mathbf{t}]P, \tag{11.81}$$

whereas for the noncontact node, neither contact nor friction exist:

$$\mathbf{F}_E^k = 0. \tag{11.82}$$

The force equilibrium equation for this node is obtained as

$$\mathbf{F}_I^k(\Delta\mathbf{u}) = \mathbf{F}_E^k(\Delta\mathbf{u}). \tag{11.83}$$

By the standard assembling procedure, Eq. (11.63) can be obtained as the global equilibrium equation from all nodal equilibria expressed by Eq. (11.83). Both the $\mathbf{F}_I$ and $\mathbf{F}_E$ vectors have the same dimension, which is also the number of nonlinear equations that must be solved.

## 11.6 Consistent Full Set Algorithm

### Overview

A consistent full set, or CFS, algorithm[13] was developed for enforcing the contact condition in sheet-forming simulations. The CFS algorithm can improve the accuracy and the numerical stability while retaining the computational efficiency needed for realistic industrial application. Because the first CFS version implemented defines the contact error along the punch travel direction (or $z$ direction), it is called Z-CFS.[14] Preliminary results obtained from the Z-CFS algorithm have shown a significant improvement over the DNP method. Even so, the Z-CFS method was found to have a

[13] D. Zhou and R. H. Wagoner, "Application of A New Algorithm in Sheet Forming Simulation," *Computer Applications in Shaping & Forming of Materials*. eds. M. Y. Demeri, TMS, Denver, USA, 1993, 41.

[14] M. J. Saran and R. H. Wagoner, "A Consistent Implicit Formulation for Nonlinear Finite Element Modeling with contact and Friction: Part 1 – Theory" and "Part 2 – Numerical Verification and Results," *J. Appl. Mech. Trans.* ASNE 58, 499–512 (1991).

convergence problem when the draw-in angle approaches 70°. Although some success was obtained by using case-by-case numerical tricks to obtain convergence, a better algorithm was needed.

A more advanced version of CFS algorithm has proved better in terms of convergence and computation time. Because this new CFS algorithm defines the contact error along the mesh-normal direction, it is called the N-CFS method. The N-CFS algorithm obtains a simultaneous solution of a set of nonlinear equations, including equilibrium and contact conditions. In the N-CFS algorithm, the mesh normal is consistently used in the measurement of contact error and in the calculation of contact pressure and external force, as well as in defining the external part of the stiffness equation.

The N-CFS algorithm exhibits better convergence and numerical stability in deep drawing simulation, when draw-in is significant and relative nodal displacement is large, when the draw-in angle is large (even approaching 90°), and when the contact region changes rapidly (during redrawing, for example).

In SHEET-S and SHEET-3, the tool is assumed to consist of a fixed die and punch moving in the $z$ direction of a fixed Cartesian coordinate system. Additional tooling, such as a blank holder, is also needed sometimes. The $z$ coordinate of a certain point on the original punch surface is defined by a single-valued continuous function[15] of $x$ and $y$ coordinates:

$$ {}^{0}z_{p} = {}^{0}S(x, y). \tag{11.84} $$

The following nonpenetration geometric constraint for any contact node must be satisfied:

$$ D(x, y, z, t + \Delta t) = \alpha |\mathbf{X}_{\text{tool}} - \mathbf{X}_{\text{node}}| = 0, \tag{11.85} $$

where $\mathbf{X}_{\text{node}}$ is the nodal position vector and $\mathbf{X}_{\text{tool}}$ is the position vector at the point on the tool surface corresponding to the node, where contact is presumed to exist; $\alpha$ is either $+1$ or $-1$, depending on the relative positions of tool and sheet. More specifically, the general contact error $D$ is defined differently in the Z-CFS and the N-CFS methods. We denote the specific error measures by $D_Z$ and $D_N$, respectively. The distinction between using $D_Z$ and $D_N$ as measures of contact error defines the essential difference between the Z-CFS and the N-CFS methods, as shown in Fig. 11.10.

The sign of $D_N$, denoted by $\alpha$ in Eq. (11.85), depends on the relationship of $\mathbf{X}_{\text{tool}}$ to $\mathbf{X}_{\text{node}}$ along the measuring direction. The $D_N$ is positive if it is measured along the mesh-normal direction from the contact tool point to the node. It is negative if the order of measurement is reversed. A positive or negative value tells us whether the node is located outside or inside of the tool, either the punch or the die, as shown in Fig. 11.11. In order to fully understand the N-CFS method, one has to first understand how the sign of the mesh-normal direction is chosen.

There are two possible mesh-normal directions toward either side of the mesh surface. Since the tool's normal is defined always outward from the tool surface, the mesh normal closest to the tool-normal vector is chosen as the mesh-normal direction. For example, we know that the tool normal at the anticipated contact

[15] We assume that there are no re-entrant shapes allowed when viewed from the $z$ direction.

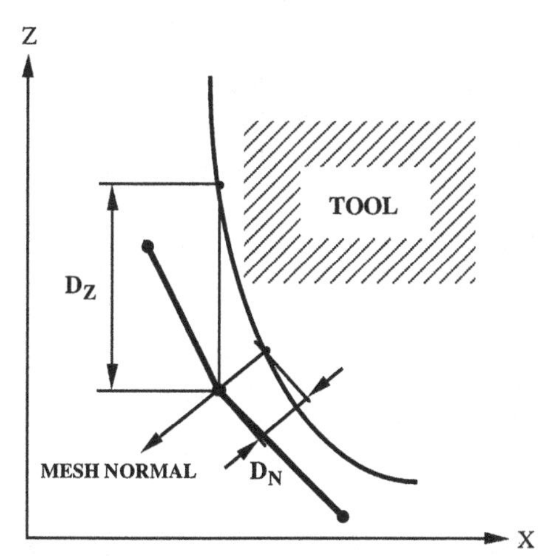


Figure 11.10. Contact errors defined by the distance along the $z$ direction in the Z-CFS and the distance along the mesh-normal direction in the N-CFS.

point in Fig. 11.11 is pointing to the lower left-hand corner of the picture, so the mesh normal is selected in such a way as is shown in the figure.

In the N-CFS algorithm, a full set of governing equations includes both the force equilibrium relations, Eq. (11.63), and the geometric constraints condition, Eq. (11.85), for all nodes. Because each node can become a contact node, the CFS algorithm adds one more degree of freedom, the contact pressure $P$, to each node.[16] N-CFS assigns three degrees of freedom to a two-dimensional node and four degrees of freedom to a three-dimensional node. Combining both the equilibrium condition and contact condition, we have the following full set consistent governing equation:

$$\begin{cases} \mathbf{F}_I(\Delta\mathbf{u}) = \mathbf{F}_E(\Delta\mathbf{u}) \\ D(x, y, z, t + \Delta t) = 0 \end{cases}. \tag{11.86}$$

When the loop of equilibrium and contact iteration converge in a Newtonian iteration, both force equilibrium and geometric contact conditions are satisfied.

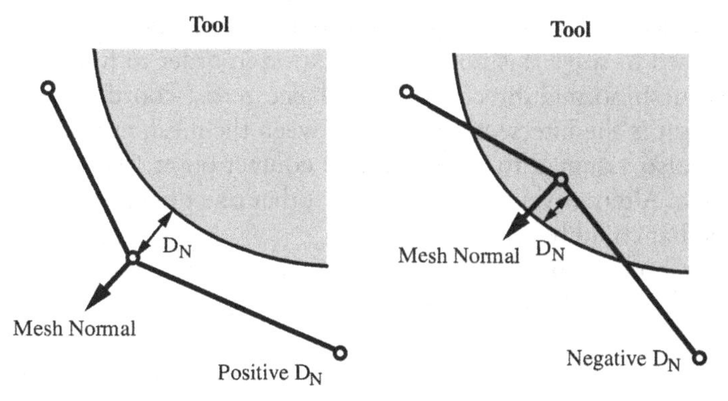


Figure 11.11. Method for determining the sign of the $D_N$ algorithm.

[16] $P$ actually represents the normal contact force at the representative node.

### Implementation of the Z-CFS Method

The most serious obstacle for industrial application of static implicit FEM in the 1980s was the difficulty of obtaining convergence during draw-in. No real solution was in sight for general three-dimensional cases. Each geometry required special techniques and ad hoc procedures. This deficiency provided the greatest impetus to the development of static explicit and dynamic explicit methods.

The contact error in Z-CFS is readily determined: for a node with coordinates ($x_n$, $y_n$, $z_n$), one can directly find the corresponding point on the tool surface with the same coordinates of $x$ and $y$ ($x_{\text{tool}} = x_n$ $y_{\text{tool}} = y_n$, $z_{\text{tool}}$). The $D_z$ is simply the $z$-coordinate difference of the two points with an appropriate sign. The derivative of $D_z$ is the tool surface derivative defined either by the mesh normal or the tool normal. The basic equations for Z-CFS are as follows:

$$\begin{bmatrix} \mathbf{K}_I - \mathbf{K}_E & \mathbf{N} \\ \mathbf{C} & 0 \end{bmatrix} \begin{bmatrix} \delta \mathbf{u} \\ \delta P \end{bmatrix} = \begin{bmatrix} \mathbf{F}_E - \mathbf{F}_I \\ D_z \end{bmatrix}, \tag{11.87}$$

where $\delta P$ is the contact pressure correction increment (which tends to zero as iteration proceeds), and

$$\mathbf{N} = \mathbf{n} - \mu\phi\mathbf{t}, \tag{11.88}$$

$$\mathbf{C} = \begin{bmatrix} \dfrac{\partial(D_z)}{\partial x} & \dfrac{\partial(D_z)}{\partial y} & -1 \end{bmatrix} = \begin{bmatrix} \dfrac{\partial^0 S}{\partial x} & \dfrac{\partial^0 S}{\partial y} & -1 \end{bmatrix}. \tag{11.89}$$

Equation (11.87) is the elemental equation, which can then be assembled to form the global equations.

### Implementation of the N-CFS Algorithm

The critical aspects of the N-CFS algorithm are how to determine the mesh-normal direction contact error and the derivatives with respect to nodal positions required by the stiffness equation. The $D_N$ must be found by a numerical method carried out along the mesh normal direction.

In our implementation, a few test points are taken along the mesh-normal direction (Figs. 11.12 and 11.13), while the $z$-coordinate distance from these test points to the tool surface is determined and compared. The $z$-coordinate distance is a one-dimensional nonlinear function of the distance from the test point to the node.

A Newtonian iteration is used to solve this nonlinear equation in order to find a special point ($T$ point) in the mesh-normal direction, which has a zero $z$-coordinate distance to the tool. The $T$ point is the intersection point between the mesh normal and the tool surface, and it is also assumed to be the desired contact point. Usually only a few iterations are enough. Alternatively, there are many other one-dimensional numerical searching methods that could be used.

The distance $D_N$ depends on both the position of node $N$, $x_N$, and its mesh normal N:

$$D_N = \mathbf{f}(x_N, \mathbf{N}). \tag{11.90}$$

The distance $D_N$ between $N$ and $T$ should reduce continually during iteration, and finally they numerically coincide at some point on the tool surface, not necessarily at the original point $T$, because $T$ changes at each iteration. In general, the tool-normal direction at point $T$ will not be parallel to the mesh normal at $N$. However,

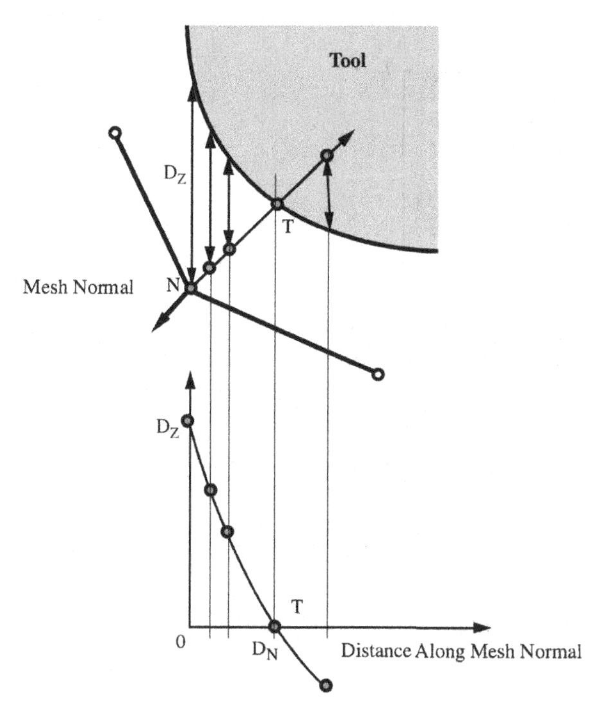


Figure 11.12. Searching for $D_N$ along the mesh-normal direction.

by assuming that the tool normal, which is never determined or used in our program, is parallel to the mesh normal (which is known), we can implement the N-CFS algorithm more simply. In numerous tests, we found that this simplification was always at least as stable and as time efficient as the more complicated algorithm, which makes use of both tool and mesh normal in defining the derivatives of $D_N$ with respect to the nodal positions.

To construct the stiffness matrix where the basic unknown is $\Delta\mathbf{u}$, we need to find the partial derivatives of $D_N$ with respect to all related $\Delta\mathbf{u}$, as expressed by the chain rule:

$$\mathbf{C}_N = \frac{\partial D_N}{\partial(\Delta\mathbf{u})} = \left.\frac{\partial D_N}{\partial(x_N)}\right|_N \frac{\partial(x_N)}{\partial(\Delta\mathbf{u})} + \left.\frac{\partial D_N}{\partial\mathbf{N}}\right|_{X_N} \frac{\partial\mathbf{N}}{\partial(\Delta\mathbf{u})}. \tag{11.91}$$

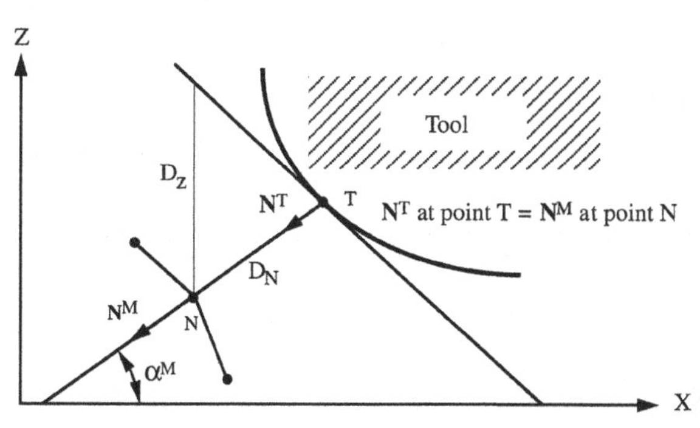


Figure 11.13. Contact distance from the nodal position to the tool surface along the mesh-normal direction.

In matrix form this is

$$\mathrm{C}_N = \begin{bmatrix} \dfrac{\partial D_N}{\partial \Delta u_x} \\ \dfrac{\partial D_N}{\partial \Delta u_y} \\ \dfrac{\partial D_N}{\partial \Delta u_z} \end{bmatrix}^T = \begin{bmatrix} \left.\dfrac{\partial D_N}{\partial \Delta u_x}\right|_N + \left.\dfrac{\partial D_N}{\partial \mathbf{N}}\right|_{X_N} \dfrac{\partial \mathbf{N}}{\partial \Delta u_x} \\ \left.\dfrac{\partial D_N}{\partial \Delta u_y}\right|_N + \left.\dfrac{\partial D_N}{\partial \mathbf{N}}\right|_{X_N} \dfrac{\partial \mathbf{N}}{\partial \Delta u_y} \\ \left.\dfrac{\partial D_N}{\partial \Delta u_z}\right|_N + \left.\dfrac{\partial D_N}{\partial \mathbf{N}}\right|_{X_N} \dfrac{\partial \mathbf{N}}{\partial \Delta u_z} \end{bmatrix}^T . \tag{11.92}$$

The first term on the right is the partial derivative of $D$ when the mesh normal direction is held constant and the position of the principal node is varied infinitesimally. (Geometrically, this implies that all connecting nodes move in unison with the principal node in order to maintain the same mesh normal direction, but this detail need not be considered further.) The second term is the partial derivative when the principal nodal position is held constant and the mesh-normal direction has an infinitesimal variation. (Geometrically, this means that the connecting nodes are moved infinitesimally to obtain the infinitesimal change in the mesh normal while the principal node is held at a fixed position.)

The contact distance $D_N$ depends on the nodal position, the $z$-direction error, and on the mesh normal. It can be expressed as

$$D_N = \frac{D_z}{\sqrt{1+S_x^2}}, \tag{11.93}$$

where the derivative $S_x$ is denoted in Eq. (11.39) for the mesh-normal direction and the $D_z$ is not the same as the one used in the Z-CFS algorithm, as shown in Fig. 11.10. The $D_z$ defined here is the distance from the node to the tangential plane, not to the real tool surface. Because the tangential plane is considered as the local tool surface at point $T$, the $D_z$ will be regarded as the distance between the *local tool surface* (lts) and the node $N$ in the $z$ direction.

$$D_z = z^{\mathrm{lts}} - z^{\mathrm{node}}. \tag{11.94}$$

The derivatives of $D_N$ to the fundamental variable, $\Delta\mathbf{u}$, in two dimensions is quite straightforward.

$$\mathrm{C}_N = \frac{\partial D_N}{\partial(\Delta\mathbf{u})} = \frac{1}{\sqrt{1+S_x^2}}\frac{\partial D_z}{\partial(\Delta\mathbf{u})} + \frac{\partial\left(1/\sqrt{1+S_x^2}\right)}{\partial(\Delta\mathbf{u})} D_z, \tag{11.95}$$

or in matrix form

$$\mathrm{C}_N = \begin{bmatrix} \dfrac{\partial D_N}{\partial \Delta u_x} \\ \dfrac{\partial D_N}{\partial \Delta u_z} \end{bmatrix}^T = \begin{bmatrix} \dfrac{1}{\sqrt{1+S_x^2}}\dfrac{\partial D_z}{\partial \Delta u_x} + \dfrac{\partial\left(1/\sqrt{1+S_x^2}\right)}{\partial \Delta u_x} D_z \\ \dfrac{1}{\sqrt{1+S_x^2}}\dfrac{\partial D_z}{\partial \Delta u_z} + \dfrac{\partial\left(1/\sqrt{1+S_x^2}\right)}{\partial \Delta u_z} D_z \end{bmatrix}^T , \tag{11.96}$$

where

$$\frac{\partial\left(1/\sqrt{1+S_x^2}\right)}{\partial(\Delta\mathbf{u})} = -\frac{1}{\left(1+S_x^2\right)^{\frac{3}{2}}}\begin{bmatrix} S_x \cdot S_{xx} \\ S_x \cdot S_{xz} \end{bmatrix}. \tag{11.97}$$

The definition of the $S_x$, $S_{xx}$, and $S_{xz}$ are the first- and second-order derivatives, which again are calculated from the element mesh and saved at the beginning of each step. The three-dimensional formulation of the N-CFS method is obtained in a similar manner to the two-dimensional case.

### Comparison Between the N-CFS and the Z-CFS Methods

Three different algorithms have been proposed, implemented, and tested for the stretching–drawing FE calculation in SHEET-3 and SHEET-S: The direct nodal projection method (DNP), the Z-CFS method (with modification), and the N-CFS method. Figure 11.14 shows schematically the iteration number of these three methods when a typical stretching–drawing problem is simulated.

A brief summary of our experience with these methods is as follows. First, the DNP method, without special numerical tricks, works only for the first part of stamping, in which stretching is dominant and frictional contact nonlinearities are less significant. When draw-in starts, resulting in little stretching and large nodal displacement relative to the tools, the DNP method ceases to converge. Second, the Z-CFS method can overcome the difficulty of the DNP method and shows better convergence performance. However, it has been found that the convergence difficulty is severe when the wall angle is greater than approximately 70° in a drawing simulation. This limits the usefulness significantly for all deep-drawn parts. Third, the modified Z-CFS method with various numerical tricks can pass the "70° limit" after greatly increased iteration number and lengthened computation times. Some problems cannot be solved even by using every possible convergence-enhancing device, and in spite of very large

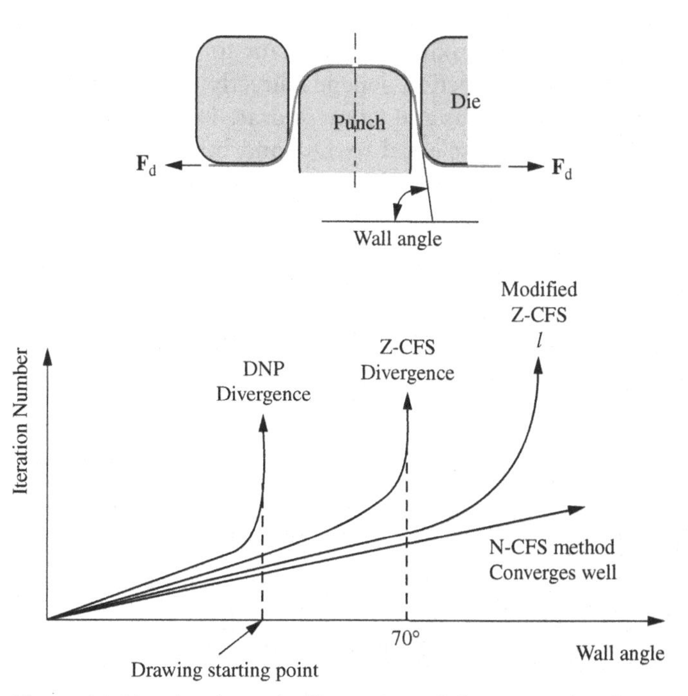


Figure 11.14. A schematic illustration of the accumulated iteration number of four methods when a typical stretching–drawing problem is simulated with an increasing wall angle from 0° to 90°.

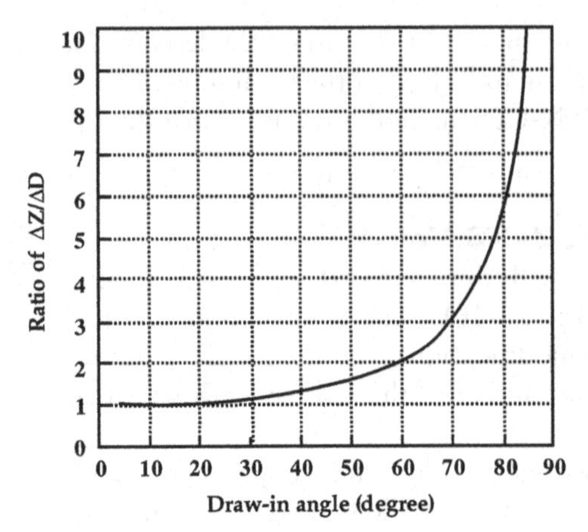


Figure 11.15. $D_Z/D_N$ ratio increasing as a function of draw-in angle.

iteration numbers, which sometimes are of the order of 200 per step. Fourth, the N-CFS method exhibits remarkable stability and strong convergence.

In the N-CFS method, the effect of the contact term on the nodal position updating is along the mesh-normal direction, instead of the $z$ direction. Updating along this normal minimizes the additional nonequilibrium force, therefore resulting in a reduced conflict between contact and equilibrium. In terms of the relation between equilibrium and contact, the N-CFS algorithm is more consistent, as both the contact error and contact pressure use the same mesh-normal direction.

Figure 11.15 illustrates why the N-CFS has better convergence. The contact error along the mesh-normal direction is approximately constant for any draw-in angle, whereas the contact error along the $z$ direction depends directly on this angle. Figure 11.15 shows the $D_Z/D_N$ ratio increasing as a function of draw-in angle. When the draw-in angle is less than 20°, $D_Z$ is almost equal to $D_N$, and both the Z-CFS and the N-CFS methods work equally well; at 60°, $D_Z$ is twice $D_N$; at 70°, $D_Z$ is three times of $D_N$; at 80°, six times. Beyond 70°, the ratio of $D_Z$ over $D_N$ increases exponentially with the draw-in angle toward 90°. Experience shows that beyond 70°, convergence is almost unobtainable for the Z-CFS method.

Prediction of the final nodal position is given in Fig. 11.16 for the simulation of a drawing operation with an arbitrary punch with eight radii (8-R), designed only for code-testing purposes. The sheet is initially flat and will be drawn into the die cavity from both sides with the same drawbead restraining force, $F_d = 300$ N/mm. A friction coefficient of $\mu = 0.1$ was used. The material parameters employed are those for an IF steel. In this example, there are more critical points where new contact areas appear and nodal displacement will change direction abruptly. This complex geometry was used to demonstrate the draw and redraw performance of the N-CFS algorithm.

Figure 11.16 shows the final nodal positions at a final punch height of 36 mm. Material starts to be drawn in from both sides when the punch height is 13 mm.

A comparison of various algorithms of SHEET-S shows the following.

1. The direct nodal projection algorithm does not converge beyond the stretch–draw transition.
2. The Z-CFS algorithm converges with an average of 27 iterations per step up to a punch travel of 19 mm, at which point the solution diverges.

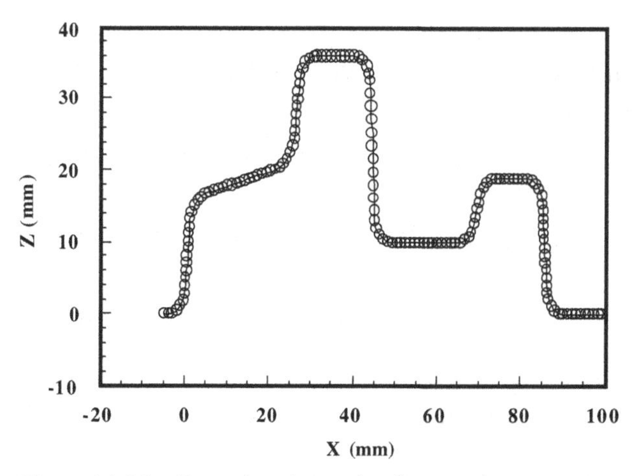


Figure 11.16. Deep drawing and redrawing by an 8-R punch.

3. The N-CFS algorithm converges to the final punch height, 36 mm, without difficulty. The CPU time used for this simulation in a VAX-8500 is 1280 s and the average iteration number is nine times, with 1 mm/step punch travel.

## 11.7 Numerical Solution Procedure

### Newton Iteration and Line Search Algorithm

Chapter 3 introduced the Newton–Raphson method for solving a nonlinear system. Equation (11.86) defines a set of nonlinear equations to be linearized for solution by this method.

The N-CFS algorithm introduces one more variable to the basic variable at each node. Now the generalized nodal variable is represented by a vector $\Delta\mathbf{r}$:

$$\Delta\mathbf{r} = \begin{bmatrix} \Delta u_x \\ \Delta u_y \\ \Delta u_z \\ \Delta P \end{bmatrix}. \tag{11.98}$$

By using this generalized variable, we can express the global equilibrium and contact condition in a compact form:

$$\mathbf{R}_I(\Delta\mathbf{r}) = \mathbf{R}_E(\Delta\mathbf{r}). \tag{11.99}$$

The generalized internal force vector $\mathbf{R}_I$ and $\mathbf{R}_E$ for a general node $k$ can be defined respectively as

$$\mathbf{R}_I^k = \begin{bmatrix} \mathbf{F}_I(\Delta u_x) \\ \mathbf{F}_I(\Delta u_y) \\ \mathbf{F}_I(\Delta u_z) \\ D_N \end{bmatrix}^k, \qquad \mathbf{R}_E^k = \begin{bmatrix} \mathbf{F}_E(\Delta u_x) \\ \mathbf{F}_E(\Delta u_y) \\ \mathbf{F}_E(\Delta u_z) \\ 0 \end{bmatrix}^k. \tag{11.100}$$

With the use of the Newton method, the stiffness equation becomes

$$\mathbf{K}\delta\Delta\mathbf{r} = (\mathbf{R}_E - \mathbf{R}_I)|_{\Delta\mathbf{r}^*}, \tag{11.101}$$

where the $\mathbf{K}$ is the tangent stiffness matrix and $\delta\Delta\mathbf{r}$ is a correction vector that will be zero at the end of a time step if convergence is achieved. Letting superscript $i$ denote the current iteration number, we express the stiffness equation in iteration $i$ by rewriting Eq. (11.101) as

$$\mathbf{K}^{i-1}\delta\Delta\mathbf{r}^i = (\mathbf{R}_E - \mathbf{R}_I)|_{\Delta\mathbf{r}^{i-1}}, \tag{11.102}$$

where only the $\delta\Delta\mathbf{r}$ is unknown. The correction rule during the iteration is expressed as

$$\Delta\mathbf{r}^i = \Delta\mathbf{r}^{i-1} + \gamma\delta\Delta\mathbf{r}^i, \tag{11.103}$$

where $0 \le \gamma \le 1$, and the accumulation rule after convergence at the end of a time step is (note that the step number is denoted by a presuperscript, while the iteration number by a postsuperscript)

$$^{n+1}\mathbf{r} = {}^{n}\mathbf{r} + {}^{n}\Delta\mathbf{r}. \tag{11.104}$$

In the N-CFS algorithm, if the contact status of a node is changed from contact to free, no accumulation for the $P$ in Eq. (11.98) is needed. This $P$ should be set to zero. If this node becomes a contact node later, the $P$ should be accumulated from zero again.

The terms in Eq. (11.102) may be expressed in matrix form as

$$\begin{bmatrix} \mathbf{K}_I + \mathbf{K}_E & \mathbf{N} \\ \mathbf{C}_N & 0 \end{bmatrix}^{i-1} \begin{bmatrix} \delta\mathbf{u} \\ \delta\mathbf{P} \end{bmatrix}^{i} = \begin{bmatrix} \mathbf{F}_E \\ D_N \end{bmatrix}^{i-1} - \begin{bmatrix} \mathbf{F}_I \\ 0 \end{bmatrix}^{i-1}, \tag{11.105}$$

where $i$ and $i-1$ denote the current and previous iteration number, respectively. The following detailed expressions are presented explicitly as follows:

$$\mathbf{K}_i = \frac{\partial \mathbf{F}_i(\Delta\mathbf{u})}{\partial \Delta\mathbf{u}} = h_0 \int_{A_0} \left[ \bar{\sigma}(\bar{\varepsilon} + \Delta\bar{\varepsilon}) \frac{\partial^2(\Delta\bar{\varepsilon})}{\partial\Delta\mathbf{u}\,\partial\Delta\mathbf{u}} + \left( \frac{\partial\bar{\sigma}}{\partial\bar{\varepsilon}} + \frac{1}{\Delta t}\frac{\partial\bar{\sigma}}{\partial\dot{\bar{\varepsilon}}} \right) \frac{\partial\Delta\bar{\varepsilon}}{\partial\Delta\mathbf{u}} \frac{\partial\Delta\bar{\varepsilon}}{\partial\Delta\mathbf{u}} \right] \mathrm{d}A, \tag{11.106}$$

$$\mathbf{K}_E = \frac{\partial \mathbf{F}_E}{\partial\Delta\mathbf{u}} = P\left[ \frac{\partial\mathbf{n}}{\partial\Delta\mathbf{u}} - \mu\left( \frac{\partial\mathbf{t}}{\partial\Delta\mathbf{u}}\phi + \frac{\partial\phi}{\partial\Delta\mathbf{u}}\mathbf{t} \right) \right], \tag{11.107}$$

$$\mathbf{N} = \mathbf{n} - \mu\phi\mathbf{t}, \tag{11.108}$$

$$\frac{\partial\phi}{\partial\Delta\mathbf{u}} = \frac{3}{\delta\cosh^2(3|\Delta\mathbf{u}_t|/\delta)} \frac{\partial|\Delta\mathbf{u}_t|}{\partial\Delta\mathbf{u}}, \tag{11.109}$$

where $h_0$ and $A_0$ are the element thickness and area at the beginning of the time step.

$\mathbf{C}_N$ is the expression of derivatives of $D_N$ with respect to $\Delta\mathbf{u}$ given as follows:

$$\mathbf{C}_N = \left[ \frac{\partial(D_N)}{\partial x} \quad \frac{\partial(D_N)}{\partial y} \quad \frac{\partial(D_N)}{\partial z} \right]. \tag{11.110}$$

The detailed expressions for $D_N$ and $\mathbf{C}_N$ have been given in a previous section.

In Eqs. (11.106) and (11.107), the integrands are calculated at the incremental effective strain produced by the trial solution. A line element for two dimensions and a constant strain triangular element for three dimensions have been chosen for their simplicity. With the use of Hill's new yield criterion or Hosford's yield function, a Hollomon hardening law or a Voce hardening law, the resulting stiffness and force-related matrices and arrays can be calculated.

### Trial Solution, Convergence, and Update

For each incremental time step, an initial trial solution $\Delta\mathbf{u}^*$ is required to start the iterative procedure. For the very first step in the two-dimensional code, the trial solution is assumed to be a zero displacement field. In the three-dimensional code, the trial solution is obtained by solving the problem with a nonlinear elastic theory. For subsequent steps, the obtained solution $\Delta\mathbf{u}$ in the previous time step is used to initiate the iteration. If the time step length in the current step $\Delta t_i$ and in the previous step $\Delta t_{i-1}$ is different, the trial solution has to be scaled and is simply equal to

$$\Delta\mathbf{u}^{*i} = \Delta\mathbf{u}^{i-1}\frac{\Delta t_i}{\Delta t_{i-1}}. \tag{11.111}$$

A fractional norm Re is defined as the "residual" force calculated after each iteration by the following equation:

$$\mathrm{Re}^{i-1} = \|[\mathbf{F}_E]^{i-1} - [\mathbf{F}_I]^{i-1}\|. \tag{11.112}$$

It was found that direct use of the norm on the right side of Eq. (11.105) as the residual force is not necessary, because the difference between the dimension of force terms and the contact error term is usually 3–5 orders of magnitude. The residual force Re is a sufficient indicator of equilibrium. Theoretically, a "true" convergence can be declared only when the residual force is equal to zero. Alternatively, another fractional norm $N_f$, defined by

$$N_f = \frac{\|\delta\Delta\mathbf{u}\|}{\|\Delta\mathbf{u}\|}, \tag{11.113}$$

is used to monitor the convergence of iteration process. Numerically, if the $N_f$ is zero, it means a local minimum has been found for the residual force, while this local minimum is not necessarily the global minimum. Consequently, a residual force check is preferred to guarantee that an equilibrium result can be obtained. Convergence is completed if $N_f$ is smaller than a given tolerance, typically $10^{-5}$, and Re is smaller than $10^{-3}$ times the norm of the external forces.

Chapter 3 introduced the Newton method for solving a nonlinear system. Equation (3.45) shows one of the methods to find a positive scalar $\gamma$, Eq. (11.103), for better convergence. An alternative method for finding the $\gamma$ during iteration is presented here.

Using the previous solution $\Delta\mathbf{u}$ to initiate the iteration is a standard method. However, if the contact condition or the boundary condition has a sudden change, for example,

- when the deformation changes from stretching to drawing,
- when the draw-in restraining force is changed,
- when a new contact region appears somewhere inside of a contact-free area,

the initial solution will be far removed from the actual solution. Whenever the trial solution is not in the close neighborhood of the actual solution, convergence is very difficult to achieve. An attempt to reduce this difficulty is made by using a *line search* technique, shown in Fig. 11.17, to determine a useful value of $\gamma$ in Eq. (11.103). The determination of the value of $\gamma$ is a critical phase of the iterative scheme. The efficiency

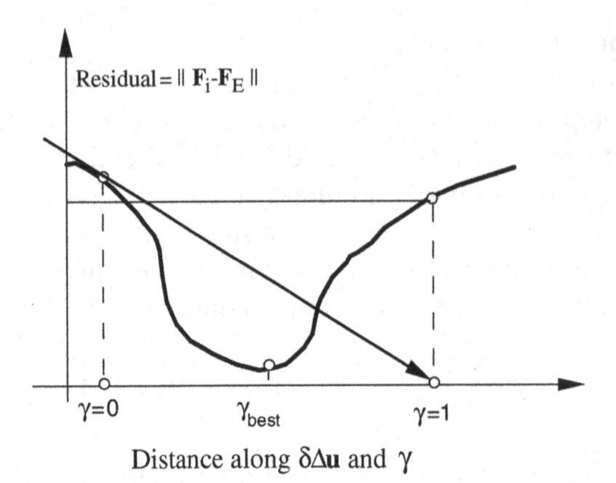


Figure 11.17. Line search along the Newton direction.

of the method to find the best $\gamma$ value is directly related to the stability of the scheme with respect to convergence, stability, and robustness.

In our implementation of line searching, values of $\gamma$ equal to 0.1, 0.2,. . ., 1.0 are used to calculate Re and the "best" $\gamma$, that is, the one that minimizes Re, is found. This calculation is fast because it does not require solution of the stiffness equation. A numerical study showed that the residual force curve as a function of $\gamma$ value has a very complex shape, thus the minimum residual force can be anywhere in the interval (0, 1). Generally speaking, in a typical time step, the best $\gamma$ value changes gradually from the left end to the right end in the interval (0, 1) during the whole iteration. The time step is automatically divided by 2 if no convergence is found after a certain number of iterations. The iteration of the Newton–Raphson method is then started again from a rescaled initial solution.

### Contact Checking

At the beginning and end of each time increment, the contact status has to be checked for each node. If inconsistencies are found, the contact conditions are updated and iteration proceeds. No contact checking is made during an equilibrium iteration.

At the beginning of each time step, the punch moves a certain distance parallel to the $z$ axis. Before the appropriate force vectors and the stiffness matrices are calculated, it is decided, in a node-by-node manner, whether a particular node should change its contact situation, either from a contact node to a free node or vice versa. Each node located on or inside the tool is considered to be a contact node. Each contact node with a tensile force on the tool (i.e., an adhesive force) is set free and its contact pressure, $P$, is set to zero.

At the end of an incremental step, the N-CFS algorithm fulfills both the geometrical and the equilibrium constraints. If a free node is found located inside the tool, the node is updated to be a contact node and another equilibrium iteration sequence is required. Similarly, if the contact force is adhesive, the node is released and another equilibrium iteration sequence is initiated. While it is tempting to combine contact checking and equilibrium iteration in the same loops, experience shows that, at the best, such procedures incur additional CPU time and, at the worst, destroy convergence.

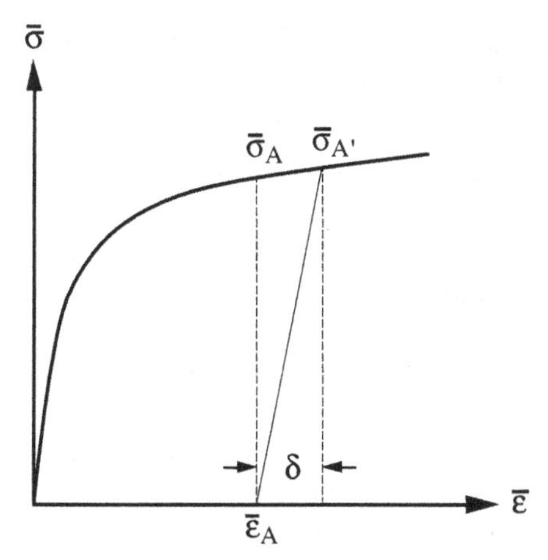


Figure 11.18. Modification for dealing with an unloading problem in a rigid-plastic FE.

### Local Unloading Problem with Rigid Plasticity[17]

During deformation, the effective stress follows the hardening curve until a material element faces unloading. Unloading typically occurs at the beginning of the punch stroke, when the deformation changes from stretching to drawing, and when some elements decrease their load-bearing capacity because of geometric changes. The rigid-plastic constitutive relation allows unloading to happen with no deformation; thus the one-to-one correspondence between stress and strain needed for a solution is destroyed.

A two-stage plastic hardening law, consisting of a linear hardening curve[18] and a standard nonlinear hardening curve, can be used for the unloading element. At each time step, a straight line is introduced between $\bar{\varepsilon}_A$ and $\bar{\sigma}_{A'}$, as shown in Fig. 11.18, so that the hardening law for each step consists of two parts, divided by an incremental effective strain tolerance, $\delta$, typically 0.001. As long as $\delta$ is small relative to the strain increment for a given step, the linear part of the curve will not affect the solution. In other cases, when only limited regions of the sheet undergo elastic unloading, the errors in the loading regions will be insignificant, and in the near-unloading regions of the order of $n\delta$, where $n$ is the number of near-unloading steps for a given element.

This quasi-elastic unloading is not physically realistic, because plastic normality is used to determine the direction of straining, instead of using Hooke's law for elasticity. During one-dimensional unloading, for example, this unloading procedure delivers small strain increments opposite to those of true elastic unloading.

Since deformation within the first region obeys plastic constitutive equations, the plastic hardening would be accumulated and permanent even under unloading. In a stretch-forming operation, such an error is relatively small because few unloading steps can be accumulated in a small region before failure occurs. The total error may be of the order of elastic strain. With a drawing simulation with low draw-in restraining forces and a deep and complex-shaped die cavity, the number of unloading elements and steps may be large and, consequently, the accumulated error may

[17] Y. Germain, K. Chung, and R. H. Wagoner, "A Rigid-Viscoplastic Finite Element Program for Sheet Metal Forming Analysis," *Int. J. Mech. Sci.* 31, 1–24 (1989).

[18] Other initial forms may be used. A hyperbolic tangent form has been used to eliminate the discontinuous derivative at a transition point. This difference has little effect on accuracy or efficiency in most cases.

become significant. For this reason, we do not update the material hardness in elements deforming in the quasi-elastic linear range. This method has proved to be effective in handling material unloading without introducing significant time penalties.

#### Implementation of Drawbead Positions in Two Dimensions

For a two-dimensional code, it is convenient to set a force boundary condition at the two end nodes to represent a drawbead. This approach can lead to errors because the material outside the drawbead may deform in the simulation, but not in the real operation. Therefore, we set the thickness (or alternatively, the strengths) of the elements located outside the drawbead much higher than the actual one. When an element located outside passes the drawbead, the initial actual thickness of the sheet is reassigned to it. This treatment ensures that there is virtually no plastic deformation in the elements outside of the drawbead under the given restraining force. This approximation may not be accurate in three-dimensional cases, because plastic deformation may occur in the sheet beyond the drawbead as draw-in occurs, particularly in curved areas of drawbead where in-plane compression is required.

#### Implementation of Blank-Holder Positions in Two Dimensions

A blank holder is commonly used in practice to constrain the blank against the die outside the die cavity. The blank may contact both the tool and the blank holder, or either one alone. During normal simulation practice, the code will check all contacting nodes to determine if the contact pressure is nonnegative; that is, the code must ensure that the external pressures are all compressive. If a node is found with a negative pressure, it is set free of contact and another iteration loop is initiated. However, between the blank holder and the die, there is no need to check this condition or to update the node constraint. A blank-holder boundary can be specified so that the pressure check will not be performed for nodes beyond the draw bead. This modification improves the computation speed and convergence rate greatly.

#### Draw-in Boundary Condition in Two Dimensions

In a two-dimensional linear element code, the two end nodes initially have fixed positions, until the internal force exceeds a specified value, the draw-in restraining force, $\mathbf{F}_d$. For draw simulation, a force boundary condition ($\mathbf{F}_i = \mathbf{F}_d$) is applied when the internal force exceeds the limit.

For a flat blank holder without drawbead, we set the draw-in restraining force depending on the blank-holder pressure ($P_{\text{bh}}$), the area under the blank holder ($A$), and the friction coefficient ($\mu$) as follows:

$$|\mathbf{F}_d| = 2\mu P_{\text{bh}} A. \qquad (11.114)$$

Here $\mathbf{F}_d$ is antiparallel to the internal force $\mathbf{F}_i$ acting on the end node. The internal draw-in force $\mathbf{F}_i$ is expressed by its components as follows:

$$\mathbf{F}_i = \mathbf{F}_x + \mathbf{F}_z. \qquad (11.115)$$

$\mathbf{F}_d$ may be set separately for the two end nodes, thus allowing for one- or two-ended draw or stretch, or differential draw, as desired.

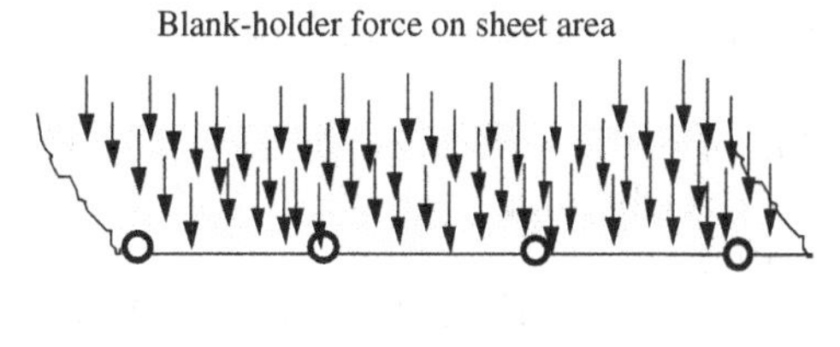


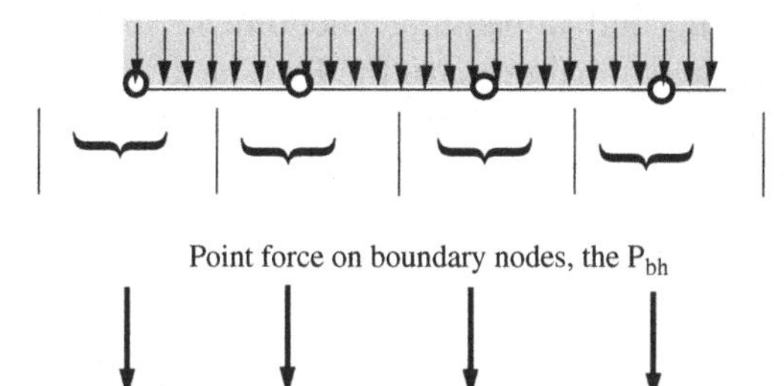


Figure 11.19. Transfer of the area-force type of blank-holder force to the point-force type of boundary node press.

### Draw-in Boundary Condition in Three Dimensions

In a three-dimensional membrane code, the following simplistic assumptions can be used for determining the boundary condition for draw-in, assuming that a draw-bead is absent:

1. The blank-holder pressure distribution on the contact area under the blank holder is known.
2. The area force can be transformed equivalently to a distributed line force acting on the periphery, and finally to a set of discretized boundary nodal forces along the peripheral nodes.
3. Coulomb's friction exists between the sheet and the tool.

In the case of uniform blank-holder pressure, for example, the basic procedure to find the draw-in constraining force $\mathbf{F}_d$ on each boundary node is proposed as follows.

1. Calculate the total length of the periphery of the sheet blank.
2. Calculate the line force per unit length acting on the periphery by dividing the total blank-holder force by the total length of the periphery.
3. Discretize the line force to each individual boundary node according to the adjacent element length. The blank-holder pressure can be transferred to a nodal pressure, $P_{bh}$, on each boundary node, as shown in Fig. 11.19.
4. Calculate the draw-in restraining force. The draw-in restraining force depends on the nodal pressure ($P_{bh}$) and the friction coefficient ($\mu$) as follows:

$$\mathbf{F}_d = 2\mu P. \tag{11.116}$$

Unlike the two-dimensional case, the $\mathbf{F}_d$ on a three-dimensional node does not have a certain direction but does have a definite magnitude. The inward draw-in force $\mathbf{F}_i$ is obtained by the components of the nodal internal force ($\mathbf{F}_x$, $\mathbf{F}_y$, and $\mathbf{F}_z$) and is expressed as follows:

$$\mathbf{F}_i = \mathbf{F}_x + \mathbf{F}_y + \mathbf{F}_z. \tag{11.117}$$

When the magnitude of the inward draw-in force $\mathbf{F}_i$ exceeds the prescribed draw-in restraining force $\mathbf{F}_d$, deformation will be changed from stretching to drawing. The draw-in restraining force $\mathbf{F}_d$ will take the value of the inward draw-in force at this critical moment. If the $\mathbf{F}_i$ later drops below the $\mathbf{F}_d$, the boundary condition is changed back to a stretch one; that is, a fixed nodal position.

In the case in which a drawbead is used, the major part of the draw-in restraining force is not caused by the friction in the interface between sheet and the blank holder or the die surface; instead, the friction, bending, and unbending phenomena that occur in the drawbead area make the predominate contribution to the $\mathbf{F}_d$. The draw-in restraining force can be determined either by testing on a drawbead experimental device, or by a solid-element simulation.

## 11.8 Example Simulations

The SHEET-S and SHEET-3 programs described in some detail above have been used for a variety of applications during the past 10 years. While the individual references should be consulted for details of the simulation techniques and the results, a few brief examples may be useful in illustrating the range of capability. In addition, several examples of in-plane simulations of forming tests were presented. These simulations were carried out by using an in-plane version, SHEET-2, of a program with features similar to SHEET-S and SHEET-3.

### Section Analysis of a Hood Inner Panel[19]

One of the important choices in FEM analysis centers on the assumed symmetry of the geometry, such as tooling, workpiece properties, and boundary conditions. It is possible in many cases to reduce the size and complexity of a problem greatly by using special assumptions such as axisymmetry or plane strain. Axisymmetric and plane-strain symmetries are very similar in formulation, implementation, and execution time.

On one hand, most stamping operations require a three-dimensional model. On the other hand, there are local sections in a complex stamping (see sections A-A or D-D in Fig. 11.20) where the forming operation can be usefully simulated by a two-dimensional plane-strain sectional analysis. The two-dimensional procedure is strictly accurate only for the forming operations that have long regions of constant cross section.

Numerically stable two-dimensional computer codes have been used for routine production. The simulations performed with these codes have provided valuable information that can be obtained before soft tool tryout. The effects of changes in the part geometry, material properties, and boundary conditions can be evaluated at substantially lower cost and faster time. Two-dimensional simulations provide an effective means for the design optimization of tooling.

For an inner panel of DQSK (drawing-quality, specially killed) sheet steel, the mechanical properties are as follows:

- plastic anisotropy parameter, average $R = 1.46$;
- Hill's yield function parameter, $M = 2.0$;

[19] Y. T. Keum, C.-T. Wang, M. J. Saran, and R. H. Wagoner, "Practical Die Design via Section Analysis," *J. Mat. Process. Tech.* 35, 1–36 (1992).

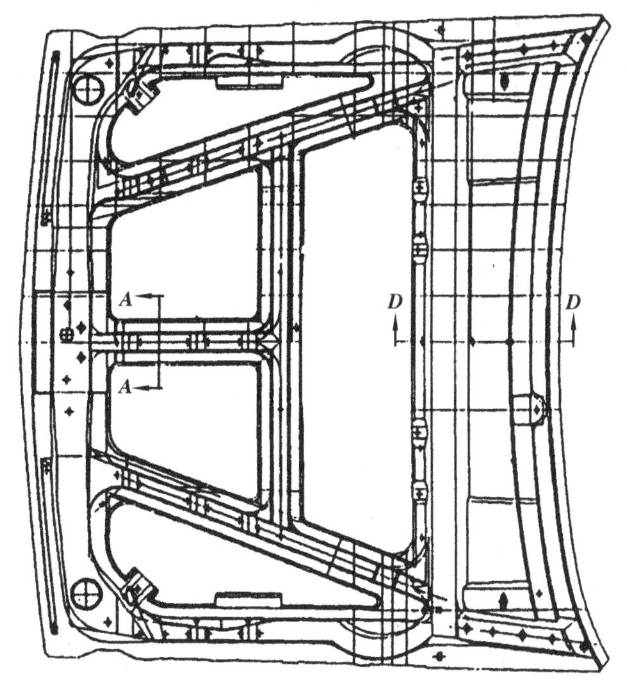


Figure 11.20. Top view of a hood inner panel, showing the positions that can be analyzed by a plane-strain code.

- hardening law, $\bar{\sigma} = K(\bar{\varepsilon} + \varepsilon_0)^n$, where $K = 503$ MPa, $\varepsilon_0 = 0.004$, and $n = 0.2$;
- Coulomb's friction coefficient, $\mu = 0.09$;
- Young's modulus, $E = 207$ GPa;
- thickness of sheet blank, $t = 0.64$ mm.

### Section A-A

Figure 11.21(a) shows the tooling geometry used in the analysis of Section A-A. In fact, since this section is symmetric with respect to the center line, only the right half was modeled. The thickness strains for the final and original tool shapes at full punch depth are shown in Fig. 11.21(b); measured thickness strains are also shown as circles. A good agreement between the FEM analysis and measurement is revealed. (The measurements were done on production panels made with the "final tools.") In the analysis of the final and original tool shapes, the maximum thickness strains are 12% and 43%, respectively. Considering the maximum allowable thinning of approximately 23% for the target material, we find the original tool shape is expected to fail, by a wide margin, at the highly strained area. The shape optimized during die tryout, however, is predicted by the FEM to be a safe design.

### Section B-B

Figure 11.22(a) shows the tooling geometry used in the analysis of Section B-B. The inclusion of the surface feature adjacent to the original Section B-B was found necessary in order to achieve a satisfactory modeling result; 98 line elements were used. The predicted thickness strains for the final and original tool shapes at the full punch depth are shown in Fig. 11.22(b); measured thickness strains are also shown as circles. A good agreement between the FEM analysis and measurement is revealed.

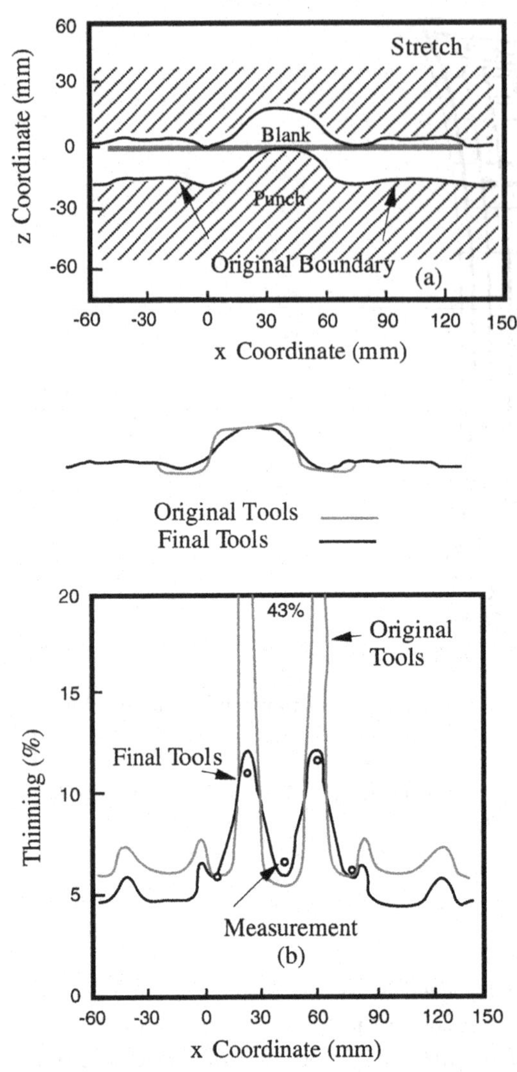


Figure 11.21. (a) Tooling geometry for analyzing Section A-A and tool profiles showing the final and original shapes. (b) Comparison of percentage of thinning in Section A-A.

In the analysis of the final and original tool shapes, the maximum thickness strains are 16% and 42%, respectively. The maximum allowable thinning of approximately 23% for the target material and the original tooling shape results in greatly exceeding the forming limit.

### Ohio State University Formability Test[20]

A sheet-metal formability test measures the degree to which a metal sheet can be stretched or drawn before failure. Examples include tensile tests, limiting dome height (LDH) tests, and Erichsen/Olsen cup tests. Of these, the LDH test has shown good correlation with press performance.

[20] F. I. Saunders and R. H. Wagoner, "Finite Element Modeling of a New Formability Test," in *Computer Applications in Shaping & Forming of Materials*, M. Y. Demeri, ed. (TMS, Denver, 1993), pp. 205–220.

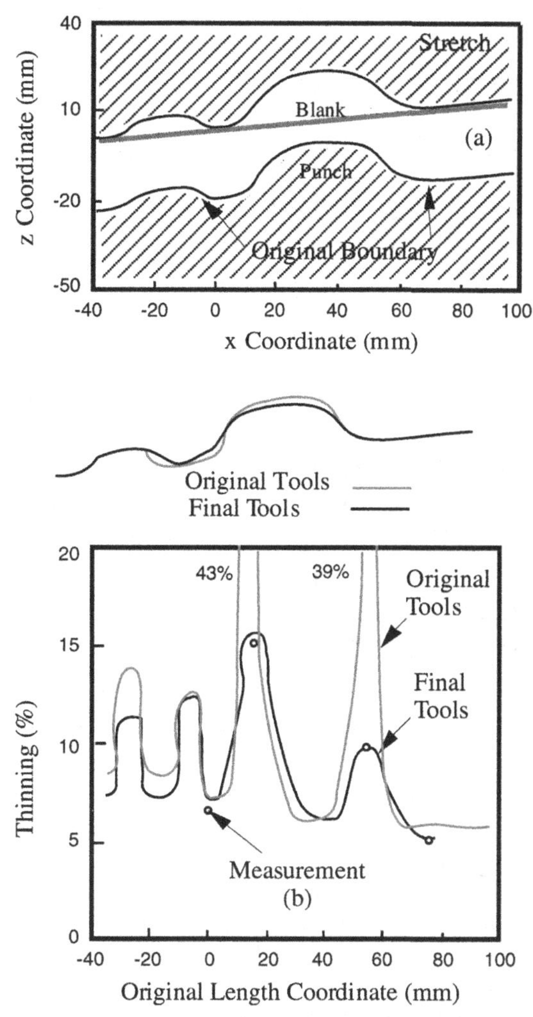


Figure 11.22. (a) Tooling geometry for analyzing Section B-B and tool profiles showing the final and original shapes. (b) Comparison of percentage of thinning in Section B-B.

In order to improve on the LDH test, the **Ohio State University (OSU) formability test** was proposed. The design was based on a cylindrical version of the LDH geometry in order to retain its press performance correlation and to eliminate problems of poor reproducibility and long testing times. The OSU formability test was designed to promote stable plane-strain conditions in the failure region for various materials and processing conditions.

Originally three cylindrical type geometries were proposed for the new test (Fig. 11.23). Initial experimental studies showed the flat-elliptical type geometry produced the strain state at failure closest to plane strain, so this geometry was used for verification of the new test. The scatter of test results for the OSU test was about half, on average, of the LDH test and it was 5–10 times faster to perform.

However, it was desirable to optimize the final geometry of the OSU formability test so that it was suitable for a wide range of plastic anisotropy parameters, to minimize scatter, and to minimize the load capacity. For these goals to be achieved, a FEM model was constructed. Figure 11.24 shows the FE mesh for the OSU formability

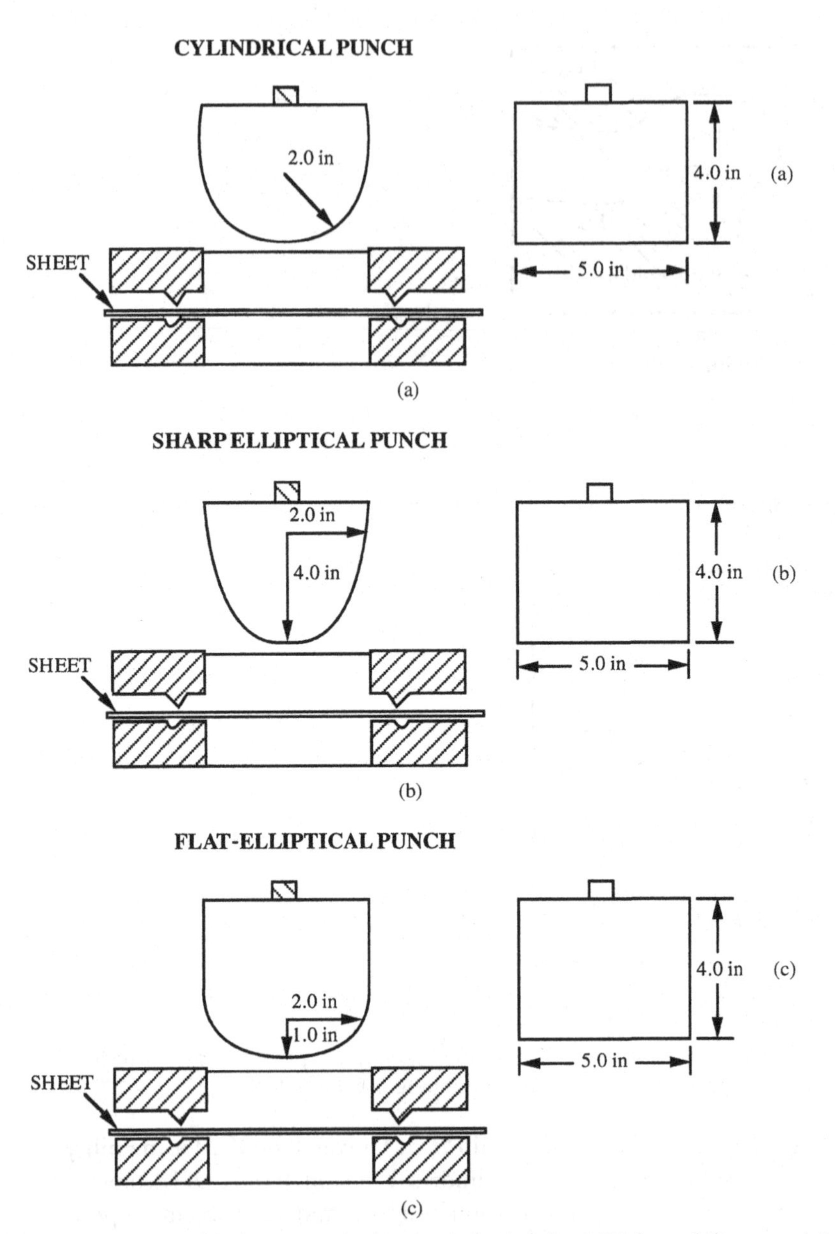


Figure 11.23. The three proposed geometries for the OSU formability test: (a) the cylindrical, (b) the sharp elliptical, and (c) the flat-elliptical geometries.

test. Because of symmetry, the mesh only represents one quarter of the blank. The mesh has 256 nodes and 450 elements.

A three-dimensional rigid-viscoplastic membrane element was used in the simulation. Figure 11.25 compares the simulated strains to experimental ones. The simulations were carried out by using Hill's 1948 (old) and 1979 (new) yield functions. The new criterion features an experimentally determined $M$ value of 2.5, based on uniaxial tension and plane-strain tension test for the IF steel. The friction coefficient used is the value that produced best-fit simulated strains. The results show that the 1979 theory gives more accurate results than the 1948 criterion. For the simulated

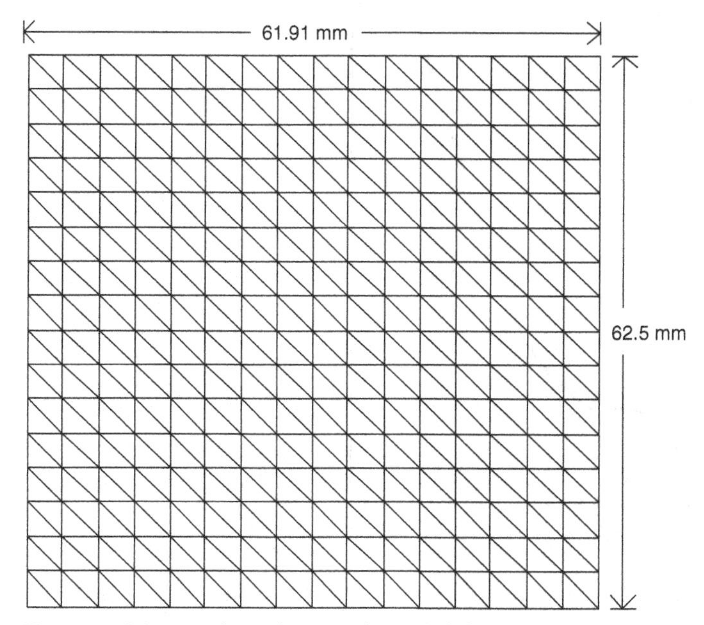


Figure 11.24. Mesh used in simulation of the OSU formability test.

strains using Hill's old theory, the major strain is too large over the contact region while it is too small in the unsupported region. Because the amount of stretch that is prescribed by the boundary conditions has to be constant, the major strain can be modified by adjusting the friction coefficient to redistribute the strains over the punch and fit the experimental values. However, the minor strains are uniformly greater for Hill's 1948 theory than the experimental strains and the 1979 theory prediction. The strain ratios for Hill's 1979 theory are also much closer to experimental results.

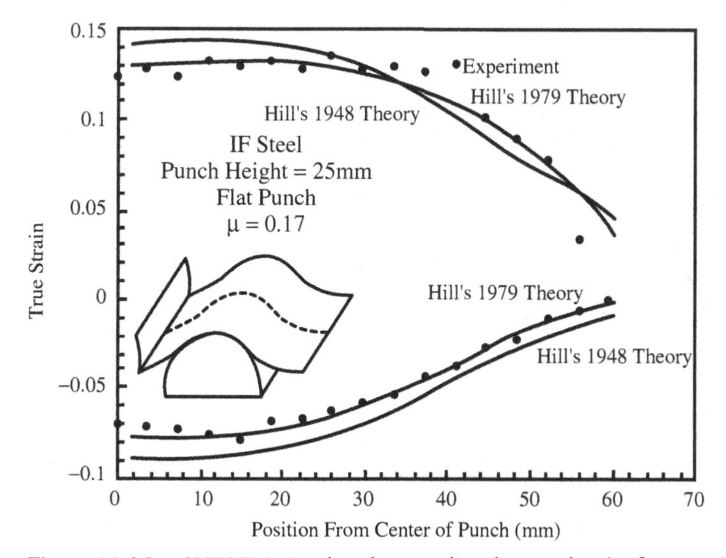


Figure 11.25. SHEET-3 simulated strain distribution for the flat punch compared to experiments using Hill's 1948 and 1979 theories for IF steel.

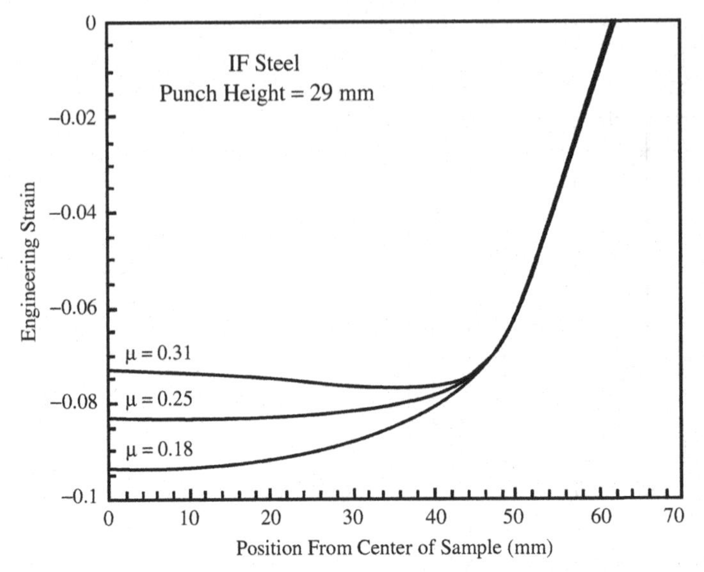


Figure 11.26. Sensitivity of the lateral strain of the blank to the friction coefficient.

To observe how the friction coefficient influences the minor strain, we show the sensitivity of the lateral strain of the blank to friction for the flat elliptical geometry in Fig. 11.26. The strain distribution changes dramatically in the contact region, but it is virtually unaffected in the region between the tools.

Figure 11.27 shows the simulated major to minor strain ratios for three $r$ values for IF steel (punch height = 26 mm, $\mu = 0.18$) plotted as a function of position normal to the major stretch axis as shown on the schematic in the figure. The strain state in the unsupported region of the blank is very sensitive to $r$, with a small value favoring the plane strain. If the major strain is held constant, decreasing the $r$ value decreases the amount of minor strain, increasing the amount of plane strain in the unsupported region.

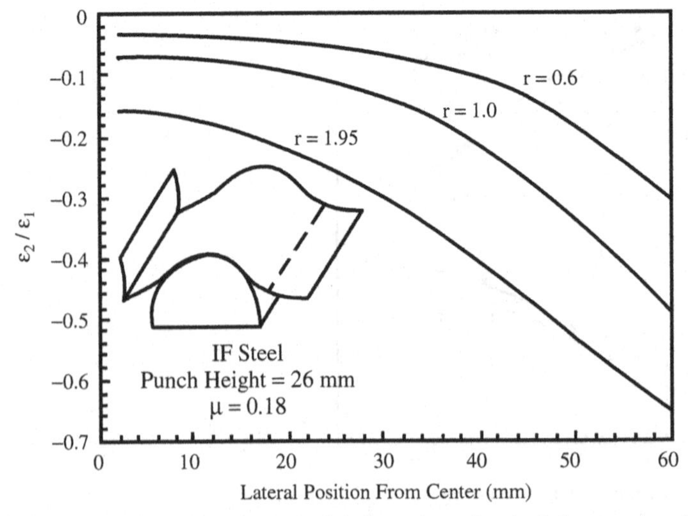


Figure 11.27. 3-D FEM simulated strain ratios in failure region for flat punch for different $r$ values.

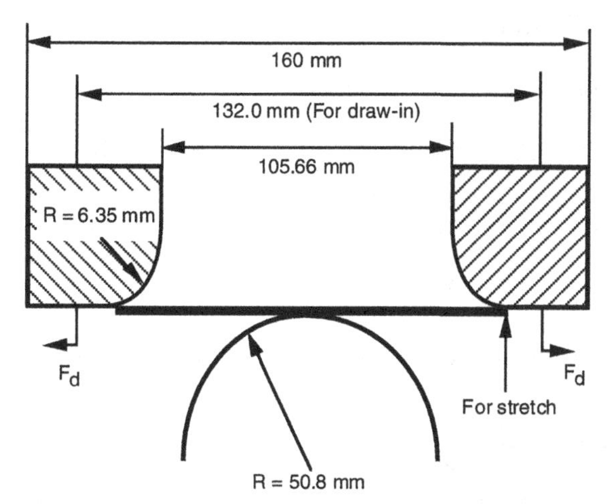


Figure 11.28. The OSU benchmark test: the overall geometry is either axisymmetric or plane strain.

## 11.9 Performance of SHEET-3 in International Benchmark Tests[21]

Benchmarking is a procedure by which a diverse set of simulations or measurements can be measured or compared. This procedure is useful for comparing the performance of sheet-forming FEM programs. Along with the growth of numerical simulation of sheet-metal forming is the growth of interest in the benchmark tests of simulation and experimental techniques.

### OSU Benchmark Test[22]

By 1989, at the time of the first sheet-forming benchmark test, sheet-forming analysis had attracted a great deal of interest, and preliminary investigations were underway with regard to element formulations, testing of commercial programs, friction laws, constitutive equations, and formability limits. The ability to simulate general two- or three-dimensional sheet-forming operations reliably, accurately, and consistently had not been demonstrated. The emphasis was on simple shapes for testing, and most programs depended on the membrane element. Nearly all programs were based on the static implicit method, although a few researchers were developing static explicit programs, usually based directly or indirectly on DYNA 3-D. Programs had not been linked to industrial CAD/CAM packages to enable routine use with production parts and die configurations.

The OSU benchmark test of 1989 revealed the status of the field. The test itself was constructed to provide the simplest geometries, material properties, and boundary conditions representing sheet-forming operations. Standard meshes were provided. Both plane-strain and axisymmetric configurations were modeled, although the test called for three-dimensional simulations of the axisymmetric case, if this capability was available. Surprisingly, 25 sets of data were submitted for this problem.

[21] R. H. Wagoner, and D. Zhou, "Overview: Benchmarking of Sheet Metal FEM Programs," presented at the Near Net Shape Manufacturing Conferences, ASM, Sept. 1993, pp. 1–12.

[22] J. K. Lee, R. H. Wagoner, and E. Nakamachi, "Summary of a Benchmark Test for Sheet Forming Analysis," in *Computational Mechanics '91*, S. N. Atluri, D. E. Beskos, R. Jones, and G. Yagawa, eds. (W. H. Wolfe, Alpharatta, GA, 1991), pp. 588–593.

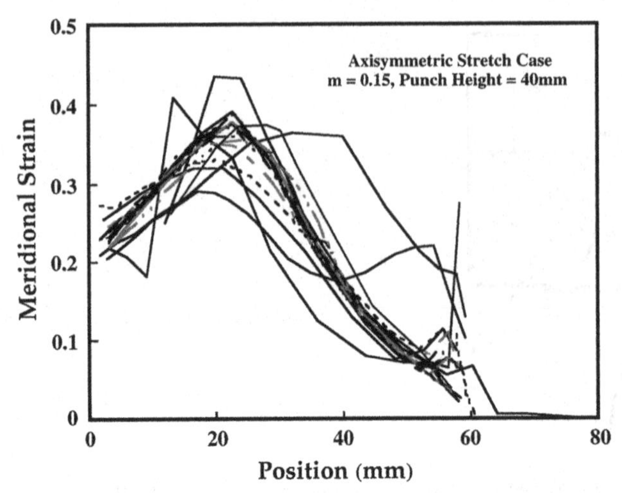


Figure 11.29. The OSU benchmark test results obtained by various FEM codes, with axisymmetric stretching.

Two typical results are shown in Figs. 11.29 and 11.30. The axisymmetric stretch case showed encouraging agreement, the best of all cases tested. The other cases had far more scatter. The scatter in the plane-strain draw case (Fig. 11.30) was particularly troubling. Even more telling, the uniform strain for frictionless plane-strain cases, which may be calculated analytically, was not reproduced by most of the contributors.

As an example of the FEM work at OSU[23] during this period, Fig. 11.31 shows typical deformed meshes, with initial widths of 180 mm and 25 mm, respectively, for the hemispherical punch stretching of various steels with various coatings, with the radial and circumferential strain direction labeled for the two blank shapes modeled. Examples of the fit between experimental and computer-generated strain distributions for the 25-mm-wide samples are presented in Fig. 11.32.

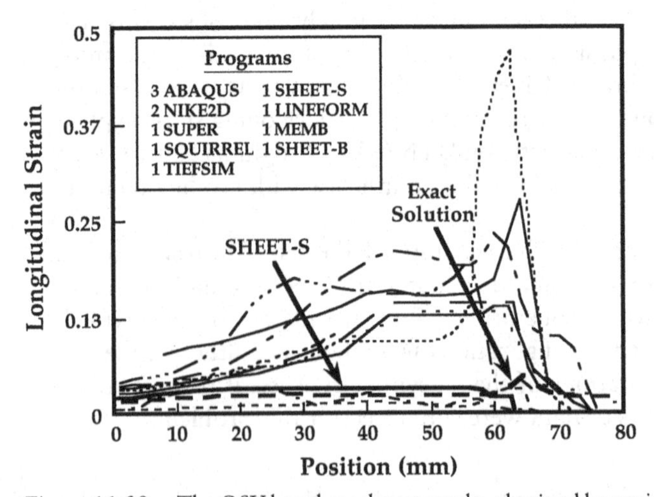


Figure 11.30. The OSU benchmark test results obtained by various FEM codes, with plane-strain drawing.

[23] J. R. Knibloe and R. H. Wagoner, "Experimental Investigation and Finite Element Modeling of Hemispherically Stretched Steel Sheet," *Metal. Trans. A* **20A**, 1509–1521 (1989).

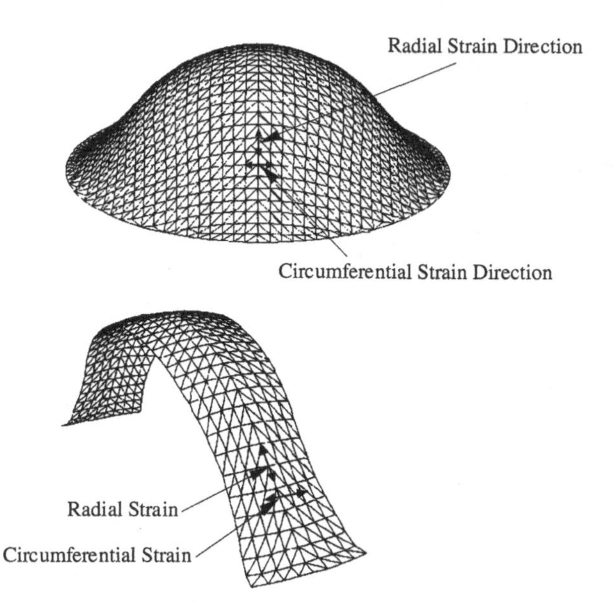


Figure 11.31. Deformed FEM meshes, hemispherical punch-stretch forming.

By comparing simulations with tedious experiments for both kinds of blanks, we were able to clarify the relationship between yield surface and friction coefficient in producing simulated strain distributions. By fitting to experimental data, Fig. 11.33, the best-fit $\mu$ was shown to depend on yield surface shape, or the best choice of $M$ (Hill's new parameter) on choice of $\mu$. This result verified that the coefficients of friction used in forming analysis can be, in fact, model dependent. That is, the choice of friction coefficient masks fundamental inaccuracies, caused by errors in constitutive equation, bending-versus-membrane element choice, and other such choices.

By 1989, the mesh-normal formulation had been introduced in an attempt to solve persistent convergence problems during draw-in simulation with SHEET-S and

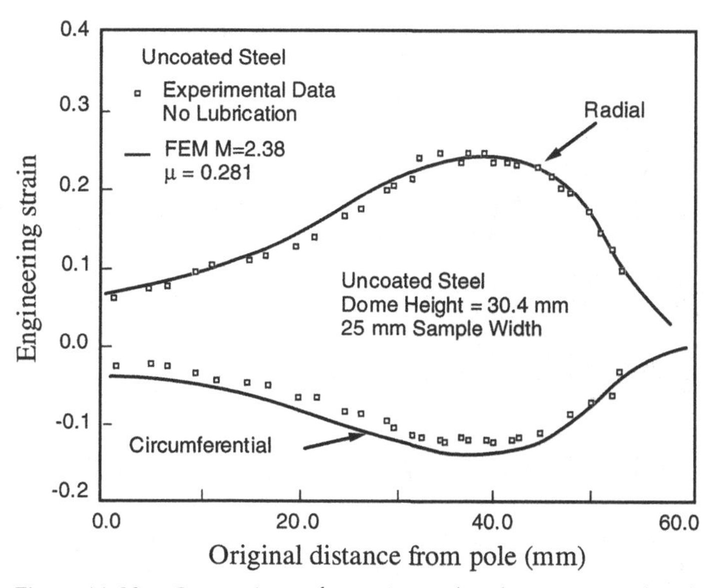


Figure 11.32. Comparison of experimental and FEM strain distributions for 25-mm-wide uncoated steel samples with no lubrication.

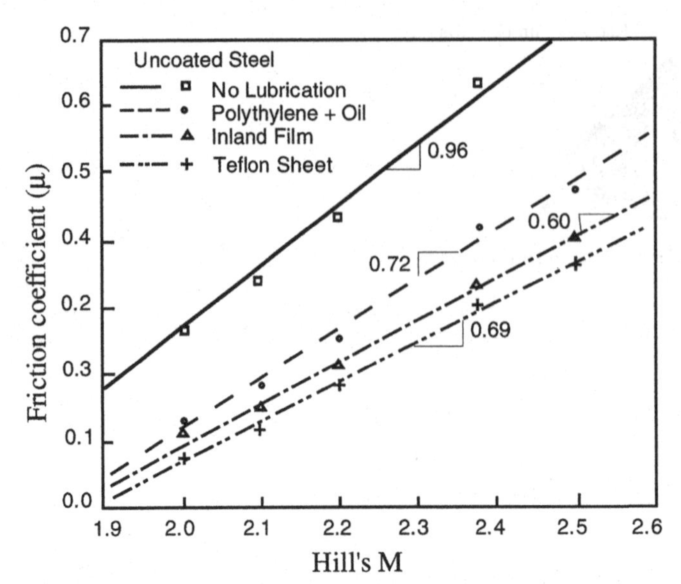


Figure 11.33. Relationship between best-fit friction coefficient ($\mu$) and Hill's anisotropy parameter ($M$).

SHEET-3. While it improved aspects of the simulation, it did not solve the fundamental instability of the DNP contact enforcement method, which was not changed until about 2 years later, when the CFS method was introduced. (Contact algorithms were described earlier in this chapter.)

### VDI Benchmark Test[24]

Much of the progress between 1989 and 1991 can be revealed by reference to the VDI benchmark geometry, shown in Fig. 11.34.

The VDI Conference demonstrated that the popularity and importance of the research area continued to grow. Very complex parts were being simulated, and manufacturing decisions were being made based on these simulations. Some companies

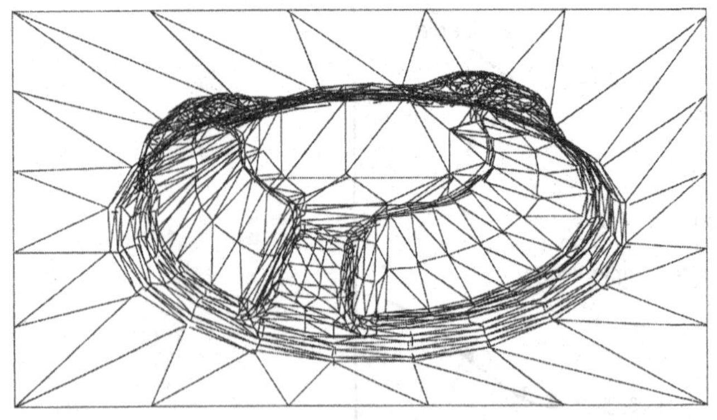

Figure 11.34. The VDI benchmark test configuration generated by a CAD system and supplied from a conference organizer.

[24] FE-Simulation of 3D Sheet Metal Forming Processes in Automotive Industry (Proceedings Conference held May 14–16, 1991, Zurich, Switzerland), VDI Berichet 894, VDI-Verlag GmbH, Dusseldorf.

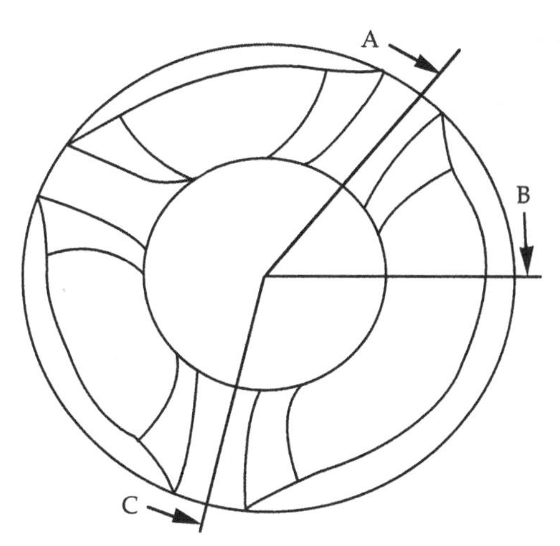


Figure 11.35. Position for the thicknesses strains of the VDI benchmark test to be measured and compared with FEM results, shown as lines A, B, and C.

had linked their CAD/CAM systems to analysis programs and could therefore test dies of immediate interest. Some wrinkling has been simulated, but consistent prediction of wrinkling remained a major challenge.

Although oral presentations at this conference were made by 11 groups in this connection, only seven groups submitted strain data for comparison. A variety of approaches were used, including the static implicit, dynamic explicit, and static explicit approaches. Both membrane and shell elements were used. The computation times for programs with friction capabilities ranged from approximately 1.5 h to 7 h, expressed relative to estimated Cray-YMP processing speeds. A typical time was 2.5 h. Thickness strains along scan lines A, B, and C are shown in Fig. 11.35.

A comparison with the similar plots in Fig. 11.36 reveals the magnitude of the progress made during the intervening 2 years since the 1989 OSU benchmark test. The scatter among the various contributions is far less for the VDI benchmark, in spite of the more complex geometry and boundary conditions. However, it should also be noted that the VDI benchmark is close to axisymmetric, which represented the closest agreement between simulation and experiment in 1989 (see Fig. 11.29).

The VDI conference and benchmark also introduced comparison of simulations with experimental forming, although the choice of friction coefficients and detailed tooling shapes were not specified precisely.

### NUMISHEET '93

The NUMISHEET '93 conference[25,26] illustrates the rapid advances in the field of computer simulation of sheet-metal forming between 1991 and 1993. Two three-dimensional benchmark problems were proposed by this conference in order to compare the various FEM simulation codes throughout the world. The two

[25] *Numerical Simulation of 3-D Sheet Metal Forming Processes – Verification of Simulation with Experiment*, A. Makinouchi, E. Nakamachi, E. Onate, and R. H. Wagoner, eds., proceedings of NUMISHEET '93 Tokyo, Japan, Aug. 30, 1993.

[26] D. Zhou and R. H. Wagoner, "Development and Application of Sheet Forming Simulation," presented at NUMISHEET '93 Tokyo, Japan, Aug. 30, 1993, pp. 3–17.

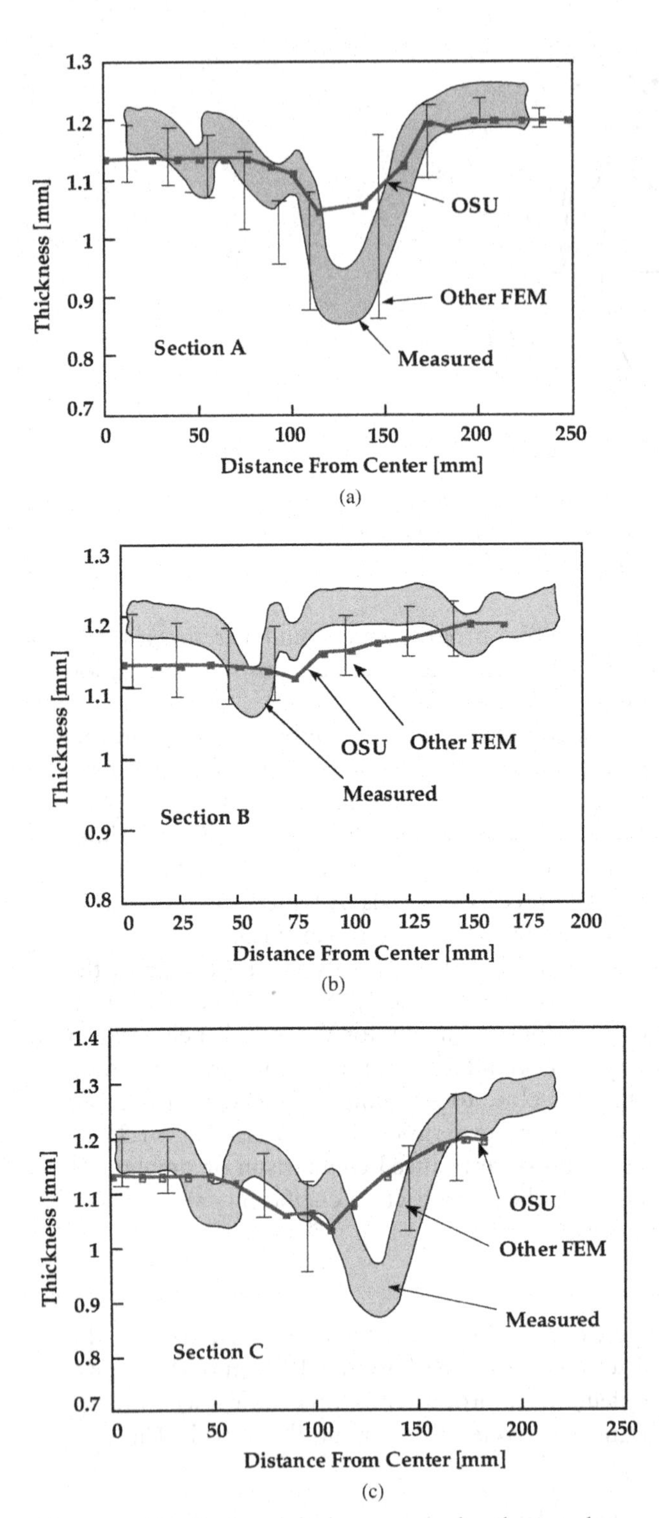


Figure 11.36. Predicted thickness strain distribution along sections (a) A, (b) B, and (c) C from seven groups of FEM programs for the VDI benchmark test.

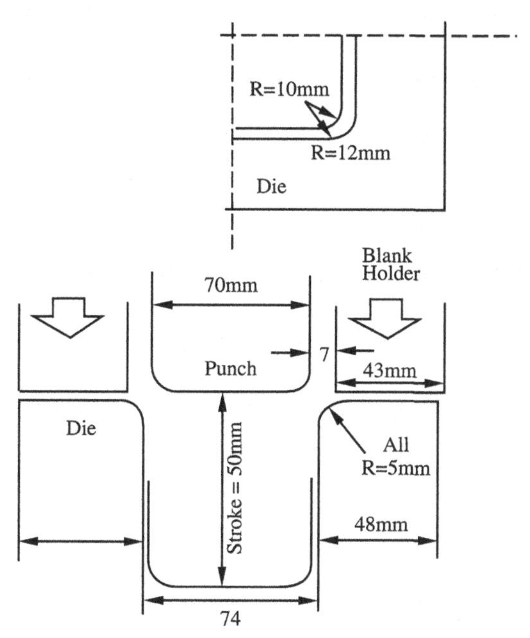


Figure 11.37. The tool geometry of the NUMISHEET '93 square punch, deep drawing, benchmark problem.

three-dimensional problems are described as follows.

1. Square cup deep drawing: This problem emphasized a program's ability to simulate a large amount of material draw-in with a relatively small die clearance and a nearly vertical wall. Ten sets of experimental and 27 sets of simulation results were presented at the conference.
2. Front fender stamping: This benchmark tests the ability to simulate a typical, complex autobody panel. The blank holder has a double curved, undevelopable surface. The forming was performed in a double-action press at Nissan, Japan. Thirteen sets of simulation results were presented at the conference. The springback capability of simulation programs was tested by a two-dimensional strip drawing benchmark, using tooling similar to the square punch. This test was the first for springback simulation capability.

### *Square Punch Benchmark*

The square punch tool is shown in Fig. 11.37. Figure 11.38 shows the draw-in amount measured from the three directions for two materials, aluminum alloy and mild steel. Comparisons were made between the average experimental measurement and average simulation results. The results of the OSU three-dimensional code, SHEET-3, are consistent with them. The mesh of an initial flat sheet is made of 338 triangular constant strain elements for a quarter of the geometry that is due to a four-fold symmetry. The materials are assumed to be hardened by the Hollomon hardening law and Hill's 1948 planar anisotropic yield function.

### *Front Fender Benchmark*

The part shape is shown in Fig. 11.39. More than 66,000 equispaced three-dimensional points, separated by a distance of 5 mm in the $x$ and $y$ directions, were

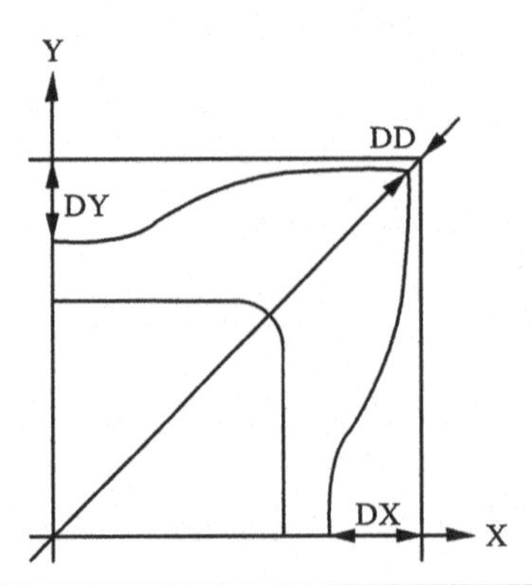


| Material | | DX | DD | DY |
|---|---|---|---|---|
| Aluminum | SHEET-3 | 25.57 | 11.51 | 25.57 |
| | Av. Simu. | 27.75 | 15.86 | 27.61 |
| Mild Steel | SHEET -3 | 27.09 | 13.79 | 27.09 |
| | Av. Exp. | 27.96 | 15.36 | 27.95 |
| | Av. Simu. | 28.08 | 15.86 | 28.16 |

Figure 11.38. Draw-in amount measured from the three directions for two materials, an aluminum alloy and a mild steel, for the square punch drawing problem in NUMISHEET '93.

used to describe the punch surface. The blank mesh is made by 957 triangular constant strain elements. The related material properties and blank information are given as follows: elastic modulus, $E = 2.06\text{E}+05$ (N/mm$^2$); Hollomon hardening law, $\bar{\sigma} = 557.66(0.01276 + \bar{\varepsilon})^{0.2488}$ (N/mm$^2$).

Hill's 1979 yield function (plane stress case) is used with an anisotropic index of $M = 2.0$ and an average normal anisotropy of $r = 1.71$. The friction coefficient is $\mu = 0.128$. The thickness is 0.81 mm.

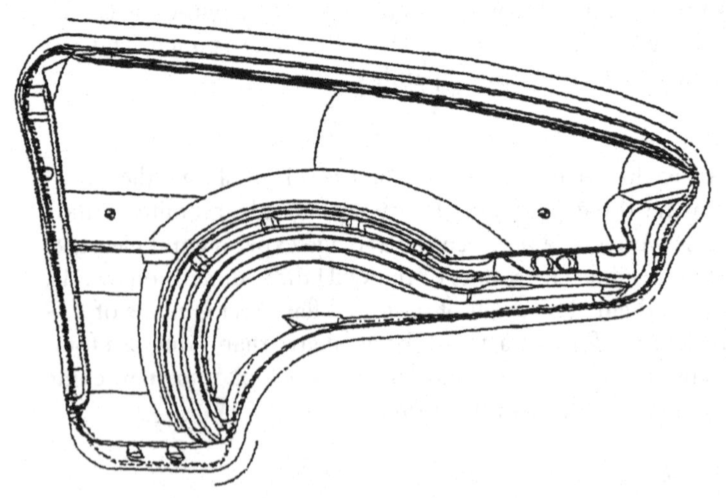

Figure 11.39. The front fender part geometry used in the NUMISHEET '93 benchmark stamping test.

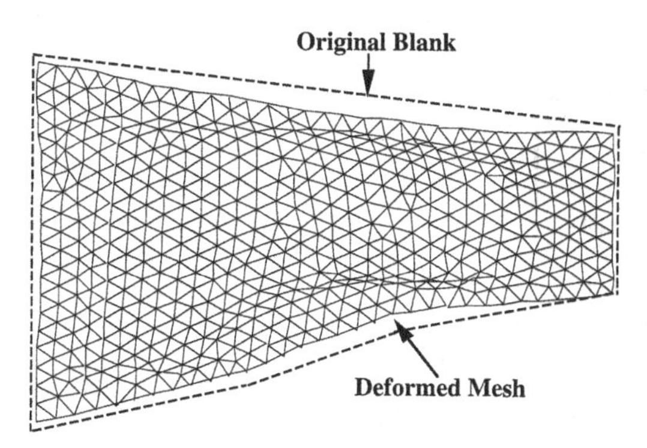


Figure 11.40. SHEET-3 mesh used in simulating the NUMISHEET '93 front fender.

Because SHEET-3 is a rigid-viscoplastic membrane element code, the binder setting could not be simulated because it involves an undevelopable, doubly curved surface involving mainly elastic straining and more bending than tensile deformation. Our simulation started from the binder set position, using the blank surface data provided by the conference organizers. However, convergence was impossible during the first 40 mm of punch travel because of the "loose metal," which eliminates the possibility of stable equilibrium in the membrane approximation. After this punch travel, however, the simulation proceeded normally.

Figure 11.40 shows the plan view of the original blank shape and the deformed mesh configuration for our simulation using SHEET-3, and Fig. 11.41 shows the simulated final thickness strain profiles. The thickness strain distribution shows that the thinnest section is located at the eyebrow area above the front wheel. Comparison of the final profile with the initial blank profile in Fig. 11.40 indicates that a significant amount of draw-in is involved. The N-CFS method was successful at simulating this complex geometry with draw-in, except for the first 40 mm of punch height, in which loose metal makes convergence impossible with a membrane approach.

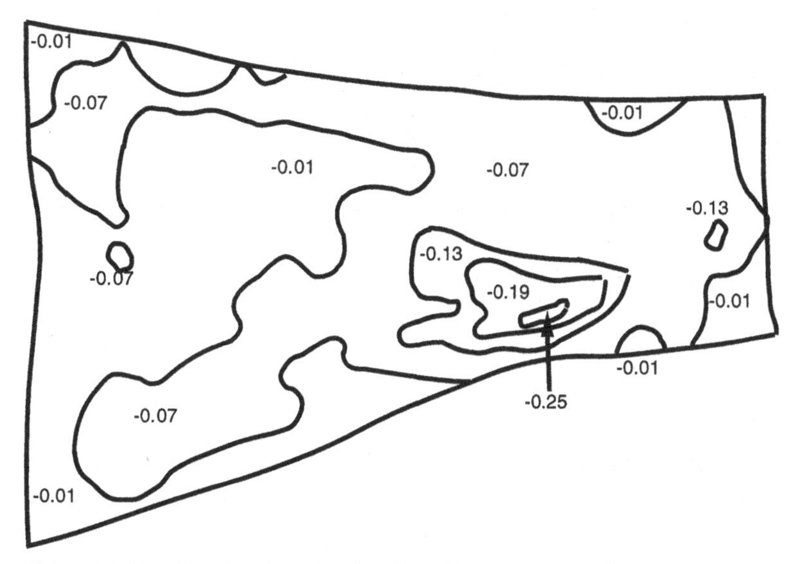


Figure 11.41. Postforming simulated thickness contours from SHEET-3 for the front fender of NUMISHEET '93.

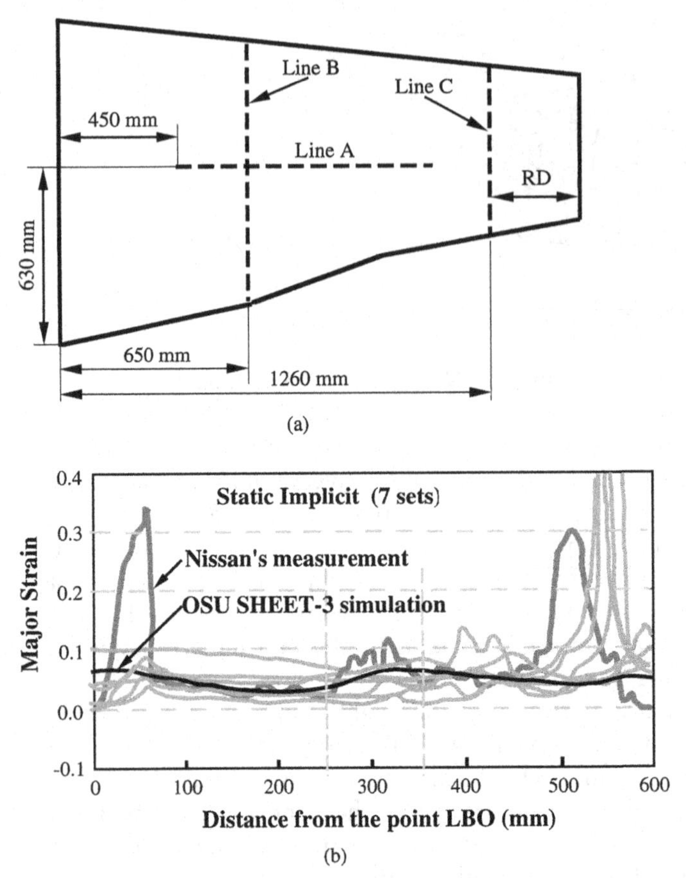


Figure 11.42. (a) Original blank geometry with scan lines for comparison of strains determined by measurement and simulation. (b) Comparison of seven simulated strain distributions from static implicit FEM programs with measurements from a formed panel.

The NUMISHEET '93 organizers presented comparisons of simulated strains along three lines of the original blank, as shown in Fig. 11.42(a). Line B was considered the most interesting because of the complex deformation patterns there. Figure 11.42(b) compares results presented in the proceedings volume for participating programs based on the static implicit method, which was the most prevalent method and which showed the smallest scatter among the simulations. Because the boundary conditions were not clearly established, comparisons with the experimental results, also shown in Fig. 11.42(b), are not conclusive.

## PROBLEMS

### A. Proficiency Problems

1. Calculate the strain distribution for the 1989 OSU benchmark test (Fig. 11.28), plane-strain version at punch heights of 20, 40, and 60 mm. Compare with benchmark results.
2. Assuming a Considere-type localization condition (i.e., necking begins at maximum load), what is the maximum punch height attainable in the plane-strain OSU benchmark test, shown in Fig. 11.28?

3. Imagine a uniform mesh of squares arranged in a rectangular array $n \times m$, where $m > n$. Assume that solver times and storage sizes (implicit, direct solution) can be approximated by

   $$\mathrm{CPU(s)} = K^{1/250} db^2 \,(\mathrm{s}),$$

   where $d$ is the degrees of freedom (DOF), and $b$ is the bandwidth (equal to the maximum number of degrees of freedom between adjacent nodes, however numbered); and

   $$\mathrm{MB} = 8/1000 db (\mathrm{mb}).$$

   (a) For fixed mesh dimensions of $n = 50, m = 200$, compare the CPU times and storage requirements for the following elements:

   membrane: 3 DOF per node
   or 4 DOF per node, CFS
   shell: 6 DOF per node
   or 7 DOF per node, CFS
   solid: 3 DOF per node, 4 DOF for CFS
   in 1 element through-thickness
   solid: 3 DOF per node, 4 DOF for CFS
   in 25 elements through-thickness

   (b) How would the CPU times and storage requirements change for each case in Part (a) if the nodes were numbered alternately?
   (c) Plot the CPU times and storage requirements for Part (a) if mesh refinement is carried out:

   (1) Part (a): $n = 50, m = 200$
   (2) 1/2 size elements: $n = 100$ m, $m = 400$
   (3) 1/4 size elements: $n = 200$ m, $m = 800$
   (4) 1/8 size elements: $n = 400$ m, $m = 1600$
   (5) 1/16 size elements: $n = 800$ m, $m = 3200$

   (d) Assuming a practical limit of CPU time of 100 h, and a storage capacity of 1 Gb, what is the finest mesh that can be used?
   (e) Assuming Moore's law, which states that computers double in speed and storage every 18 months, how long will it be until a full automotive panel (including drawbead) can be simulated with shell elements? To estimate $n$ and $m$, the panel size is approximately 1 m $\times$ 2 m, and the resolution needed for the drawbead and die radius is approximately 0.1r, where a typical radius is 10 mm (so the element size should be approximately 1 mm).
   (f) For the automotive problem in Part (e), what would the CPU time be for a section analysis (i.e., along either the $n$ or $m$ direction) for

   membrane elements: 2 DOF per node
   shell elements: 3 DOF per node
   solid elements: 2 DOF per node,
   25 elements through-thickness

   Given the results for Parts (e) and (f), what is the better approach for early design simulations?

4. With reference to Fig. 11.4, hold nodes 1 and 3 in fixed positions and allow node 2 to traverse along the tool surface. Compute the slope of both the tool normal and

mesh normal as node 2 moves. Plot these two normals and the curvature obtained from them. Which variation is smoother?

5. Calculate and approximate time for the line search part of the solution algorithm and compare it to the linear solver times (see Proficiency Problem 3). How important is the line search time for large problems?
6. For a simple cup-drawing operation, the original blank diameter is 250 mm and the punch diameter is 125 mm. A blank-holder force of 200 kN is applied through a lubricant on both sides of the sheet with friction coefficient of $\mu = 0.15$. What equivalent drawbead force should be applied per length of blank circumference? How should this force vary throughout punch stroke?

## B. Depth Problems

7. Why is the time to solve a set of linear equations directly approximated by $db^2$, where $d$ is the number of equations, and $b$ is the bandwidth? (See Proficiency Problem 3.) Discuss ways to reduce this rapid growth of solution time with DOF and bandwidth.
8. Why does the DNP contact algorithm provide little assurance of convergence? Why is this problem much worse for draw problems, which are principally force controlled, than for stretch problems, which are principally displacement controlled?
9. With reference to Eq. (11.65), discuss the strain path during a single step. What material directions are followed? What is the trajectory of the nodes during a proportional strain path? Does this lead to incompatible nodal motion; that is, is it inconsistent with the assumed paths in connecting elements?
10. Why is regularization of the Coulomb friction law required for numerical convergence?
11. Why does Z-CFS fail for nearly vertical draw-in cases?
12. Discuss the use of tolerance measures Re, Eq. (11.112), and $N_f$, Eq. (11.113). How will these vary with material strength, mesh refinement, and step size? Propose a simple method for normalizing these measures to avoid scaling problems.
13. In some cases, it is impossible to find a consistent set of contact nodes and an equilibrated solution. The solution will oscillate between two states of contact and equilibrium. Discuss the various causes of this and what can be done to proceed with reliable solutions.
14. What approximations are made in the section analysis presented for the hood sections, shown in Figs. 11.20–11.22?
15. Two additional NUMISHEET benchmark conferences have been held since the drafting of the manuscript for this book. The references for these are as follows:
    - Proceedings of NUMISHEET '99, University of Franche-Compte, Besançon, France, 1999.
    - NUMISHEET '96, The Ohio State University, 1996.

    What progress has been made in sheet-forming simulations during the intervening years?

CHAPTER TWELVE

# Recent Research Topics

Once the physical problem is taken into account properly, the major concern of the engineer is reliability of the numerical code, and cost and delay to run an application. The growing need of very complicated problems, for which meshing "by hand" is less and less thinkable, urged the development of friendly automatic meshing and remeshing tools.

In contrast, it is now possible to evaluate quantitatively the confidence we can put into an actual computation with the a posteriori error estimation. The combination of both approaches has been achieved: after a first trial computation, the most advanced methodology permits us to generate a new mesh automatically, which is locally refined in order that the error is always below an acceptable value for practical purpose.

Finally, when very large problems are in three dimensions, both computer time and memory allocation may become prohibitive. This concern can be taken into account by utilizing iterative methods for solving large linear systems, in order to save computer time, reduce the memory requirement, and facilitate the optimal use of parallel computers.

## 12.1 Meshing and Remeshing

The meshing capability of a FE code is now very important, as the manual generation of the complex meshes that are treated is almost impossible and always dissuading. When the incremental simulation of a forming sequence is considered, automatic remeshing is still more important. We shall briefly review the main methods for meshing and discuss their possible generalization to remeshing.

### Meshing

Different steps in the development of meshing software can be distinguished, which can be combined in some respect.

#### *Structured Meshing*

Structured meshing requires the operator to subdivide the domain into simple subdomains, which can be easily meshed separately and assembled to produce the complete mesh (this is schematically illustrated in Fig. 12.1). This method is still

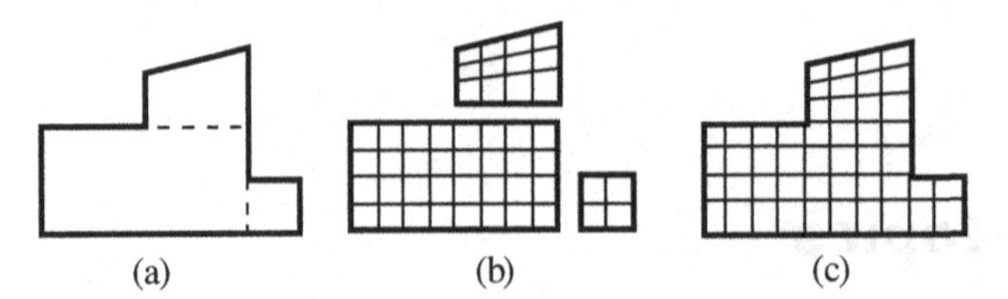


Figure 12.1. Structured meshing: (a) initial part divided into subdomains; (b) meshing of each subdomain; and (c) assembly of the meshes.

widely used in three-dimensional meshing codes, despite some disadvantages.

1. The division into subdomains is not always straightforward.
2. The connections between subdomains can be a constraint that results in a low-quality mesh.
3. The process is very difficult to transform into a fully automatic procedure, so it is not appropriate for the remeshing of very complicated shapes without human intervention.

### *Dichotomic Meshing*

The basic idea of dichotomic meshing is to part the domain into two subdomains, and then part each subdomain into two smaller subdomains until only triangles (or tetrahedrons) are obtained. This general procedure is illustrated schematically in the example of Fig. 12.2. This method is powerful and does not suffer the limitations of the previous one.

### *The Generalized Delaunay Tesselation Method*

With its original form, this method is based on a precise mathematical approach that we shall describe only very briefly here (for more details the interested reader can refer to Delaunay[1]). Let us first consider a convex domain, the boundary of which is given and approximated by a series of straight segments: the general objective is

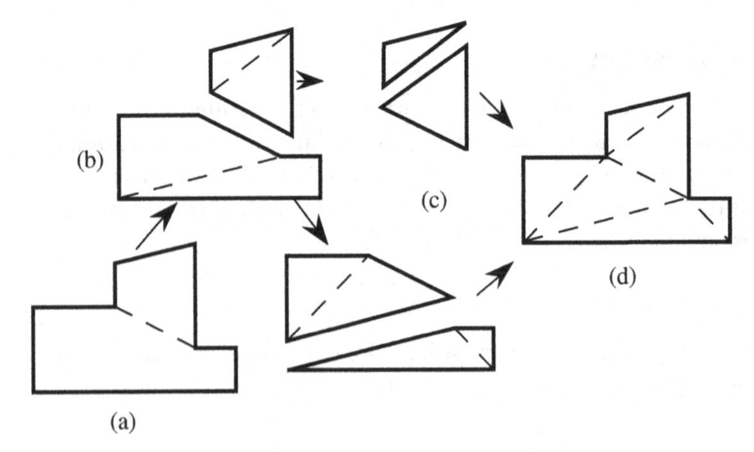


Figure 12.2. Dichotomic meshing: (a) initial domain and first parting; (b) two subdomains are parted; (c) last parting; and (d) assembly of the triangles and final mesh.

[1] C. Borgers, "Generalized Delauney triangulation of non convex domains," *Comput. Math. Applie.*, 20, no. 7, 45–49 (1990).

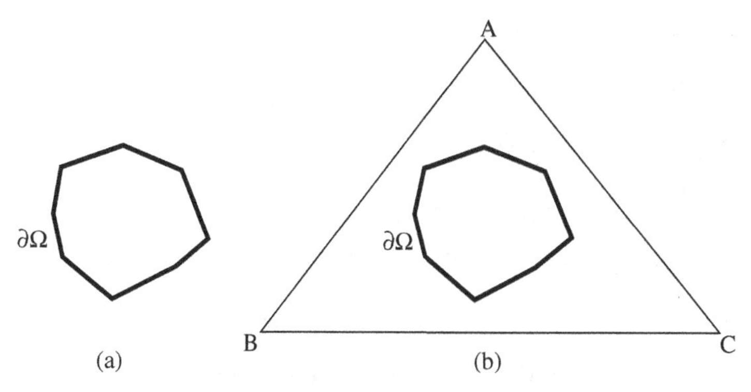


Figure 12.3. Step 1 of the Delaunay tesselation: (a) definition of the boundary $\partial\Omega$, and (b) addition of a triangle $ABC$ containing the domain.

to produce a triangular mesh by using the sides of the boundary. Several steps are introduced according to the following.

In the first step, there is determination of a (fictitious) triangle, which will contain the domain boundary (Fig. 12.3).

In the second step, each of the nodes of the domain boundary is added successively to the already formed mesh, including the extra (fictitious) triangle. For each addition of a node the mesh is modified with the following substeps:

a. The existing mesh is pictured in Fig. 12.4(a), to which the node $N$ is added.
b. A set of triangles is determined, for which the encircling circle contains the added node $N$; they are hatched in Fig. 12.4(b).
c. The exterior boundary $\Gamma(N)$ of the zone made up by the hatched triangles is determined, while the sides of the inner triangle to $\Gamma(N)$ are destroyed; see Fig. 12.4(c).
d. Finally, the new mesh is obtained by joining the new node $N$ to each vertex of $\Gamma(N)$; a mathematical property ensures that it is always possible.

In the third step, the final mesh of the domain is obtained by deleting all the triangles that contain at least one node of the fictitious triangle we added initially, as it is illustrated in Fig. 12.5.

When the domain is no longer convex, which is the more general case, the above "classical" Delaunay triangulation is not able to generate a proper mesh. By the previous procedure we obtain the smallest convex domain that contains the actual domain, so that it must be generalized. In Fig. 12.6 the part to be meshed is represented by its boundary with thick lines. The triangulation is based on all the nodes of the boundary,

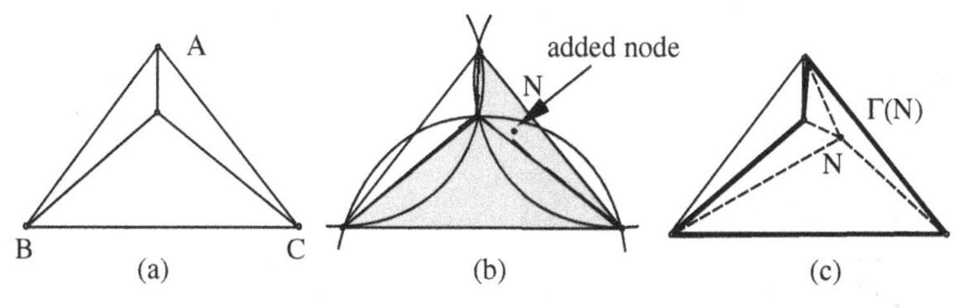


Figure 12.4. Step 2 of the Delaunay tesselation: (a) mesh after the introduction of the first boundary node; (b) addition of the second boundary node; and (c) the new mesh is generated after suppression of the hatched triangles.

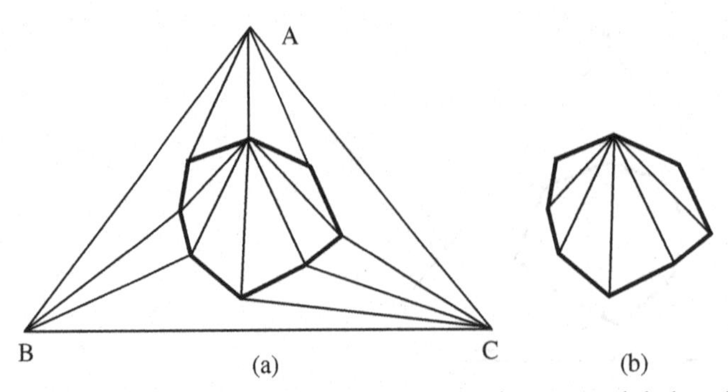


Figure 12.5. Step 3 of the Delaunay triangulation: (a) global mesh of the domain and the triangle; and (b) mesh of the domain.

and it is made of thin lines and some of the thick segments of the boundary. Obviously, however, this mesh cannot be considered as a consistent mesh of the domain.

Two situations are mixed in the previous example.

a. The segments of the boundary are sides of triangles that do not belong to the domain; then the consistent mesh is obtained by suppressing only some outside triangles; see Fig. 12.7(a).
b. Some segments of the boundary do not belong to any triangle, so that the triangulation must be transformed before a consistent mesh can be obtained. It can be shown that, when a segment of the boundary is not included in the sides of the triangles of the current mesh, the iterative algorithm, which consists of adding a node at the middle of the segments of the boundary that are not sides of the triangles, is able to produce a consistent mesh. This is schematically indicated in Fig. 12.8.

In the fourth step, there is improvement of the mesh.

Many (iterative) algorithms can be invented to produce a more suitable mesh, when we keep in mind that further FE computation is of course the final goal. Precise criteria must be set in order to evaluate the effect of any change of the mesh. Without becoming involved with too many technical details, we can quote three criteria.

1. There is sufficient density of the triangles in order that each element is small enough to represent local evolutions reasonably well.
2. There is nondistortion of the elements so that each of them is "close enough" to an equilateral triangle.

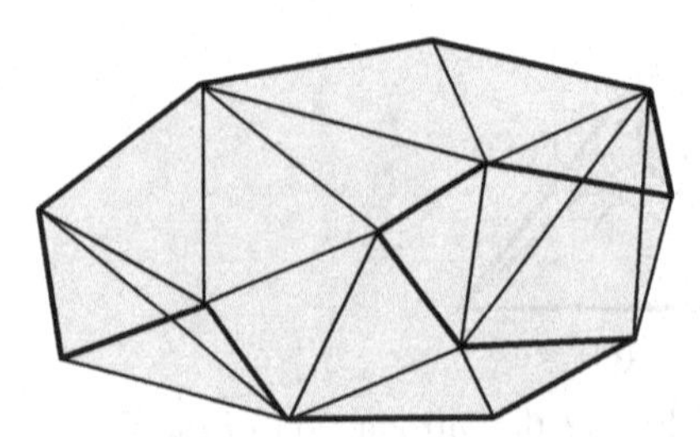

Figure 12.6. Meshing of a nonconvex domain by classical Delaunay triangulation.

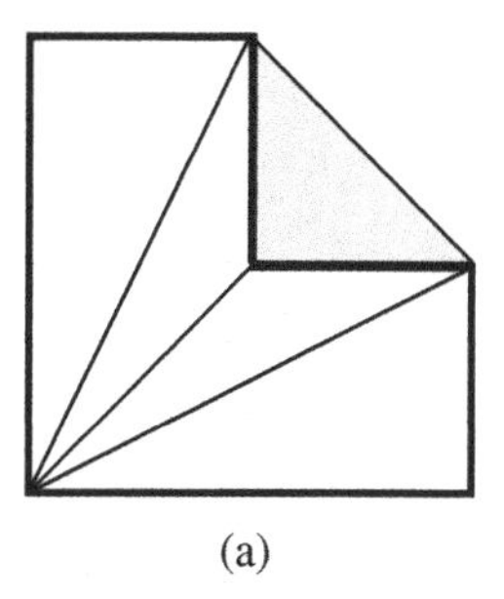

(a)

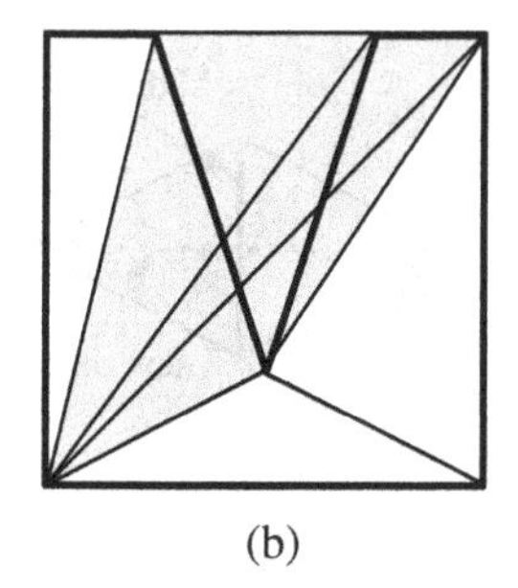

(b)

Figure 12.7. Meshing of nonconvex domains: (a) a mesh can be obtained by deleting the hatched triangle; (b) if the hatched triangles are deleted, the mesh is not acceptable.

3. If the size of elements is allowed to vary greatly within the whole domain, this evolution must be progressive in order that a small triangle will not be adjacent to a large one.

It is obvious that the first and third criteria can be reached by adding progressively internal and/or boundary nodes, using the same procedure as before. The second criterion can be improved locally by the exchange of diagonal procedure, whenever it is possible, as explained in Fig. 12.9. The process is performed iteratively for each pair of adjacent triangles and can be repeated several times.

Finally, a third algorithm is often introduced, called barycentric smoothing. The goal of this treatment is also to produce more uniform and more regular elements: this is illustrated by Fig. 12.10.

More precisely, an iterative algorithm is used so that each node $M_i$, with vector coordinates $\mathbf{X}_i$, is moved into $M'_i$, with vector coordinates $\mathbf{X}'_i$ given by the equation

$$\mathbf{X}'_i = \frac{1}{c(i)} \sum_{j \in C(i)} \mathbf{X}_j, \tag{12.1}$$

where $C(i)$ is the set of indices of nodes connected to node number $i$ by the side of a triangle, and $c(i)$ is the number of indices in $C(i)$.

The procedure is repeated until convergence, that is, until Eq. (12.1) produces no change in the nodes' positions: this is generally done with few iterations.

A more efficient barycentric smoothing makes use of the geometric aspect of the triangles. A quality factor is defined for an element $e$ by the ratio between $d^e_{\text{ins}}$, the

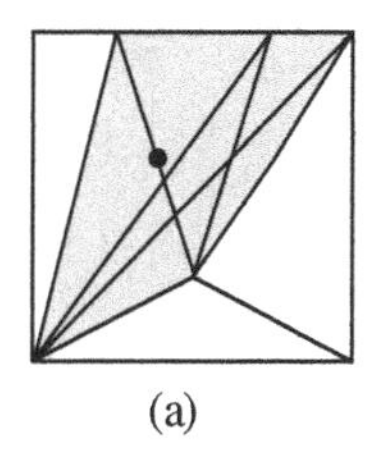

(a)

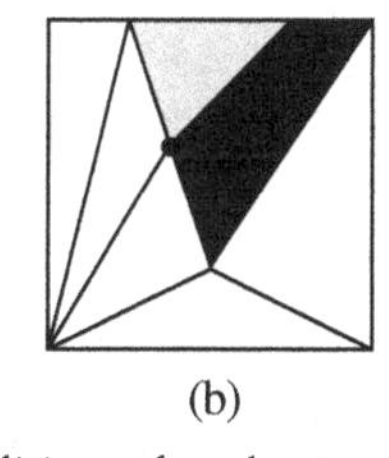

(b)

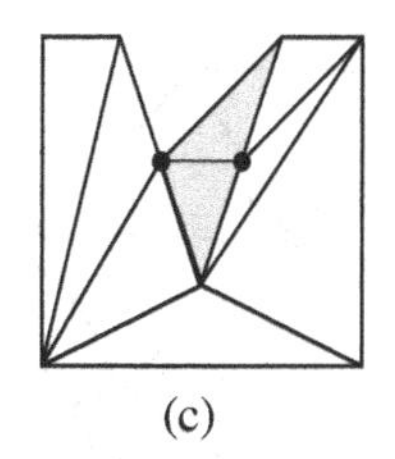

(c)

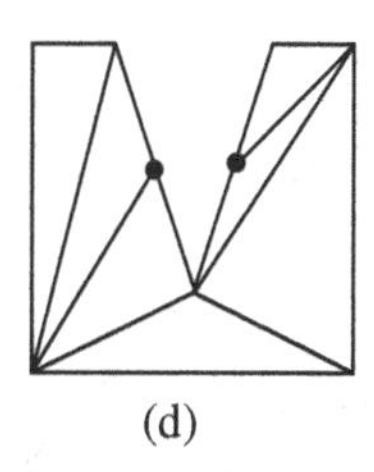

(d)

Figure 12.8. Addition of nodes to obtain a consistent mesh: (a) addition of a node on a side of the boundary that is not a side of a triangle; (b) remeshing, in which the hatched triangle is eliminated, and a second node is added; (c) remeshing and elimination of the hatched triangle; and (d) final mesh.

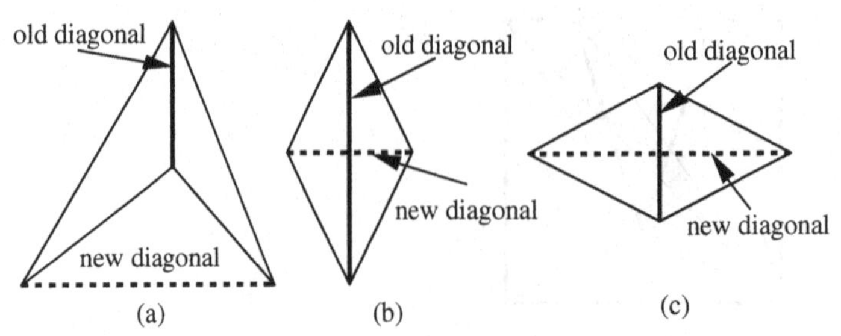


Figure 12.9. Exchange of diagonal procedure for two adjacent triangles: (a) the transform is not possible (the resulting mesh would be inconsistent); (b) the transform is possible and improves the mesh quality; and (c) it is possible but the quality is lowered.

diameter of the inscribed circle, and $d^e_{\text{enc}}$, the diameter of the encircling diameter:

$$q^e = d^e_{\text{ins}}/d^e_{\text{enc}}. \tag{12.2}$$

We remark that the quality parameter is equal to 0.5 for an equilateral triangle and is less than 0.5 otherwise. Now the barycentric smoothing (see Fig. 12.11) is defined with the centroids of surrounding triangles, and with the weights

$$w^e = 1/(1 + q^e). \tag{12.3}$$

This weighting factor is higher when the quality of the corresponding element is poor, that is, when the element is distorted.

The barycentric smoothing is defined by

$$\mathbf{X}'_i = \frac{1}{S_i}\sum_e w^e \mathbf{X}^e, \tag{12.4}$$

where $\mathbf{X}^e$ is the coordinate vector of the centroid of element $e$, $S_i$ is the sum of the element weights,

$$S_i = \sum_e w^e, \tag{12.5}$$

and the summation is extended to the surrounding element having one node as node $i$.

### Remeshing

When the updated Lagrangian method is used, all the nodes of the mesh are moved according to the material velocity and displacement. After several steps, large

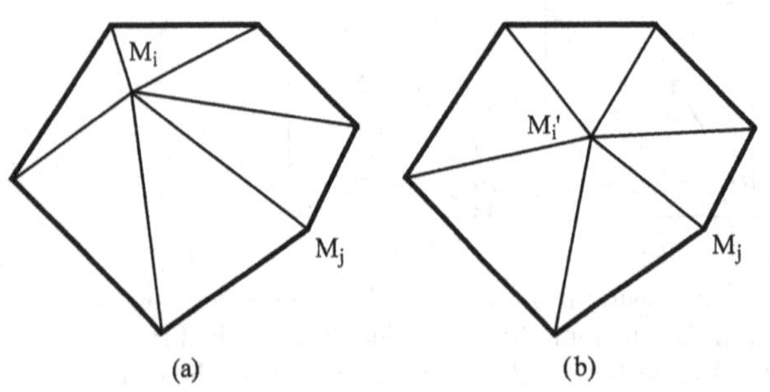


Figure 12.10. Nodal barycentric smoothing: (a) before smoothing; and (b) the interior node $M_i$ is moved into $M'_i$ as in Eq. 12.1.

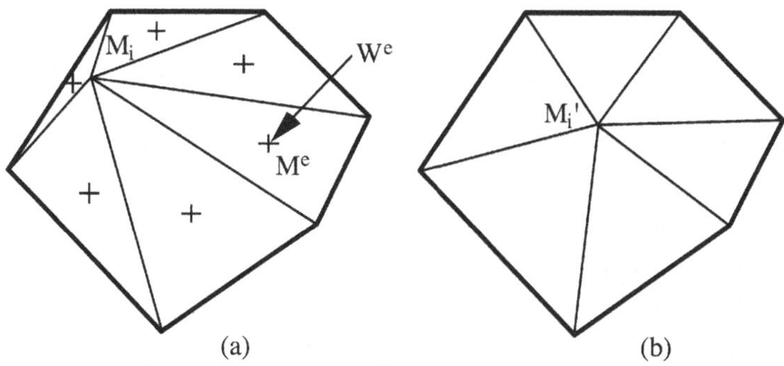


Figure 12.11. Element barycentric smoothing: (a) before smoothing; and (b) the interior node $M_i$ is moved into $M_i'$, the weights being located at the centroids of the initial triangles.

element distortions can occur so that the precision of the FE computation is no more satisfactory, and even the calculation can be stopped as a result of the degeneracy of some elements. Moreover, the displacement of the tools produces new contact areas, which can result in a very important change of the topology of the part. These events must be recognized, using element quality tests.

For isoparametric elements, degeneracy must be tested, at least at integration points, by evaluating the Jacobian determinant of the transform between the local space and the physical space (see Section 3.2). For example, in two dimensions, we can use the test

$$\det(\mathbf{A})/|\mathbf{A}|^2 > r^d, \tag{12.6}$$

where $r^d$ is a small positive number (e.g., $10^{-3}$), and $|\mathbf{A}|$ is the norm of the $\mathbf{A}$ matrix defined by

$$|\mathbf{A}|^2 = \sum_{ij} \mathbf{A}_{ij}^2. \tag{12.7}$$

When the test defined by Eq. (12.6) is not satisfied, the mechanical computation is stopped and the remeshing procedure is started. Classically one can distinguish three different stages for remeshing applied to the metal-forming process.

In the first step, there is definition of the new boundary of the deformed part, which is itself obtained generally after several time steps of FE computations and nodal updates. When curved elements are used, for example, quadratic six-node triangles, the boundary can be approximated by an "overdiscretization" with linear segments as pictured in Fig. 12.12.

Because of approximation, this definition can be slightly inconsistent with the tool definition. Two local problems can arise: the new overdiscretized boundary can penetrate inside the tool domain, or a zone of the part, previously in contact with the tool, may loose contact. In both cases a normal reprojection on the tool surface is performed.

In the second step, the remeshing process is then analogous to the initial meshing.

In the third step, the memory variables (equivalent strain, temperature, etc.) must be transported from the old mesh to the new one by some interpolation. At first sight, the easiest method is to use the shape functions of the old mesh to find the values of the variable at the nodes of the new mesh. Some difficulties may arise: for example,

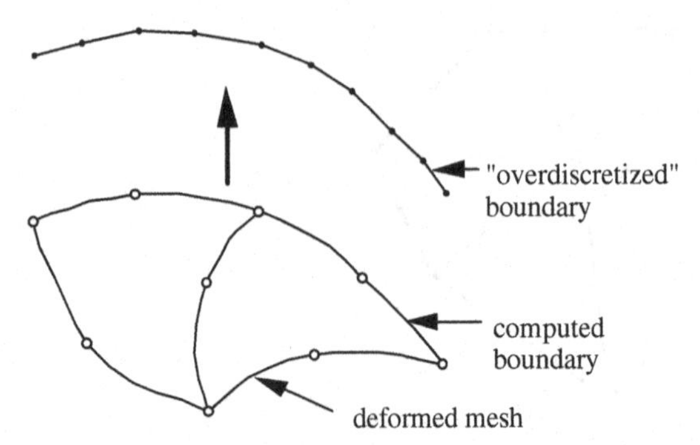


Figure 12.12. Overdiscretization of a part of the boundary.

when a stress field satisfying the discretized virtual work principle is interpolated in this way, the new stress field does not satisfy it anymore. This is specially important for elastoplastic analysis for which the new interpolated stress field will satisfy neither the integral form of the equilibrium equation, nor the plasticity yield criterion. In this case, the remedy is to perform a fictitious displacement increment, from the interpolated stess field, in order to verify the equilibrium plasticity equations on the new mesh.

## 12.2 Error Estimation

The a priori evaluation of the error caused by FE discretization by a function of the size $h$ of the mesh, when it tends to zero, was studied for a rather long time: it involves a sophisticated mathematical theory, which is outside the scope of this book, and generally does not provide practical information for a given mesh. More recently the a posteriori error estimation of the error of discretization has become popular, as it allows us to have more confidence in engineering computations. This latter approach will be outlined here, and we shall see in the next section that the mathematical results are also necessary for adaptive mesh refinement.

### The Principle of Error Estimation for Elastic Problems

The elastic problem in a domain $\Omega$ was defined in Section 5.8.[2] For any displacement field $\mathbf{u}$, the associated strain $\varepsilon$, and the stress tensor $\boldsymbol{\sigma}$ derived with the Hooke law, we can associate the energy norm:

$$|\mathbf{u}| = \left( \int_{\Omega} \sigma : \varepsilon \mathrm{d}V \right)^{1/2}. \tag{12.8}$$

Now $(\mathbf{u}, \boldsymbol{\sigma}, \varepsilon)$ will represent the exact solution of the elastic problem,[3] while $(\mathbf{u}_h, \boldsymbol{\sigma}_h, \varepsilon_h)$ will be the finite-element solution for a given mesh, which is denoted by $h$ for simplicity. The error of discretization is defined by the difference[4]:

$$\mathbf{e}_h = \mathbf{u} - \mathbf{u}_h. \tag{12.9}$$

[2] R. H. Wagoner and J.-L. Chenot, *Fundamentals of Metal Forming*, (Wiley, New York, 1997).

[3] This is generally unknown.

[4] The notation $\mathbf{e}_h$ will not introduce any confusion here, as here we shall not use the deviatoric strain tensor.

According to Eq. (12.8), the energy norm of the error is written as

$$|\mathbf{e}_h| = \left[ \int_\Omega \boldsymbol{\sigma}(\mathbf{u} - \mathbf{u}_h) : \boldsymbol{\varepsilon}(\mathbf{u} - \mathbf{u}_h) \, \mathrm{d}V \right]^{1/2}. \tag{12.10}$$

As a result of the linearity of $\boldsymbol{\varepsilon}$ and $\boldsymbol{\sigma}$ with respect to $\mathbf{u}$, Eq. (12.10) can be transformed into

$$|\mathbf{e}_h| = \left[ \int_\Omega (\boldsymbol{\sigma} - \boldsymbol{\sigma}_h) : (\boldsymbol{\varepsilon} - \boldsymbol{\varepsilon}_h) \, \mathrm{d}V \right]^{1/2}. \tag{12.11}$$

If the linear elasticity law is expressed with the fourth-rank tensor $\mathbf{D}$ by

$$\boldsymbol{\sigma} = \mathbf{D} : \boldsymbol{\varepsilon}, \tag{12.12}$$

it is convenient to further transform Eq. (12.11) into

$$|\mathbf{e}_h| = \left[ \int_\Omega (\boldsymbol{\sigma} - \boldsymbol{\sigma}_h) : \mathbf{D}^{-1} : (\boldsymbol{\sigma} - \boldsymbol{\sigma}_h) \, \mathrm{d}V \right]^{1/2}. \tag{12.13}$$

Generally it is impossible to compute the norm of the error exactly, as $\mathbf{u}$ is not known. Thus we shall restrict ourselves to the calculation of an estimation of $|\mathbf{e}_h|$: Several methods are proposed in the literature, but for simplicity we shall present only the Zienkiewicz and Zhu estimator.[5] The basic idea is to remark that the FE solution verifies all the equations of the elastic problem,[6] but that usually the strain and stress tensor fields, which are derived from the continuous displacement field by differentiation, are not continuous in the domain $\Omega$. More specifically the discontinuities of $\boldsymbol{\varepsilon}_h$ and $\boldsymbol{\sigma}_h$ appear at the element's interfaces, and some measure of these discontinuities represents the error. The methodology developed by Zienkiewicz and Zhu consists of building, by a linear transformation, a continuous strain field $\tilde{\boldsymbol{\varepsilon}}_h$ and a continuous stress field $\tilde{\boldsymbol{\sigma}}_h$, which are close to the finite approximations $\boldsymbol{\varepsilon}_h$ and $\boldsymbol{\sigma}_h$, respectively. They are interpolated by the same shape functions as the displacement field $\mathbf{u}_h$:

$$\tilde{\boldsymbol{\varepsilon}}_h = \sum_n \tilde{\boldsymbol{\varepsilon}}_{h,n} N_n, \tag{12.14a}$$

$$\tilde{\boldsymbol{\sigma}}_h = \sum_n \tilde{\boldsymbol{\sigma}}_{h,n} N_n. \tag{12.14b}$$

The error estimation is then defined by the expression

$$\theta_h = \left[ \int_\Omega (\tilde{\boldsymbol{\sigma}}_h - \boldsymbol{\sigma}_h) : (\tilde{\boldsymbol{\varepsilon}}_h - \boldsymbol{\varepsilon}_h) \, \mathrm{d}V \right]^{1/2}. \tag{12.15}$$

This latter expression can be transformed exactly into

$$\theta_h = \left[ \int_\Omega (\tilde{\boldsymbol{\sigma}}_h - \boldsymbol{\sigma}_h) : \mathbf{D}^{-1} : (\tilde{\boldsymbol{\sigma}}_h - \boldsymbol{\sigma}_h) \, \mathrm{d}V \right]^{1/2}, \tag{12.16}$$

[5] O. C. Zienkiewicz and J. Z. Zhu, "A Simple Error Estimator and Adaptive Procedure for Practical Engineering Analysis," *Int. J. Num. Meth. Eng.* **24**, 337–357 (1987).

[6] In fact the equilibrium equation is satisfied only in the weak (integral) sense. Moreover, as numerical integration is only approximate, with the Gaussian method, even the weak equations are expressed approximately. However, it is generally assumed that the major contribution to the error arises from the discontinuity of the stress and strain tensors.

when one remembers that the linearity of Eq. (12.12) and of the smoothing procedure allows us to write

$$\sigma(\tilde{\varepsilon}_h) = \tilde{\sigma}(\varepsilon_h). \tag{12.17}$$

For a more explicit representation of the error, it is convenient to introduce the relative error indicator, the form of which is

$$\eta_h = \frac{\theta_h}{\left(|\mathbf{u}_h|^2 + \theta_h^2\right)^{1/2}}. \tag{12.18}$$

This is usually given in percent.

The "smoothed" fields $\tilde{\varepsilon}_h$ and $\tilde{\sigma}_h$ can be obtained by different schemes. The simplest one is to compute the arithmetic average of the contributions of each element having a given node in common; this purely local procedure was outlined in Section 5.1 (see Fig. 5.1).

A more satisfactory approach is obtained when a global least-squares smoothing is performed. When $\sigma_h$ is known, $\tilde{\sigma}_h$ is expressed by Eq. (12.14b) and its nodal values are determined by the minimization of the mean-square difference:

$$Q(\tilde{\sigma}_h) = \int_\Omega (\tilde{\sigma}_h - \sigma_h) : (\tilde{\sigma}_h - \sigma_h)\,\mathrm{d}V. \tag{12.19a}$$

A similar result can be obtained when the energy norm is used instead of Eq. (12.19a), leading to the minimization of

$$Q'(\tilde{\sigma}_h) = \int_\Omega (\tilde{\sigma}_h - \sigma_h) : \mathbf{D}^{-1} : (\tilde{\sigma}_h - \sigma_h)\,\mathrm{d}V. \tag{12.19b}$$

It is easy to verify that the minimization of Eq. (12.19), which is a quadratic form, gives an independent linear system for each component of $\sigma_h$ with the same matrix:

$$\sum_n \tilde{\sigma}_{h,ijn} \int_\Omega N_n N_m\,\mathrm{d}V = \int_\Omega \sigma_{h,ij} N_m\,\mathrm{d}V. \tag{12.20}$$

The general procedure is outlined schematically in Fig. 12.13: with a linear element the stress FE solution is piecewise constant and becomes continuous after smoothing. The error estimate is a function of the hatched difference between the two stresses.

A more efficient smoothing technique is the Orkisz method, which is a generalization of the classical finite-difference method.[7] The general idea is to compute at each node the first and second derivatives of the displacement field in order to determine the strain and stress tensors. For a node $n$, the computation of the derivatives will be made with the nodal displacements of the neighbors, referred to by the $p$ index (see Fig 12.14).

For any component $i$ of the displacement vector $\mathbf{u}$, the second-order Taylor expansion, expressed at the point with coordinated vector $\mathbf{x}$, is written as

$$u_i(\mathbf{x}) = U_{in} + \sum_k \frac{\partial u_{in}}{\partial x_k}\Delta x_k + \frac{1}{2}\sum_{k,l} \frac{\partial^2 u_{in}}{\partial x_k \partial x_l}\Delta x_k \Delta x_l, \tag{12.21}$$

where $\Delta\mathbf{x} = \mathbf{x} - \mathbf{X}_n$.

[7] T. Liszka and J. Orkisz, "The Finite Difference Method at Arbitrary Irregular Grids and Its Application in Applied Mechanics," *Comp. Struct.* 11, 83–95 (1980).

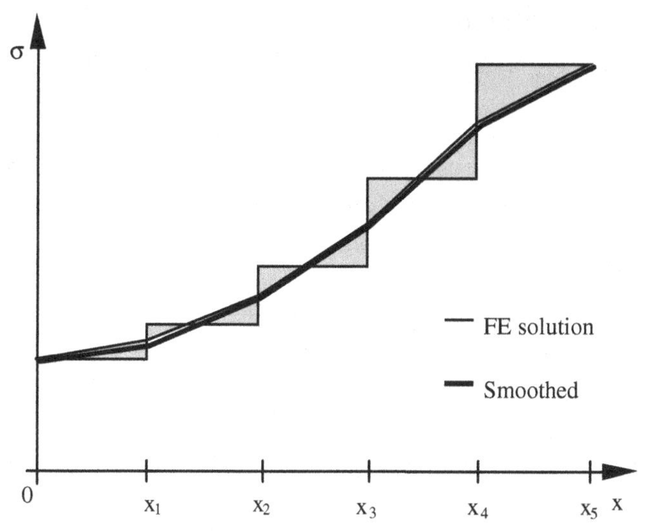


Figure 12.13. Smoothing of the FE stress discontinuous solution.

Equation (12.21) can be written for each neighboring node with coordinate vector $\mathbf{X}_p$. The least-squares fitting between the Taylor polynomial of Eq. (12.21) and the actual values of the displacement at the neighbors can be expressed (in two dimensions) by

$$S\left(\frac{\partial u_{in}}{\partial x_1}, \frac{\partial u_{in}}{\partial x_2}, \frac{\partial^2 u_{in}}{\partial x_1^2}, \frac{\partial^2 u_{in}}{\partial x_1 \partial x_2}, \frac{\partial u_{in}^2}{\partial x_2^2}\right) = \sum_p \frac{[u_i(\mathbf{X}_p) - U_{ip}]^2}{|\Delta \mathbf{x}_p|^6}. \tag{12.22}$$

Here we have $\Delta \mathbf{x}_p = \mathbf{X}_p - \mathbf{X}_n$.

The best fitting is obtained when the function $S$ is minimized with respect to the first- and second-order derivatives of the displacement field at node $n$; that is, $\partial u_{in}/\partial x_k$ and $\partial^2 u_{in}/\partial x_k \partial x_1$. It is clear that, for each node and for each component of the displacement field, $S$ being a quadratic function that must be minimized, we obtain a set of linear systems of order 5, the unknowns of which are the nodal approximations of the first- and second-order derivatives. For a given node, one can easily see that the linear systems corresponding to each components of the displacement have the same matrix. They differ only by their right-hand side, which depends on the actual values of the displacement components.

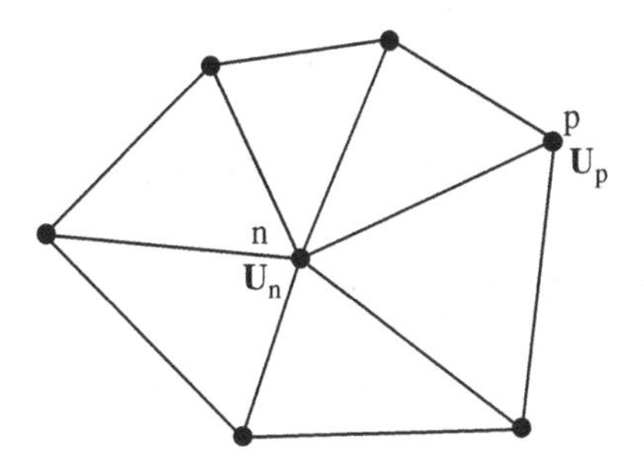


Figure 12.14. A node $n$ and its neighbors $p$ for the Orkisz method.

### Error Estimation for a Viscoplastic Material

The viscoplastic constitutive equation can be written with a viscosity depending on the equivalent strain rate:

$$\mathbf{s} = 2\mu(\dot{\bar{\varepsilon}})\dot{\varepsilon}. \tag{12.23}$$

Here we put

$$\mu(\dot{\bar{\varepsilon}}) = K(\sqrt{3}\dot{\bar{\varepsilon}})^{m-1}. \tag{12.24}$$

With this notation, the density of energy of the viscoplastic material is

$$\dot{w} = \mathbf{s} : \dot{\varepsilon} = \frac{1}{2\mu(\dot{\bar{\varepsilon}})}\mathbf{s} : \mathbf{s}. \tag{12.25}$$

The energy norm can be written in the form

$$|\mathbf{v}| = \int_{\Omega} \frac{1}{2\mu(\dot{\bar{\varepsilon}})}\mathbf{s} : \mathbf{s}\,\mathrm{d}V. \tag{12.26}$$

An approach similar to that used for the elastic problem can be developed in the following way.[8] The velocity field $\mathbf{v}_h$, computed by the finite-element method with a given mesh, is used to derive the strain rate tensor $\dot{\varepsilon}_h$ by differentiation. Then the FE deviatoric stress tensor $\mathbf{s}_h$ is determined from $\dot{\varepsilon}_h$ by utilizing the constitutive equation:

$$\mathbf{s}_h = 2\mu(\dot{\bar{\varepsilon}}_h)\dot{\varepsilon}_h. \tag{12.27}$$

At this stage the error is introduced according to

$$|\mathbf{e}_h| = \left[\int_{\Omega} \frac{1}{2\mu(\dot{\bar{\varepsilon}})}(\mathbf{s} - \mathbf{s}_h) : (\mathbf{s} - \mathbf{s}_h)\,\mathrm{d}V\right]^{1/2}. \tag{12.28}$$

Now the smoothed deviatoric stress tensor $\tilde{\mathbf{s}}_h$ is computed from $\mathbf{s}_h$ either by a local smoothing or from a global least-squares approximation by equations similar to Eq. (12.20). The error indicator is computed by the formula

$$\theta_h = \left[\int_{\Omega} \frac{1}{2\mu(\dot{\bar{\varepsilon}})}(\tilde{\mathbf{s}}_h - \mathbf{s}_h) : (\tilde{\mathbf{s}}_h - \mathbf{s}_h)\,\mathrm{d}V\right]^{1/2}, \tag{12.29}$$

and the relative error indicator $\eta_h$ is also defined according to Eq. (12.18).

## 12.3 Adaptive Remeshing

The problem is to build a mesh so that the number of elements is as small as possible and the error is less than or equal to a prescribed value $e_0$. We suppose that a first computation is done with a trial mesh, which can be relatively coarse. The adaptive remeshing will allow us to generate a new mesh, which will provide a prescribed

[8] In fact, it is easy to see that different choices can be made to write the energy norm. All of these choices are equivalent for the linear elastic material, but they give different results for the viscoplastic behavior. This choice, Eq. (12.26), seems to be the more reliable; it corresponds to O. C. Zienkiewicz, J. Z. Zhu, and G. C. Huang, "Error Estimation and Adaptivity in Flow Formulation for Forming Problems," *Int. J. Num. Meth. Eng.* 25, 23–42 (1988).

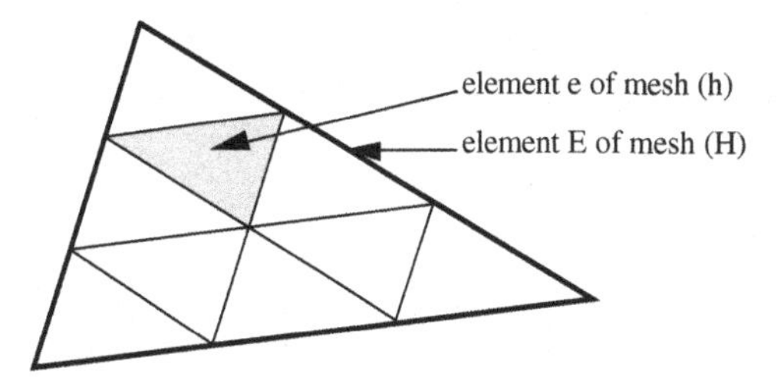


Figure 12.15. Subdivision of a coarse element into refined elements.

accuracy. Without complicated mathematical developments, we shall briefly indicate the different ingredients that are necessary, and the main steps for the practical determination of the mesh refinement that will fulfill the accuracy requirement.

More precisely, we define two embedded meshes.

1. There is an initial or coarse mesh, which does not give the desired error $e_0$; it will be denoted by $(H)$ with elements $\Omega_E$, the local size of which is $H_E$.
2. There is a final mesh $(h)$, which is obtained by subdividing each element $E$ of $H$ into $n_E$ smaller elements $\Omega_e$ of local size $h_e$, as indicated in Fig 12.15.

For any element $\Omega_e$ included in element $\Omega_E$, the refinement factor[9] is introduced according to

$$r_E = \frac{h_e}{H_E}. \tag{12.30}$$

We see immediately that each element of $H$ is subdivided into $1/r_E$ in each space direction, so that we build $n_E$ elements of the fine mesh with

$$n_E = \frac{1}{r_E^2} = \left(\frac{H_E}{h_e}\right)^2 \tag{12.31}$$

in two dimensions.

### Local Error and Local Error Estimation

Let us first consider again the simple linear elastic problem. By analogy with Eq. (12.13), it is reasonable to assume that the local contribution to the error of the element number $E$ can be written as

$$|\mathbf{e}_H|_E = \left[\int_{\Omega^E} (\boldsymbol{\sigma} - \boldsymbol{\sigma}_H) : \mathbf{D}^{-1} : (\boldsymbol{\sigma} - \boldsymbol{\sigma}_H)\,\mathrm{d}V\right]^{1/2}, \tag{12.32}$$

where we have simply restricted the domain of integration to the element $\Omega^E$.

To this local error we associate similarly a local error indicator, which is written according to Eq. (12.16) as

$$\theta_H^E = \left[\int_{\Omega^E} (\tilde{\boldsymbol{\sigma}}_H - \boldsymbol{\sigma}_H) : \mathbf{D}^{-1} : (\tilde{\boldsymbol{\sigma}}_H - \boldsymbol{\sigma}_H)\,\mathrm{d}V\right]^{1/2}. \tag{12.33}$$

[9] This is supposed to be an integer here for simplicity, but this is not essential in the following.

The same expressions for the refined mesh can be defined; we shall also need the error (or error indicator) contribution of the refined mesh to an element $E$ of the coarse mesh:

$$|\mathrm{e}_h|_E = \left[ \int_{\Omega^E} (\sigma - \sigma_h) : \mathrm{D}^{-1} : (\sigma - \sigma_h)\, \mathrm{d}V \right]^{1/2}. \tag{12.34}$$

Similar expressions can also be defined for viscoplastic materials in which the integral of Eq. (12.28), or Eq. (12.29), is restricted to an element.

### Evaluation of the Local Refinement Factor

In order to predict the local refinement accurately, we shall suppose that the behavior of the error is known when the mesh is refined. More precisely, the order of convergence is assumed equal to $p$,[10] that is, we can write

$$|\mathrm{e}_h| \cong C(\Omega) h^p, \tag{12.35}$$

where $C(\Omega)$ is a positive constant. Equation (12.35) can also be used locally on $\Omega^E$, the element number $E$ for the coarse mesh; with the definition we have

$$|\mathrm{e}_H|_E \cong C(\Omega_E) H_E^p. \tag{12.36}$$

The same result applies to the refined mesh; with Eq. (12.34) we get

$$|\mathrm{e}_h|_E \cong C(\Omega_E) h_E^p. \tag{12.37}$$

Combining Eqs. (12.36), (12.37), and (12.35), we conclude that

$$\frac{|\mathrm{e}_h|_E}{|\mathrm{e}_H|_E} \cong \frac{h_e^p}{H_E^p} = r_E^p. \tag{12.38}$$

Now if we use a uniform refinement principle, which guarantees an optimal number of refined elements, we shall assume that within each refined element we impose a constant error $e_u$:

$$|\mathrm{e}_h|_e \cong e_u. \tag{12.39}$$

If we remember the definitions of local errors, we observe that within a coarse element,

$$|e_h|_E^2 = n_E |e_h|_h^2 = \frac{1}{r_E^2} e_u^2. \tag{12.40}$$

With the help of Eqs. (12.38) and (12.40), we obtain

$$\frac{e_u}{|\mathrm{e}_H|_E} \cong r_E^{p+1}. \tag{12.41}$$

Now if Eq. (12.11) is computed with the element contributions of the fine mesh, we shall write

$$|\mathrm{e}_h|^2 = \sum_e |\mathrm{e}_h|_e^2 = \sum_E \left( \sum_{\substack{e \\ \Omega_e \subset \Omega_E}} |\mathrm{e}_h|_e^2 \right). \tag{12.42}$$

[10] With six-node triangular elements, we have $p = 2$ for the elastic problem and $p \approx 1$ for the viscoplastic problem.

Now we target an error $e_0$ on the fine mesh, with a hypothesis of uniform error distribution according to Eq. (12.39), so that Eq. (12.42) can be transformed into

$$e_0^2 \cong |\mathbf{e}_h|^2 \cong \sum_E \frac{1}{r_E^2} e_u^2. \tag{12.43}$$

With Eq. (12.41), $r_E$ is substituted in Eq. (12.43) so that

$$e_0^2 \cong e_u^2 \sum_E \left( \frac{\mathbf{e}_u}{|\mathbf{e}_H|_E} \right)^{-2/p+1}. \tag{12.44}$$

This last equation allows us to determine the value $e_u$ of the uniform error on the fine mesh:

$$e_u \cong \frac{e_0^{p+1/p}}{\left( \sum_E |\mathbf{e}_H|_E^{2/p+1} \right)^{p+1/2p}}. \tag{12.45}$$

The refinement factor $r_E$ is computed when Eqs. (12.41) and (12.45) are combined:

$$r_E \cong \frac{|\mathbf{e}_H|_E^{-1/p+1}}{\left( \sum_{E'} |\mathbf{e}_H|_{E'}^{2/p+1} \right)^{1/2p}} e_0^{1/p}. \tag{12.46a}$$

The practical value of the refinement factor is obtained by replacing the error, which is not known, by the error indicator, leading to the approximation

$$r_E \cong \frac{|\theta_H^E|^{-1/p+1}}{\left( \sum_{E'} |\theta_H^{E'}|^{2/p+1} \right)^{1/2p}} e_0^{1/p}. \tag{12.46b}$$

### Adaptive Remeshing

We suppose here that the first coarse mesh is built and that we have computed in each point a refinement factor that gives a new desired size for local elements. The adaptive remeshing procedure must generate new elements in order that the new mesh size requirement is satisfied everywhere, while keeping a good geometrical quality of the elements. At first a precise definition of the size of an element $e$ must be chosen. It is here the maximum of the lengths of its sides:

$$h^e = \max\left( |\mathbf{X}_1^e - \mathbf{X}_2^e|,\ |\mathbf{X}_2^e - \mathbf{X}_3^e|,\ |\mathbf{X}_3^e - \mathbf{X}_1^e| \right). \tag{12.47}$$

Now we shall consider three major steps for adaptive remeshing.

In the first step, the sides of the boundary are successively checked with respect to the required optimal local size $h_{\text{opt}}^e$. For each side of the boundary that is greater than $h_{\text{opt}}^e$, additional nodes are added on the boundary until the condition is fulfilled. After that transformation of the boundary, a new mesh is generated with the Delaunay procedure.

In the second step, for each internal side, which belongs to two adjacent elements, the more severe size condition is imposed. Nodes are added on the side in order to fulfill the local size condition, and the mesh is updated according to the Delaunay method. The procedure is repeated until no side is greater than $h^e_{\mathrm{opt}}$. Another procedure is to add internal nodes at the centroids of all the elements that have at least one side that does not fulfill the local condition on the size.

In the third step, a smoothing procedure is again necessary in order to have no distorted elements, but, as we must keep the local adaptation, the algorithms of Section 12.1 must take into account the local sizes. One possible solution is to move each node by a barycentric smoothing involving the middles of the adjacent sides with the weights

$$w_{ij} = (1 + a_{ij})/(1 + q_{ij}), \tag{12.48}$$

where the parameters in Eq. (12.48) are determined according to the following.

1. The quality parameter $q_{ij}$ is defined by the arithmetic average of the two adjacent elements $e$ and $e'$ sharing the same side $(ij)$:

$$q_{ij} = \frac{1}{2}(q_e + q_{e'}). \tag{12.49}$$

2. The adaptation parameter $a_{ij}$ is equal to zero when the side $(ij)$ is less than or equal to $h^e_{\mathrm{opt}}$, and otherwise it is equal to

$$a_{ij} = \frac{1}{2}\left(\frac{|\mathbf{X}_i - \mathbf{X}_j|}{h_{ij\ \mathrm{opt}}} - 1\right). \tag{12.50}$$

3. For the elements $e$ and $e'$ adjacent to side $ij$, the targeted local size of the side is defined by

$$h_{ij\ \mathrm{opt}} = \min_{e,e'}\left(h^e_{\mathrm{opt}},\ h^{e'}_{\mathrm{opt}}\right). \tag{12.51}$$

The procedure is schematically pictured in Fig. 12.16, and the explicit barycentric formula is quite similar to that in Eq. (12.4), when the middles $M_{ij}$ of the adjacent sides are introduced and the weighting factors are updated.

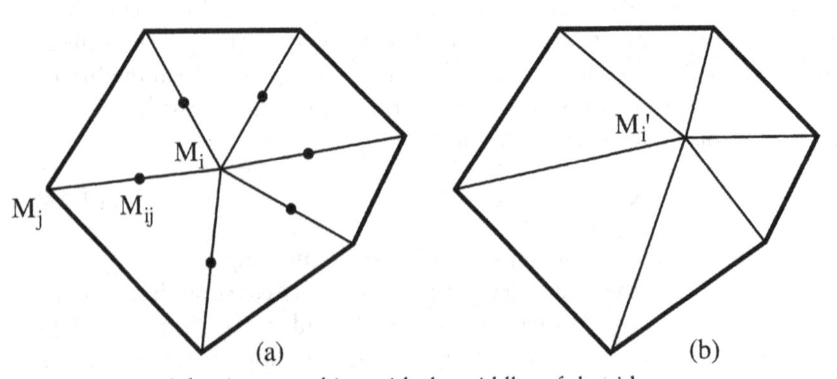


Figure 12.16. Adaptive smoothing with the middles of the sides.

## 12.4 Application to Orthogonal Machining

The problem of metal cutting has attracted considerable research interest in the past 20 years. As a result of extreme localization of deformation in the vicinity of the rake face of the tool, it is particularly difficult to obtain reliable experimental results on metal flow. Most of the numerical models are not entirely predictive, as some empirical parameters must be introduced in order to be able to perform the computation. Even when a sophisticated FE numerical model is used,[11] the results can be questioned with regard to the accuracy of the prediction of maximum strain rate and temperature. Thus the example of orthogonal machining, which is schematically represented in Fig. 12.17, is a very severe test for an error estimation procedure and adaptive remeshing.

The FORGE2 viscoplastic code with thermal coupling was adapted to the orthogonal, two-dimensional process of machining.* Though stationary in principle, the process was simulated by using an incremental formulation, until a steady-state regime was reached. The most important parameters were chosen according to the following.

Geometry:

| | |
|---|---|
| rake angle: | $\gamma = 5^\circ$ |
| relief angle: | $\eta = 8^\circ$ |
| radius of cutting edge: | $r_c = 0.025$ mm |
| depth of cut: | $v_1 = 0.25$ mm |

Mechanical parameters:

$$m = p = 0.1, \quad \alpha = 0.5$$

$$K = K_0(1 + a\bar{\varepsilon})\exp(\beta/T)$$

with $K_0 = 1.25 \times 10^6$ Pa s$^{-m}$, $a = 0.1$, $\beta = 200\ ^\circ\mathrm{K}^{-1}$.

Thermal parameters:

$$\rho = 7.8 \times 10^3 \text{ kg m}^{-3}, \quad c = 700 \text{ J kg}^{-1}\ ^\circ\mathrm{K}^{-1}, \quad k = 23 \text{ W m}^{-1}\ ^\circ\mathrm{K}^{-1}.$$

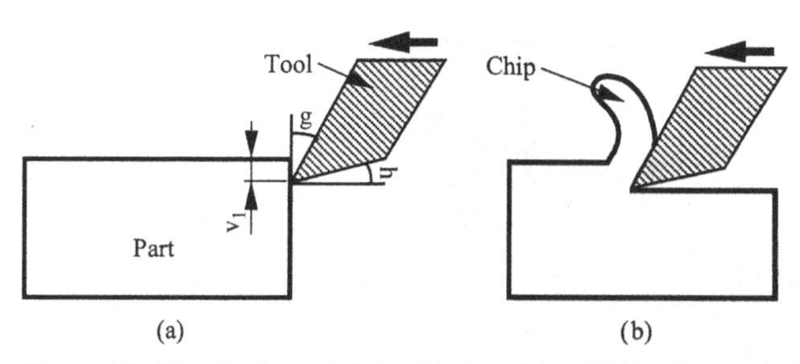


Figure 12.17. Orthogonal machining: (a) initialization and (b) steady state.

[11] G. S. Sekhon and J.-L. Chenot, "Numerical Simulation of Continuous Chip Formation During Non-steady Orthogonal Cutting," *Eng. Comput.* 10, 31–48 (1993).

* L. Fourment and J.-L. Chenot, Adaptive remeshing and error control for forming processes, Revue europeenne des elements finis, 3, no. 2, 247–280 (1994).

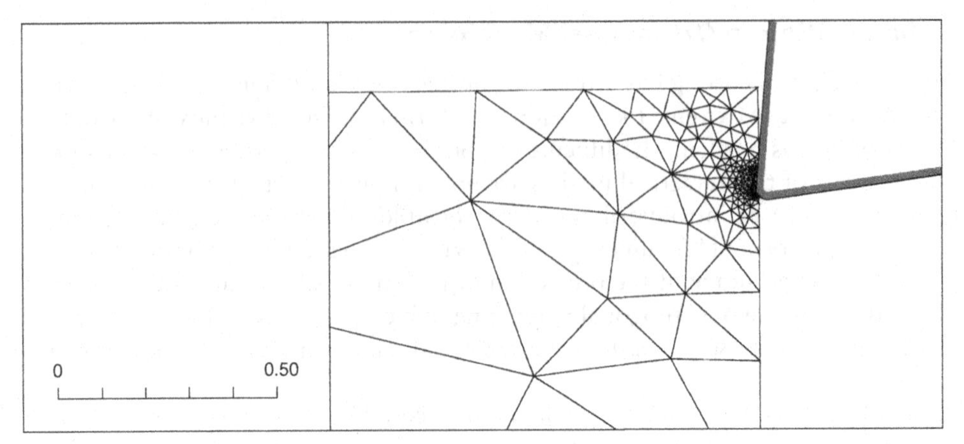


Figure 12.18. Mesh refinement at the beginning of orthogonal machining.

In Fig. 12.18 the mesh is represented after adaptive remeshing in the vicinity of the cutting edge, showing a very important refinement. Such a mesh refinement was necessary to reach a prescribed relative energy norm accuracy of 10%. When the chip is formed, the optimal mesh size estimation to reach the same accuracy is plotted in Fig. 12.19. We observe that the targeted mesh size is very small again close to the tool edge, and also in the zone where it is known from laboratory experiments that a very intense shear occurs. Here we can see that the maximation size ratio between the maximum mesh size $h_M$ and the minimum mesh size $h_m$ is $h_M/h_m \cong 35$.

At this stage of the machining process it is informative to observe that the map of iso equivalent strain rate exhibits a maximum value as high as $10^5$ s$^{-1}$, while it is also high in a very narrow zone which can be represented only when very small elements are produced in this area (Fig. 9.20a). The map of the isotemperatures is not the same

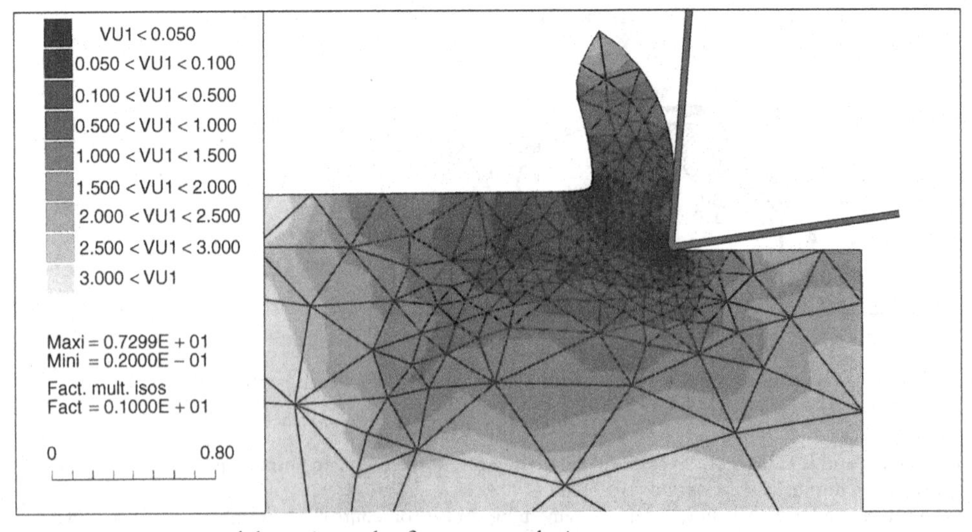


Figure 12.19. Map of the estimated refinement mesh size.

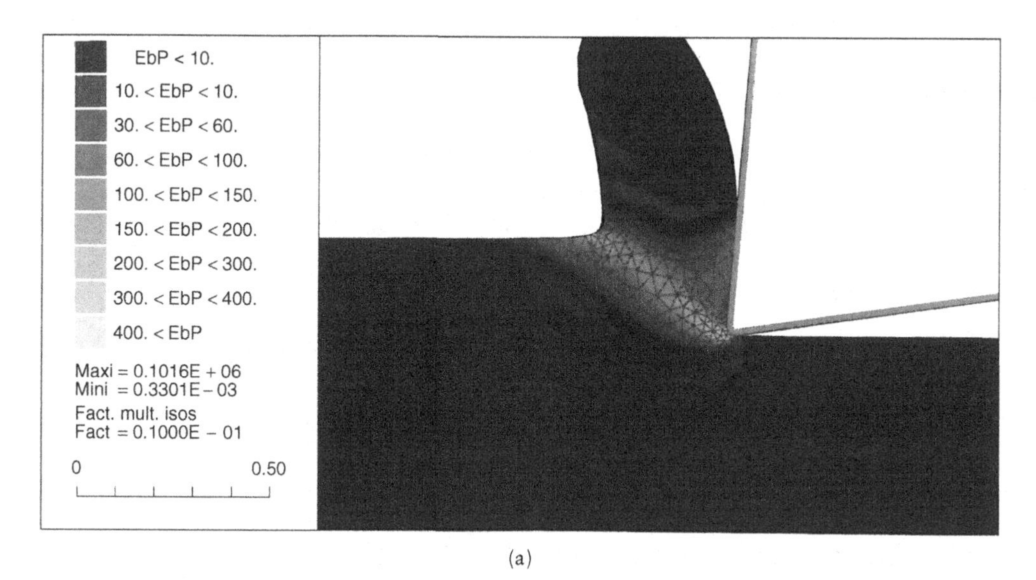


(a)

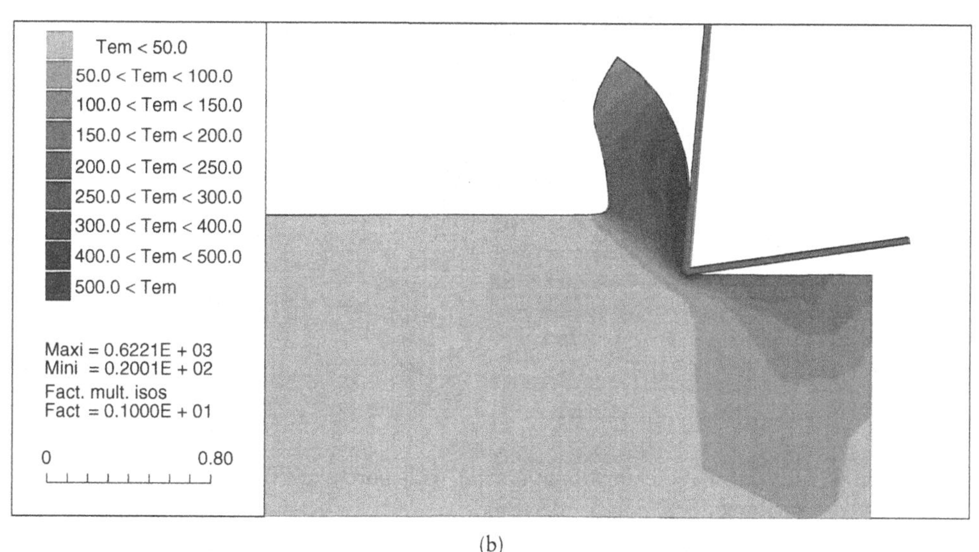


(b)

Figure 12.20. Prediction of orthogonal machining: (a) map of isoequivalent strain rate ($s^{-1}$) and (b) map of isotemperatures (°C).

due to the friction contribution to heat generation on the tool: the maximum value is $T = 622$ °C (Fig 12.20b).

A technological parameter that is important to predict is the shape and the curvature of the chip; this can be done when a quasi-steady state is achieved. In Fig. 12.21(a) the stationary state is reached for the computation with only a geometrical adaptation when the mesh is regenerated. In Fig. 12.21(b) the fully adaptive procedure is used, which leads to a more accurate result with many more elements generated where they are necessary. It can be observed by the comparisons of the two computations of Fig. 12.21 that even the shape and curvature of the chip are not exactly the same.

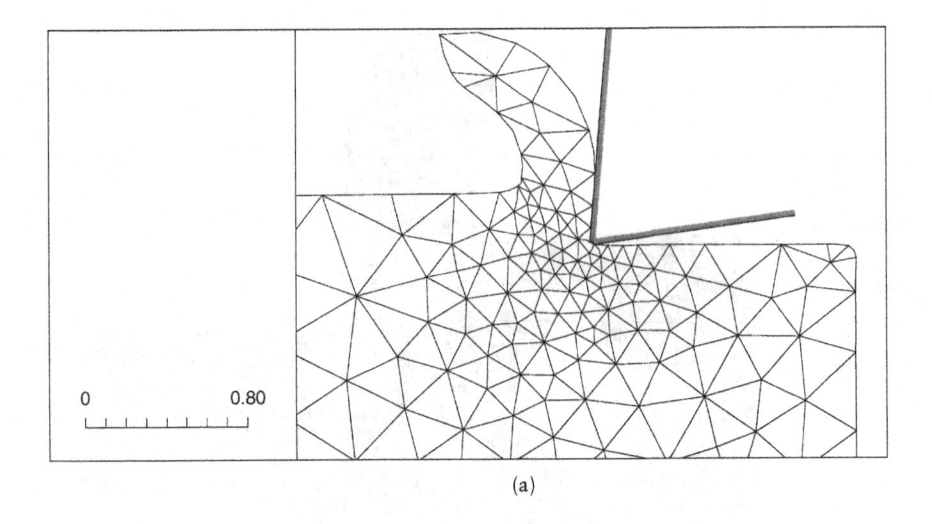


(a)

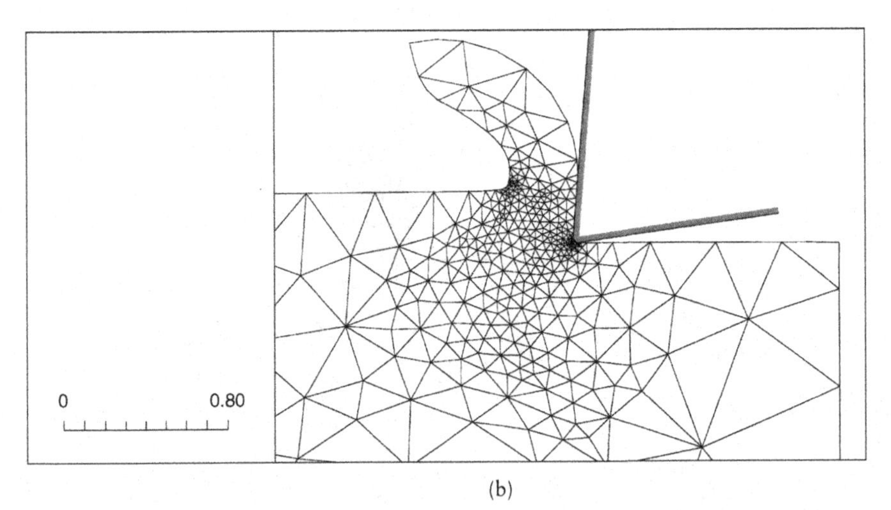


(b)

Figure 12.21. Prediction of the chip formation: (a) with purely geometrical remeshing and (b) with adaptive remeshing.

## 12.5 Advanced Solution Methods

Generally, in FE computation, the more CPU time-consuming step is the numerical resolution of linear systems. For nonlinear incremental problems, a linear system must be solved at each iteration of the Newton–Raphson linearization method, and of course at each time step. Direct methods were popular in the past, and are still widely used, as they are quite robust for most problems and easy to code for sequential computation. However, for three-dimensional problems with $n$ degrees of freedom, the associated CPU time grows as $n^{7/3}$ if a classical Crout method is used with a skyline storage. In contrast, the CPU memory requirement grows rather fast as it is proportional to $n^{5/3}$. Finally, it seems to be impossible to take advantage of the architecture of the parallel computers, as the Crout algorithm necessitates very important communications between processors. These reasons have motivated many authors interested in very large three-dimensional problems to improve or develop iterative methods for solving linear systems. One of the most common algorithms is the conjugate gradient method, which is quite effective for well-conditioned positive linear systems[12] and is the origin of several other iterative methods.

### The Minimal Residual Method

When the penalty method is used for viscoplastic materials, the penalty factor must be large for enforcing the incompressibility condition with a satisfactory approximation, but the linear systems we obtain after linearization are very ill conditioned. With the mixed formulation described in Section 6.3, the matrices are much better conditioned, but they are not positive, so that the conjugate gradient method cannot be used directly. The residual of the linear system to solve is written as

$$\mathbf{r} = \mathbf{R} - \mathbf{AY}, \tag{12.52}$$

where

**A** is the matrix of the linear system, or the Newton–Raphson matrix after linearization,
**Y** represents the vector of the unknown, for example, the nodal component vectors **V** and **P**, or their increment $\Delta\mathbf{V}$ and $\Delta\mathbf{P}$, which are computed during an iteration of the Newton–Raphson procedure, and
**R** is the right-hand side of the linear system.

The problem can be also stated as the minimization of the error function $E$, with

$$E(\mathbf{r}) = (\mathbf{M}^{-1}\mathbf{r})\mathbf{r}, \tag{12.53}$$

where **M** is a definite positive preconditioning matrix, of which the inverse is easy to compute. Here we can chose the absolute value of the diagonal, as the special form of the mini-element does not give zero terms on the diagonal.

The algorithm can be summarized briefly in following form.

1. Select a starting value $\mathbf{Y}_0$ (which is often taken to be null), to which we can associate the initial direction of descent $\mathbf{p}_0 = \mathbf{r}_0$.
2. At iteration number $k$:
   - the new iterate is given by

$$\mathbf{Y}^{k+1} = \mathbf{Y}^k + \alpha_k \mathbf{p}^k, \tag{12.54}$$

where $\alpha_k$ minimizes the scalar function $E$ so that it is easy to verify that it is given by the expression

$$\alpha_k = \frac{\mathbf{r}^k(\mathbf{M}^{-1}\mathbf{A}\mathbf{p}^k)}{(\mathbf{M}^{-1}\mathbf{A}\mathbf{p}^k)(\mathbf{A}\mathbf{p}^k)}; \tag{12.55}$$

[12] The conditioning of a matrix is defined by the ratio:

$\max_i |\lambda_i| / \min_i |\lambda_i|$

where $\lambda_i$ is an eigenvalue of the matrix. The matrix is well conditioned when the ratio is not very different from one and is ill conditioned when it is higher than $10^3$ or $10^4$.

- the next descent direction is computed according to

$$\mathbf{p}^{k+1} = \mathbf{M}^{-1}\mathbf{r}^k + \beta_{k+1}\mathbf{p}^k; \tag{12.56}$$

- $\beta_{k+1}$ is determined so that $\mathbf{p}^{k+1}$ is conjugate with respect to $\mathbf{p}^k$, giving

$$\beta_{k+1}\frac{(\mathbf{M}^{-1}\mathbf{A}\mathbf{p}^k)(\mathbf{A}\mathbf{M}^{-1}\mathbf{A}\mathbf{r}^{k+1})}{(\mathbf{M}^{-1}\mathbf{A}\mathbf{p}^k)(\mathbf{A}\mathbf{p}^k)}. \tag{12.57}$$

3. Update the error by

$$E(\mathbf{r}^{k+1}) = E(\mathbf{r}^k) - \frac{(\mathbf{r}^k\mathbf{M}^{-1}\mathbf{A}\mathbf{p}^k)^2}{(\mathbf{M}^{-1}\mathbf{A}\mathbf{p}^k)(\mathbf{A}\mathbf{p}^k)} \tag{12.58}$$

and check if the error is small enough to stop the iteration procedure.

4. If stagnation[13] of $E$ occurs, a restart procedure is necessary; that is, we go to step 1 without keeping a memory of the previous directions of descent.

The advantages of the minimum residual method are a decrease of memory requirement as the zero term in the matrix need not be stored, a decrease also of the CPU time when large problems are analyzed, and the possibility of efficient parallelization as outlined in the next subsection.

#### Parallel Computing

An iterative procedure, as described in the previous subsection, is desirable in order to develop more easily and effectively a parallel version of a FE code. We can see that most of the computational effort to evaluate Eqs. (12.54) to (12.58), can be done on separate processors. To achieve this goal, we must build a domain decomposition; that is, there is a partition of the domain $\Omega$ into nonoverlapping subdomains $\Omega_{(r)}$, with each subdomain corresponding to a different processor.

The efficiency of the coding depends not only on the coding but also on the effectiveness of the data transmission between the different processors. The results for an example run on an IBM SP2 computer are presented in Table 12.1. In Table 12.1 the speed up $S_u$ is the ratio of the computer times corresponding to one and $n_{\text{proc}}$ processors respectively, the efficiency being the ratio $E_f = S_u/n_{\text{proc}}$ (expressed in percent).

#### The Multigrid Method

The multigrid method is widely used at the research level, and in some industrial codes, for many linear physical problems; a general presentation is given by Brandt.[14]

**Table 12.1. Example Results**

| $n_{\text{proc}}$ | $S_u$ | $E_f$ |
|---|---|---|
| 16 | 15 | 95 |
| 24 | 21 | 90 |
| 34 | 28 | 80 |

[13] Stagnation can occur for large linear systems as a result of an accumulation of rounding arithmetic errors.

[14] A. Brandt, "Multi-Level Adaptative Solutions to Boundary-Value Problems," *Math. Comput.* 31, 333–390 (1977).

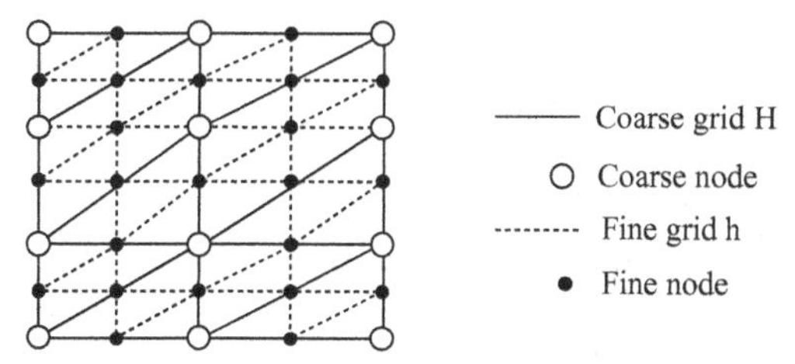


Figure 12.22. Coarse and fine grids for a 2-grid method.

It is also used for nonlinear problems such as incompressible viscoplastic material flow; for an example, see Hadj et al.[15] (using a direct method of resolution for the linear system in velocity). The present formulation corresponds to a mixed formulation and the use of the minimum residual method for solving linear systems corresponding to $\mathbf{r} = 0$ in Eq. (12.52).

In a two-grid method an iterative algorithm is constructed, which uses a coarse and a fine mesh and the corresponding discretizations, in order to compute the solution on the finer mesh. At each step of the two-grid method, several iterations of the minimum residual method are performed on each level of grid, and if the results are transmitted by specific operators. This process permits significant improvement of computer time by accelerating the convergence rate of the minimum residual method.

In order to be able to define the two-grid algorithm more precisely, we introduce the following notation. The fine grid is denoted by $h$, and the coarse one by $H$. We suppose that the fine grid is obtained by a subdivision of the coarse mesh $h$, for which each side of the elements are subdivided into two equal segments. A simple example of such a structure is schematically pictured in Fig. 12.22 for a discretization into triangles.[16] The nodes of the coarse mesh are also nodes of the fine mesh. In the numbering of nodes of the fine mesh, the nodes that also belong to the coarse mesh will have the same numbers as in the coarse mesh. Thus for such a common node with number $I$ in the coarse mesh, and number $i$ in the fine mesh, we have $i = I$.

1. On the fine grid, let $\mathbf{Y}_h$, $\mathbf{A}_h$, $\mathbf{r}_h$, $\mathbf{S}_h$ be a global vector, the discretized matrix, the residual and the minimum residual method smoothing operator, respectively.
2. Similarly, on the coarse mesh, let $\mathbf{Y}_H$, $\mathbf{A}_H$, $\mathbf{r}_H$, $\mathbf{S}_H$ be a global vector, the discretized matrix, the residual and the minimum residual method smoothing operator.
3. The extrapolation operator $\mathbf{I}_H^h$ allows the approximation of a vector on the fine mesh from its components on the coarse mesh; it is defined by

$$\mathbf{Y}_{ih} = \left[\mathbf{I}_H^h(\mathbf{Y}_H)\right] = \mathbf{Y}_{IH}, \tag{12.59}$$

where $I$ is the number of a node belonging to the coarse mesh; if $i$ is the number of a node of the fine mesh only,

$$\mathbf{Y}_{ih} = \left[\mathbf{I}_H^h(\mathbf{Y}_H)\right]_i = \frac{1}{2}(\mathbf{Y}_{I_1H} + \mathbf{Y}_{I_2H}), \tag{12.60}$$

[15] M. El Hadj, Y. Demay, and J. L. Chenot, "Application of the Multi-Grid Method to the Finite Element Computation of Incompressible Visco-Plastic Flows," in *Computational Plasticity: Models, Software and Applications*, D. R. J. Owen et al., eds. (Pineridge Press, Swansea, 1987), pp. 425–434.

[16] The multigrid method can be generalized to other elements, in two and three dimensions.

Figure 12.23. Coarse neighbors of a fine node.

where $I_1$ and $I_2$ are the numbers of the two neighbors of node $i$ (as pictured in Fig. 12.23).

4. The projection operator $\mathbf{I}_h^H$ restricts a fine right-hand side to a coarse vector; it is the transpose of the extrapolation operator $\mathbf{I}_H^h$ (denoted by the superscript$^T$):

$$\mathbf{I}_h^H = \left(\mathbf{I}_H^h\right)^T. \tag{12.61}$$

When applied to a fine residual $\mathbf{r}_h$, it gives

$$\mathbf{r}_{IH} = \left[\mathbf{I}_h^H(\mathbf{R}_h)\right]_I = \mathbf{r}_{Ih} + \frac{1}{2}\sum_{i\in nb(I)} \mathbf{r}_{ih}, \tag{12.62}$$

where $nb(I)$ is the set of fine node numbers that are neighbors of coarse node $I$ (see Fig. 12.24).

The two-grid algorithm is then described in term of iterative steps as follows. The solution vector is initialized on the fine grid according to $\mathbf{Y}_h^{(0)} = \mathbf{0}$ and, at the end of step $(k)$, the approximate fine solution of the linear system is denoted $\mathbf{Y}_h^{(k)}$.

1. Perform $\mu$ iterations of smoothing on the fine grid by the minimum residual method,

$$\mathbf{Y}_h^{(k)*} = \mathbf{S}_h^{\mu}\left[\mathbf{Y}_h^{(k)}\right], \tag{12.63}$$

and update the residual,

$$\mathbf{r}_h^{(k)*} = \mathbf{b}_h - \mathbf{A}_h\mathbf{Y}_h^{(k)*}. \tag{12.64}$$

2. Project the residual on the coarse grid:

$$\mathbf{r}_H^{(k)} = \mathbf{I}_h^H\left(\mathbf{r}_h^{(k)*}\right). \tag{12.65}$$

3. Perform $\nu$ iterations of the minimum residual method on the coarse grid to obtain a coarse correction:

$$\Delta\mathbf{Y}_H^{(k)} = \mathbf{S}_H^{\nu}\left[\mathbf{r}_H^{(k)*}\right]. \tag{12.66}$$

4. Extrapolate the coarse correction from the coarse grid to the fine mesh,

$$\Delta\mathbf{Y}_h^{(k)} = \mathbf{I}_H^h\left[\Delta\mathbf{Y}_H^{(k)}\right], \tag{12.67}$$

and update the solution on the fine mesh by

$$\mathbf{Y}_h^{(k+1)} = \mathbf{Y}_h^{(k)} + \Delta\mathbf{Y}_h^{(k)}. \tag{12.68}$$

5. Compute the new residual on the fine mesh:

$$\mathbf{r}_h^{(k+1)} = \mathbf{b}_h - \mathbf{A}_h\mathbf{Y}_h^{(k+1)}. \tag{12.69}$$

If $|\mathbf{r}_h^{(k+1)}|$, the norm of the residual, is greater than a given tolerance, then repeat steps 1–5.

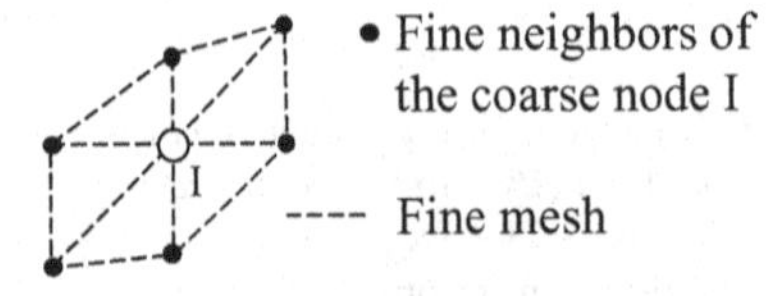


Figure 12.24. Fine neighbors of a coarse node.

The $\mu$ and $\nu$ parameters must be optimized to achieve fast convergence for a given class of problems. The two-grid method is able to produce acceleration factors of 2–5 on practical examples with a viscoplastic behavior.[17] Moreover, the method can be generalized and applied to several levels of grids.

## PROBLEMS

### A. Proficiency Problems

1. How can the segments of the boundary of a polygonal meshed domain be distinguished from the other (inner) segments?
2. Define a test to check whether an exchange of diagonals in a quadrilateral (see Fig. 12.9) is possible. When this mesh transformation is possible, define a quantitative criterion to determine if the exchange of diagonals improves the mesh quality locally.
3. Determine the transformation that maps the equilateral triangle with sides equal to unity onto the reference element. Compute the ratio of the Jacobian determinant over the norm of the Jacobian matrix of the transformation.
4. Show that the refinement factor is equal to 1 when the error is equal to the desired error $e_0$ and is uniformly distributed over the elements.

### B. Depth Problems

5. Explain why an error indicator given by Eq. (12.15) should not be recommended for a viscoplastic material. Find another form that could be possibly substituted into Eq. (12.29).
6. We consider again the problem of a simple rod subjected to inertial forces that are due to rotation. This problem was presented and solved by a simple FE discretization in Section 2.11. Compute explicitly the constant by element stress field, and determine the smoothed stress field obtained by a simple arithmetic average at nodes. Compute the error indicator and the relative error indicator. Compare it with the exact error energy norm.
7. Perform the same computations as in Problem 6, using a least-squares smoothing of the stress field for the error indicator evaluation.
8. Evaluate the local refinement parameter with a hypothesis of uniform relative error indicator (instead of uniform error indicator).

### C. Numerical and Computational Problems

9. Write a FORTRAN computer routine for the triangulation of a convex domain, using the Delaunay method. A test algorithm should be added to check that the domain is convex, that is, that all the geometrical points lie on the same side of any straight line containing a segment of the boundary.

[17] K. Mocellin, L. Fourment and J.-L. Chenot, "An efficient multigrid solver for incompressible fluid problems: Application to the 3D hot forgoing process," Simulation of Materials Processing: Theory, Methods and Applications, Proceedings of Numiform '98, ed. by J. Huétink and F. P. T. Baaijens, A. A. Balkema, Rotterdom, 233–238 (1998).

10. Write a FORTRAN code for the interpolation of a scalar field that is defined on one mesh and must be transported at the nodes of a new mesh. We suppose that the two meshes are both composed of linear or quadratic triangles.
11. Write a FORTRAN code for the two-dimensional error estimation of a viscoplastic problem, using four node elements and a local smoothing. Input parameters are consistency, rate sensitivity index, node coordinates, mesh connection table, and nodal velocity vectors. Output values are the error estimation in each element, and the global error indicator.
12. Write a FORTRAN routine for the computation of the refinement factor in each element. Input data are the mesh connection table, node coordinates, error indicator on each element, rate $p$ of convergence of the error (when it is evaluated in the energy norm), and the targeted global error.
13. Write a FORTRAN routine for solving a general symmetric linear system by the minimum residual method.
14. Write a FORTRAN program to solve a one-dimensional thermal problem, using the two-grid method. The conduction factor $k$ is a function of the space variable $x$.

# Index